Toyota Land Cruiser Automotive Repair Manual

by Jeff Kibler, Robert Maddox and John H Haynes

Member of the Guild of Motoring Writers

Models covered:
FJ60, FJ62, FJ80 and FZJ80 Series Land Cruiser
1980 through 1996

(92056-5E1)

Haynes Group Limited
Haynes North America, Inc.
www.haynes.com

Acknowledgements

We are grateful for the help and cooperation of the Toyota Motor Company for their assistance with technical information and certain illustrations. Technical consultants who contributed to this project include James Cota and Jamie Sarté. Special thanks for technical assistance from Collin Andersen representing Man-A-Fre Land Cruiser in Agoura, California and Jonathan Ward representing TLC, Inc. in Van Nuys, California.

A book in the Haynes Automotive Repair Manual Series

ISBN-10: 1-56392-301-7
ISBN-13: 978-1-56392-301-2

Library of Congress Catalog Card Number 97-80597

While every attempt is made to ensure that the information in this manual is correct, no liability can be accepted by the authors or publishers for loss, damage or injury caused by any errors in, or omissions from, the information given.

Contents

Introductory pages

About this manual	0-5
Introduction to the Toyota Land Cruiser	0-5
Vehicle identification numbers	0-6
Buying parts	0-8
Maintenance techniques, tools and working facilities	0-8
Jacking and towing	0-16
Booster battery (jump) starting	0-16
Automotive chemicals and lubricants	0-17
Conversion factors	0-18
Safety first!	0-19
Troubleshooting	0-20

Chapter 1
Tune-up and routine maintenance — **1-1**

Chapter 2 Part A
Pushrod engines — **2A-1**

Chapter 2 Part B
Overhead camshaft engine — **2B-1**

Chapter 2 Part C
General engine overhaul procedures — **2C-1**

Chapter 3
Cooling, heating and air conditioning systems — **3-1**

Chapter 4 Part A
Fuel and exhaust systems - carbureted engines — **4A-1**

Chapter 4 Part B
Fuel and exhaust systems - fuel-injected engines — **4B-1**

Chapter 5
Engine electrical systems — **5-1**

Chapter 6
Emissions and engine control systems — **6-1**

Chapter 7 Part A
Manual transmission — **7A-1**

Chapter 7 Part B
Automatic transmission — **7B-1**

Chapter 7 Part C
Transfer case — **7C-1**

Chapter 8
Clutch and driveline — **8-1**

Chapter 9
Brakes — **9-1**

Chapter 10
Suspension and steering systems — **10-1**

Chapter 11
Body — **11-1**

Chapter 12
Chassis electrical system — **12-1**

Wiring diagrams — **12-21**

Index — **IND-1**

Haynes mechanic, author and photographer with 1994 Toyota Land Cruiser

About this manual

Its purpose

The purpose of this manual is to help you get the best value from your vehicle. It can do so in several ways. It can help you decide what work must be done, even if you choose to have it done by a dealer service department or a repair shop; it provides information and procedures for routine maintenance and servicing; and it offers diagnostic and repair procedures to follow when trouble occurs.

We hope you use the manual to tackle the work yourself. For many simpler jobs, doing it yourself may be quicker than arranging an appointment to get the vehicle into a shop and making the trips to leave it and pick it up. More importantly, a lot of money can be saved by avoiding the expense the shop must pass on to you to cover its labor and overhead costs. An added benefit is the sense of satisfaction and accomplishment that you feel after doing the job yourself.

Using the manual

The manual is divided into Chapters. Each Chapter is divided into numbered Sections, which are headed in bold type between horizontal lines. Each Section consists of consecutively numbered paragraphs.

At the beginning of each numbered Section you will be referred to any illustrations which apply to the procedures in that Section. The reference numbers used in illustration captions pinpoint the pertinent Section and the Step within that Section. That is, illustration 3.2 means the illustration refers to Section 3 and Step (or paragraph) 2 within that Section.

Procedures, once described in the text, are not normally repeated. When it's necessary to refer to another Chapter, the reference will be given as Chapter and Section number. Cross references given without use of the word "Chapter" apply to Sections and/or paragraphs in the same Chapter. For example, "see Section 8" means in the same Chapter.

References to the left or right side of the vehicle assume you are sitting in the driver's seat, facing forward.

Even though we have prepared this manual with extreme care, neither the publisher nor the author can accept responsibility for any errors in, or omissions from, the information given.

NOTE

A **Note** provides information necessary to properly complete a procedure or information which will make the procedure easier to understand.

CAUTION

A **Caution** provides a special procedure or special steps which must be taken while completing the procedure where the Caution is found. Not heeding a Caution can result in damage to the assembly being worked on.

WARNING

A **Warning** provides a special procedure or special steps which must be taken while completing the procedure where the Warning is found. Not heeding a Warning can result in personal injury.

Introduction to the Toyota Land Cruiser

FJ60, FJ62, FJ80 and FZJ80 Toyota Land Cruiser models are available in four-door station wagon body styles.

The FJ60 was manufactured from 1980 through 1987 and is equipped with the 2F engine. The FJ62 was manufactured from 1988 through 1990 and is equipped with the 3F-E engine. The FJ80 was manufactured from 1991 through 1992 and is also equipped with the 3F-E engine. The FZJ80 series was manufactured from 1993 through 1996 and is equipped with the 1FZ-FE DOHC engine.

2F pushrod engines are equipped with a carburetor. 3F-E pushrod and 1FZ-FE DOHC engines are equipped with port fuel injection.

The engine drives the rear wheels through either a manual or automatic transmission via a driveshaft and solid rear axle. A transfer case and driveshaft are used to drive the front axle. All models are equipped with 4WD.

The solid axle front suspension features either leaf springs (FJ60, 62) or coil springs and control arms (FJ80), and shock absorbers. A solid axle at the rear is suspended by leaf springs (FJ60, 62) or coil springs (FJ80) and shock absorbers.

The steering box is mounted to the left of the engine and is connected to the steering arms through a series of rods. All models are equipped with power steering.

The brakes are disc or drum at the front and disc or drums at the rear, with power assist standard.

Vehicle identification numbers

Modifications are a continuing and unpublicized process in vehicle manufacturing. Since spare parts manuals and lists are compiled on a numerical basis, the individual vehicle numbers are essential to correctly identify the component required.

Vehicle Identification Number (VIN)

This very important identification number is stamped on a plate attached to the left side of the dashboard just inside the windshield on the driver's side of the vehicle (see illustration). The VIN also appears on the Vehicle Certificate of Title and Registration. It contains information such as where and when the vehicle was manufactured, the model year and the body style.

Engine identification number

The engine ID number on 2F and 3F-E engines is located on a machined surface on the right side of the block (see illustration). On 1FZ-FE engines, the ID number is located on the left side of the engine block (see illustration).

Manufacturer Certification label

The Manufacturer Certification label is affixed to the front door pillar. The plate contains the name of the manufacturer, the month and year of production, the Gross Vehicle Weight Rating (GVWR) and the certification statement (see illustration).

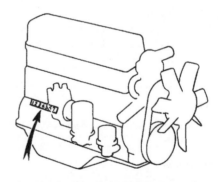

The Engine Identification Number is stamped on the right side of the engine block on 2F and 3F-E engines

The Vehicle Identification Number (VIN) is visible through the driver's side of the windshield

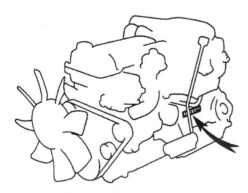

The Engine Identification Number is stamped on the left side of the engine block on 1FZ-FE engines

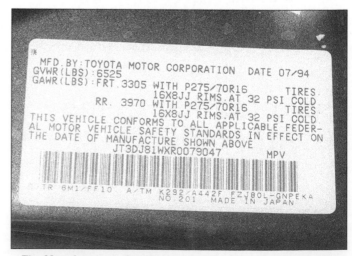

The Manufacturer's Certification label is affixed to the driver's side door end or post

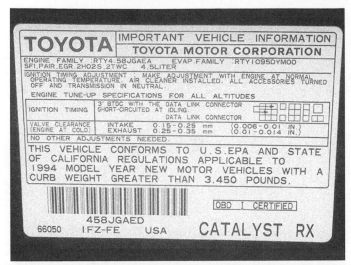

The Vehicle Emissions Control Information (VECI) label is located on the bottom side of the hood

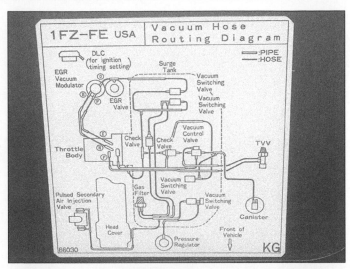

The Vacuum schematic label is located on the bottom side of the hood

The transfer case identification label is located on the side of the case

The transmission identification label is located on the side of the transmission body

Vehicle Emissions Control Information (VECI) label and Vacuum Schematic

The emissions control information label and vacuum schematic is found under the hood. This label contains information on the emissions control equipment installed on the vehicle, as well as tune-up specifications (see illustrations).

Transfer case and transmission identification Number

The transfer case and/or the transmission identification number is located on the side of the component. Transfer case numbers are stamped directly into the transfer case (see illustration) while the transmission numbers are stamped onto ID plates (see illustrations).

Buying parts

Replacement parts are available from many sources, which generally fall into one of two categories - authorized dealer parts departments and independent retail auto parts stores. Our advice concerning these parts is as follows:

Retail auto parts stores: Good auto parts stores will stock frequently needed components which wear out relatively fast, such as clutch components, exhaust systems, brake parts, tune-up parts, etc. These stores often supply new or reconditioned parts on an exchange basis, which can save a considerable amount of money. Discount auto parts stores are often very good places to buy materials and parts needed for general vehicle maintenance such as oil, grease, filters, spark plugs, belts, touch-up paint, bulbs, etc. They also usually sell tools and general accessories, have convenient hours, charge lower prices and can often be found not far from home.

Authorized dealer parts department: This is the best source for parts which are unique to the vehicle and not generally available elsewhere (such as major engine parts, transmission parts, trim pieces, etc.).

Warranty information: If the vehicle is still covered under warranty, be sure that any replacement parts purchased - regardless of the source - do not invalidate the warranty!

To be sure of obtaining the correct parts, have engine and chassis numbers available and, if possible, take the old parts along for positive identification.

Maintenance techniques, tools and working facilities

Maintenance techniques

There are a number of techniques involved in maintenance and repair that will be referred to throughout this manual. Application of these techniques will enable the home mechanic to be more efficient, better organized and capable of performing the various tasks properly, which will ensure that the repair job is thorough and complete.

Fasteners

Fasteners are nuts, bolts, studs and screws used to hold two or more parts together. There are a few things to keep in mind when working with fasteners. Almost all of them use a locking device of some type, either a lockwasher, locknut, locking tab or thread adhesive. All threaded fasteners should be clean and straight, with undamaged threads and undamaged corners on the hex head where the wrench fits. Develop the habit of replacing all damaged nuts and bolts with new ones. Special locknuts with nylon or fiber inserts can only be used once. If they are removed, they lose their locking ability and must be replaced with new ones.

Rusted nuts and bolts should be treated with a penetrating fluid to ease removal and prevent breakage. Some mechanics use turpentine in a spout-type oil can, which works quite well. After applying the rust penetrant, let it work for a few minutes before trying to loosen the nut or bolt. Badly rusted fasteners may have to be chiseled or sawed off or removed with a special nut breaker, available at tool stores.

If a bolt or stud breaks off in an assembly, it can be drilled and removed with a special tool commonly available for this purpose. Most automotive machine shops can perform this task, as well as other repair procedures, such as the repair of threaded holes that have been stripped out.

Flat washers and lockwashers, when removed from an assembly, should always be replaced exactly as removed. Replace any damaged washers with new ones. Never use a lockwasher on any soft metal surface (such as aluminum), thin sheet metal or plastic.

Fastener sizes

For a number of reasons, automobile manufacturers are making wider and wider use of metric fasteners. Therefore, it is important to be able to tell the difference between standard (sometimes called U.S. or SAE) and metric hardware, since they cannot be interchanged.

All bolts, whether standard or metric, are sized according to diameter, thread pitch and length. For example, a standard 1/2 - 13 x 1 bolt is 1/2 inch in diameter, has 13 threads per inch and is 1 inch long. An M12 - 1.75 x 25 metric bolt is 12 mm in diameter, has a thread pitch of 1.75 mm (the distance between threads) and is 25 mm long. The two bolts are nearly identical, and easily confused, but they are not interchangeable.

In addition to the differences in diameter, thread pitch and length, metric and standard bolts can also be distinguished by examining the bolt heads. To begin with, the distance across the flats on a standard bolt head is measured in inches, while the same dimension on a metric bolt is sized in millimeters (the same is true for nuts). As a result, a standard wrench should not be used on a metric bolt and a metric wrench should not be used on a standard bolt. Also, most standard bolts have slashes radiating out from the center of the head to denote the grade or strength of the bolt, which is an indication of the amount of torque that can be applied to it. The greater the number of slashes, the greater the strength of the bolt. Grades 0 through 5 are commonly used on automobiles. Metric bolts have a property class (grade) number, rather than a slash, molded into their heads to indicate bolt strength. In this case, the higher the number, the stronger the bolt. Property class numbers 8.8, 9.8 and 10.9 are commonly used on automobiles.

Strength markings can also be used to distinguish standard hex nuts from metric hex nuts. Many standard nuts have dots stamped into one side, while metric nuts are marked with a number. The greater the number of dots, or the higher the number, the greater the strength of the nut.

Metric studs are also marked on their ends according to property class (grade). Larger studs are numbered (the same as metric bolts), while smaller studs carry a geometric code to denote grade.

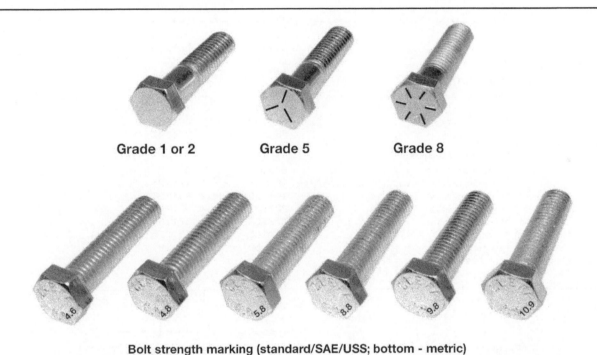

Grade 1 or 2   Grade 5   Grade 8

Bolt strength marking (standard/SAE/USS; bottom - metric)

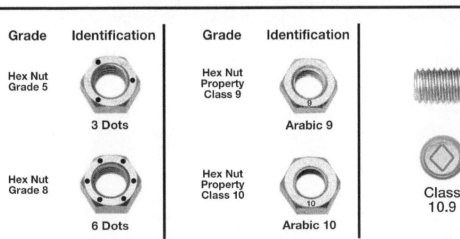

Grade	Identification	Grade	Identification
Hex Nut Grade 5	3 Dots	Hex Nut Property Class 9	Arabic 9
Hex Nut Grade 8	6 Dots	Hex Nut Property Class 10	Arabic 10

Standard hex nut strength markings

Metric hex nut strength markings

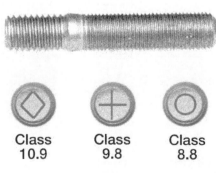

Class 10.9   Class 9.8   Class 8.8

Metric stud strength markings

00-1 HAYNES

It should be noted that many fasteners, especially Grades 0 through 2, have no distinguishing marks on them. When such is the case, the only way to determine whether it is standard or metric is to measure the thread pitch or compare it to a known fastener of the same size.

Standard fasteners are often referred to as SAE, as opposed to metric. However, it should be noted that SAE technically refers to a non-metric fine thread fastener only. Coarse thread non-metric fasteners are referred to as USS sizes.

Since fasteners of the same size (both standard and metric) may have different strength ratings, be sure to reinstall any bolts, studs or nuts removed from your vehicle in their original locations. Also, when replacing a fastener with a new one, make sure that the new one has a strength rating equal to or greater than the original.

Tightening sequences and procedures

Most threaded fasteners should be tightened to a specific torque value (torque is the twisting force applied to a threaded component such as a nut or bolt). Overtightening the fastener can weaken it and cause it to break, while undertightening can cause it to eventually come loose. Bolts, screws and studs, depending on the material they are made of and their thread diameters, have specific torque values, many of which are noted in the Specifications at the beginning of each Chapter. Be sure to follow the torque recommendations closely. For fasteners not assigned a specific torque, a general torque value chart is presented here as a guide. These torque values are for dry (unlubricated) fasteners threaded into steel or cast iron (not aluminum). As was previously mentioned, the size and grade of a fastener determine the amount of torque that can safely be applied to it. The figures listed

Metric thread sizes	Ft-lbs	Nm
M-6	6 to 9	9 to 12
M-8	14 to 21	19 to 28
M-10	28 to 40	38 to 54
M-12	50 to 71	68 to 96
M-14	80 to 140	109 to 154

Pipe thread sizes		
1/8	5 to 8	7 to 10
1/4	12 to 18	17 to 24
3/8	22 to 33	30 to 44
1/2	25 to 35	34 to 47

U.S. thread sizes		
1/4 - 20	6 to 9	9 to 12
5/16 - 18	12 to 18	17 to 24
5/16 - 24	14 to 20	19 to 27
3/8 - 16	22 to 32	30 to 43
3/8 - 24	27 to 38	37 to 51
7/16 - 14	40 to 55	55 to 74
7/16 - 20	40 to 60	55 to 81
1/2 - 13	55 to 80	75 to 108

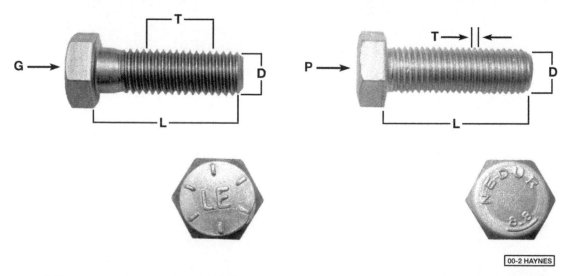

00-2 HAYNES

Standard (SAE and USS) bolt dimensions/grade marks

- G   Grade marks (bolt strength)
- L   Length (in inches)
- T   Thread pitch (number of threads per inch)
- D   Nominal diameter (in inches)

Metric bolt dimensions/grade marks

- P   Property class (bolt strength)
- L   Length (in millimeters)
- T   Thread pitch (distance between threads in millimeters)
- D   Diameter

here are approximate for Grade 2 and Grade 3 fasteners. Higher grades can tolerate higher torque values.

Fasteners laid out in a pattern, such as cylinder head bolts, oil pan bolts, differential cover bolts, etc., must be loosened or tightened in sequence to avoid warping the component. This sequence will normally be shown in the appropriate Chapter. If a specific pattern is not given, the following procedures can be used to prevent warping.

Initially, the bolts or nuts should be assembled finger-tight only. Next, they should be tightened one full turn each, in a criss-cross or diagonal pattern. After each one has been tightened one full turn, return to the first one and tighten them all one-half turn, following the same pattern. Finally, tighten each of them one-quarter turn at a time until each fastener has been tightened to the proper torque. To loosen and remove the fasteners, the procedure would be reversed.

Component disassembly

Component disassembly should be done with care and purpose to help ensure that the parts go back together properly. Always keep track of the sequence in which parts are removed. Make note of special characteristics or marks on parts that can be installed more than one way, such as a grooved thrust washer on a shaft. It is a good idea to lay the disassembled parts out on a clean surface in the order that they were removed. It may also be helpful to make sketches or take instant photos of components before removal.

When removing fasteners from a component, keep track of their locations. Sometimes threading a bolt back in a part, or putting the washers and nut back on a stud, can prevent mix-ups later. If nuts and bolts cannot be returned to their original locations, they should be kept in a compartmented box or a series of small boxes. A cupcake or muffin tin is ideal for this purpose, since each cavity can hold the bolts and nuts from a particular area (i.e. oil pan bolts, valve cover bolts, engine mount bolts, etc.). A pan of this type is especially helpful when working on assemblies with very small parts, such as the carburetor, alternator, valve train or interior dash and trim pieces. The cavities can be marked with paint or tape to identify the contents.

Whenever wiring looms, harnesses or connectors are separated, it is a good idea to identify the two halves with numbered pieces of masking tape so they can be easily reconnected.

Gasket sealing surfaces

Throughout any vehicle, gaskets are used to seal the mating surfaces between two parts and keep lubricants, fluids, vacuum or pressure contained in an assembly.

Many times these gaskets are coated with a liquid or paste-type gasket sealing compound before assembly. Age, heat and pressure can sometimes cause the two parts to stick together so tightly that they are very difficult to separate. Often, the assembly can be loosened by striking it with a soft-face hammer near the mating surfaces. A regular hammer can be used if a block of wood is placed between the hammer and the part. Do not hammer on cast parts or parts that could be easily damaged. With any particularly stubborn part, always recheck to make sure that every fastener has been removed.

Avoid using a screwdriver or bar to pry apart an assembly, as they can easily mar the gasket sealing surfaces of the parts, which must remain smooth. If prying is absolutely necessary, use an old broom handle, but keep in mind that extra clean up will be necessary if the wood splinters.

After the parts are separated, the old gasket must be carefully scraped off and the gasket surfaces cleaned. Stubborn gasket material can be soaked with rust penetrant or treated with a special chemical to soften it so it can be easily scraped off. A scraper can be fashioned from a piece of copper tubing by flattening and sharpening one end. Copper is recommended because it is usually softer than the surfaces to be scraped, which reduces the chance of gouging the part. Some gaskets can be removed with a wire brush, but regardless of the method used, the mating surfaces must be left clean and smooth. If for some reason the gasket surface is gouged, then a gasket sealer thick enough to fill scratches will have to be used during reassembly of the components. For most applications, a non-drying (or semi-drying) gasket sealer should be used.

Hose removal tips

Warning: *If the vehicle is equipped with air conditioning, do not disconnect any of the A/C hoses without first having the system depressurized by a dealer service department or a service station.*

Hose removal precautions closely parallel gasket removal precautions. Avoid scratching or gouging the surface that the hose mates against or the connection may leak. This is especially true for radiator hoses. Because of various chemical reactions, the rubber in hoses can bond itself to the metal spigot that the hose fits over. To remove a hose, first loosen the hose clamps that secure it to the spigot. Then, with slip-joint pliers, grab the hose at the clamp and rotate it around the spigot. Work it back and forth until it is completely free, then pull it off. Silicone or other lubricants will ease removal if they can be applied between the hose and the outside of the spigot. Apply the same lubricant to the inside of the hose and the outside of the spigot to simplify installation.

As a last resort (and if the hose is to be replaced with a new one anyway), the rubber can be slit with a knife and the hose peeled from the spigot. If this must be done, be careful that the metal connection is not damaged.

If a hose clamp is broken or damaged, do not reuse it. Wire-type clamps usually weaken with age, so it is a good idea to replace them with screw-type clamps whenever a hose is removed.

Tools

A selection of good tools is a basic requirement for anyone who plans to maintain and repair his or her own vehicle. For the owner who has few tools, the initial investment might seem high, but when compared to the spiraling costs of professional auto maintenance and repair, it is a wise one.

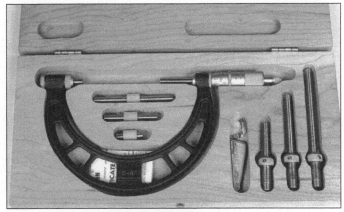

Micrometer set

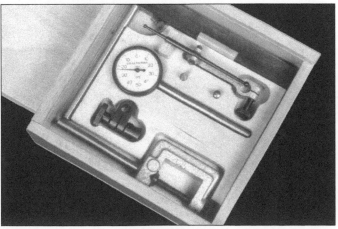

Dial indicator set

Dial caliper

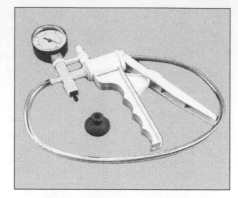

Hand-operated vacuum pump

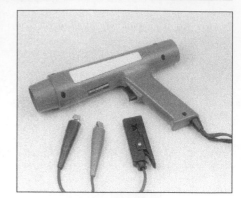

Timing light

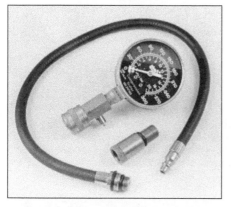

Compression gauge with spark plug
hole adapter

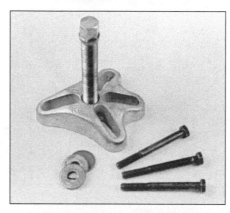

Damper/steering wheel puller

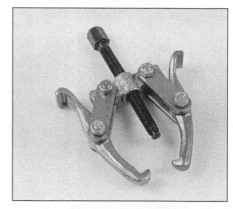

General purpose puller

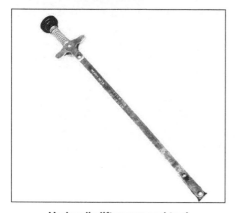

Hydraulic lifter removal tool

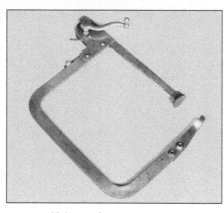

Valve spring compressor

Valve spring compressor

Ridge reamer

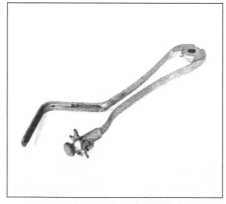

Piston ring groove cleaning tool

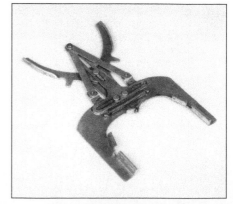

Ring removal/installation tool

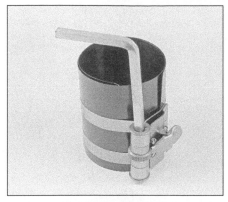

Ring compressor

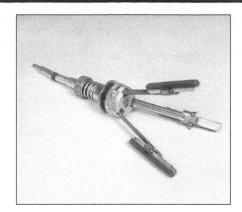

Cylinder hone

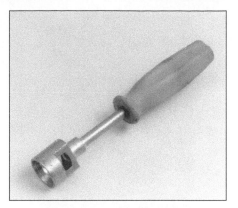

Brake hold-down spring tool

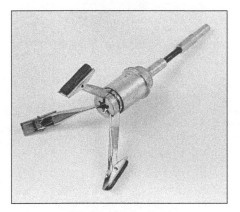

Brake cylinder hone

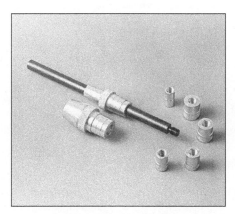

Clutch plate alignment tool

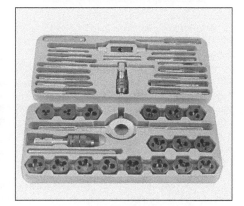

Tap and die set

To help the owner decide which tools are needed to perform the tasks detailed in this manual, the following tool lists are offered: *Maintenance and minor repair, Repair/overhaul* and *Special.*

The newcomer to practical mechanics should start off with the *maintenance and minor repair* tool kit, which is adequate for the simpler jobs performed on a vehicle. Then, as confidence and experience grow, the owner can tackle more difficult tasks, buying additional tools as they are needed. Eventually the basic kit will be expanded into the *repair and overhaul* tool set. Over a period of time, the experienced do-it-yourselfer will assemble a tool set complete enough for most repair and overhaul procedures and will add tools from the special category when it is felt that the expense is justified by the frequency of use.

Maintenance and minor repair tool kit

The tools in this list should be considered the minimum required for performance of routine maintenance, servicing and minor repair work. We recommend the purchase of combination wrenches (box-end and open-end combined in one wrench). While more expensive than open end wrenches, they offer the advantages of both types of wrench.

Combination wrench set (1/4-inch to 1 inch or 6 mm to 19 mm)
Adjustable wrench, 8 inch
Spark plug wrench with rubber insert
Spark plug gap adjusting tool
Feeler gauge set
Brake bleeder wrench
Standard screwdriver (5/16-inch x 6 inch)
Phillips screwdriver (No. 2 x 6 inch)
Combination pliers - 6 inch
Hacksaw and assortment of blades
Tire pressure gauge
Grease gun

Oil can
Fine emery cloth
Wire brush
Battery post and cable cleaning tool
Oil filter wrench
Funnel (medium size)
Safety goggles
Jackstands (2)
Drain pan

Note: *If basic tune-ups are going to be part of routine maintenance, it will be necessary to purchase a good quality stroboscopic timing light and combination tachometer/dwell meter. Although they are included in the list of special tools, it is mentioned here because they are absolutely necessary for tuning most vehicles properly.*

Repair and overhaul tool set

These tools are essential for anyone who plans to perform major repairs and are in addition to those in the maintenance and minor repair tool kit. Included is a comprehensive set of sockets which, though expensive, are invaluable because of their versatility, especially when various extensions and drives are available. We recommend the 1/2-inch drive over the 3/8-inch drive. Although the larger drive is bulky and more expensive, it has the capacity of accepting a very wide range of large sockets. Ideally, however, the mechanic should have a 3/8-inch drive set and a 1/2-inch drive set.

Socket set(s)
Reversible ratchet
Extension - 10 inch
Universal joint
Torque wrench (same size drive as sockets)
Ball peen hammer - 8 ounce
Soft-face hammer (plastic/rubber)
Standard screwdriver (1/4-inch x 6 inch)

Standard screwdriver (stubby - 5/16-inch)
Phillips screwdriver (No. 3 x 8 inch)
Phillips screwdriver (stubby - No. 2)
Pliers - vise grip
Pliers - lineman's
Pliers - needle nose
Pliers - snap-ring (internal and external)
Cold chisel - 1/2-inch
Scribe
Scraper (made from flattened copper tubing)
Centerpunch
Pin punches (1/16, 1/8, 3/16-inch)
Steel rule/straightedge - 12 inch
Allen wrench set (1/8 to 3/8-inch or 4 mm to 10 mm)
A selection of files
Wire brush (large)
Jackstands (second set)
Jack (scissor or hydraulic type)

Note: *Another tool which is often useful is an electric drill with a chuck capacity of 3/8-inch and a set of good quality drill bits.*

Special tools

The tools in this list include those which are not used regularly, are expensive to buy, or which need to be used in accordance with their manufacturer's instructions. Unless these tools will be used frequently, it is not very economical to purchase many of them. A consideration would be to split the cost and use between yourself and a friend or friends. In addition, most of these tools can be obtained from a tool rental shop on a temporary basis.

This list primarily contains only those tools and instruments widely available to the public, and not those special tools produced by the vehicle manufacturer for distribution to dealer service departments. Occasionally, references to the manufacturer's special tools are included in the text of this manual. Generally, an alternative method of doing the job without the special tool is offered. However, sometimes there is no alternative to their use. Where this is the case, and the tool cannot be purchased or borrowed, the work should be turned over to the dealer service department or an automotive repair shop.

Valve spring compressor
Piston ring groove cleaning tool
Piston ring compressor
Piston ring installation tool
Cylinder compression gauge
Cylinder ridge reamer
Cylinder surfacing hone
Cylinder bore gauge
Micrometers and/or dial calipers
Hydraulic lifter removal tool
Balljoint separator
Universal-type puller
Impact screwdriver
Dial indicator set
Stroboscopic timing light (inductive pick-up)
Hand operated vacuum/pressure pump
Tachometer/dwell meter
Universal electrical multimeter
Cable hoist
Brake spring removal and installation tools
Floor jack

Buying tools

For the do-it-yourselfer who is just starting to get involved in vehicle maintenance and repair, there are a number of options available when purchasing tools. If maintenance and minor repair is the extent of the work to be done, the purchase of individual tools is satisfactory. If, on the other hand, extensive work is planned, it would be a good idea to purchase a modest tool set from one of the large retail chain stores. A set can usually be bought at a substantial savings over the individual tool prices, and they often come with a tool box. As additional tools are

needed, add-on sets, individual tools and a larger tool box can be purchased to expand the tool selection. Building a tool set gradually allows the cost of the tools to be spread over a longer period of time and gives the mechanic the freedom to choose only those tools that will actually be used.

Tool stores will often be the only source of some of the special tools that are needed, but regardless of where tools are bought, try to avoid cheap ones, especially when buying screwdrivers and sockets, because they won't last very long. The expense involved in replacing cheap tools will eventually be greater than the initial cost of quality tools.

Care and maintenance of tools

Good tools are expensive, so it makes sense to treat them with respect. Keep them clean and in usable condition and store them properly when not in use. Always wipe off any dirt, grease or metal chips before putting them away. Never leave tools lying around in the work area. Upon completion of a job, always check closely under the hood for tools that may have been left there so they won't get lost during a test drive.

Some tools, such as screwdrivers, pliers, wrenches and sockets, can be hung on a panel mounted on the garage or workshop wall, while others should be kept in a tool box or tray. Measuring instruments, gauges, meters, etc. must be carefully stored where they cannot be damaged by weather or impact from other tools.

When tools are used with care and stored properly, they will last a very long time. Even with the best of care, though, tools will wear out if used frequently. When a tool is damaged or worn out, replace it. Subsequent jobs will be safer and more enjoyable if you do.

How to repair damaged threads

Sometimes, the internal threads of a nut or bolt hole can become stripped, usually from overtightening. Stripping threads is an all-too-common occurrence, especially when working with aluminum parts, because aluminum is so soft that it easily strips out.

Usually, external or internal threads are only partially stripped. After they've been cleaned up with a tap or die, they'll still work. Sometimes, however, threads are badly damaged. When this happens, you've got three choices:

1) Drill and tap the hole to the next suitable oversize and install a larger diameter bolt, screw or stud.
2) Drill and tap the hole to accept a threaded plug, then drill and tap the plug to the original screw size. You can also buy a plug already threaded to the original size. Then you simply drill a hole to the specified size, then run the threaded plug into the hole with a bolt and jam nut. Once the plug is fully seated, remove the jam nut and bolt.
3) The third method uses a patented thread repair kit like Heli-Coil or Slimsert. These easy-to-use kits are designed to repair damaged threads in straight-through holes and blind holes. Both are available as kits which can handle a variety of sizes and thread patterns. Drill the hole, then tap it with the special included tap. Install the Heli-Coil and the hole is back to its original diameter and thread pitch.

Regardless of which method you use, be sure to proceed calmly and carefully. A little impatience or carelessness during one of these relatively simple procedures can ruin your whole day's work and cost you a bundle if you wreck an expensive part.

Working facilities

Not to be overlooked when discussing tools is the workshop. If anything more than routine maintenance is to be carried out, some sort of suitable work area is essential.

It is understood, and appreciated, that many home mechanics do not have a good workshop or garage available, and end up removing an engine or doing major repairs outside. It is recommended, however, that the overhaul or repair be completed under the cover of a roof.

A clean, flat workbench or table of comfortable working height is an absolute necessity. The workbench should be equipped with a vise that has a jaw opening of at least four inches.

As mentioned previously, some clean, dry storage space is also required for tools, as well as the lubricants, fluids, cleaning solvents, etc. which soon become necessary.

Sometimes waste oil and fluids, drained from the engine or cooling system during normal maintenance or repairs, present a disposal problem. To avoid pouring them on the ground or into a sewage system, pour the used fluids into large containers, seal them with caps and take them to an authorized disposal site or recycling center. Plastic jugs, such as old antifreeze containers, are ideal for this purpose.

Always keep a supply of old newspapers and clean rags available. Old towels are excellent for mopping up spills. Many mechanics use rolls of paper towels for most work because they are readily available and disposable. To help keep the area under the vehicle clean, a large cardboard box can be cut open and flattened to protect the garage or shop floor.

Whenever working over a painted surface, such as when leaning over a fender to service something under the hood, always cover it with an old blanket or bedspread to protect the finish. Vinyl covered pads, made especially for this purpose, are available at auto parts stores.

Jacking and towing

Jacking

The jack supplied with the vehicle should only be used for raising the vehicle when changing a tire or placing jackstands under the frame. **Warning:** *Never work under the vehicle or start the engine while this jack is being used as the only means of support.*

The vehicle should be on level ground with the hazard flashers on, the wheels blocked, the parking brake applied and the transmission in Park (automatic) or Reverse (manual). If a tire is being changed, loosen the lug nuts one-half turn and leave them in place until the wheel is raised off the ground.

Place the jack under the vehicle suspension in the indicated position **(see illustration)**. Operate the jack with a slow, smooth motion until the wheel is raised off the ground. Remove the lug nuts, pull off the wheel, install the spare and thread the lug nuts back on with the beveled sides facing in. Tighten them snugly, but wait until the vehicle is lowered to tighten them completely.

Lower the vehicle, remove the jack and tighten the nuts (if loosened or removed) in a criss-cross pattern.

Towing

As a general rule, the vehicle should be towed with professional towing equipment. If towed from the front, the rear wheels should be placed on a towing dolly. If towed from the rear, the front wheels should be placed on a towing dolly. If towed with either two or four wheels on the ground, disconnect the driveshaft(s) from the differential(s).

When a vehicle is towed with the rear wheels raised, the steering wheel must be clamped in the straight ahead position with a special device designed for use during towing. The ignition key must be in the OFF position, since the steering lock mechanism isn't strong enough to hold the front wheels straight while towing.

Equipment specifically designed for towing should be used. It should be attached to the main structural members of the vehicle, not the bumpers or brackets. Safety is a major consideration when towing and all applicable state and local laws must be obeyed. A safety chain system must be used at all times. Remember that power steering and power brakes will not work with the engine off.

Front and rear jacking point - Place the jack on the side of the vehicle under the axle housing

Booster battery (jump) starting

Observe these precautions when using a booster battery to start a vehicle:

a) *Before connecting the booster battery, make sure the ignition switch is in the Off position.*
b) *Turn off the lights, heater and other electrical loads.*
c) *Your eyes should be shielded. Safety goggles are a good idea.*
d) *Make sure the booster battery is the same voltage as the dead one in the vehicle.*
e) *The two vehicles MUST NOT TOUCH each other!*
f) *Make sure the transaxle is in Neutral (manual) or Park (automatic).*
g) *If the booster battery is not a maintenance-free type, remove the vent caps and lay a cloth over the vent holes.*

Connect the red jumper cable to the positive (+) terminals of each battery **(see illustration)**.

Connect one end of the black jumper cable to the negative (-) terminal of the booster battery. The other end of this cable should be connected to a good ground on the vehicle to be started, such as a bolt or bracket on the body.

Start the engine using the booster battery, then, with the engine running at idle speed, disconnect the jumper cables in the reverse order of connection.

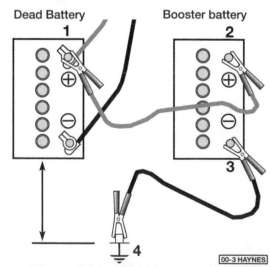

Make the booster battery cable connections in the numerical order shown (note that the negative cable of the booster battery is NOT attached to the negative terminal of the dead battery)

Automotive chemicals and lubricants

A number of automotive chemicals and lubricants are available for use during vehicle maintenance and repair. They include a wide variety of products ranging from cleaning solvents and degreasers to lubricants and protective sprays for rubber, plastic and vinyl.

Cleaners

Carburetor cleaner and choke cleaner is a strong solvent for gum, varnish and carbon. Most carburetor cleaners leave a dry-type lubricant film which will not harden or gum up. Because of this film it is not recommended for use on electrical components.

Brake system cleaner is used to remove grease and brake fluid from the brake system, where clean surfaces are absolutely necessary. It leaves no residue and often eliminates brake squeal caused by contaminants.

Electrical cleaner removes oxidation, corrosion and carbon deposits from electrical contacts, restoring full current flow. It can also be used to clean spark plugs, carburetor jets, voltage regulators and other parts where an oil-free surface is desired.

Demoisturants remove water and moisture from electrical components such as alternators, voltage regulators, electrical connectors and fuse blocks. They are non-conductive, non-corrosive and non-flammable.

Degreasers are heavy-duty solvents used to remove grease from the outside of the engine and from chassis components. They can be sprayed or brushed on and, depending on the type, are rinsed off either with water or solvent.

Lubricants

Motor oil is the lubricant formulated for use in engines. It normally contains a wide variety of additives to prevent corrosion and reduce foaming and wear. Motor oil comes in various weights (viscosity ratings) from 5 to 50. The recommended weight of the oil depends on the season, temperature and the demands on the engine. Light oil is used in cold climates and under light load conditions. Heavy oil is used in hot climates and where high loads are encountered. Multi-viscosity oils are designed to have characteristics of both light and heavy oils and are available in a number of weights from 5W-20 to 20W-50.

Gear oil is designed to be used in differentials, manual transmissions and other areas where high-temperature lubrication is required.

Chassis and wheel bearing grease is a heavy grease used where increased loads and friction are encountered, such as for wheel bearings, balljoints, tie-rod ends and universal joints.

High-temperature wheel bearing grease is designed to withstand the extreme temperatures encountered by wheel bearings in disc brake equipped vehicles. It usually contains molybdenum disulfide (moly), which is a dry-type lubricant.

White grease is a heavy grease for metal-to-metal applications where water is a problem. White grease stays soft under both low and high temperatures (usually from -100 to +190-degrees F), and will not wash off or dilute in the presence of water.

Assembly lube is a special extreme pressure lubricant, usually containing moly, used to lubricate high-load parts (such as main and rod bearings and cam lobes) for initial start-up of a new engine. The assembly lube lubricates the parts without being squeezed out or washed away until the engine oiling system begins to function.

Silicone lubricants are used to protect rubber, plastic, vinyl and nylon parts.

Graphite lubricants are used where oils cannot be used due to contamination problems, such as in locks. The dry graphite will lubricate metal parts while remaining uncontaminated by dirt, water, oil or acids. It is electrically conductive and will not foul electrical contacts in locks such as the ignition switch.

Moly penetrants loosen and lubricate frozen, rusted and corroded fasteners and prevent future rusting or freezing.

Heat-sink grease is a special electrically non-conductive grease that is used for mounting electronic ignition modules where it is essential that heat is transferred away from the module.

Sealants

RTV sealant is one of the most widely used gasket compounds. Made from silicone, RTV is air curing, it seals, bonds, waterproofs, fills surface irregularities, remains flexible, doesn't shrink, is relatively easy to remove, and is used as a supplementary sealer with almost all low and medium temperature gaskets.

Anaerobic sealant is much like RTV in that it can be used either to seal gaskets or to form gaskets by itself. It remains flexible, is solvent resistant and fills surface imperfections. The difference between an anaerobic sealant and an RTV-type sealant is in the curing. RTV cures when exposed to air, while an anaerobic sealant cures only in the absence of air. This means that an anaerobic sealant cures only after the assembly of parts, sealing them together.

Thread and pipe sealant is used for sealing hydraulic and pneumatic fittings and vacuum lines. It is usually made from a Teflon compound, and comes in a spray, a paint-on liquid and as a wrap-around tape.

Chemicals

Anti-seize compound prevents seizing, galling, cold welding, rust and corrosion in fasteners. High-temperature ant-seize, usually made with copper and graphite lubricants, is used for exhaust system and exhaust manifold bolts.

Anaerobic locking compounds are used to keep fasteners from vibrating or working loose and cure only after installation, in the absence of air. Medium strength locking compound is used for small nuts, bolts and screws that may be removed later. High-strength locking compound is for large nuts, bolts and studs which aren't removed on a regular basis.

Oil additives range from viscosity index improvers to chemical treatments that claim to reduce internal engine friction. It should be noted that most oil manufacturers caution against using additives with their oils.

Gas additives perform several functions, depending on their chemical makeup. They usually contain solvents that help dissolve gum and varnish that build up on carburetor, fuel injection and intake parts. They also serve to break down carbon deposits that form on the inside surfaces of the combustion chambers. Some additives contain upper cylinder lubricants for valves and piston rings, and others contain chemicals to remove condensation from the gas tank.

Miscellaneous

Brake fluid is specially formulated hydraulic fluid that can withstand the heat and pressure encountered in brake systems. Care must be taken so this fluid does not come in contact with painted surfaces or plastics. An opened container should always be resealed to prevent contamination by water or dirt.

Weatherstrip adhesive is used to bond weatherstripping around doors, windows and trunk lids. It is sometimes used to attach trim pieces.

Undercoating is a petroleum-based, tar-like substance that is designed to protect metal surfaces on the underside of the vehicle from corrosion. It also acts as a sound-deadening agent by insulating the bottom of the vehicle.

Waxes and polishes are used to help protect painted and plated surfaces from the weather. Different types of paint may require the use of different types of wax and polish. Some polishes utilize a chemical or abrasive cleaner to help remove the top layer of oxidized (dull) paint on older vehicles. In recent years many non-wax polishes that contain a wide variety of chemicals such as polymers and silicones have been introduced. These non-wax polishes are usually easier to apply and last longer than conventional waxes and polishes.

Conversion factors

Length (distance)

Inches (in)	X	25.4	= Millimetres (mm)	X 0.0394	= Inches (in)
Feet (ft)	X	0.305	= Metres (m)	X 3.281	= Feet (ft)
Miles	X	1.609	= Kilometres (km)	X 0.621	= Miles

Volume (capacity)

Cubic inches (cu in; in³)	X	16.387	= Cubic centimetres (cc; cm³)	X 0.061	= Cubic inches (cu in; in³)
Imperial pints (Imp pt)	X	0.568	= Litres (l)	X 1.76	= Imperial pints (Imp pt)
Imperial quarts (Imp qt)	X	1.137	= Litres (l)	X 0.88	= Imperial quarts (Imp qt)
Imperial quarts (Imp qt)	X	1.201	= US quarts (US qt)	X 0.833	= Imperial quarts (Imp qt)
US quarts (US qt)	X	0.946	= Litres (l)	X 1.057	= US quarts (US qt)
Imperial gallons (Imp gal)	X	4.546	= Litres (l)	X 0.22	= Imperial gallons (Imp gal)
Imperial gallons (Imp gal)	X	1.201	= US gallons (US gal)	X 0.833	= Imperial gallons (Imp gal)
US gallons (US gal)	X	3.785	= Litres (l)	X 0.264	= US gallons (US gal)

Mass (weight)

Ounces (oz)	X	28.35	= Grams (g)	X 0.035	= Ounces (oz)
Pounds (lb)	X	0.454	= Kilograms (kg)	X 2.205	= Pounds (lb)

Force

Ounces-force (ozf; oz)	X	0.278	= Newtons (N)	X 3.6	= Ounces-force (ozf; oz)
Pounds-force (lbf; lb)	X	4.448	= Newtons (N)	X 0.225	= Pounds-force (lbf; lb)
Newtons (N)	X	0.1	= Kilograms-force (kgf; kg)	X 9.81	= Newtons (N)

Pressure

Pounds-force per square inch (psi; lbf/in²; lb/in²)	X	0.070	= Kilograms-force per square centimetre (kgf/cm²; kg/cm²)	X 14.223	= Pounds-force per square inch (psi; lbf/in²; lb/in²)
Pounds-force per square inch (psi; lbf/in²; lb/in²)	X	0.068	= Atmospheres (atm)	X 14.696	= Pounds-force per square inch (psi; lbf/in²; lb/in²)
Pounds-force per square inch (psi; lbf/in²; lb/in²)	X	0.069	= Bars	X 14.5	= Pounds-force per square inch (psi; lbf/in²; lb/in²)
Pounds-force per square inch (psi; lbf/in²; lb/in²)	X	6.895	= Kilopascals (kPa)	X 0.145	= Pounds-force per square inch (psi; lbf/in²; lb/in²)
Kilopascals (kPa)	X	0.01	= Kilograms-force per square centimetre (kgf/cm²; kg/cm²)	X 98.1	= Kilopascals (kPa)

Torque (moment of force)

Pounds-force inches (lbf in; lb in)	X	1.152	= Kilograms-force centimetre (kgf cm; kg cm)	X 0.868	= Pounds-force inches (lbf in; lb in)
Pounds-force inches (lbf in; lb in)	X	0.113	= Newton metres (Nm)	X 8.85	= Pounds-force inches (lbf in; lb in)
Pounds-force inches (lbf in; lb in)	X	0.083	= Pounds-force feet (lbf ft; lb ft)	X 12	= Pounds-force inches (lbf in; lb in)
Pounds-force feet (lbf ft; lb ft)	X	0.138	= Kilograms-force metres (kgf m; kg m)	X 7.233	= Pounds-force feet (lbf ft; lb ft)
Pounds-force feet (lbf ft; lb ft)	X	1.356	= Newton metres (Nm)	X 0.738	= Pounds-force feet (lbf ft; lb ft)
Newton metres (Nm)	X	0.102	= Kilograms-force metres (kgf m; kg m)	X 9.804	= Newton metres (Nm)

Vacuum

Inches mercury (in. Hg)	X	3.377	= Kilopascals (kPa)	X 0.2961	= Inches mercury
Inches mercury (in. Hg)	X	25.4	= Millimeters mercury (mm Hg)	X 0.0394	= Inches mercury

Power

Horsepower (hp)	X	745.7	= Watts (W)	X 0.0013	= Horsepower (hp)

Velocity (speed)

Miles per hour (miles/hr; mph)	X	1.609	= Kilometres per hour (km/hr; kph)	X 0.621	= Miles per hour (miles/hr; mph)

Fuel consumption*

Miles per gallon, Imperial (mpg)	X	0.354	= Kilometres per litre (km/l)	X 2.825	= Miles per gallon, Imperial (mpg)
Miles per gallon, US (mpg)	X	0.425	= Kilometres per litre (km/l)	X 2.352	= Miles per gallon, US (mpg)

Temperature

Degrees Fahrenheit = (°C x 1.8) + 32

Degrees Celsius (Degrees Centigrade; °C) = (°F - 32) x 0.56

*It is common practice to convert from miles per gallon (mpg) to litres/100 kilometres (l/100km), where mpg (Imperial) x l/100 km = 282 and mpg (US) x l/100 km = 235

Safety first!

Regardless of how enthusiastic you may be about getting on with the job at hand, take the time to ensure that your safety is not jeopardized. A moment's lack of attention can result in an accident, as can failure to observe certain simple safety precautions. The possibility of an accident will always exist, and the following points should not be considered a comprehensive list of all dangers. Rather, they are intended to make you aware of the risks and to encourage a safety conscious approach to all work you carry out on your vehicle.

Essential DOs and DON'Ts

DON'T rely on a jack when working under the vehicle. Always use approved jackstands to support the weight of the vehicle and place them under the recommended lift or support points.

DON'T attempt to loosen extremely tight fasteners (i.e. wheel lug nuts) while the vehicle is on a jack - it may fall.

DON'T start the engine without first making sure that the transmission is in Neutral (or Park where applicable) and the parking brake is set.

DON'T remove the radiator cap from a hot cooling system - let it cool or cover it with a cloth and release the pressure gradually.

DON'T attempt to drain the engine oil until you are sure it has cooled to the point that it will not burn you.

DON'T touch any part of the engine or exhaust system until it has cooled sufficiently to avoid burns.

DON'T siphon toxic liquids such as gasoline, antifreeze and brake fluid by mouth, or allow them to remain on your skin.

DON'T inhale brake lining dust - it is potentially hazardous (see *Asbestos* below).

DON'T allow spilled oil or grease to remain on the floor - wipe it up before someone slips on it.

DON'T use loose fitting wrenches or other tools which may slip and cause injury.

DON'T push on wrenches when loosening or tightening nuts or bolts. Always try to pull the wrench toward you. If the situation calls for pushing the wrench away, push with an open hand to avoid scraped knuckles if the wrench should slip.

DON'T attempt to lift a heavy component alone - get someone to help you.

DON'T rush or take unsafe shortcuts to finish a job.

DON'T allow children or animals in or around the vehicle while you are working on it.

DO wear eye protection when using power tools such as a drill, sander, bench grinder, etc. and when working under a vehicle.

DO keep loose clothing and long hair well out of the way of moving parts.

DO make sure that any hoist used has a safe working load rating adequate for the job.

DO get someone to check on you periodically when working alone on a vehicle.

DO carry out work in a logical sequence and make sure that everything is correctly assembled and tightened.

DO keep chemicals and fluids tightly capped and out of the reach of children and pets.

DO remember that your vehicle's safety affects that of yourself and others. If in doubt on any point, get professional advice.

Asbestos

Certain friction, insulating, sealing, and other products - such as brake linings, brake bands, clutch linings, torque converters, gaskets, etc. - may contain asbestos. Extreme care must be taken to avoid inhalation of dust from such products, since it is hazardous to health. If in doubt, assume that they do contain asbestos.

Fire

Remember at all times that gasoline is highly flammable. Never smoke or have any kind of open flame around when working on a vehicle. But the risk does not end there. A spark caused by an electrical short circuit, by two metal surfaces contacting each other, or even by static electricity built up in your body under certain conditions, can ignite gasoline vapors, which in a confined space are highly explosive. Do not, under any circumstances, use gasoline for cleaning parts. Use an approved safety solvent.

Always disconnect the battery ground (-) cable at the battery before working on any part of the fuel system or electrical system. Never risk spilling fuel on a hot engine or exhaust component. It is strongly recommended that a fire extinguisher suitable for use on fuel and electrical fires be kept handy in the garage or workshop at all times. Never try to extinguish a fuel or electrical fire with water.

Fumes

Certain fumes are highly toxic and can quickly cause unconsciousness and even death if inhaled to any extent. Gasoline vapor falls into this category, as do the vapors from some cleaning solvents. Any draining or pouring of such volatile fluids should be done in a well ventilated area.

When using cleaning fluids and solvents, read the instructions on the container carefully. Never use materials from unmarked containers.

Never run the engine in an enclosed space, such as a garage. Exhaust fumes contain carbon monoxide, which is extremely poisonous. If you need to run the engine, always do so in the open air, or at least have the rear of the vehicle outside the work area.

If you are fortunate enough to have the use of an inspection pit, never drain or pour gasoline and never run the engine while the vehicle is over the pit. The fumes, being heavier than air, will concentrate in the pit with possibly lethal results.

The battery

Never create a spark or allow a bare light bulb near a battery. They normally give off a certain amount of hydrogen gas, which is highly explosive.

Always disconnect the battery ground (-) cable at the battery before working on the fuel or electrical systems.

If possible, loosen the filler caps or cover when charging the battery from an external source (this does not apply to sealed or maintenance-free batteries). Do not charge at an excessive rate or the battery may burst.

Take care when adding water to a non maintenance-free battery and when carrying a battery. The electrolyte, even when diluted, is very corrosive and should not be allowed to contact clothing or skin.

Always wear eye protection when cleaning the battery to prevent the caustic deposits from entering your eyes.

Household current

When using an electric power tool, inspection light, etc., which operates on household current, always make sure that the tool is correctly connected to its plug and that, where necessary, it is properly grounded. Do not use such items in damp conditions and, again, do not create a spark or apply excessive heat in the vicinity of fuel or fuel vapor.

Secondary ignition system voltage

A severe electric shock can result from touching certain parts of the ignition system (such as the spark plug wires) when the engine is running or being cranked, particularly if components are damp or the insulation is defective. In the case of an electronic ignition system, the secondary system voltage is much higher and could prove fatal.

Troubleshooting

Contents

Symptom	Section

Engine and performance

Alternator light fails to come on when key is turned on 13
Alternator light stays on 12
Battery will not hold a charge 11
Engine backfires 18
Engine diesels (continues to run) after being turned off 21
Engine hard to start when cold 4
Engine hard to start when hot 5
Engine lacks power 17
Engine 'lopes' while idling or idles erratically 8
Engine misses at idle speed 9
Engine misses throughout driving speed range 4
Engine rotates but will not start 2
Engine stalls 16
Engine starts but stops immediately 7
Engine surges while holding accelerator steady 19
Engine will not rotate when attempting to start 1
Excessive fuel consumption 24
Excessively high idle speed 10
Excessive oil consumption 23
Fuel odor 25
Hesitation or stumble during acceleration 15
Low oil pressure 22
Miscellaneous engine noises 26
Pinging or knocking engine sounds when engine is under load 20
Starter motor noisy or engages roughly 6
Starter motor operates without turning engine 3

Cooling system

Abnormal coolant loss 31
Corrosion 33
External coolant leakage 29
Internal coolant leakage 30
Overcooling 28
Overheating 27
Poor coolant circulation 32

Clutch

Clutch pedal stays on floor when disengaged 39
Clutch slips (engine speed increases
    with no increase in vehicle speed) 35
Fails to release (pedal pressed to the floor - shift lever
    does not move freely in and out of Reverse) 34
Grabbing (chattering) as clutch is engaged 36
Squeal or rumble with clutch disengaged (pedal depressed) 38
Squeal or rumble with clutch engaged (pedal released) 37

Manual transmission

Difficulty engaging gears 45
Noise occurs while shifting gears 46
Noisy in all gears 41
Noisy in Neutral with engine running 40
Noisy in one particular gear 42
Oil leaks 44
Slips out of gear 43

Automatic transmission

Engine will start in gears other than Park or Neutral 50

Symptom	Section

Fluid leakage 47
General shift mechanism problems 48
Transmission slips, shifts rough, is noisy or has
    no drive in forward or Reverse gears 51
Transmission will not downshift with the
    accelerator pedal pressed to the floor 49

Driveshaft

Knock or clunk when transmission is under initial
    load (just after transmission is put into gear) 53
Leaks at front of driveshaft 52
Metallic grating sound consistent with vehicle speed 54
Scraping noise 56
Vibration 55
Whining or whistling noise 57

Rear axle and differential

Knocking sound when starting or shifting gears 59
Noise - same when in drive as when vehicle is coasting 58
Noise when turning 60
Oil leaks 62
Vibration 61

Transfer case (4WD models)

Difficult shifting 64
Gear jumping out of mesh 63
Noise 65

Brakes

Brake pedal feels spongy when depressed 69
Brake pedal pulsates during brake application 72
Brakes drag (indicated by sluggish engine performance
    or wheels being very hot after driving) 73
Excessive brake pedal travel 68
Excessive effort required to stop vehicle 70
Noise (high-pitched squeal) 67
Pedal travels to the floor with little resistance 71
Rear brakes lock up under heavy brake application 75
Rear brakes lock up under light brake application 74
Vehicle pulls to one side during braking 66

Suspension and steering

Excessively stiff steering 80
Excessive pitching and/or rolling around
    corners or during braking 78
Excessive play in steering 81
Excessive tire wear (not specific to one area) 87
Excessive tire wear on inside edge 89
Excessive tire wear on outside edge 88
Lack of power assistance 82
Miscellaneous noises 86
Noisy power steering pump 85
Shimmy, shake or vibration 77
Steering effort not the same in both
    directions (power system) 84
Steering wheel fails to return to straight-ahead position 83
Tire tread worn in one place 90
Vehicle pulls to one side 76
Wandering or general instability 79

Engine and performance

1   Engine will not rotate when attempting to start

1   Battery terminal connections loose or corroded. Check the cable terminals at the battery; tighten cable clamp and/or clean off corrosion as necessary (see Chapter 1).

2   Battery discharged or faulty. If the cable ends are clean and tight on the battery posts, turn the key to the On position and switch on the headlights or windshield wipers. If they won't run, the battery is discharged.

3   Automatic transmission not engaged in park (P) or Neutral (N).

4   Broken, loose or disconnected wires in the starting circuit. Inspect all wires and connectors at the battery, starter solenoid and ignition switch (on steering column).

5   Starter motor pinion jammed in flywheel ring gear. If manual transmission, place transmission in gear and rock the vehicle to manually turn the engine. Remove starter (Chapter 5) and inspect pinion and flywheel (Chapter 2) at earliest convenience.

6   Starter solenoid faulty (Chapter 5).

7   Starter motor faulty (Chapter 5).

8   Ignition switch faulty (Chapter 12).

9   Engine seized. Try to turn the crankshaft with a large socket and breaker bar on the pulley bolt.

2   Engine rotates but will not start

1   Fuel tank empty.

2   Battery discharged (engine rotates slowly). Check the operation of electrical components as described in previous Section.

3   Battery terminal connections loose or corroded. See previous Section.

4   Fuel not reaching carburetor or fuel injector. Check for clogged fuel filter or lines and defective fuel pump. Also make sure the tank vent lines aren't clogged (Chapter 4A or 4B).

5   Choke not operating properly (Chapter 1).

6   Faulty distributor components. Check the cap and rotor (Chapter 1).

7   Low cylinder compression. Check as described in Chapter 2.

8   Valve clearances not properly adjusted (Chapter 1).

9   Water in fuel. Drain tank and fill with new fuel.

10   Defective ignition coil (Chapter 5).

11   Dirty or clogged carburetor jets or fuel injector (Chapter 4B).

12   Carburetor out of adjustment. Check the float level (Chapter 4A).

13   Wet or damaged ignition components (Chapters 1 and 5).

14   Worn, faulty or incorrectly gapped spark plugs (Chapter 1).

15   Broken, loose or disconnected wires in the starting circuit (see previous Section).

16   Loose distributor (changing ignition timing). Turn the distributor body as necessary to start the engine, then adjust the ignition timing as soon as possible (Chapter 1).

17   Broken, loose or disconnected wires at the ignition coil or faulty coil (Chapter 5).

18   Timing chain or gear failure or wear affecting valve timing (Chapter 2).

3   Starter motor operates without turning engine

1   Starter pinion sticking. Remove the starter (Chapter 5) and inspect.

2   Starter pinion or flywheel/driveplate teeth worn or broken. Remove the inspection cover and inspect.

4   Engine hard to start when cold

1   Battery discharged or low. Check as described in Chapter 1.

2   Fuel not reaching the carburetor or fuel injectors. Check the fuel filter, lines and fuel pump (Chapters 1 and 4A, 4B).

3   Choke inoperative (Chapters 1 and 4).

4   Defective spark plugs (Chapter 1).

5   Fault with the fuel injection or engine management system (Chapter 4B or 6).

5   Engine hard to start when hot

1   Air filter dirty (Chapter 1).

2   Fuel not reaching carburetor or fuel injectors (see Chapter 4A or 4B). Check for a vapor lock situation, brought about by clogged fuel tank vent lines.

3   Bad engine ground connection.

4   Choke sticking (Chapter 1).

5   Defective pick-up coil in distributor (Chapter 5).

6   Float level too high (Chapter 4A).

7   Fault with the fuel injection or engine management system (Chapter 4B or 6).

6   Starter motor noisy or engages roughly

1   Pinion or flywheel/driveplate teeth worn or broken. Remove the inspection cover on the left side of the engine and inspect.

2   Starter motor mounting bolts loose or missing.

7   Engine starts but stops immediately

1   Loose or damaged wire harness connections at distributor, coil or alternator.

2   Intake manifold vacuum leaks. Make sure all mounting bolts/nuts are tight and all vacuum hoses connected to the manifold are attached properly and in good condition (Chapters 2A, 4A, 4B).

3   Insufficient fuel flow to carburetor or fuel injectors (Chapters 4A and 4B).

4   Idle speed incorrect (Chapter 1).

8   Engine 'lopes' while idling or idles erratically

1   Vacuum leaks. Check mounting bolts at the intake manifold for tightness. Make sure that all vacuum hoses are connected and in good condition. Use a stethoscope or a length of fuel hose held against your ear to listen for vacuum leaks while the engine is running. A hissing sound will be heard. A soapy water solution will also detect leaks. Check the intake manifold gasket surfaces.

2   Leaking EGR valve or plugged PCV valve (see Chapters 1 and 6).

3   Air filter clogged (Chapter 1).

4   Fuel pump not delivering sufficient fuel (Chapter 4A, 4B).

5   Leaking head gasket. Perform a cylinder compression check (Chapter 2).

6   Timing gear or chain worn (Chapter 2C).

7   Camshaft lobes worn (Chapter 2).

8   Valve clearance out of adjustment - Chapter 1 .

9   Valves burned or otherwise leaking (Chapter 2).

10   Ignition timing out of adjustment (Chapter 1).

11   Ignition system not operating properly (Chapters 1 and 5).

12   Thermostatic air cleaner not operating properly (Chapter 1).

13   Choke not operating properly (Chapters 1 and 4).

14   Dirty or clogged injector(s). Carburetor dirty, clogged or out of adjustment. Check the float level (Chapter 4A, 4B).

15   Idle speed out of adjustment (Chapter 1).

9   Engine misses at idle speed

1   Spark plugs faulty or not gapped properly (Chapter 1).

2   Faulty spark plug wires (Chapter 1).
3   Wet or damaged distributor components (Chapter 1).
4   Short circuits in ignition, coil or spark plug wires.
5   Sticking or faulty emissions systems (see Chapter 6).
6   Clogged fuel filter and/or foreign matter in fuel. Remove the fuel filter (Chapter 1) and inspect.
7   Vacuum leaks at intake manifold or hose connections. Check as described in Section 8.
8   Incorrect idle speed (Chapter 1) or idle mixture (Chapter 4).
9   Incorrect ignition timing (Chapter 1).
10   Low or uneven cylinder compression. Check as described in Chapter 2.
11   Choke not operating properly (Chapter 1).
12   Clogged or dirty fuel injectors (Chapter 4).

10   Excessively high idle speed

1   Sticking throttle linkage (Chapter 4A).
2   Choke opened excessively at idle (Chapter 4A).
3   Idle speed incorrectly adjusted (Chapter 1).
4   Valve clearance incorrectly adjusted (Chapter 1).

11   Battery will not hold a charge

1   Alternator drivebelt defective or not adjusted properly (Chapter 1).
2   Battery cables loose or corroded (Chapter 1).
3   Alternator not charging properly (Chapter 5).
4   Loose, broken or faulty wires in the charging circuit (Chapter 5).
5   Short circuit causing a continuous drain on the battery.
6   Battery defective internally.

12   Alternator light stays on

1   Fault in alternator or charging circuit (Chapter 5).
2   Alternator drivebelt defective or not properly adjusted (Chapter 1).

13   Alternator light fails to come on when key is turned on

1   Faulty bulb (Chapter 12).
2   Defective alternator (Chapter 5).
3   Fault in the printed circuit, dash wiring or bulb holder (Chapter 12).

14   Engine misses throughout driving speed range

1   Fuel filter clogged and/or impurities in the fuel system. Check fuel filter (Chapter 1) or clean system (Chapter 4).
2   Faulty or incorrectly gapped spark plugs (Chapter 1).
3   Incorrect ignition timing (Chapter 1).
4   Cracked distributor cap, disconnected distributor wires or damaged distributor components (Chapter 1).
5   Defective spark plug wires (Chapter 1).
6   Emissions system components faulty (Chapter 6).
7   Low or uneven cylinder compression pressures. Check as described in Chapter 2.
8   Weak or faulty ignition coil (Chapter 5).
9   Weak or faulty ignition system (Chapter 5).
10   Vacuum leaks at intake manifold or vacuum hoses (see Section 8).
11   Dirty or clogged carburetor or fuel injector (Chapter 4A, 4B).
12   Leaky EGR valve (Chapter 6).
13   Carburetor out of adjustment (Chapter 4A).
14   Idle speed out of adjustment (Chapter 1).

15   Hesitation or stumble during acceleration

1   Ignition timing incorrect (Chapter 1).
2   Ignition system not operating properly (Chapter 5).
3   Dirty or clogged carburetor or fuel injector (Chapter 4A, 4B).
4   Low fuel pressure. Check for proper operation of the fuel pump and for restrictions in the fuel filter and lines (Chapter 4A).
5   Carburetor out of adjustment (Chapter 4A).
6   Fault with the fuel injection or engine management system (Chapter 4B or 6).

16   Engine stalls

1   Idle speed incorrect (Chapter 1).
2   Fuel filter clogged and/or water and impurities in the fuel system (Chapter 1).
3   Choke not operating properly (Chapter 1).
4   Damaged or wet distributor cap and wires.
5   Emissions system components faulty (Chapter 6).
6   Faulty or incorrectly gapped spark plugs (Chapter 1). Also check the spark plug wires (Chapter 1).
7   Vacuum leak at the carburetor, intake manifold or vacuum hoses. Check as described in Section 8.
8   Valve clearances incorrect (Chapter 1).
9   Fault with the fuel injection or engine management system (Chapter 4B or 6).

17   Engine lacks power

1   Incorrect ignition timing (Chapter 1).
2   Excessive play in distributor shaft. At the same time check for faulty distributor cap, wires, etc. (Chapter 1).
3   Faulty or incorrectly gapped spark plugs (Chapter 1).
4   Air filter dirty (Chapter 1).
5   Faulty ignition coil (Chapter 5).
6   Brakes binding (Chapters 1 and 10).
7   Automatic transmission fluid level incorrect, causing slippage (Chapter 1).
8   Clutch slipping (Chapter 8).
9   Fuel filter clogged and/or impurities in the fuel system (Chapters 1 and 4).
10   EGR system not functioning properly (Chapter 6).
11   Use of sub-standard fuel. Fill tank with proper octane fuel.
12   Low or uneven cylinder compression pressures. Check as described in Chapter 2D.
13   Air leak at carburetor or intake manifold (check as described in Section 8).
14   Dirty or clogged carburetor jets or malfunctioning choke (Chapters 1 and 4).
15   Fault with the fuel injection or engine management system (Chapter 4B or 6).

18   Engine backfires

1   EGR system not functioning properly (Chapter 6).
2   Ignition timing incorrect (Chapter 1).
3   Thermostatic air cleaner system not operating properly (Chapter 6).
4   Vacuum leak (refer to Section 8).
5   Valve clearances incorrect (Chapter 1).
6   Damaged valve springs or sticking valves (Chapter 2).
7   Intake air leak (see Section 8).
8   Carburetor float level out of adjustment (Chapter 4A).

19 Engine surges while holding accelerator steady

1 Intake air leak (see Section 8).
2 Fuel pump not working properly (Chapter 4A, 4B).
3 Fault with the fuel injection or engine management system (Chapter 4B or 6).

20 Pinging or knocking engine sounds when engine is under load

1 Incorrect grade of fuel. Fill tank with fuel of the proper octane rating.
2 Ignition timing incorrect (Chapter 1).
3 Carbon build-up in combustion chambers. Remove cylinder heads and clean combustion chambers (Chapter 2).
4 Incorrect spark plugs (Chapter 1).

21 Engine diesels (continues to run) after being turned off

1 Idle speed too high (Chapter 1).
2 Ignition timing incorrect (Chapter 1).
3 Incorrect spark plug heat range (Chapter 1).
4 Intake air leak (see Section 8).
5 Carbon build-up in combustion chambers. Remove the cylinder heads and clean the combustion chambers (Chapter 2).
6 Valves sticking (Chapter 2).
7 Valve clearances incorrect (Chapter 1).
8 EGR system not operating properly (Chapter 6).
9 Fuel shut-off system not operating properly (Chapter 6).
10 Check for causes of overheating (Section 27).

22 Low oil pressure

1 Improper grade of oil.
2 Oil pump worn or damaged (Chapter 2).
3 Engine overheating (refer to Section 27).
4 Clogged oil filter (Chapter 1).
5 Clogged oil strainer (Chapter 2).
6 Oil pressure gauge not working properly (Chapter 2D).

23 Excessive oil consumption

1 Loose oil drain plug.
2 Loose bolts or damaged oil pan gasket (Chapter 2).
3 Loose bolts or damaged front cover gasket (Chapter 2).
4 Front or rear crankshaft oil seal leaking (Chapter 2).
5 Loose bolts or damaged rocker arm cover gasket (Chapter 2).
6 Loose oil filter (Chapter 1).
7 Loose or damaged oil pressure switch (Chapter 2).
8 Pistons and cylinders excessively worn (Chapter 2).
9 Piston rings not installed correctly on pistons (Chapter 2).
10 Worn or damaged piston rings (Chapter 2).
11 Intake and/or exhaust valve oil seals worn or damaged (Chapter 2).
12 Worn valve stems.
13 Worn or damaged valves/guides (Chapter 2).

24 Excessive fuel consumption

1 Dirty or clogged air filter element (Chapter 1).
2 Incorrect ignition timing (Chapter 1).
3 Incorrect idle speed (Chapter 1).
4 Low tire pressure or incorrect tire size (Chapter 11).

5 Fuel leakage. Check all connections, lines and components in the fuel system (Chapter 4A).
6 Choke not operating properly (Chapter 1).
7 Dirty or clogged carburetor jets or fuel injectors (Chapter 4A, 4B).
8 Fault with the fuel injection or engine management system (Chapter 4B or 6).

25 Fuel odor

1 Fuel leakage. Check all connections, lines and components in the fuel system (Chapter 4A).
2 Fuel tank overfilled. Fill only to automatic shut-off.
3 Charcoal canister filter in Evaporative Emissions Control system clogged (Chapter 1).
4 Vapor leaks from Evaporative Emissions Control system lines (Chapter 6).

26 Miscellaneous engine noises

1 A strong dull noise that becomes more rapid as the engine accelerates indicates worn or damaged crankshaft bearings or an unevenly worn crankshaft. To pinpoint the trouble spot, remove the spark plug wire from one plug at a time and crank the engine over. If the noise stops, the cylinder with the removed plug wire indicates the problem area. Replace the bearing and/or service or replace the crankshaft (Chapter 2).
2 A similar (yet slightly higher pitched) noise to the crankshaft knocking described in the previous paragraph, that becomes more rapid as the engine accelerates, indicates worn or damaged connecting rod bearings (Chapter 2). The procedure for locating the problem cylinder is the same as described in Paragraph 1.
3 An overlapping metallic noise that increases in intensity as the engine speed increases, yet diminishes as the engine warms up indicates abnormal piston and cylinder wear (Chapter 2). To locate the problem cylinder, use the procedure described in Paragraph 1.
4 A rapid clicking noise that becomes faster as the engine accelerates indicates a worn piston pin or piston pin hole. This sound will happen each time the piston hits the highest and lowest points in the stroke (Chapter 2). The procedure for locating the problem piston is described in Paragraph 1.
5 A metallic clicking noise coming from the water pump indicates worn or damaged water pump bearings or pump. Replace the water pump with a new one (Chapter 3).
6 A rapid tapping sound or clicking sound that becomes faster as the engine speed increases indicates "valve tapping" or improperly adjusted valve clearances. This can be identified by holding one end of a section of hose to your ear and placing the other end at different spots along the rocker arm cover. The point where the sound is loudest indicates the problem valve. Adjust the valve clearance (Chapter 1 or 2B). If the problem persists, you likely have a collapsed valve lifter or other damaged valve train component. Changing the engine oil and adding a high viscosity oil treatment will sometimes cure a stuck lifter problem. If the problem still persists, the lifters, pushrods and rocker arms must be removed for inspection (see Chapter 2).
7 A steady metallic rattling or rapping sound coming from the area of the timing chain cover indicates a worn, damaged or out-of-adjustment timing chain. Service or replace the chain and related components (Chapter 2).

Cooling system

27 Overheating

1 Insufficient coolant in system (Chapter 1).
2 Drivebelt defective or not adjusted properly (Chapter 1).

3   Radiator core blocked or radiator grille dirty or restricted (Chapter 3).
4   Thermostat faulty (Chapter 3).
5   Fan not functioning properly (Chapter 3).
6   Radiator cap not maintaining proper pressure. Have cap pressure tested by gas station or repair shop.
7   Ignition timing incorrect (Chapter 1).
8   Defective water pump (Chapter 3).
9   Improper grade of engine oil.
10   Inaccurate temperature gauge (Chapter 12).

28   Overcooling

1   Thermostat faulty (Chapter 3).
2   Inaccurate temperature gauge (Chapter 12).

29   External coolant leakage

1   Deteriorated or damaged hoses. Loose clamps at hose connections (Chapter 1).
2   Water pump seals defective. If this is the case, water will drip from the weep hole in the water pump body (Chapter 3).
3   Leakage from radiator core or header tank. This will require the radiator to be professionally repaired (see Chapter 3 for removal procedures).
4   Engine drain plugs or water jacket freeze plugs leaking (see Chapters 1 and 2).
5   Leak from coolant temperature switch (Chapter 3).
6   Leak from damaged gaskets or small cracks (Chapter 2).
7   Damaged head gasket. This can be verified by checking the condition of the engine oil as noted in Section 30.

30   Internal coolant leakage

Note: *Internal coolant leaks can usually be detected by examining the oil. Check the dipstick and inside the rocker arm cover for water deposits and an oil consistency like that of a milkshake.*
1   Leaking cylinder head gasket. Have the system pressure tested or remove the cylinder head (Chapter 2) and inspect.
2   Cracked cylinder bore or cylinder head. Dismantle engine and inspect (Chapter 2).
3   Loose cylinder head bolts (tighten as described in Chapter 2).

31   Abnormal coolant loss

1   Overfilling system (Chapter 1).
2   Coolant boiling away due to overheating (see causes in Section 27).
3   Internal or external leakage (see Sections 29 and 30).
4   Faulty radiator cap. Have the cap pressure tested.
5   Cooling system being pressurized by engine compression. This could be due to a cracked head or block or leaking head gaskets.

32   Poor coolant circulation

1   Inoperative water pump. A quick test is to pinch the top radiator hose closed with your hand while the engine is idling, then release it. You should feel a surge of coolant if the pump is working properly (Chapter 3).
2   Restriction in cooling system. Drain, flush and refill the system (Chapter 1). If necessary, remove the radiator (Chapter 3) and have it reverse flushed or professionally cleaned.

3   Loose water pump drivebelt (Chapter 1).
4   Thermostat sticking (Chapter 3).
5   Insufficient coolant (Chapter 1).

33   Corrosion

1   Excessive impurities in the water. Soft, clean water is recommended. Distilled or rainwater is satisfactory.
2   Insufficient antifreeze solution (refer to Chapter 1 for the proper ratio of water to antifreeze).
3   Infrequent flushing and draining of system. Regular flushing of the cooling system should be carried out at the specified intervals as described in (Chapter 1).

Clutch

Note: *All clutch related service information is located in Chapter 8, unless otherwise noted.*

34   Fails to release (pedal pressed to the floor - shift lever does not move freely in and out of Reverse)

1   Freeplay incorrectly adjusted.
2   Clutch contaminated with oil. Remove clutch plate and inspect.
3   Clutch plate warped, distorted or otherwise damaged.
4   Diaphragm spring fatigued. Remove clutch cover/pressure plate assembly and inspect.
5   Broken, binding or damaged release cable (cable-operated clutch).
6   Leakage of fluid from clutch hydraulic system. Inspect master cylinder, operating cylinder and connecting lines.
7   Air in clutch hydraulic system. Bleed the system.
8   Insufficient pedal stroke. Check and adjust as necessary.
9   Piston seal in operating cylinder deformed or damaged.
10   Lack of grease on pilot bearing.

35   Clutch slips (engine speed increases with no increase in vehicle speed)

1   Worn or oil-soaked clutch plate.
2   Clutch plate not broken in. It may take 30 or 40 normal starts for a new clutch to seat.
3   Diaphragm spring weak or damaged. Remove clutch cover/pressure plate assembly and inspect.
4   Flywheel warped (Chapter 2).
5   Debris in master cylinder preventing the piston from returning to its normal position.
6   Clutch hydraulic line damaged.
7   Binding in the release mechanism.

36   Grabbing (chattering) as clutch is engaged

1   Oil on clutch plate. Remove and inspect. Repair any leaks.
2   Worn or loose engine or transmission mounts. They may move slightly when clutch is released. Inspect mounts and bolts.
3   Worn splines on transmission input shaft. Remove clutch components and inspect.
4   Warped pressure plate or flywheel. Remove clutch components and inspect.
5   Diaphragm spring fatigued. Remove clutch cover/pressure plate assembly and inspect.
6   Clutch linings hardened or warped.
7   Clutch lining rivets loose.

37 Squeal or rumble with clutch engaged (pedal released)

1   Improper pedal adjustment. Adjust pedal freeplay.
2   Release bearing binding on transmission shaft. Remove clutch components and check bearing. Remove any burrs or nicks, clean and relubricate before reinstallation.
3   Pilot bearing worn or damaged.
4   Clutch rivets loose.
5   Clutch plate cracked.
6   Fatigued clutch plate torsion springs. Replace clutch plate.

38 Squeal or rumble with clutch disengaged (pedal depressed)

1   Worn or damaged release bearing.
2   Worn or broken pressure plate diaphragm fingers.

39 Clutch pedal stays on floor when disengaged

    Binding linkage or release bearing. Inspect linkage or remove clutch components as necessary.

Manual transmission

Note: *All manual transmission service information is located in Chapter 7A, unless otherwise noted.*

40 Noisy in Neutral with engine running

1   Input shaft bearing worn.
2   Damaged main drive gear bearing.
3   Insufficient transmission oil (Chapter 1).
4   Transmission oil in poor condition. Drain and fill with proper grade oil. Check old oil for water and debris (Chapter 1).
5   Noise can be caused by variations in engine torque. Change the idle speed and see if noise disappears.

41 Noisy in all gears

1   Any of the above causes, and/or:
2   Worn or damaged output gear bearings or shaft.

42 Noisy in one particular gear

1   Worn, damaged or chipped gear teeth.
2   Worn or damaged synchronizer.

43 Slips out of gear

1   Transmission loose on clutch housing.
2   Stiff shift lever seal.
3   Shift linkage binding.
4   Broken or loose input gear bearing retainer.
5   Dirt between clutch lever and engine housing.
6   Worn linkage.
7   Damaged or worn check balls, fork rod ball grooves or check springs.

8   Worn mainshaft or countershaft bearings.
9   Loose engine mounts (Chapter 2).
10   Excessive gear end play.
11   Worn synchronizers.

44 Oil leaks

1   Excessive amount of lubricant in transmission (see Chapter 1 for correct checking procedures). Drain lubricant as required.
2   Rear oil seal or speedometer oil seal damaged.
3   To pinpoint a leak, first remove all built-up dirt and grime from the transmission. Degreasing agents and/or steam cleaning will achieve this. With the underside clean, drive the vehicle at low speeds so the air flow will not blow the leak far from its source. Raise the vehicle and determine where the leak is located.

45 Difficulty engaging gears

1   Clutch not releasing completely.
2   Loose or damaged shift linkage. Make a thorough inspection, replacing parts as necessary.
3   Insufficient transmission oil (Chapter 1).
4   Transmission oil in poor condition. Drain and fill with proper grade oil. Check oil for water and debris (Chapter 1).
5   Worn or damaged striking rod.
6   Sticking or jamming gears.

46 Noise occurs while shifting gears

1   Check for proper operation of the clutch (Chapter 8).
2   Faulty synchronizer assemblies.

Automatic transmission

Note: *Due to the complexity of the automatic transmission, it's difficult for the home mechanic to properly diagnose and service. For problems other than the following, the vehicle should be taken to a reputable mechanic.*

47 Fluid leakage

1   Automatic transmission fluid is a deep red color, and fluid leaks should not be confused with engine oil which can easily be blown by air flow to the transmission.
2   To pinpoint a leak, first remove all built-up dirt and grime from the transmission. Degreasing agents and/or steam cleaning will achieve this. With the underside clean, drive the vehicle at low speeds so the air flow will not blow the leak far from its source. Raise the vehicle and determine where the leak is located. Common areas of leakage are:
   a) *Fluid pan: tighten mounting bolts and/or replace pan gasket as necessary (Chapter 1).*
   b) *Rear extension: tighten bolts and/or replace oil seal as necessary.*
   c) *Filler pipe: replace the rubber oil seal where pipe enters transmission case.*
   d) *Transmission oil lines: tighten fittings where lines enter transmission case and/or replace lines.*
   e) *Vent pipe: transmission overfilled and/or water in fluid (see checking procedures, Chapter 1).*
   f) *Speedometer connector: replace the O-ring where speedometer cable enters transmission case.*

48 General shift mechanism problems

Chapter 7B deals with checking and adjusting the shift linkage on automatic transmissions. Common problems which may be caused by out of adjustment linkage are:

a) *Engine starting in gears other than P (park) or N (Neutral).*
b) *Indicator pointing to a gear other than the one actually engaged.*
c) *Vehicle moves with transmission in P (Park) position.*

49 Transmission will not downshift with the accelerator pedal pressed to the floor

Chapter 7B deals with adjusting the TV linkage to enable the transmission to downshift properly.

50 Engine will start in gears other than Park or Neutral

Chapter 7B deals with adjusting the Neutral start switch installed on automatic transmissions.

51 Transmission slips, shifts rough, is noisy or has no drive in forward or Reverse gears

1   There are many probable causes for the above problems, but the home mechanic should concern himself only with one possibility; fluid level.
2   Before taking the vehicle to a shop, check the fluid level and condition as described in Chapter 1. Add fluid, if necessary, or change the fluid and filter if needed. If problems persist, have a professional diagnose the transmission.

Driveshaft

Note: *Refer to Chapter 8, unless otherwise specified, for service information.*

52 Leaks at front of driveshaft

Defective transmission rear seal. See Chapter 7 for replacement procedure. As this is done, check the splined yoke for burrs or roughness that could damage the new seal. Remove burrs with a fine file or whetstone.

53 Knock or clunk when transmission is under initial load (just after transmission is put into gear)

1   Loose or disconnected rear suspension components. Check all mounting bolts and bushings (Chapters 7 and 10).
2   Loose driveshaft bolts. Inspect all bolts and nuts and tighten them securely.
3   Worn or damaged universal joint bearings. Inspect the universal joints (Chapter 8).
4   Worn sleeve yoke and mainshaft spline.

54 Metallic grating sound consistent with vehicle speed

Pronounced wear in the universal joint bearings. Replace U-joints or driveshafts, as necessary.

55 Vibration

Note: *Before blaming the driveshaft, make sure the tires are perfectly balanced and perform the following test.*
1   Install a tachometer inside the vehicle to monitor engine speed as the vehicle is driven. Drive the vehicle and note the engine speed at which the vibration (roughness) is most pronounced. Now shift the transmission to a different gear and bring the engine speed to the same point.
2   If the vibration occurs at the same engine speed (rpm) regardless of which gear the transmission is in, the driveshaft is NOT at fault since the driveshaft speed varies.
3   If the vibration decreases or is eliminated when the transmission is in a different gear at the same engine speed, refer to the following probable causes.
4   Bent or dented driveshaft. Inspect and replace as necessary.
5   Undercoating or built-up dirt, etc. on the driveshaft. Clean the shaft thoroughly.
6   Worn universal joint bearings. Replace the U-joints or driveshaft as necessary.
7   Driveshaft and/or companion flange out of balance. Check for missing weights on the shaft. Remove driveshaft and reinstall 180-degrees from original position, then recheck. Have the driveshaft balanced if problem persists.
8   Loose driveshaft mounting bolts/nuts.
9   Defective center bearing, if so equipped.
10   Worn transmission rear bushing (Chapter 7).

56 Scraping noise

Make sure the dust cover on the sleeve yoke isn't rubbing on the transmission extension housing.

57 Whining or whistling noise

Defective center bearing, if so equipped.

Rear axle and differential

Note: *For differential servicing information, refer to Chapter 8, unless otherwise specified.*

58 Noise - same when in drive as when vehicle is coasting

1   Road noise. No corrective action available.
2   Tire noise. Inspect tires and check tire pressures (Chapter 1).
3   Front wheel bearings loose, worn or damaged (Chapter 1).
4   Insufficient differential oil (Chapter 1).
5   Defective differential.

59 Knocking sound when starting or shifting gears

Defective or incorrectly adjusted differential.

60 Noise when turning

Defective differential.

61 Vibration

See probable causes under Driveshaft. Proceed under the guidelines listed for the driveshaft. If the problem persists, check the rear wheel bearings by raising the rear of the vehicle and spinning the wheels by hand. Listen for evidence of rough (noisy) bearings. Remove and inspect (Chapter 8).

62 Oil leaks

1   Pinion oil seal damaged (Chapter 8).
2   Axleshaft oil seals damaged (Chapter 8).
3   Differential cover leaking. Tighten mounting bolts or replace the gasket as required.
4   Loose filler or drain plug on differential (Chapter 1).
5   Clogged or damaged breather on differential.

Transfer case

Note: *Unless otherwise specified, refer to Chapter 7C for service and repair information.*

63 Gear jumping out of mesh

1   Incorrect control lever freeplay
2   Interference between the control lever and the console.
3   Play or fatigue in the transfer case mounts.
4   Internal wear or incorrect adjustments.

64 Difficult shifting

1   Lack of oil.
2   Internal wear, damage or incorrect adjustment.

65 Noise

1   Lack of oil in transfer case.
2   Noise in 4H and 4L, but not in 2H indicates cause is in the front differential or front axle.
3   Noise in 2H, 4H and 4L indicates cause is in rear differential or rear axle.
4   Noise in 2H and 4H but not in 4L, or in 4L only, indicates internal wear or damage in transfer case.

Brakes

Note: *Before assuming a brake problem exists, make sure the tires are in good condition and inflated properly, the front end alignment is correct and the vehicle is not loaded with weight in an unequal manner. All service procedures for the brakes are included in Chapter 9, unless otherwise noted.*

66 Vehicle pulls to one side during braking

1   Defective, damaged or oil contaminated brake pad on one side. Inspect as described in Chapter 1. Refer to Chapter 10 if replacement is required.
2   Excessive wear of brake pad material or disc on one side. Inspect and repair as necessary.

3   Loose or disconnected front suspension components. Inspect and tighten all bolts securely (Chapters 1 and 11).
4   Defective caliper assembly. Remove caliper and inspect for stuck piston or damage.
5   Scored or out of round rotor.
6   Loose caliper mounting bolts.
7   Incorrect wheel bearing adjustment.

67 Noise (high-pitched squeal)

1   Front brake pads worn out. This noise comes from the wear sensor rubbing against the disc. Replace pads with new ones immediately!
2   Glazed or contaminated pads.
3   Dirty or scored disc.
4   Bent support plate.

68 Excessive brake pedal travel

1   Partial brake system failure. Inspect entire system (Chapter 1) and correct as required.
2   Insufficient fluid in master cylinder. Check (Chapter 1) and add fluid - bleed system if necessary.
3   Air in system. Bleed system.
4   Brakes out of adjustment. Check the operation of the automatic adjusters.
5   Defective proportioning valve. Replace valve and bleed system.

69 Brake pedal feels spongy when depressed

1   Air in brake lines. Bleed the brake system.
2   Deteriorated rubber brake hoses. Inspect all system hoses and lines. Replace parts as necessary.
3   Master cylinder mounting nuts loose. Inspect master cylinder bolts (nuts) and tighten them securely.
4   Master cylinder faulty.
5   Incorrect shoe or pad clearance.
6   Defective check valve. Replace valve and bleed system.
7   Clogged reservoir cap vent hole.
8   Deformed rubber brake lines.
9   Soft or swollen caliper seals.
10 Poor quality brake fluid. Bleed entire system and fill with new approved fluid.

70 Excessive effort required to stop vehicle

1   Power brake booster not operating properly.
2   Excessively worn linings or pads. Check and replace if necessary.
3   One or more caliper pistons seized or sticking. Inspect and rebuild as required.
4   Brake pads or linings contaminated with oil or grease. Inspect and replace as required.
5   New pads or linings installed and not yet seated. It'll take a while for the new material to seat against the disc or drum.
6   Worn or damaged master cylinder or caliper assemblies. Check particularly for frozen pistons.
7   Also see causes listed under Section 69.

71 Pedal travels to the floor with little resistance

Little or no fluid in the master cylinder reservoir caused by leaking caliper piston(s) or loose, damaged or disconnected brake lines. Inspect entire system and repair as necessary.

72   Brake pedal pulsates during brake application

1   Wheel bearings damaged, worn or out of adjustment (Chapter 1).
2   Caliper not sliding properly due to improper installation or obstructions. Remove and inspect.
3   Disc not within specifications. Remove the disc and check for excessive lateral runout and parallelism. Have the discs resurfaced or replace them with new ones. Also make sure that all discs are the same thickness.
4   Out-of-round rear brake drums. Remove the drums and have them resurfaced or replace them with new ones.

73   Brakes drag (indicated by sluggish engine performance or wheels being very hot after driving)

1   Output rod adjustment incorrect at the brake pedal.
2   Obstructed master cylinder compensator. Disassemble master cylinder and clean.
3   Master cylinder piston seized in bore. Overhaul master cylinder.
4   Caliper assembly in need of overhaul.
5   Brake pads or shoes worn out.
6   Piston cups in master cylinder or caliper assembly deformed. Overhaul master cylinder.
7   Parking brake assembly will not release.
8   Clogged brake lines.
9   Wheel bearings out of adjustment (Chapter 1).
10   Brake pedal height improperly adjusted.
11   Wheel cylinder needs overhaul.
12   Improper shoe-to-drum clearance. Adjust as necessary.

74   Rear brakes lock up under light brake application

1   Tire pressures too high.
2   Tires excessively worn (Chapter 1).

75   Rear brakes lock up under heavy brake application

1   Tire pressures too high.
2   Tires excessively worn (Chapter 1).
3   Front brake pads contaminated with oil, mud or water. Clean or replace the pads.
4   Front brake pads excessively worn.
5   Defective master cylinder or caliper assembly.

Suspension and steering

Note: *All service procedures for the suspension and steering systems are included in Chapter 10, unless otherwise noted.*

76   Vehicle pulls to one side

1   Tire pressures uneven (Chapter 1).
2   Defective tire (Chapter 1).
3   Excessive wear in suspension or steering components (Chapter 1).
4   Wheel alignment incorrect.
5   Front brakes dragging. Inspect as described in Section 73.
6   Wheel bearings improperly adjusted (Chapter 1 or 8).
7   Wheel lug nuts loose.

77   Shimmy, shake or vibration

1   Tire or wheel out of balance or out of round. Have them balanced on the vehicle.
2   Loose, worn or out of adjustment wheel bearings (Chapter 1 or 8).
3   Shock absorbers and/or suspension components worn or damaged. Check for worn bushings in the upper and lower links.
4   Wheel lug nuts loose.
5   Incorrect tire pressures.
6   Excessively worn or damaged tire.
7   Loosely mounted steering gear housing.
8   Steering gear improperly adjusted.
9   Loose, worn or damaged steering components.
10   Damaged idler arm.
11   Worn balljoint.

78   Excessive pitching and/or rolling around corners or during braking

1   Defective shock absorbers. Replace as a set.
2   Broken or weak leaf springs and/or suspension components.
3   Worn or damaged stabilizer bar or bushings.

79   Wandering or general instability

1   Improper tire pressures.
2   Worn or damaged upper and lower link or tension rod bushings.
3   Incorrect front end alignment.
4   Worn or damaged steering linkage or suspension components.
5   Improperly adjusted steering gear.
6   Out-of-balance wheels.
7   Loose wheel lug nuts.
8   Worn rear shock absorbers.
9   Fatigued or damaged rear leaf springs.

80   Excessively stiff steering

1   Lack of lubricant in power steering fluid reservoir, where appropriate (Chapter 1).
2   Incorrect tire pressures (Chapter 1).
3   Lack of lubrication at balljoints (Chapter 1).
4   Front end out of alignment.
5   Steering gear out of adjustment or lacking lubrication.
6   Improperly adjusted wheel bearings.
7   Worn or damaged steering gear.
8   Interference of steering column with turn signal switch.
9   Low tire pressures.
10   Worn or damaged balljoints.
11   Worn or damaged steering linkage.
12   See also Section 79.

81   Excessive play in steering

1   Loose wheel bearings (Chapter 1 or 8).
2   Excessive wear in suspension bushings (Chapter 1).
3   Steering gear improperly adjusted.
4   Incorrect wheel alignment.
5   Steering gear mounting bolts loose.
6   Worn steering linkage.

82 Lack of power assistance

1 Steering pump drivebelt faulty or not adjusted properly (Chapter 1).
2 Fluid level low (Chapter 1).
3 Hoses or pipes restricting the flow. Inspect and replace parts as necessary.
4 Air in power steering system. Bleed system.
5 Defective power steering pump.

83 Steering wheel fails to return to straight-ahead position

1 Incorrect front end alignment.
2 Tire pressures low.
3 Steering gear worn or damaged.
4 Steering column out of alignment.
5 Worn or damaged balljoint.
6 Worn or damaged steering linkage.
7 Steering linkage in need of lubrication.
8 Insufficient oil in steering gear.
9 Lack of fluid in power steering pump.

84 Steering effort not the same in both directions (power system)

1 Leaks in steering gear.
2 Clogged fluid passage in steering gear.

85 Noisy power steering pump

1 Insufficient oil in pump.
2 Clogged hoses or oil filter in pump.
3 Loose pulley.
4 Improperly adjusted drivebelt (Chapter 1).
5 Defective pump.

86 Miscellaneous noises

1 Improper tire pressures.

2 Insufficiently lubricated balljoint or steering linkage.
3 Loose or worn steering gear, steering linkage or suspension components.
4 Defective shock absorber.
5 Defective wheel bearing.
6 Worn or damaged suspension bushings.
7 Damaged leaf spring.
8 Loose wheel lug nuts.
9 Worn or damaged rear axleshaft spline.
10 Worn or damaged rear shock absorber mounting bushing.
11 Excessive rear axle end play.
12 See also causes of noises at the rear axle and driveshaft.

87 Excessive tire wear (not specific to one area)

1 Incorrect tire pressures.
2 Tires out of balance. Have them balanced on the vehicle.
3 Wheels damaged. Inspect and replace as necessary.
4 Suspension or steering components worn (Chapter 1).

88 Excessive tire wear on outside edge

1 Incorrect tire pressure.
2 Excessive speed in turns.
3 Front end alignment incorrect (excessive toe-in).

89 Excessive tire wear on inside edge

1 Incorrect tire pressure.
2 Front end alignment incorrect (toe-out).
3 Loose or damaged steering components (Chapter 1).

90 Tire tread worn in one place

1 Tires out of balance. Have them balanced on the vehicle.
2 Damaged or buckled wheel. Inspect and replace if necessary.
3 Defective tire.

Notes

Chapter 1
Tune-up and routine maintenance

Contents

	Section
Air filter check and replacement	27
Automatic transmission fluid and filter change	40
Automatic transmission fluid level check	6
Battery check, maintenance and charging	9
Brake check	33
Carburetor choke check	25
Chassis lubrication	15
Clutch/brake pedal height and freeplay adjustment	34
Cooling system check	10
Cooling system servicing (draining, flushing and refilling)	38
Differential lubricant change	43
Differential lubricant level check	19
Drivebelt check, adjustment and replacement	31
Engine oil and filter change	8
Evaporative emissions control system check	23
Exhaust control valve (heat riser) lubrication and check	22
Exhaust Gas Recirculation (EGR) system check	24
Exhaust system check	16
Fluid level checks	4
Front hub and wheel bearing check, repack and adjustment	39
Fuel filter replacement	35
Fuel system check	32
Idle speed check and adjustment	30
Ignition timing check and adjustment	37

	Section
Introduction	2
Maintenance schedule	1
Manual transmission lubricant change	41
Manual transmission lubricant level check	17
Positive Crankcase Ventilation (PCV) valve check and replacement	26
Power steering fluid level check	7
Seat belt check	20
Spark plug replacement	28
Spark plug wire, distributor cap and rotor check and replacement	36
Suspension and steering check	14
Thermostatically controlled air cleaner check	21
Tire and tire pressure checks	5
Tire rotation	13
Transfer case lubricant change	42
Transfer case lubricant level check	18
Tune-up general information	3
Underhood hose check and replacement	11
Valve clearance check and adjustment (1FZ-FE engines)	44
Valve clearance check and adjustment (2F and 3F-E pushrod engines)	29
Wiper blade inspection and replacement	12

Specifications

Recommended lubricants and fluids

Engine oil
  Type .... API grade SG or SH energy conserving oil
  Viscosity .... See accompanying chart

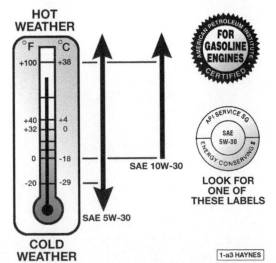

Engine oil viscosity chart - for best fuel economy and cold starting, select the lowest SAE viscosity grade for the expected temperature range

1-a3 HAYNES

Coolant .... Ethylene glycol based anti-freeze
Brake fluid.. DOT 3 brake fluid
Clutch fluid... DOT 3 brake fluid
Power steering fluid .... DEXRON II automatic transmission fluid
Automatic transmission fluid .... DEXRON II automatic transmission fluid
Manual transmission lubricant... API GL-4 or GL-5 SAE 80W-90 gear oil
Transfer case lubricant .... API GL-4 or GL-5 SAE 80W-90 gear oil
Differential lubricant... API GL-5 SAE 80W-90 hypoid gear oil
Chassis grease .... NLGI No. 2 lithium base chassis grease

Capacities*

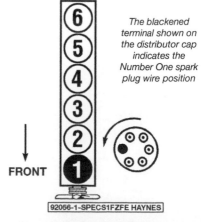

Engine oil (with filter change)
    2F engine.. 8.2 quarts (7.8 liters)
    3F-E engine.. 8.2 quarts (7.8 liters)
    1FZ-FE engine... 7.8 quarts (7.4 liters)
Cooling system
    FJ60 and FJ62 .... 17.4 quarts (16.5 liters)
    FJ80 .... 15.0 quarts (14.2 liters)
Automatic transmission (drain and refill)
    FJ62 .... 5.3 quarts (5.0 liters)
    FJ80 .... 6.3 quarts (6.0 liters)
Manual transmission (drain and refill)
    H41 4-speed .... 3.3 quarts (3.5 liters)
    H55F 5-speed .... 4.7 quarts (4.9 liters)
Transfer case (drain and refill)
    Without ABS... 1.8 quarts (1.7 liters)
    With ABS .... 1.4 quarts (1.3 liters)
Differential
    Front axle
        Without differential lock .... 2.9 quarts (2.8 liters)
        With differential lock .... 2.8 quarts (2.6 liters)
    Rear axle .... 3.4 quarts (3.2 liters)

All capacities approximate. Add as necessary to bring to appropriate level.

The blackened terminal shown on the distributor cap indicates the Number One spark plug wire position

92056-1-SPECS2F3F HAYNES

Cylinder location and distributor rotation diagram - 2F and 3F-E engines

Ignition system

Spark plug type and gap
2F engine.. Champion RN14YC or equivalent @ 0.032 inch
3F-E engine .... Champion RN14YC or equivalent @ 0.032 inch
1FZ-FE engine .... Champion RC12YC or equivalent @ 0.032 inch

Ignition timing and idle speed
2F engine.. 7 degrees BTDC @ 650 rpm in Neutral*
3F-E engine .... 7 degrees BTDC @ 650 rpm in Neutral**
1FZ-FE engine .... 3 degrees BTDC @ 650 rpm in Neutral**

Note: *Use the information printed on the Vehicle Emissions Control Information label, if different than the Specifications listed here.*
* *With the distributor vacuum line disconnected and plugged.*
** *With a jumper wire connected between terminals TE1 and E1 of the engine compartment service connector.*

Firing order and distributor rotation
2F and 3F-E engines
    Firing order 1-5-3-6-2-4
    Distributor rotation .... Clockwise
1FZ-FE engine
    Firing order 1-5-3-6-2-4
    Distributor rotation .... Counterclockwise

Valve clearance
2F engine (engine hot)
    Intake valve .... 0.008 inch (0.20 mm)
    Exhaust valve .... 0.014 inch (0.35 mm)
3F-E engine (engine hot)
    Intake valve .... 0.008 inch (0.20 mm)
    Exhaust valve .... 0.014 inch (0.35 mm)
1FZ-FE engine (engine cold)
    Intake valve .... 0.006 to 0.010 inch (0.15 to 0.25 mm)
    Exhaust valve .... 0.010 to 0.014 inch (0.25 to 0.35 mm)

The blackened terminal shown on the distributor cap indicates the Number One spark plug wire position

92056-1-SPECS1FZFE HAYNES

Cylinder location and distributor rotation diagram - 1FZ-FE engines

Clutch pedal
FJ60 and FJ62
    Height .... 7.13 inch (181 mm)
    Freeplay .... 0.51 to 0.91 inch (13 to 23 mm)

Brakes
Disc brake pad lining thickness (minimum) .... See Chapter 9
Drum brake shoe lining thickness (minimum) .... See Chapter 9
Brake pedal
    FJ60 and FJ62
        Height .... 7.09 inches (180 mm)
        Freeplay .... 0.12 to 0.24 inch (3 to 6 mm)
    FJ80
        Height .... 6.59 inches (167 mm)
        Freeplay .... 0.12 to 0.24 inch (3 to 6 mm)
Parking brake adjustment
    Drum brake .... 5 to 6 clicks
    Disc brake .... 5 to 6 clicks

Suspension and steering
Steering wheel freeplay limit .... 1.58 inches (40 mm)
Balljoint allowable movement .... 0.0 inch (0.0 mm)

Torque specifications   **Ft-lbs** unless otherwise indicated
Note: One foot-pound (ft-lb) of torque is equivalent to 12 inch-pounds (in-lbs) of torque. Torque values below approximately 15 ft-lbs are expressed in inch-pounds, because most foot-pound torque wrenches are not accurate at these smaller values.

Automatic transmission pan bolts .... 61 in-lbs
Automatic transmission drain plug 20
Transfer case fill/drain plug 27
Differential fill/drain plug 36
Spark plugs 20
Engine oil drain plug 20
Wheel lug nuts 90

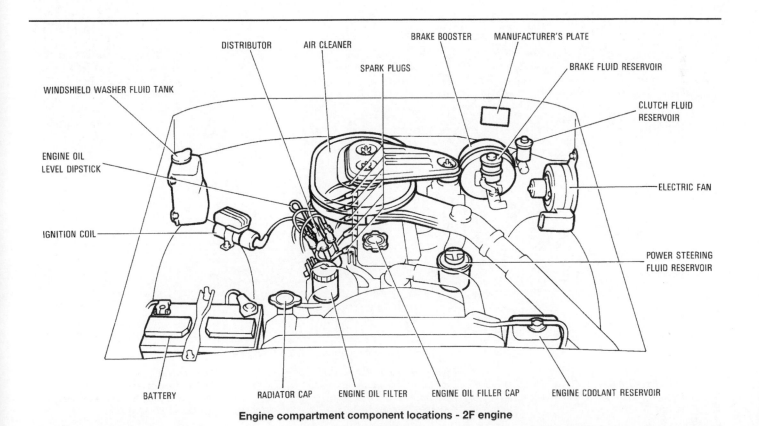

Engine compartment component locations - 2F engine

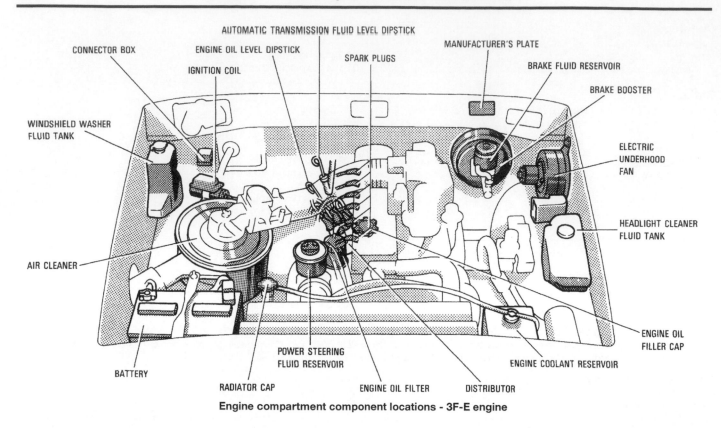

Engine compartment component locations - 3F-E engine

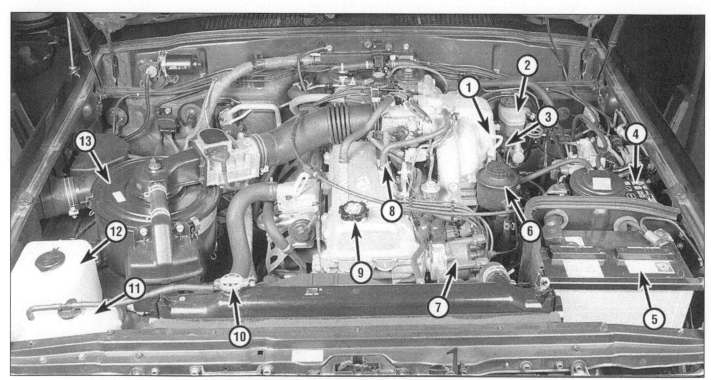

Engine compartment component locations - 1FZ-FE engine

1	Engine oil dipstick	6	Power steering fluid reservoir	10	Radiator cap
2	Brake fluid reservoir	7	Distributor cap and wires	11	Engine coolant reservoir
3	Automatic transmission fluid dipstick	8	PCV valve	12	Windshield washer fluid reservoir
4	Fuse box	9	Engine oil filler cap	13	Air cleaner housing
5	Battery				

Typical engine compartment underside component locations

1	Front differential check/fill plug	5	Coil spring	9	Automatic transmission fluid pan
2	Front differential drain plug	6	Shock absorber	10	Front driveshaft
3	Track bar	7	Stabilizer bar bracket	11	Exhaust system
4	Steering gearbox	8	Engine oil drain plug	12	Steering stop

Typical rear underside component locations

1	Track bar	4	Rear shock absorber	7	Muffler
2	Tail pipe	5	Differential check/fill plug	8	Rear driveshaft
3	Rear brake caliper	6	Differential drain plug		

1 Toyota Land Cruiser Maintenance schedule

Every 250 miles or weekly, whichever comes first

Check the engine oil level (Section 4)
Check the engine coolant level (Section 4)
Check the windshield washer fluid level (Section 4)
Check the brake and clutch fluid level (Section 4)
Check the tires and tire pressures (Section 5)

Every 3000 miles or 3 months, whichever comes first

All items listed above, plus . . .
Check the automatic transmission fluid level (Section 6)
Check the power steering fluid level (Section 7)
Change the engine oil and filter (Section 8)

Every 7500 miles or 6 months, whichever comes first

Check and service the battery (Section 9)
Check the cooling system (Section 10)
Inspect and replace, if necessary, all underhood hoses (Section 11)
Inspect and replace, if necessary, the windshield wiper blades (Section 12)
Rotate the tires (Section 13)
Inspect the suspension and steering components (Section 14)
Lubricate the chassis components (Section 15)*
Inspect the exhaust system (Section 16)
Check the manual transmission lubricant (Section 17)
Check the transfer case lubricant level (Section 18)
Check the differential lubricant level (Section 19)
Check the seat belts (Section 20)

Every 15,000 miles or 12 months, whichever comes first

All items listed above, plus . . .
Inspect the air cleaner thermostatic control valve (Section 21)
Check the exhaust control valve (heat riser) (Section 22)
Inspect the evaporative emissions control system (Section 23)
Check the EGR valve (Section 24)
Check the carburetor choke operation (Section 25)
Check the Positive Crankcase Ventilation (PCV) system (Section 26)

Replace the air filter and PCV filter (Section 27)
Replace the spark plugs (Section 28)
Check and adjust if necessary, the valve clearance (2F and 3F-E engines) (Section 29)
Check the idle speed adjustment (Section 30)
Check the engine drivebelts (Section 31)
Inspect the fuel system (Section 32)
Check the brakes (Section 33)*
Check the clutch and brake pedal for proper height and freeplay (Section 34)
Replace the fuel filter (Section 35)

Every 30,000 miles or 24 months, whichever comes first

All items listed above, plus . . .
Inspect the spark plug wires, distributor cap and rotor (Section 36)
Check the ignition timing (Section 37)
Service the cooling system (drain, flush and refill) (Section 38)
Inspect and repack the front wheel bearings (Section 39)
Change the automatic transmission fluid and filter (Section 40)**
Change the manual transmission lubricant (Section 41)
Change the transfer case lubricant (Section 42)
Change the differential lubricant (Section 43)

Every 60,000 miles or 48 months, whichever comes first

Check and adjust if necessary, the valve clearance (1FZ FE engines) (Section 44)

Every 80,000 miles

Replace the oxygen sensor (see Chapter 6)
This item is affected by "severe" operating conditions, as described below. If the vehicle is operated under severe conditions, perform all maintenance indicated with an asterisk (*) at 3000 mile/three-month intervals. Severe conditions exist if you mainly operate the vehicle . . .
in dusty areas
towing a trailer
idling for extended periods and/or driving at low speeds when outside temperatures remain below freezing and most trips are less than four miles long
**If operated under one or more of the following conditions, change the automatic transmission fluid every 15,000 miles:*
in heavy city traffic where the outside temperature regularly reaches 90-degrees F or higher
in hilly or mountainous terrain
frequent trailer pulling

2 Introduction

This Chapter is designed to help the home mechanic maintain the Toyota Land Cruiser for peak performance, economy, safety and long life.

On the following pages is a master maintenance schedule, followed by Sections dealing specifically with each item on the schedule. Visual checks, adjustments, component replacement and other helpful items are included. Refer to the **accompanying illustrations** of the engine compartment and the underside of the vehicle for the location of various components.

Servicing your Land Cruiser in accordance with the mileage/time maintenance schedule and the following Sections will provide it with a planned maintenance program that should result in a long and reliable service life. This is a comprehensive plan, so maintaining some items but not others at the specified service intervals will not produce the same results.

As you service your vehicle, you will discover that many of the procedures can, and should, be grouped together because of the nature of the particular procedure you're performing or because of the close proximity of two otherwise unrelated components to one another.

For example, if the vehicle is raised for any reason, you should inspect the exhaust, suspension, steering and fuel systems while you're under the vehicle. When you're rotating the tires, it makes good sense to check the brakes and wheel bearings since the wheels are already removed.

Finally, let's suppose you have to borrow or rent a torque wrench. Even if you only need to tighten the spark plugs, you might as well check the torque of as many critical fasteners as time allows.

The first step of this maintenance program is to prepare yourself before the actual work begins. Read through all Sections pertinent to the procedures you're planning to do, then make a list of and gather together all the parts and tools you will need to do the job. If it looks as if you might run into problems during a particular segment of some procedure, seek advice from your local parts man or dealer service department.

3 Tune-up general information

The term tune-up is used in this manual to represent a combination of individual operations rather than one specific procedure.

If, from the time the vehicle is new, the routine maintenance schedule is followed closely and frequent checks are made of fluid levels and high wear items, as suggested throughout this manual, the engine will be kept in relatively good running condition and the need for additional work will be minimized.

More likely than not, however, there will be times when the engine is running poorly due to lack of regular maintenance. This is even more likely if a used vehicle, which has not received regular and frequent maintenance checks, is purchased. In such cases, an engine tune-up will be needed outside of the regular routine maintenance intervals.

The first step in any tune-up or diagnostic procedure to help correct a poor running engine is a cylinder compression check. A compression check (see Chapter 2, Part C) will help determine the condition of internal engine components and should be used as a guide for tune-up and repair procedures. If, for instance, the compression check indicates serious internal engine wear, a conventional tune-up won't improve the performance of the engine and would be a waste of time and money. Because of its importance, the compression check should be done by someone with the right equipment and the knowledge to use it properly.

The following procedures are those most often needed to bring a generally poor running engine back into a proper state of tune.

Minor tune-up

Check all engine related fluids (Section 4)
Clean, inspect and test the battery (Section 9)
Check the cooling system (Section 10)
Check all underhood hoses (Section 11)
Check the PCV valve (Section 26)
Check the air filter and the PCV filter (Section 27)

Replace the spark plugs (Section 28)
Inspect the spark plug wires, distributor cap and rotor (Section 36)
Check and adjust the valve clearances (Section 29 and 44)
Check and adjust the idle speed (Section 30)
Check and adjust the ignition timing (Section 37)
Check and adjust the drivebelts (Section 31)

Major tune-up

All items listed under Minor tune-up, plus . . .

Check the EGR system (Section 24)
Replace the PCV valve (Section 26)
Replace the air filter and the PCV filter (Section 27)
Check the fuel system (Section 32)
Replace the fuel filter (Section 35)
Replace the distributor cap, rotor and spark plug wires (Section 36)
Check the ignition system (Chapter 5)
Check the charging system (Chapter 5)

4 Fluid level checks (every 250 miles or weekly)

Note: *The following are fluid level checks to be done on a 250 mile or weekly basis. Additional fluid level checks can be found in specific maintenance procedures which follow. Regardless of intervals, be alert to fluid leaks under the vehicle which would indicate a fault to be corrected immediately.*

1   Fluids are an essential part of the lubrication, cooling, brake and windshield washer systems. Because the fluids gradually become depleted and/or contaminated during normal operation of the vehicle, they must be periodically replenished. See *Recommended lubricants and fluids* at the beginning of this Chapter before adding fluid to any of the following components. **Note:** *The vehicle must be on level ground when fluid levels are checked.*

Engine oil

Refer to illustrations 4.2a, 4.2b, 4.4 and 4.6

2   The engine oil level is checked with a dipstick that extends through a tube and into the oil pan at the bottom of the engine **(see illustrations)**.

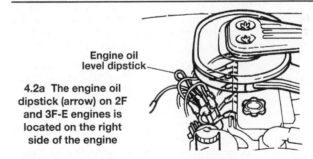

Engine oil level dipstick

4.2a The engine oil dipstick (arrow) on 2F and 3F-E engines is located on the right side of the engine

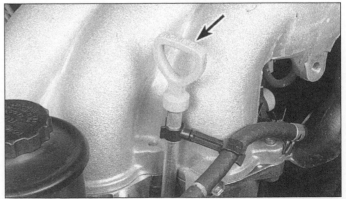

4.2b On 1FZ-FE engines the oil dipstick (arrow) is located on the left side of the engine

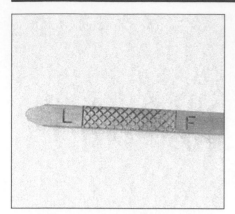

4.4 The oil level must be maintained between the marks at all times - it takes one quart of oil to raise the level from the LOW mark to the FULL mark

4.6 Oil is added to the engine after removing the twist off cap (arrow) located on the valve cover

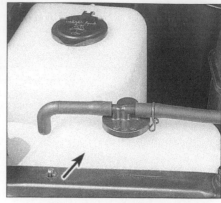

4.8 Check the coolant level in the reservoir (arrow) with the engine hot - it should be visible through the translucent reservoir

3   The oil level should be checked before the vehicle has been driven, or about 15 minutes after the engine has been shut off. If the oil is checked immediately after driving the vehicle, some of the oil will remain in the upper engine components, resulting in an inaccurate reading on the dipstick.

4   Pull the dipstick out of the tube and wipe all the oil from the end with a clean rag or paper towel. Insert the clean dipstick all the way back into the tube, then pull it out again. Note the oil at the end of the dipstick. Add oil as necessary to keep the level between the ADD and FULL marks on the dipstick **(see illustration)**.

5   Do not overfill the engine by adding too much oil since this may result in oil-fouled spark plugs, oil leaks or oil seal failures.

6   Oil is added to the engine after removing the threaded cap from the valve cover **(see illustration)**. A funnel may help to reduce spills.

7   Checking the oil level is an important preventive maintenance step. A consistently low oil level indicates oil leakage through damaged seals, defective gaskets or past worn rings or valve guides. If the oil looks milky or has water droplets in it, the cylinder head gasket(s) may be blown or the head(s) or block may be cracked. The engine should be checked immediately. The condition of the oil should also be checked. Whenever you check the oil level, slide your thumb and index finger up the dipstick before wiping off the oil. If you see small dirt or metal particles clinging to the dipstick, the oil should be changed (see Section 8).

Engine coolant

Refer to illustration 4.8

Warning: *Do not allow antifreeze to come in contact with your skin or painted surfaces of the vehicle. Flush contaminated areas immediately with plenty of water. Don't store new coolant or leave old coolant lying around where it's accessible to children or pets – they're attracted by its sweet smell. Ingestion of even a small amount of coolant can be fatal! Wipe up garage floor and drip pan spills immediately. Keep antifreeze containers covered and repair cooling system leaks as soon as they're noticed.*

8   All vehicles covered by this manual are equipped with a pressurized coolant recovery system. A coolant reservoir which is located next to the radiator in the engine compartment is connected by a hose to the base of the coolant filler cap **(see illustration)**. If the coolant gets too hot during engine operation, coolant can escape through a pressurized filler cap, then through a connecting hose into the reservoir. As the engine cools, the coolant is automatically drawn back into the cooling system to maintain the correct level.

9   The coolant level should be checked regularly. It must be between the FULL and LOW lines on the tank. The level will vary with the temperature of the engine. When the engine is cold, the coolant level should be at or slightly above the LOW mark on the tank. Once the engine has warmed up, the level should be at or near the FULL mark. If it isn't, allow the fluid in the tank to cool, then remove the cap

from the reservoir and add coolant to bring the level up to the FULL line. Use only ethylene/glycol type coolant and water in the mixture ratio recommended by your owner's manual. Do not use supplemental inhibitor additives. If only a small amount of coolant is required to bring the system up to the proper level, water can be used. However, repeated additions of water will dilute the recommended antifreeze and water solution. In order to maintain the proper ratio of antifreeze and water, it is advisable to top up the coolant level with the correct mixture. Refer to your owner's manual for the recommended ratio.

10   If the coolant level drops within a short time after replenishment, there may be a leak in the system. Inspect the radiator, hoses, engine coolant filler cap, drain plugs and water pump. If no leak is evident, have the radiator cap pressure tested by your dealer. **Warning:** *Never remove the radiator cap or the coolant recovery reservoir cap when the engine is running or has just been shut down, because the cooling system is hot. Escaping steam and scalding liquid could cause serious injury.*

11   If it is necessary to open the radiator cap, wait until the system has cooled completely, then wrap a thick cloth around the cap and turn it to the first stop. If any steam escapes, wait until the system has cooled further, then remove the cap.

12   When checking the coolant level, always note its condition. It should be relatively clear. If it is brown or rust colored, the system should be drained, flushed and refilled. Even if the coolant appears to be normal, the corrosion inhibitors wear out with use, so it must be replaced at the specified intervals.

13   Do not allow antifreeze to come in contact with your skin or painted surfaces of the vehicle. Flush contacted areas immediately with plenty of water.

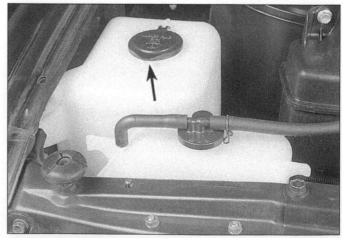

4.14 The windshield washer fluid reservoir (arrow) is located at the front of the engine compartment

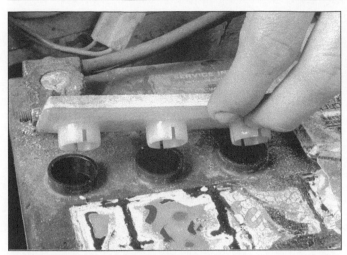

4.15a On earlier style batteries the cell caps have to be removed to check the water level in the battery - if the level is too low, add distilled water only

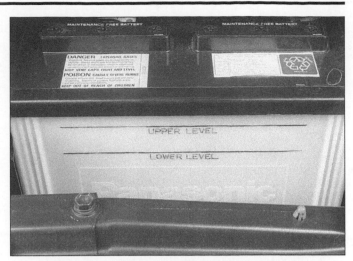

14.15b On later models the battery electrolyte level can be visually checked through the translucent battery case - if the level is to low, add distilled water only

Windshield washer fluid

Refer to illustration 4.14

14   Fluid for the windshield washer system may be stored in two places depending on model year of the vehicle. The primary reservoir for the front windshield/headlight washer system is located on the right (passenger) side of the engine compartment **(see illustration)**. On some models a second reservoir for the rear window washer system (if equipped) is located in the interior of the left rear passenger compartment. In milder climates, plain water can be used to top up the reservoir, but the reservoir should be kept no more than two-thirds full to allow for expansion should the water freeze. In colder climates, the use of a specially designed windshield washer fluid, available at your dealer and any auto parts store, will help lower the freezing point of the fluid. Mix the solution with water in accordance with the manufacturer's directions on the container. Do not use regular antifreeze. It will damage the vehicle's paint.

Battery electrolyte

Refer to illustrations 4.15a and 4.15b

15   On models not equipped with a sealed battery, check the electrolyte level **(see illustrations)** of all six battery cells. It must be between the upper and lower levels. If the level is low, remove the

filler/vent cap and add distilled water. Install and securely re-tighten the cap. **Caution:** *Overfilling the cells may cause electrolyte to spill over during periods of heavy charging, causing corrosion or damage.*

Brake and clutch fluid

Refer to illustrations 4.17a and 4.17b

16   The brake master cylinder is mounted on the front of the power booster unit in the engine compartment. The hydraulic clutch master cylinder used on manual transmission vehicles is located next to the brake master cylinder.

17   To check the fluid level of the brake and clutch master cylinders, simply look at the MAX and MIN marks on the reservoir **(see illustrations)**. The level should be 1/4 inch below the maximum fill line.

18   If the level is low, wipe the top of the reservoir cover with a clean rag to prevent contamination of the brake system before lifting the cover.

19   Add only the specified brake fluid to the brake and clutch reservoirs (refer to *Recommended lubricants and fluids* at the front of this Chapter or to your owner's manual). Mixing different types of brake fluid can damage the system. **Warning:** *Use caution when filling either reservoir - brake fluid can harm your eyes and damage painted surfaces. Do not use brake fluid that has been opened for more than one year or has been left open. Brake fluid absorbs moisture from the air. Excess moisture can cause a dangerous loss of braking.*

20   While the reservoir cap is removed, inspect the master cylinder reservoir for contamination. If deposits, dirt particles or water droplets are present, the system should be drained and refilled.

4.17a The fluid level inside the brake reservoir can easily be checked by observing the level from the outside - fluid can be added to the reservoir after the cover is removed by prying up on the cap (later model shown)

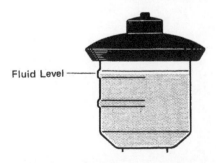

Fluid Level —

4.17b The fluid level inside the clutch reservoir can easily be checked by observing the level from the outside - fluid can be added to the reservoir after the cover is removed by prying up on the cap (early model shown)

21   After filling the reservoir to the proper level, make sure the lid is properly seated to prevent fluid leakage and/or system pressure loss.

22   The fluid in the brake master cylinder will drop slightly as the brake pads at each wheel wear down during normal operation. If either master cylinder requires repeated replenishing to keep it at the proper level, this is an indication of leakage in the brake or clutch system, which should be corrected immediately. If the brake system shows an indication of leakage check all brake lines and connections, along with the calipers, wheel cylinders and booster (see Section 33 for more information). If the hydraulic clutch system shows an indication of leakage check all clutch lines and connections, along with the clutch release cylinder (see Chapter 8 for more information).

23   If, upon checking the brake or clutch master cylinder fluid level, you discover one or both reservoirs empty or nearly empty, the systems should be bled (see Chapter 9).

---

5   Tire and tire pressure checks (every 250 miles or weekly)

Refer to illustrations 5.2, 5.3, 5.4a, 5.4b and 5.8

1   Periodic inspection of the tires may spare you the inconvenience of being stranded with a flat tire. It can also provide you with vital information regarding possible problems in the steering and suspension systems before major damage occurs.

2   The original tires on this vehicle are equipped with 1/2-inch wear bands that will appear when tread depth reaches 1/16-inch. Tread wear can be monitored with a simple, inexpensive device known as a tread depth gauge **(see illustration)**.

3   Note any abnormal tread wear **(see illustration)**. Tread pattern irregularities such as cupping, flat spots and more wear on one side than the other are indications of front end alignment and/or balance problems. If any of these conditions are noted, take the vehicle to a tire

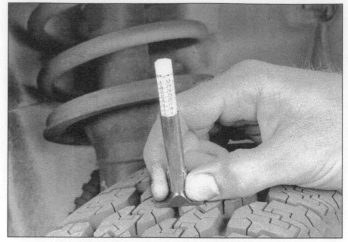

5.2 Use a tire tread depth gauge to monitor tire wear - they are available at auto parts stores and service stations and cost very little

shop or service station to correct the problem.

4   Look closely for cuts, punctures and embedded nails or tacks. Sometimes a tire will hold air pressure for a short time or leak down very slowly after a nail has embedded itself in the tread. If a slow leak persists, check the valve stem core to make sure it's tight **(see illustration)**. Examine the tread for an object that may have embedded itself in the tire or for a "plug" that may have begun to leak (radial tire punctures are repaired with a rubber plug that's installed in the hole). If a puncture is suspected, it can be easily verified by spraying a solution of soapy water onto the puncture area **(see illustration)**. The soapy solution will bubble if there's a leak. Unless the puncture is unusually

UNDERINFLATION

CUPPING

Cupping may be caused by:
- Underinflation and/or mechanical irregularities such as out-of-balance condition of wheel and/or tire, and bent or damaged wheel.
- Loose or worn steering tie-rod or steering idler arm.
- Loose, damaged or worn front suspension parts.

OVERINFLATION

INCORRECT TOE-IN OR EXTREME CAMBER

FEATHERING DUE TO MISALIGNMENT

5.3 This chart will help you determine the condition of the tires, the probable cause(s) of abnormal wear and the corrective action necessary

5.4a If a tire loses air on a steady basis, check the valve core first to make sure it's snug (special inexpensive wrenches are commonly available at auto parts stores)

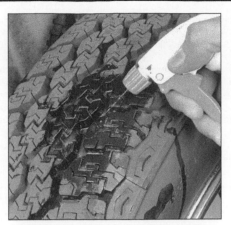

5.4b If the valve core is tight, raise the corner of the vehicle with the low tire and spray a soapy water solution onto the tread as the tire is turned slowly - leaks will cause small bubbles to appear

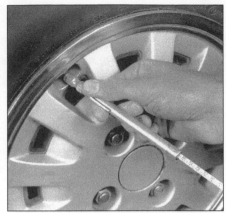

5.8 To extend the life of the tires, check the air pressure at least once a week with an accurate gauge (don't forget the spare)

large, a tire shop or service station can usually repair the tire.

5   Carefully inspect the inner sidewall of each tire for evidence of brake fluid leakage. If you see any, inspect the brakes immediately.

6   Correct air pressure adds miles to the lifespan of the tires, improves mileage and enhances overall ride quality. Tire pressure cannot be accurately estimated by looking at a tire, especially if it's a radial. A tire pressure gauge is essential. Keep an accurate gauge in the vehicle. The pressure gauges attached to the nozzles of air hoses at gas stations are often inaccurate.

7   Always check tire pressure when the tires are cold. Cold, in this case, means the vehicle has not been driven over a mile in the three hours preceding a tire pressure check. A pressure rise of four to eight pounds is not uncommon once the tires are warm.

8   Unscrew the valve cap protruding from the wheel or hubcap and push the gauge firmly onto the valve stem (see illustration). Note the reading on the gauge and compare the figure to the recommended tire pressure shown on the placard on the driver's side door pillar. Be sure to reinstall the valve cap to keep dirt and moisture out of the valve stem mechanism. Check all four tires and, if necessary, add enough air to bring them up to the recommended pressure.

9   Don't forget to keep the spare tire inflated to the specified pressure (consult your owner's manual). Note that the air pressure specified for the compact spare is significantly higher than the pressure of the regular tires.

6   Automatic transmission fluid level check (every 3000 miles or 3 months)

Refer to illustrations 6.4 and 6.6

1   The level of the automatic transmission fluid should be carefully maintained. Low fluid level can lead to slipping or loss of drive, while overfilling can cause foaming, loss of fluid and transmission damage.

2   The transmission fluid level should only be checked when the transmission is hot (at its normal operating temperature). If the vehicle has just been driven over 10 miles (15 miles in a frigid climate), and the fluid temperature is 160 to 175-degrees F, the transmission is hot. **Caution:** *If the vehicle has just been driven for a long time at high speed or in city traffic in hot weather, or if it has been pulling a trailer, an accurate fluid level reading cannot be obtained. Allow the fluid to cool down for about 30 minutes.*

3   If the vehicle has not been driven, park the vehicle on level ground, set the parking brake, then start the engine and bring it to operating temperature. While the engine is idling, depress the brake pedal and move the selector lever through all the gear ranges, beginning and ending in Park.

4   With the engine still idling, remove the dipstick from its tube (see

6.4 The automatic transmission fluid dipstick (arrow) is located at the rear of the engine compartment

illustration). Check the level of the fluid on the dipstick and note its condition.

5   Wipe the fluid from the dipstick with a clean rag and reinsert it back into the filler tube until the cap seats.

6   Pull the dipstick out again and note the fluid level (see illustration). If the transmission is cold, the level should be in the COLD or COOL range on the dipstick. If it is hot, the fluid level should be in the HOT range. If the level is at the low side of either range, add the specified automatic transmission fluid through the dipstick tube with a funnel.

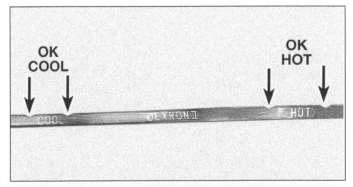

6.6 The automatic transmission fluid level must be maintained between the ADD mark and the FULL mark (arrows) at the indicated operating temperature

7   Add just enough of the recommended fluid to fill the transmission to the proper level. It takes about one pint to raise the level from the low mark to the high mark when the fluid is hot, so add the fluid a little at a time and keep checking the level until it is correct.

8   The condition of the fluid should also be checked along with the level. If the fluid at the end of the dipstick is black or a dark reddish brown color, or if it emits a burned smell, the fluid should be changed (see Section 40). If you are in doubt about the condition of the fluid, purchase some new fluid and compare the two for color and smell.

7   Power steering fluid level check (every 3000 miles or 3 months)

Refer to illustrations 7.2 and 7.6

1   Unlike manual steering, the power steering system relies on fluid which may, over a period of time, require replenishing.

2   The fluid reservoir for the power steering pump on most models is located on the pump body at the front of the engine. Later models are equipped with a remote fluid reservoir which is mounted on the left side of the engine **(see illustration)**.

3   For the check, the front wheels should be pointed straight ahead and the engine should be off.

4   Use a clean rag to wipe off the reservoir cap and the area around the cap. This will help prevent any foreign matter from entering the reservoir during the check.

7.2 The power steering fluid dipstick is located in the power steering pump reservoir - turn the cap (arrow) counterclockwise to remove it

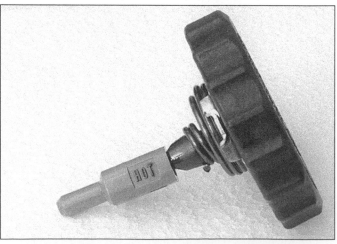

7.6 The marks on the dipstick indicate the safe fluid range

5   Twist off the cap and determine the temperature of the fluid at the end of the dipstick with your finger.

6   Wipe off the fluid with a clean rag, reinsert the dipstick, then withdraw it and read the fluid level. The fluid should be at the proper level, depending on whether it was checked hot or cold **(see illustration)**. Never allow the fluid level to drop below the lower mark on the dipstick.

7   If additional fluid is required, pour the specified type directly into the reservoir, using a funnel to prevent spills.

8   If the reservoir requires frequent fluid additions, all power steering hoses, hose connections and the power steering pump should be carefully checked for leaks.

8   Engine oil and filter change (every 3000 miles or 3 months)

Refer to illustrations 8.2, 8.7, 8.13 and 8.15

1   Frequent oil changes are the best preventive maintenance the home mechanic can give the engine, because aging oil becomes diluted and contaminated, which leads to premature engine wear.

2   Make sure that you have all the necessary tools before you begin this procedure **(see illustration)**. You should also have plenty of rags or newspapers handy for mopping up any spills.

3   Access to the underside of the vehicle is greatly improved if the vehicle can be lifted on a hoist, driven onto ramps or supported by jackstands.

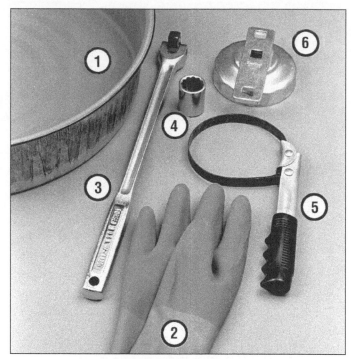

8.2 These tools are required when changing the engine oil and filter

1   *Drain pan - It should be fairly shallow in depth, but wide to prevent spills*

2   *Rubber gloves - When removing the drain plug and filter, you will get oil on your hands (the gloves will prevent burns)*

3   *Breaker bar - Sometimes the oil drain plug is tight, and a long breaker bar is needed to loosen it*

4   *Socket - To be used with the breaker bar or a ratchet (must be the correct size to fit the drain plug)*

5   *Filter wrench - This is a metal band-type wrench, which requires clearance around the filter to be effective*

6   *Filter wrench - This type fits on the bottom of the filter and can be turned with a ratchet or breaker bar (different size wrenches are available for different types of filters)*

4   If this is your first oil change, get under the vehicle and familiarize yourself with the location of the oil drain plug. The engine and exhaust components will be warm during the actual work, so try to anticipate any potential problems before the engine and accessories are hot.

5   Park the vehicle on a level spot. Start the engine and allow it to reach its normal operating temperature (the needle on the temperature gauge should be at least above the bottom mark). Warm oil and contaminates will flow out more easily. Turn off the engine when it's warmed up. Remove the filler cap in the valve cover.

6   Raise the vehicle and support it on jackstands. **Warning:** *To avoid personal injury, never get beneath the vehicle when it is supported by only by a jack. The jack provided with your vehicle is designed solely for raising the vehicle to remove and replace the wheels. Always use jackstands to support the vehicle when it becomes necessary to place your body underneath the vehicle.*

7   Being careful not to touch the hot exhaust components, place the drain pan under the drain plug in the bottom of the pan and remove the plug **(see illustration)**. You may want to wear gloves while unscrewing the plug the final few turns if the engine is really hot.

8   Allow the old oil to drain into the pan. It may be necessary to move the pan farther under the engine as the oil flow slows to a trickle. Inspect the old oil for the presence of metal shavings and chips.

9   After all the oil has drained, wipe off the drain plug with a clean rag. Even minute metal particles clinging to the plug would immediately contaminate the new oil.

10   Clean the area around the drain plug opening, reinstall the plug and tighten it securely, but do not strip the threads.

11   Move the drain pan into position under the oil filter.

12   Remove all tools, rags, etc. from under the vehicle, being careful not to spill the oil in the drain pan, then lower the vehicle.

13   Loosen the oil filter by turning it counterclockwise with the filter wrench **(see illustration)**. Any standard filter wrench should work. Once the filter is loose, use your hands to unscrew it from the block. Just as the filter is detached from the block, immediately tilt the open end up to prevent the oil inside the filter from spilling out. **Warning:** *The engine exhaust manifold may still be hot, so be careful.*

14   With a clean rag, wipe off the mounting surface on the block. If a residue of old oil is allowed to remain, it will smoke when the block is heated up. It will also prevent the new filter from seating properly. Also make sure that the none of the old gasket remains stuck to the mounting surface. It can be removed with a scraper if necessary.

15   Compare the old filter with the new one to make sure they are the same type. Smear some engine oil on the rubber gasket of the new filter and screw it into place **(see illustration)**. Because over-tightening the filter will damage the gasket, do not use a filter wrench to tighten the filter. Tighten it by hand until the gasket contacts the seating surface. Then seat the filter by giving it an additional 3/4-turn.

16   Add new oil to the engine through the oil filler cap in the valve cover. Use a spout or funnel to prevent oil from spilling onto the top of the engine. Pour three quarts of fresh oil into the engine. Wait a few minutes to allow the oil to drain into the pan, then check the level on the oil dipstick (see Section 4 if necessary). If the oil level is at or near the H mark, install the filler cap hand tight, start the engine and allow the new oil to circulate.

17   Allow the engine to run for about a minute. While the engine is running, look under the vehicle and check for leaks at the oil pan drain plug and around the oil filter. If either is leaking, stop the engine and tighten the plug or filter slightly.

18   Wait a few minutes to allow the oil to trickle down into the pan, then recheck the level on the dipstick and, if necessary, add enough oil to bring the level to the H mark.

19   During the first few trips after an oil change, make it a point to check frequently for leaks and proper oil level.

20   The old oil drained from the engine cannot be reused in its present state and should be discarded. Oil reclamation centers, auto repair shops and gas stations will normally accept the oil, which can be recycled. After the oil has cooled, it can be drained into a suitable container (capped plastic jugs, topped bottles, milk cartons, etc.) for transport to one of these disposal sites.

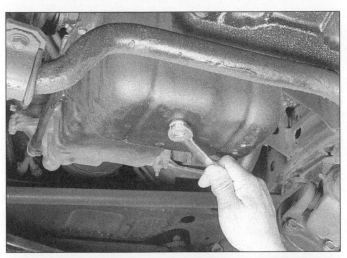

8.7 The engine oil drain plug is located on the bottom of the oil pan - it is usually very tight, so use the proper size box end wrench or socket to avoid rounding it off

8.13 The oil filter (arrow) is usually on very tight and will require a special wrench for removal - DO NOT use the wrench to tighten the new filter!

8.15 Lubricate the oil filter gasket with clean engine oil before installing the filter on the engine

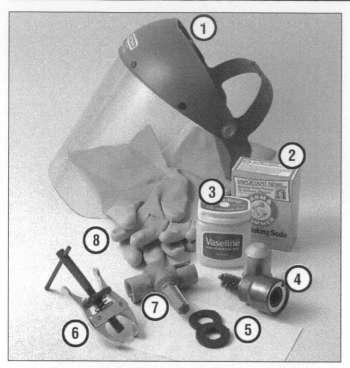

9.1 Tools and materials required for battery maintenance

1   *Face shield/safety goggles - When removing corrosion with a brush, the acidic particles can easily fly up into your eyes*
2   *Baking soda - A solution of baking soda and water can be used to neutralize corrosion*
3   *Petroleum jelly - A layer of this on the battery posts will help prevent corrosion*
4   *Battery post/cable cleaner - This wire brush cleaning tool will remove all traces of corrosion from the battery posts and cable clamps*
5   *Treated felt washers - Placing one of these on each post, directly under the cable clamps, will help prevent corrosion*
6   *Puller - Sometimes the cable clamps are very difficult to pull off the posts, even after the nut/bolt has been completely loosened. This tool pulls the clamp straight up and off the post without damage*
7   *Battery post/cable cleaner - Here is another cleaning tool which is a slightly different version of Number 4 above, but it does the same thing*
8   *Rubber gloves - Another safety item to consider when servicing the battery; remember that's acid inside the battery!*

---

9   Battery check, maintenance and charging (every 7500 miles or 6 months)

Refer to illustrations 9.1, 9.6a, 9.6b, 9.7a and 9.7b

Warning: *Certain precautions must be followed when checking and servicing the battery. Hydrogen gas, which is highly flammable, is always present in the battery cells, so keep lighted tobacco and all other open flames and sparks away from the battery. The electrolyte inside the battery is actually dilute sulfuric acid, which will cause injury if splashed on your skin or in your eyes. It will also ruin clothes and painted surfaces. When removing the battery cables, always detach the negative cable first and hook it up last!*

Caution: *If the stereo in your vehicle is equipped with an anti-theft system, make sure you have the correct activation code before disconnecting the battery.*

Maintenance

1   A routine preventive maintenance program for the battery in your vehicle is the only way to ensure quick and reliable starts. But before

9.6a Battery terminal corrosion usually appears as light, fluffy powder

9.6b Removing the cable from a battery post with a wrench - sometimes special battery pliers are required for this procedure if corrosion has caused deterioration of the nut hex (always remove the ground cable first and hook it up last!)

performing any battery maintenance, make sure that you have the proper equipment necessary to work safely around the battery **(see illustration)**.
2   There are also several precautions that should be taken whenever battery maintenance is performed. Before servicing the battery, always turn the engine and all accessories off and disconnect the cable from the negative terminal of the battery.
3   The battery produces hydrogen gas, which is both flammable and explosive. Never create a spark, smoke or light a match around the battery. Always charge the battery in a ventilated area.
4   Electrolyte contains poisonous and corrosive sulfuric acid. Do not allow it to get in your eyes, on your skin on your clothes. Never ingest it. Wear protective safety glasses when working near the battery. Keep children away from the battery.
5   Note the external condition of the battery. If the positive terminal and cable clamp on your vehicle's battery is equipped with a rubber protector, make sure that it's not torn or damaged. It should completely cover the terminal. Look for any corroded or loose connections, cracks in the case or cover or loose hold-down clamps. Also check the entire length of each cable for cracks and frayed conductors.
6   If corrosion, which looks like white, fluffy deposits **(see illustration)** is evident, particularly around the terminals, the battery should be removed for cleaning. Loosen the cable clamp bolts with a wrench, being careful to remove the ground cable first, and slide them off the terminals **(see illustration)**. Then disconnect the hold-down clamp bolt and nut, remove the clamp and lift the battery from the engine compartment.

9.7a When cleaning the cable clamps, all corrosion must be removed (the inside of the clamp is tapered to match the taper on the post, so don't remove too much material)

9.7b Regardless of the type of tool used on the battery posts, a clean, shiny surface should be the end result

7   Clean the cable clamps thoroughly with a battery brush or a terminal cleaner and a solution of warm water and baking soda **(see illustration)**. Wash the terminals and the top of the battery case with the same solution but make sure that the solution doesn't get into the battery. When cleaning the cables, terminals and battery top, wear safety goggles and rubber gloves to prevent any solution from coming in contact with your eyes or hands. Wear old clothes too - even diluted, sulfuric acid splashed onto clothes will burn holes in them. If the terminals have been extensively corroded, clean them up with a terminal cleaner **(see illustration)**. Thoroughly wash all cleaned areas with plain water.
8   Make sure that the battery tray is in good condition and the hold-down clamp bolts are tight. If the battery is removed from the tray, make sure no parts remain in the bottom of the tray when the battery is reinstalled. When reinstalling the hold-down clamp bolts, do not overtighten them.
9   Any metal parts of the vehicle damaged by corrosion should be covered with a zinc-based primer, then painted.
10   Information on removing and installing the battery can be found in Chapter 5. Information on jump starting can be found at the front of this manual. For more detailed battery checking procedures, refer to the *Haynes Automotive Electrical Manual*.

Charging

Warning: *When batteries are being charged, hydrogen gas, which is very explosive and flammable, is produced. Do not smoke or allow open flames near a battery. Wear eye protection when near the battery during charging. Also, make sure the charger is unplugged before connecting or disconnecting the battery from the charger.*
Note: *The manufacturer recommends the battery be removed from the vehicle for charging because the gas that escapes during this procedure can damage the paint. Fast charging with the battery cables connected can result in damage to the electrical system.*

11   Slow-rate charging is the best way to restore a battery that's discharged to the point where it will not start the engine. It's also a good way to maintain the battery charge in a vehicle that's only driven a few miles between starts. Maintaining the battery charge is particularly important in the winter when the battery must work harder to start the engine and electrical accessories that drain the battery are in greater use.
12   It's best to use a one or two-amp battery charger (sometimes called a "trickle" charger). They are the safest and put the least strain on the battery. They are also the least expensive. For a faster charge, you can use a higher amperage charger, but don't use one rated more than 1/10th the amp/hour rating of the battery. Rapid boost charges that claim to restore the power of the battery in one to two hours are hardest on the battery and can damage batteries not in good condition. This type of charging should only be used in emergency situations.

13   The average time necessary to charge a battery should be listed in the instructions that come with the charger. As a general rule, a trickle charger will charge a battery in 12 to 16 hours.
14   Remove all the cell caps (if equipped) and cover the holes with a clean cloth to prevent spattering electrolyte. Disconnect the negative battery cable and hook the battery charger cable clamps up to the battery posts (positive to positive, negative to negative), then plug in the charger. Make sure it is set at 12-volts if it has a selector switch.
15   If you're using a charger with a rate higher than two amps, check the battery regularly during charging to make sure it doesn't overheat. If you're using a trickle charger, you can safely let the battery charge overnight after you've checked it regularly for the first couple of hours.
16   If the battery has removable cell caps, measure the specific gravity with a hydrometer every hour during the last few hours of the charging cycle. Hydrometers are available inexpensively from auto parts stores - follow the instructions that come with the hydrometer. Consider the battery charged when there's no change in the specific gravity reading for two hours and the electrolyte in the cells is gassing (bubbling) freely. The specific gravity reading from each cell should be very close to the others. If not, the battery probably has a bad cell(s).
17   Some batteries with sealed tops have built-in hydrometers on the top that indicate the state of charge by the color displayed in the hydrometer window. Normally, a bright-colored hydrometer indicates a full charge and a dark hydrometer indicates the battery still needs charging.
18   If the battery has a sealed top and no built-in hydrometer, you can hook up a voltmeter across the battery terminals to check the charge. A fully charged battery should read 12.6 volts or higher after the surface charge has been removed.
19   Further information on the battery and jump starting can be found in Chapter 5 and at the front of this manual.

10 Cooling system check (every 7500 miles or 6 months)

Refer to illustration 10.4
1   Many major engine failures can be attributed to a faulty cooling system. If the vehicle is equipped with an automatic transmission, the cooling system also cools the transmission fluid and thus plays an important role in prolonging transmission life.
2   The cooling system should be checked with the engine cold. Do this before the vehicle is driven for the day or after it has been shut off for at least three hours.
3   Remove the radiator cap by turning it to the left until it reaches a stop. If you hear a hissing sound (indicating there is still pressure in the system), wait until this stops. Now press down on the cap with the palm of your hand and continue turning to the left until the cap can be

removed. Thoroughly clean the cap, inside and out, with clean water. Also clean the filler neck on the radiator. All traces of corrosion should be removed. The coolant inside the radiator should be relatively transparent. If it is rust colored, the system should be drained and refilled (see Section 38). If the coolant level is not up to the top, add additional antifreeze/coolant mixture (see Section 4).

4   Carefully check the large upper and lower radiator hoses along with the smaller diameter heater hoses which run from the engine to the firewall. Inspect each hose along its entire length, replacing any hose which is cracked, swollen or shows signs of deterioration. Cracks may become more apparent if the hose is squeezed **(see illustration)**. Regardless of condition, it's a good idea to replace hoses with new ones every two years.

5   Make sure all hose connections are tight. A leak in the cooling system will usually show up as white or rust colored deposits on the areas adjoining the leak. If wire-type clamps are used at the ends of the hoses, it may be a good idea to replace them with more secure screw-type clamps.

6   Use compressed air or a soft brush to remove bugs, leaves, etc. from the front of the radiator or air conditioning condenser. Be careful not to damage the delicate cooling fins or cut yourself on them.

7   Every other inspection, or at the first indication of cooling system problems, have the cap and system pressure tested. If you don't have a pressure tester, most gas stations and repair shops will do this for a minimal charge.

---

11   Underhood hose check and replacement (every 7500 miles or 6 months)

General

1   **Warning:** *Replacement of air conditioning hoses must be left to a dealer service department or air conditioning shop that has the equipment to depressurize the system safely. Never remove air conditioning components or hoses until the system has been depressurized.*

2   High temperatures in the engine compartment can cause the deterioration of the rubber and plastic hoses used for engine, accessory and emission systems operation. Periodic inspection should be made for cracks, loose clamps, material hardening and leaks. Information specific to the cooling system hoses can be found in Section 10.

3   Some, but not all, hoses are secured to the fittings with clamps. Where clamps are used, check to be sure they haven't lost their tension, allowing the hose to leak. If clamps aren't used, make sure the hose has not expanded and/or hardened where it slips over the fitting, allowing it to leak.

Vacuum hoses

4   It's quite common for vacuum hoses, especially those in the emissions system, to be color coded or identified by colored stripes molded into them. Various systems require hoses with different wall thickness, collapse resistance and temperature resistance. When replacing hoses, be sure the new ones are made of the same material.

5   Often the only effective way to check a hose is to remove it completely from the vehicle. If more than one hose is removed, be sure to label the hoses and fittings to ensure correct installation.

6   When checking vacuum hoses, be sure to include any plastic T-fittings in the check. Inspect the fittings for cracks and the hose where it fits over the fitting for distortion, which could cause leakage.

7   A small piece of vacuum hose (1/4-inch inside diameter) can be used as a stethoscope to detect vacuum leaks. Hold one end of the hose to your ear and probe around vacuum hoses and fittings, listening for the "hissing" sound characteristic of a vacuum leak. **Warning:** *When probing with the vacuum hose stethoscope, be very careful not to come into contact with moving engine components such as the drivebelt, cooling fan, etc.*

Fuel hose

Warning: *There are certain precautions which must be taken when inspecting or servicing fuel system components. Work in a well ventilated*

Check for a chafed area that could fail prematurely.

Check for a soft area indicating the hose has deteriorated inside.

Overtightening the clamp on a hardened hose will damage the hose and cause a leak.

Check each hose for swelling and oil-soaked ends. Cracks and breaks can be located by squeezing the hose.

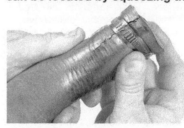

10.4 Hoses, like drivebelts, have a habit of failing at the worst possible time - to prevent the inconvenience of a blown radiator or heater hose, inspect them carefully as shown here

---

area and do not allow open flames (cigarettes, appliance pilot lights, etc.) or bare light bulbs near the work area. Mop up any spills immediately and do not store fuel soaked rags where they could ignite.

8   Check all rubber fuel lines for deterioration and chafing. Check especially for cracks in areas where the hose bends and just before fittings, such as where a hose attaches to the fuel filter.

9   Only high quality fuel line should be used for fuel line replacement. Never, under any circumstances, use unreinforced vacuum line, clear plastic tubing or water hose for fuel lines.

10   Spring-type clamps are commonly used on fuel lines. These clamps often lose their tension over a period of time, and can be "sprung" during removal. Replace all spring-type clamps with screw clamps whenever a hose is replaced.

Metal lines

11   Sections of metal line are often used in the fuel system. Check carefully to be sure the line has not been bent or crimped and that cracks have not started in the line.

12   If a section of metal fuel line must be replaced, only seamless steel tubing should be used, since copper and aluminum tubing don't have the strength necessary to withstand normal engine vibration.

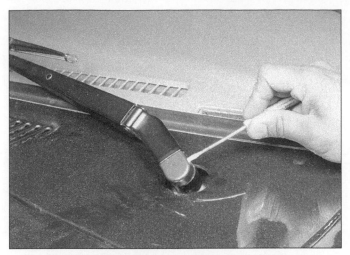

12.3 Pry open the trim cap and check the tightness of the wiper arm retaining nut

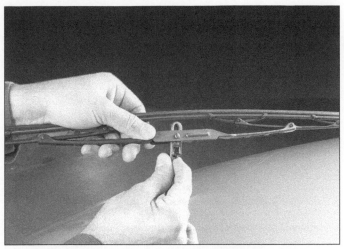

12.5 Press on the release tab, then push the blade assembly down and out of the hook in the arm

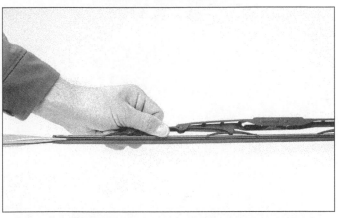

12.6 Use needle-nose pliers to compress the rubber element, then slide the element out - slide the new element in and lock the blade assembly fingers into the notches of the wiper element

13   Check the metal brake lines where they enter the master cylinder and brake proportioning unit (if used) for cracks in the lines or loose fittings. Any sign of brake fluid leakage calls for an immediate thorough inspection of the brake system.

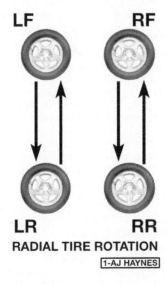

13.2 Tire rotation diagram for radial tires

12 Wiper blade inspection and replacement (every 7500 miles or 6 months)

Refer to illustrations 12.3, 12.5 and 12.6

1   The windshield wiper and blade assembly should be inspected periodically for damage, loose components and cracked or worn blade elements.
2   Road film can build up on the wiper blades and affect their efficiency, so they should be washed regularly with a mild detergent solution.
3   The action of the wiping mechanism can loosen bolts, nuts and fasteners, so they should be checked and tightened, as necessary **(see illustration),** at the same time the wiper blades are checked.
4   If the wiper blade elements are cracked, worn or warped, or no longer clean adequately, they should be replaced with new ones.
5   Lift the arm assembly away from the glass for clearance, press on the release lever, then slide the wiper blade assembly out of the hook in the end of the arm **(see illustration).**
6   Use needle-nose pliers to compress the blade element, then slide the element out of the frame and discard it **(see illustration).**
7   Installation is the reverse of removal.

13 Tire rotation (every 7500 miles or 6 months)

Refer to illustration 13.2

1   The tires should be rotated at the specified intervals and whenever uneven wear is noticed. Since the vehicle will be raised and the tires removed anyway, check the brakes (see Section 33) at this time.
2   Radial tires must be rotated in a specific pattern **(see illustration)**.
3   Refer to the information in *Jacking and towing* at the front of this manual for the proper procedures to follow when raising the vehicle and changing a tire. If the brakes are to be checked, do not apply the parking brake as stated. Make sure the tires are blocked to prevent the vehicle from rolling.
4   Preferably, the entire vehicle should be raised at the same time. This can be done on a hoist or by jacking up each corner and then lowering the vehicle onto jackstands placed under the frame rails. Always use four jackstands and make sure the vehicle is firmly supported.
5   After rotation, check and adjust the tire pressures as necessary and be sure to check the lug nut tightness.
6   For further information on the wheels and tires, refer to Chapter 10.

14 Suspension and steering check (every 7500 miles or 6 months)

Refer to illustrations 14.1, 14.6a, 14.6b and 14.7

Note: *For detailed illustrations of the steering and suspension components, refer to Chapter 10.*

With the wheels on the ground

1   With the vehicle stopped and the front wheels pointed straight ahead, rock the steering wheel gently back and forth. If freeplay is excessive, a front wheel bearing, main shaft yoke, intermediate shaft yoke, steering knuckle bearing or tie rod end is worn or the steering gear is out of adjustment or broken. Refer to Chapter 10 for the appropriate repair procedure.

2   Other symptoms, such as excessive vehicle body movement over rough roads, swaying (leaning) around corners and binding as the steering wheel is turned, may indicate faulty steering and/or suspension components.

3   Check the shock absorbers by pushing down and releasing the vehicle several times at each corner. If the vehicle does not come back to a level position within one or two bounces, the shocks/struts are worn and must be replaced. When bouncing the vehicle up and down, listen for squeaks and noises from the suspension components.

Under the vehicle

4   Raise the vehicle with a floor jack and support it securely on jackstands. See *Jacking and towing* at the front of this book for proper jacking points.

5   Check the tires for irregular wear patterns and proper inflation. See Section 5 in this Chapter for information regarding tire wear.

6   Inspect the universal joint between the steering shaft and the steering gear housing. Check the steering gear housing for grease leakage. Check the steering linkage for looseness or damage. Check the tie-rod ends for excessive play. Look for loose bolts, broken or disconnected parts and deteriorated rubber bushings on all suspension and steering components **(see illustrations)**. While an assistant turns the steering wheel from side to side, check the steering components for free movement, chafing and binding. If the steering components do not seem to be reacting with the movement of the steering wheel, try to determine where the slack is located.

7   Check the wheel bearings. Do this by spinning the front wheels. Listen for any abnormal noises and watch to make sure the wheel spins true (doesn't wobble). Grab the top and bottom of the tire and pull in-and-out on it. Notice any movement which would indicate a loose wheel bearing assembly **(see illustration)**. If the bearings are suspect, they should be checked and repacked. Refer to Section 39 for more information.

8   Check the steering knuckle, moving the knuckle up and down

14.1 Steering wheel freeplay is the amount of travel between an initial steering input and the point at which the front wheels begin to turn (indicated by a slight resistance)

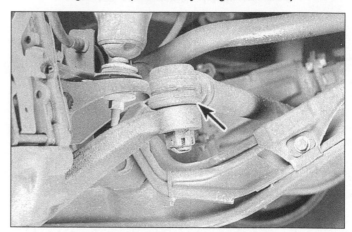

14.6a Inspect the suspension for deteriorated rubber bushings and torn grease seals (arrows)

with a pry bar to ensure that knuckle bearings have no play. If the bearings are suspect, they should be checked and repacked. Refer to Chapter 10 for more information.

9   Inspect the driveshafts for worn U-joints and for excessive play in the slip yoke and spline area (see Chapter 8).

10   Check the transfer case and differentials for evidence of fluid leakage.

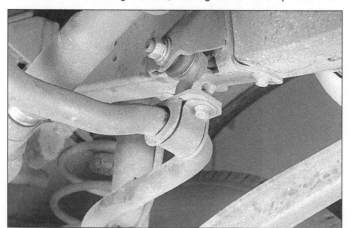

14.6b Check the stabilizer bar bushings for deterioration at the front and the rear of the vehicle

14.7 Grasp the tire as shown and check for endplay at the wheel bearings - if endplay is found the wheel bearings must be repacked and adjusted

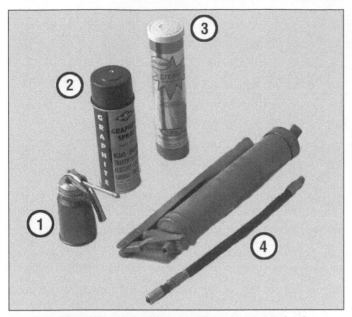

15.1 Materials required for chassis and body lubrication

1   **Engine oil** - *Light engine oil in a can like this can be used for door and hood hinges*
2   **Graphite spray** - *Used to lubricate lock cylinders*
3   **Grease** - *Grease, in a variety of types and weights, is available for use in a grease gun. Check the Specifications for your requirements*
4   **Grease gun** - *A common grease gun, shown here with a detachable hose and nozzle, is needed for chassis lubrication. After use, clean it thoroughly*

15   Chassis lubrication (every 7500 miles or 6 months)

Refer to illustrations 15.1, 15.6, 15.8, 15.9 and 15.11

1   Refer to *Recommended lubricants and fluids* at the front of this Chapter to obtain the necessary grease, etc. You will also need a grease gun **(see illustration)**. Occasionally plugs will be installed rather than grease fittings. If so, grease fittings will have to be purchased and installed.
2   Look under the vehicle for grease fittings or plugs on the steering, suspension, and driveline components. They are normally found on the tie-rod ends and universal joints. If there are plugs, remove them and install grease fittings, which will thread into the component. An automotive parts store will be able to supply the correct fittings. Straight, as well as angled, fittings are available.

15.6 Remove the pipe plug(s) to grease the steering knuckles

3   For easier access under the vehicle, raise it with a jack and place jackstands under the frame. Make sure it is safely supported by the stands. If the wheels are to be removed at this interval for tire rotation or brake inspection, loosen the lug nuts slightly while the vehicle is still on the ground.
4   Before beginning, force a little grease out of the nozzle to remove any dirt from the end of the gun. Wipe the nozzle clean with a rag.
5   With the grease gun and plenty of clean rags, crawl under the vehicle and begin lubricating the components.
6   Wipe the top of both steering knuckles free of dirt, then unscrew the pipe plug from each steering knuckle **(see illustration)**. Insert the nozzle, then squeeze the trigger on the grease gun to force grease into the component. For all other suspension and steering components, continue pumping grease into the fitting until it oozes out of the joint between the two components. If it escapes around the grease gun nozzle, the fitting is clogged or the nozzle is not completely seated on the fitting. Resecure the gun nozzle to the fitting and try again. If necessary, replace the fitting with a new one.
7   Wipe the excess grease from the components and the grease fitting. Repeat the procedure for the remaining fittings.
8   Lubricate the driveshaft slip yoke by pumping grease into the fitting until it can be seen coming out of the slip yoke seal **(see illustration)**.
9   Lubricate conventional universal joints until grease can be seen coming out of the contact points **(see illustration)**.
10   While you are under the vehicle, clean and lubricate the parking brake cable along with the cable guides and levers. This can be done by smearing some chassis grease onto the cable and its related parts with your fingers.
11   Lubricate the contact points on the steering knuckle stop and adjustment bolt **(see illustration)**.

15.8 The slip joint grease fitting is located on the yoke - pump grease into it until it comes out of the slip joint seal

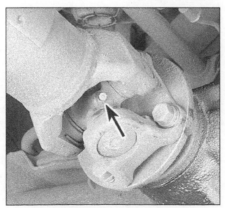

15.9 Pump grease into the universal joints until it can be seen coming out of the contact surfaces

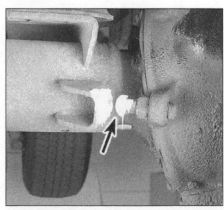

15.11 Use lithium base grease to lubricate the contact points on the steering knuckle stop and adjustment bolt

16.2a Check the exhaust pipes and connections (arrow) for signs of leakage and corrosion

16.2b Check the exhaust system rubber hangers for cracks and damage (arrow)

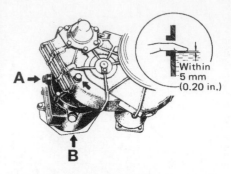

17.2 The manual transmission fill plug (A) and drain plug (B) are located on the side of the transmission case - use your finger as a dipstick to check the lubricant level

12 Open the hood and smear a little chassis grease on the hood latch mechanism. Have an assistant pull the hood release lever from inside the vehicle as you lubricate the cable at the latch.
13 Lubricate all the hinges (door, hood, etc.) with engine oil to keep them in proper working order.
14 The key lock cylinders can be lubricated with spray-on graphite or silicone lubricant, which is available at auto parts stores.
15 Lubricate the door weatherstripping with silicone spray. This will reduce chafing and retard wear.

16 Exhaust system check (every 7500 miles or 6 months)

Refer to illustrations 16.2a and 16.2b
1 With the engine cold (at least three hours after the vehicle has been driven), check the complete exhaust system from the manifold to the end of the tailpipe. Be careful around the catalytic converter (if equipped), which may be hot even after three hours. The inspection should be done with the vehicle on a hoist to permit unrestricted access. If a hoist isn't available, raise the vehicle and support it securely on jackstands.
2 Check the exhaust pipes and connections for signs of leakage and/or corrosion indicating a potential failure. Make sure that all brackets and hangers are in good condition and tight **(see illustrations)**.
3 Inspect the underside of the body for holes, corrosion, open seams, etc. which may allow exhaust gasses to enter the passenger compartment. Seal all body openings with silicone sealant or body putty.

4 Rattles and other noises can often be traced to the exhaust system, especially the hangers, mounts and heat shields. Try to move the pipes, mufflers and catalytic converter. If the components can come in contact with the body or suspension parts, secure the exhaust system with new brackets and hangers.

17 Manual transmission lubricant level check (every 7500 miles or 6 months)

Refer to illustration 17.2
1 The manual transmission has a filler plug which must be removed to check the lubricant level. If the vehicle is raised to gain access to the plug, be sure to support it safely on jackstands - DO NOT crawl under a vehicle which is supported only by a jack! Be sure the vehicle is level or the check may be inaccurate.
2 Using a wrench, unscrew the fill plug from the transmission **(see illustration)** and use a finger to reach inside the housing to determine the lubricant level. The level should be at or near the bottom of the plug hole.
3 If it isn't, add the recommended lubricant through the plug hole with a pump or squeeze bottle.
4 Install and tighten the plug and check for leaks after the first few miles of driving.

18 Transfer case lubricant level check (every 7500 miles or 6 months)

Refer to illustration 18.2
1 The transfer case lubricant level is checked by removing the upper plug located in the back of the case.
2 Use a finger to reach inside the housing to determine the lubricant level. The lubricant level should be just at the bottom of the hole **(see illustration)**. If not, add the appropriate lubricant through the opening.
3 Install and tighten the plug and check for leaks after the first few miles of driving.

19 Differential lubricant level check (every 7500 miles or 6 months)

Refer to illustration 19.2
Note: *The models covered by this manual have two differentials, be sure to check the lubricant level in both differentials.*
1 The differential has a check/fill plug which must be removed to check the lubricant level. If the vehicle must be raised to gain access to the plug, be sure to support it safely on jackstands - DO NOT crawl under the vehicle when it's supported only by the jack.

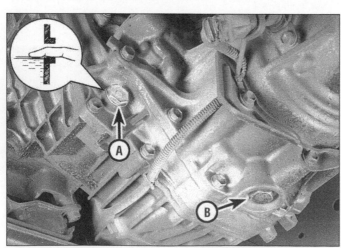

18.2 The transfer case fill plug (A) and drain plug (B) are located in the middle of the transfer case housing - use your finger as a dipstick to check the lubricant level

2   Remove the oil check/fill plug from the back of the rear differential or the front of the front differential **(see illustration)**. On some models a tag is located in the area of the plug which gives information regarding lubricant type, particularly on models equipped with a limited-slip differential.

3   Use a finger to reach inside the housing to determine the lubricant level. The oil level should be at the bottom of the plug opening. If it isn't, use a hand pump (available at auto parts stores) to add the specified lubricant until it just starts to run out of the opening.

4   Install the plug and tighten it securely.

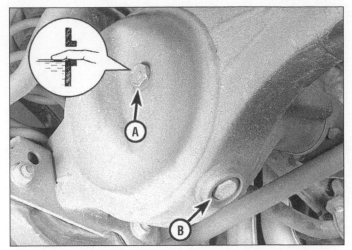

20   Seat belt check (every 7500 miles or 6 months)

1   Check the seat belts, buckles, latch plates and guide loops for any obvious damage or signs of wear.

2   Make sure the seat belt reminder light comes on when the key is turned on.

3   The seat belts are designed to lock up during a sudden stop or impact, yet allow free movement during normal driving. The retractors should hold the belt against your chest while driving and rewind the belt when the buckle is unlatched.

4   If any of the above checks reveal problems with the seat belt system, replace parts as necessary.

19.2 The differential fill plug (A) and drain plug (B) are located on the axle housing - use your finger as a dipstick to check the lubricant level

21   Thermostatically controlled air cleaner check (every 15,000 miles or 12 months)

Refer to illustrations 21.4 and 21.6

1   Carbureted engines are equipped with a thermostatically controlled air cleaner which draws air to the carburetor from different locations, depending upon engine temperature.

2   This is a visual check. If access is limited, a small mirror may have to be used.

3   Open the hood and locate the damper door inside the air cleaner assembly. It is inside the long snorkel of the metal air cleaner housing.

4   If there is a flexible air duct attached to the end of the snorkel, leading to an area behind the grille, disconnect it at the snorkel. This will enable you to look through the end of the snorkel and see the damper inside **(see illustration)**. A mirror may be needed if you can't look safely into the end of the snorkel.

5   The check should be done when the engine is cold. Start the engine and look through the snorkel at the damper, which should move to a closed position. With the damper closed, air cannot enter through the end of the snorkel, but instead enters the air cleaner through the flexible duct attached to the exhaust manifold and the heat stove passage.

6   As the engine warms up to operating temperature, the damper should open to allow air through the snorkel end **(see illustration)**. Depending on outside temperature, this may take 10 to 15 minutes. To

speed up this check you can reconnect the snorkel air duct, drive the vehicle, then check to see if the damper is completely open.

7   If the thermostatic air cleaner isn't operating properly, see Chapter 6 for more information.

22   Exhaust control valve (heat riser) lubrication and check (every 15,000 miles or 12 months)

Refer to illustration 22.2

1   Some engines covered by this manual are equipped with an exhaust heat control valve (heat riser), located between the junction of the exhaust manifold and the exhaust pipe. When the engine is cold, this valve redirects hot exhaust gases to the intake manifold for increased fuel vaporization and better low speed driveability.

2   With the engine cold to prevent burns, press on the butterfly actuator lever to make sure it opens and closes freely **(see illustration)**. If the action isn't smooth, lubricate it by applying heat-valve lubricant to the lever shaft.

3   Again with the engine cold, start the engine and watch the heat riser. Upon starting, the weight should move to the closed position. As the engine warms to normal operating temperature, the weight should move the valve to the open position, allowing a free flow of exhaust through the tailpipe. Since it could take several minutes for the system to heat up, you could mark the cold weight position, drive the vehicle, and then check the weight.

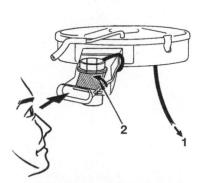

21.4 Position of the thermostatically controlled flapper door with the engine cold

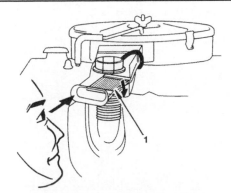

21.6 Position of the thermostatically controlled flapper door with the engine at normal operating temperature

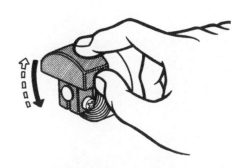

22.2 With the engine and exhaust pipe cold, try moving the heat riser valve - It should move freely

23 Evaporative emissions control system check (every 15,000 miles or 12 months)

Refer to illustration 23.2

1   The function of the evaporative emissions control system is to draw fuel vapors from the gas tank and fuel system, store them in a charcoal canister and route them to the intake manifold during normal engine operation.

2   The most common symptom of a fault in the evaporative emissions system is a strong fuel odor in the engine compartment. If a fuel odor is detected, inspect the charcoal canister, located in the front of the engine compartment **(see illustration)**. Check the canister and all hoses for damage and deterioration.

3   The evaporative emissions control system is explained in more detail in Chapter 6.

24 Exhaust Gas Recirculation (EGR) system check (every 15,000 miles or 12 months)

Refer to illustration 24.2

1   The EGR valve is usually located on the intake manifold. Most of the time when a problem develops in this emissions system, it's due to a stuck or defective EGR valve.

2   With the engine cold, to prevent burns, push on the EGR valve diaphragm. Using moderate pressure, you should be able to push the diaphragm up into the housing **(see illustration)**.

3   If the diaphragm doesn't move or is hard to move, replace the EGR valve with a new one. If in doubt about the condition of the valve, compare the free movement of your EGR valve with a new valve.

4   Refer to Chapter 6 for more information on the EGR system.

25 Carburetor choke check (every 15,000 miles or 12 months)

Refer to illustration 25.3

1   The choke operates when the engine is cold, and thus this check can only be performed before the vehicle has been started for the day.

2   Open the hood and remove the top plate of the air cleaner assembly. If any vacuum hoses must be disconnected, make sure you tag the hoses for reinstallation in their original positions. Place the top plate and wing nut aside, out of the way of moving engine components.

3   Look at the center of the air cleaner housing. You will notice a flat plate (arrow) at the carburetor opening **(see illustration)**.

4   Press the accelerator pedal to the floor. The plate should close completely. Start the engine while you watch the plate at the carburetor.

23.2 Check the evaporative emissions control canister and the hose connections for cracks and damage

Warning: *Do not position your face directly over the carburetor, as the engine could backfire, causing serious burns.* When the engine starts, the choke plate should open slightly.

5   Allow the engine to continue running at an idle speed. As the engine warms up to operating temperature, the plate should slowly open, allowing more cold air to enter through the top of the carburetor.

6   After a few minutes, the choke plate should be fully open to the vertical position. Press quickly on the accelerator to make sure the fast idle cam disengages.

7   You will notice that the engine speed corresponds with the plate opening. With the plate fully closed, the engine should run at a fast idle speed. As the plate opens and the throttle is moved to disengage the fast idle cam, the engine speed will decrease.

26 Positive Crankcase Ventilation (PCV) valve check and replacement (every 15,000 miles or 12 months)

Refer to illustration 26.2

1   The PCV valve is usually located in the valve cover.

2   With the engine idling at normal operating temperature, pull the valve (with hose attached) from the valve cover or intake manifold **(see illustration)**.

3   Place your finger over the valve opening or hose. If there's no vacuum, check for a plugged hose, manifold port, or the valve itself. Replace any plugged or deteriorated hoses.

24.2 The diaphragm in the EGR valve (which can be reached through the holes in the underside of the valve) should move easily with finger pressure

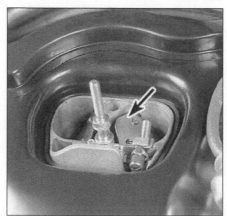

25.3 The carburetor choke plate (arrow) is visible after removing the air cleaner top plate

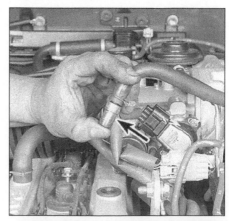

26.2 The PCV valve (arrow) is located in the valve cover - with the engine running, put your finger over the end of the PCV hose, you should feel vacuum

27.3 The first step in removing the air filter is detaching the clips on the housing, then remove the wing nut

27.5 After setting the top cover aside, the filter element can be lifted out of the housing

4   Turn off the engine and shake the PCV valve, listening for a rattle. If the valve doesn't rattle, replace it with a new one.
5   To replace the valve, pull it from the end of the hose, noting its installed position.
6   When purchasing a replacement PCV valve, make sure it's for your particular vehicle and engine size. Compare the old valve with the new one to make sure they're the same.
7   Push the valve into the end of the hose until it's seated.
8   Inspect all rubber hoses and grommets for damage and hardening. Replace them, if necessary.
9   Press the PCV valve and hose securely into position.

27   Air filter check and replacement (every 15,000 miles or 12 months)

Refer to illustrations 27.3 and 27.5
1   At the specified intervals, the air filter should be replaced with a

new one. A thorough preventive maintenance schedule would also require the filter to be inspected between filter changes.
2   On carbureted models, the air filter housing is mounted on top of the carburetor in the engine compartment.
3   On fuel injected models, the air filter housing is located at the front of the engine compartment **(see illustration)**.
4   Remove the wing nut(s) and the cover retaining clips. Then detach all hoses that would interfere with the removal of the air cleaner cover from the air cleaner housing. While the top cover is off, be careful not to drop anything down into the carburetor or air cleaner assembly.
5   Lift the air filter element out of the housing **(see illustration)** and wipe out the inside of the air cleaner housing with a clean rag.
6   Inspect the outer surface of the filter element. If it is dirty, replace it. If it is only moderately dusty, it can be reused by blowing it clean from the back to the front surface with compressed air. Because it is a pleated paper type filter, it cannot be washed or oiled. If it cannot be cleaned satisfactorily with compressed air, discard and replace it. **Caution:** *Never drive the vehicle with the air cleaner removed. Excessive engine wear could result and backfiring could even cause a fire under the hood.*
7   Place the new filter in the air cleaner housing, making sure it seats properly.
8   Installation of the cover is the reverse of removal.

28   Spark plug replacement (every 15,000 miles or 12 months)

Refer to illustrations 28.1, 28.4a and 28.4b
1   Spark plug replacement requires a spark plug socket which fits onto a ratchet wrench. This socket is lined with a rubber grommet to protect the porcelain insulator of the spark plug and to hold the plug while you insert it into the spark plug hole. You will also need a wire-type feeler gauge to check and adjust the spark plug gap and a torque wrench to tighten the new plugs to the specified torque **(see illustration)**.
2   If you are replacing the plugs, purchase the new plugs, adjust them to the proper gap and then replace each plug one at a time. **Note:** *When buying new spark plugs, it's essential that you obtain the correct plugs for your specific vehicle. This information can be found in the Specifications Section at the beginning of this Chapter, on the Vehicle Emissions Control Information (VECI) label located on the underside of the hood or in the owner's manual. If these sources specify different plugs, purchase the spark plug type specified on the VECI label because that information is provided specifically for your engine.*
3   Inspect each of the new plugs for defects. If there are any signs of cracks in the porcelain insulator of a plug, don't use it.

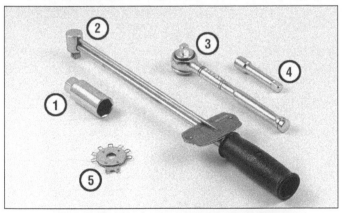

28.1 Tools required for changing spark plugs

1   *Spark plug socket - This will have special padding inside to protect the spark plug's porcelain insulator*
2   *Torque wrench - Although not mandatory, using this tool is the best way to ensure the plugs are tightened properly*
3   *Ratchet - Standard hand tool to fit the spark plug socket*
4   *Extension - Depending on model and accessories, you may need special extensions and universal joints to reach one or more of the plugs*
5   *Spark plug gap gauge - This gauge for checking the gap comes in a variety of styles. Make sure the gap for your engine is included*

28.4a Spark plug manufacturers recommend using a wire type gauge when checking the gap - if the wire does not slide between the electrodes with a slight drag, adjustment is required

28.4b To change the gap, bend the *side* electrode only, as indicated by the arrows, and be very careful not to crack or chip the porcelain insulator surrounding the center electrode

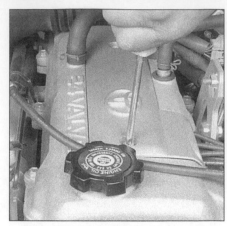

28.6a On 1FZ-FE engines, remove the trim cover on the valve cover to access the spark plugs

28.6b When removing the spark plug wires, pull only on the boot and use a twisting, pulling motion

28.8 Use a socket wrench with a long extension to unscrew the spark plugs

28.10a Apply a thin coat of anti-seize compound to the spark plug threads

4   Check the electrode gaps of the new plugs. Check the gap by inserting the wire gauge of the proper thickness between the electrodes at the tip of the plug **(see illustration)**. The gap between the electrodes should be identical to that listed in this Chapter's Specifications or on the VECI label. If the gap is incorrect, use the notched adjuster on the feeler gauge body to bend the curved side electrode slightly **(see illustration)**.

5   If the side electrode is not exactly over the center electrode, use the notched adjuster to align them.

Removal

Refer to illustrations 28.6a, 28.6b and 28.8

6   To prevent the possibility of mixing up spark plug wires, work on one spark plug at a time. Remove the wire and boot from one spark plug. Grasp the boot - not the cable - as shown, give it a half twisting motion and pull straight up **(see illustrations)**.

7   If compressed air is available, blow any dirt or foreign material away from the spark plug area before proceeding (a common bicycle pump will also work).

8   Remove the spark plug **(see illustration)**.

9   Whether you are replacing the plugs at this time or intend to reuse the old plugs, compare each old spark plug with the chart shown on the inside back cover of this manual to determine the overall running condition of the engine.

Installation

Refer to illustrations 28.10a and 28.10b

10   Prior to installation, apply a coat of anti-seize compound to the plug threads. It's often difficult to insert spark plugs into their holes without cross-threading them. To avoid this possibility, fit a short piece of rubber hose over the end of the spark plug **(see illustrations)**. The flexible hose acts as a universal joint to help align the plug with the plug hole. Should the plug begin to cross-thread, the hose will slip on the spark plug, preventing thread damage. Tighten the plug to the torque listed in this Chapter's Specifications.

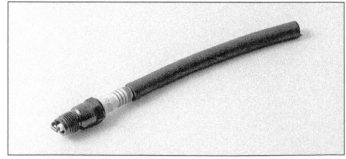

28.10b A length of rubber hose will save time and prevent damaged threads when installing the spark plugs

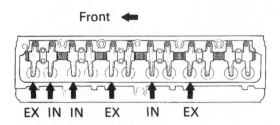

Front ⬅

EX IN IN EX IN EX

29.3 With the No. 1 piston at TDC on the compression stroke, check and adjust the clearance of the indicated valves

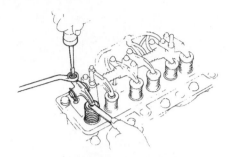

29.4 Loosen the locknut and turn the adjusting screw until the feeler gauge slips between the valve stem tip and rocker arm with a slight amount of drag

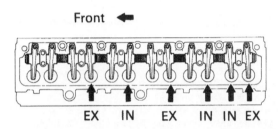

Front ⬅

EX IN EX IN IN EX

29.6 Rotate the crankshaft 360-degrees from No. 1 TDC, then check and adjust the clearance of the indicated valves

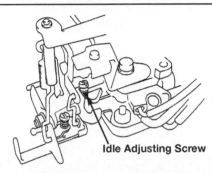

Idle Adjusting Screw

30.4 The curb idle speed can be adjusted by turning the idle stop screw (arrow) located at the base of the carburetor by the throttle lever

11   Attach the plug wire to the new spark plug, again using a twisting motion on the boot until it is firmly seated on the end of the spark plug.
12   Follow the above procedure for the remaining spark plugs, replacing them one at a time to prevent mixing up the spark plug wires.

29 Valve clearance check and adjustment (2F and 3F-E pushrod engines) (every 15,000 miles or 12 months)

Refer to illustrations 29.3, 29.4 and 29.6
Note: *The manufacturer recommends adjusting the valve clearance at the specified interval only if the valve train is making excessive noise.*
1   Make sure the engine is warmed up to operating temperature before beginning this procedure.
2   Refer to Chapter 2A and remove the valve cover, then use the distributor rotor and timing cover marks, to position the number one piston at TDC on the compression stroke (see Chapter 2A). **Note:** *Verify TDC for the number one piston by observing that both valves are closed and the rocker arms are loose.*
3   With the crankshaft at number one TDC, measure the clearance of the indicated valves **(see illustration)**. Insert a feeler gauge of the specified thickness (see this Chapter's Specifications) between the valve stem tip and the rocker arm. The feeler gauge should slip between the valve stem tip and rocker arm with a slight amount of drag.
4   If the clearance is incorrect (too loose or too tight), loosen the locknut and turn the adjusting screw slowly until you can feel a slight drag on the feeler gauge as you withdraw it from between the valve stem tip and the rocker arm **(see illustration).**
5   Once the clearance is adjusted, hold the adjusting screw with a screwdriver (to keep it from turning) and tighten the locknut to the torque listed in this Chapter's Specifications. Recheck the clearance to make sure it hasn't changed after tightening the locknut.
6   Make a mark on the crankshaft pulley and an adjacent mark on the front cover. Rotate the crankshaft one complete revolution (360-degrees) and realign the marks. Check the clearance of the remaining valves **(see illustration).**
7   If necessary, repeat the adjustment procedure described in Steps 3, 4 and 5 until all the valves are adjusted to specifications.
8   Install the valve covers.

30 Idle speed check and adjustment (every 15,000 miles or 12 months)

Refer to illustration 30.4
Note: *This Section pertains to carbureted models only. For models equipped with fuel injection, see Chapter 4B.*
1   Engine idle speed is the speed at which the engine operates when no accelerator pedal pressure is applied. The idle speed is critical to the performance of the engine as well as many engine sub-systems.
2   Make sure the parking brake is firmly set and the wheels blocked to prevent the vehicle from rolling. This is especially true if the specifications require the transmission to be in Drive. An assistant inside the vehicle pressing on the brake pedal is the safest method.
3   A hand-held tachometer must be used when adjusting idle speed to get an accurate reading. The exact hook-up for these meters varies with the manufacturer, so follow the particular directions included with the instrument.
4   For most applications, the idle speed is set by turning an adjustment screw which is located next to the throttle lever **(see illustration)**. This screw changes the amount the throttle plate is held open by the throttle linkage. The screw may be on the linkage itself or on the carburetor body.
5   For all applications, the engine must be completely warmed-up to operating temperature, which will automatically render the choke and fast idle inoperative.
6   Once you have located the idle screw, experiment with different length screwdrivers until the adjustment can be easily made without coming into contact with hot or moving engine components.
7   If the air cleaner is removed, the vacuum hose to the snorkel should be plugged.

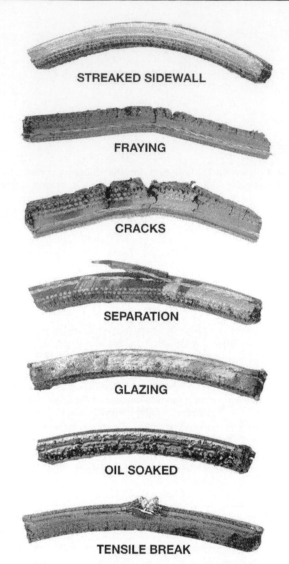

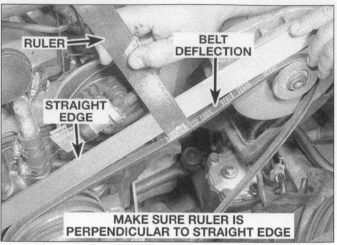

31.4 Measuring drivebelt deflection with a straightedge and ruler

31.3 Here are some of the more common problems associated with V-drivebelts (check the belts very carefully to prevent an untimely breakdown)

8   Since the manufacturer recommended many different adjustment procedures over the time period covered by this manual, it would be impractical to cover all types in this Section. Most models have a tune-up decal or Vehicle Emission Control Information (VECI) label located in the engine compartment with instructions for setting idle speed. If no VECI label is found refer to the Specifications Section at the beginning of this Chapter and to the adjustment procedures specified in Chapter 4.

---

31   Drivebelt check, adjustment and replacement (every 15,000 miles or 12 months)

Check

Refer to illustrations 31.3 and 31.4

1   The drivebelts, or V-belts as they are sometimes called, are located at the front of the engine and play an important role in the overall operation of the vehicle and its components. Due to their function and material make-up, the belts are prone to failure after a period of time and should be inspected and adjusted periodically to prevent major engine damage.
2   The number of belts used on a particular vehicle depends on the

accessories installed. Drivebelts are used to turn the alternator, power steering pump, water pump and air conditioning compressor. Depending on the pulley arrangement, more than one of these components may be driven by a single belt.
3   With the engine off, open the hood and locate the various belts at the front of the engine. Using your fingers (and a flashlight, if necessary), move along the belts checking for cracks and separation of the belt plies. Also check for fraying and glazing, which gives the belt a shiny appearance **(see illustration)**. Both sides of each belt should be inspected, which means you will have to twist the belt to check the underside.
4   The tension of each belt is checked by pushing on the belt at a distance halfway between the pulleys. Push firmly with your thumb and see how much the belt moves (deflects) **(see illustration)**. As rule of thumb, if the distance from pulley center-to-pulley center is between 7 and 11 inches, the belt should deflect 1/4-inch. If the belt travels between pulleys spaced 12 to 16 inches apart, the belt should deflect 1/2-inch for a V-belt or 1/4-inch for a multi-ribbed belt.

Adjustment

Refer to illustration 31.6

5   If it is necessary to adjust the belt tension, either to make the belt tighter or looser, it is done by moving the belt-driven accessory on the bracket.
6   For each belt on the engine there will be one component with an adjusting bolt and a pivot bolt. These bolts must be loosened slightly to enable you to move the component **(see illustration)**.
7   After the two bolts have been loosened, move the component away from the engine to tighten the belt or toward the engine to loosen the belt. Hold the accessory in position and check the belt tension. If it is correct, tighten the two bolts until just snug, then recheck the tension. If the tension is all right, tighten the bolts.
8   On earlier models it will be necessary to use some sort of prybar to move the accessory while the belt is adjusted. If this must be done to gain the proper leverage, be very careful not to damage the component being moved or the part being pried against.

Replacement

9   To replace a belt, follow the above procedures for drivebelt adjustment but slip the belt off the pulleys and remove it. Since belts tend to wear out more or less at the same time, it's a good idea to replace all of them at the same time. Mark each belt and the corresponding pulley grooves so the replacement belts can be installed properly.
10   Take the old belts with you when purchasing new ones in order to make a direct comparison for length, width and design.
11   Adjust the belts as described earlier in this Section.

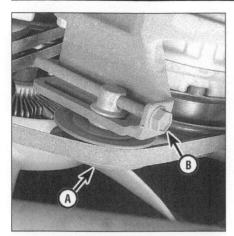

31.6 Loosen the lock bolt (A) on the front of the pulley and turn the adjustment bolt (B) in the desired direction to adjust the drive belt

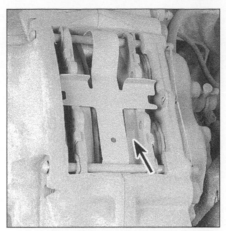

33.5 With the wheels removed, the brake pad lining (arrow) can be inspected

33.10 Check for any sign of brake fluid leakage at the brake line fittings and the brake hoses (arrow)

32 Fuel system check (every 15,000 miles or 12 months)

Warning: *Gasoline is extremely flammable, so take extra precautions when you work on any part of the fuel system. Don't smoke or allow open flames or bare light bulbs near the work area, and don't work in a garage where a natural gas-type appliance (such as a water heater or clothes dryer) with a pilot light is present. Since gasoline is carcinogenic, wear latex gloves when there's a possibility of being exposed to fuel, and, if you spill any fuel on your skin, rinse it off immediately with soap and water. Mop up any spills immediately and do not store fuel-soaked rags where they could ignite. The fuel system is under constant pressure, so, if any fuel lines are to be disconnected, the fuel pressure in the system must be relieved first (see Chapter 4 for more information). When you perform any kind of work on the fuel system, wear safety glasses and have a Class B type fire extinguisher on hand.*

1   The fuel system is most easily checked with the vehicle raised on a hoist so the components underneath the vehicle are readily visible and accessible.

2   If the smell of gasoline is noticed while driving or after the vehicle has been in the sun, the system should be thoroughly inspected immediately.

3   Remove the gas tank cap and check for damage, corrosion and an unbroken sealing imprint on the gasket. Replace the cap with a new one if necessary.

4   With the vehicle raised and safely supported, inspect the gas tank and filler neck for punctures, cracks and other damage. The connection between the filler neck and the tank is particularly critical. Sometimes a rubber filler neck will leak because of loose clamps or deteriorated rubber. These are problems a home mechanic can usually rectify. **Warning:** *Do not, under any circumstances, try to repair a fuel tank (except rubber components). A welding torch or any open flame can easily cause fuel vapors inside the tank to explode.*

5   Carefully check all rubber hoses and metal lines leading away from the fuel tank. Check for loose connections, deteriorated hoses, crimped lines and other damage. Follow the lines to the front of the vehicle, carefully inspecting them all the way to the carburetor or fuel injection system. Repair or replace damaged sections as necessary.

6   If a fuel odor is still evident after the inspection, refer to Chapter 6 and check the EVAP system.

33 Brake check (every 15,000 miles or 12 months)

Warning: *The dust created by the brake system may contain asbestos, which is harmful to your health. Never blow it out with compressed air and don't inhale any of it. An approved filtering mask should be worn when working on the brakes. Do not, under any circumstances, use petroleum-based solvents to clean brake parts. Use brake system cleaner only! Try to use non-asbestos replacement parts whenever possible.*

Note: *For detailed photographs of the brake system, refer to Chapter 9.*

1   In addition to the specified intervals, the brakes should be inspected every time the wheels are removed or whenever a defect is suspected. Any of the following symptoms could indicate a potential brake system defect: The vehicle pulls to one side when the brake pedal is depressed; the brakes make squealing or dragging noises when applied; brake pedal travel is excessive; the pedal pulsates; brake fluid leaks, usually onto the inside of the tire or wheel.

2   Loosen the wheel lug nuts.

3   Raise the vehicle and place it securely on jackstands.

4   Remove the wheels (see *Jacking and towing* at the front of this book, or your owner's manual, if necessary).

Disc brakes

Refer to illustrations 33.5 and 33.10

5   There are two pads (an outer and an inner) in each caliper. The pads are visible after the wheels are removed **(see illustration)**.

6   Measure the pad thickness. If the lining material is less than the thickness listed in this Chapter's Specifications, replace the pads. **Note:** *Keep in mind that the lining material is riveted or bonded to a metal backing plate and the metal portion is not included in this measurement.*

7   If it is difficult to determine the exact thickness of the remaining pad material by the above method, or if you are at all concerned about the condition of the pads, remove the caliper(s), then remove the pads from the calipers for further inspection (refer to Chapter 9).

8   Once the pads are removed from the calipers, clean them with brake cleaner and re-measure them with a ruler or a vernier caliper.

9   Measure the disc thickness with a micrometer to make sure that it still has service life remaining. If any disc is thinner than the specified minimum thickness, replace it (refer to Chapter 9). Even if the disc has service life remaining, check its condition. Look for scoring, gouging and burned spots. If these conditions exist, remove the disc and have it resurfaced (see Chapter 9).

10   Before installing the wheels, check all brake lines and hoses for damage, wear, deformation, cracks, corrosion, leakage, bends and twists, particularly in the vicinity of the rubber hoses at the calipers **(see illustration)**. Check the clamps for tightness and the connections for leakage. Make sure that all hoses and lines are clear of sharp edges, moving parts and the exhaust system. If any of the above conditions are noted, repair, reroute or replace the lines and/or fittings as necessary (see Chapter 9).

Drum brakes

Refer to illustrations 33.12 and 33.14

11   Refer to Chapter 9 and remove the brake drums.
12   Note the thickness of the lining material on the brake shoes **(see illustration)** and look for signs of contamination by brake fluid and grease. If the lining material is within 1/16-inch of the recessed rivets or metal shoes, replace the brake shoes with new ones. The shoes should also be replaced if they are cracked, glazed (shiny lining surfaces) or contaminated with brake fluid or grease. See Chapter 9 for the replacement procedure.
13   Check the shoe return and hold-down springs and the adjusting mechanism to make sure they're installed correctly and in good condition. Deteriorated or distorted springs, if not replaced, could allow the linings to drag and wear prematurely.
14   Check the wheel cylinders for leakage by carefully peeling back the rubber boots **(see illustration).** If brake fluid is noted behind the boots, the wheel cylinders must be replaced (see Chapter 9).
15   Check the drums for cracks, score marks, deep scratches and hard spots, which will appear as small discolored areas. If imperfections cannot be removed with emery cloth, the drums must be resurfaced by an automotive machine shop (see Chapter 9 for more detailed information).
16   Refer to Chapter 9 and install the brake drums.
17   Install the wheels and snug the wheel lug nuts finger tight.
18   Remove the jackstands and lower the vehicle.
19   Tighten the wheel lug nuts to the torque listed in this Chapter's Specifications.

Brake booster check

20   Sit in the driver's seat and perform the following sequence of tests.
21   With the engine stopped, depress the brake pedal several times - the travel distance should not change.
22   With the brake fully depressed, start the engine - the pedal should move down a little when the engine starts.
23   Depress the brake, stop the engine and hold the pedal in for about 30 seconds - the pedal should neither sink nor rise.
24   Restart the engine, run it for about a minute and turn it off. Then firmly depress the brake several times - the pedal travel should decrease with each application.
25   If your brakes do not operate as described above when the preceding tests are performed, the brake booster is either in need of repair or has failed. Refer to Chapter 9 for the removal procedure.

Parking brake

Refer to illustration 33.31

26   Slowly pull up on the parking brake and count the number of clicks you hear until the handle is up as far as it will go. The adjustment

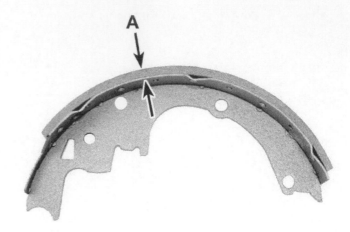

33.11 If the lining is bonded to the brake shoe, measure the lining thickness from the outer surface to the metal shoe, as shown here; if the lining is riveted to the shoe, measure from the lining outer surface to the rivet head

should be within the specified number of clicks listed in this Chapter's Specifications. If you hear more or fewer clicks, it's time to adjust the parking brake (refer to Chapter 9).
27   An alternative method of checking the parking brake is to park the vehicle on a steep hill with the parking brake set and the transmission in Neutral (be sure to stay in the vehicle during this check!). If the parking brake cannot prevent the vehicle from rolling, it is in need of adjustment (see Chapter 9).
28   However, on vehicles with rear disc brakes the parking brake assembly itself should be visually inspected every 12 months (or whenever a fault is suspected).
29   With the vehicle raised and supported on jackstands, remove the rear wheels.
30   Remove the rear discs as described in Chapter 9. Support the caliper assemblies with a coat hanger or heavy wire and do not disconnect the brake line from the caliper.
31   With the disc removed, the parking brake components are visible and can be inspected for wear and damage. The linings should last the life of the vehicle. However, they can wear down if the parking brake system has been improperly adjusted. There is no minimum thickness specification for the parking brake shoes, but as a rule of thumb, if the shoe material is less 1/32-inch (0.8 mm) thick, you should replace them **(see illustration)**. Also check the springs and adjuster mechanism and inspect the disc for deep scratches and other damage. For more information on the brake system see Chapter 9.

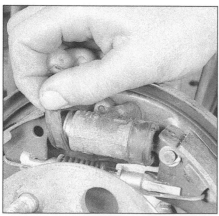

33.14 Check for fluid leakage at both ends of the wheel cylinder dust covers (arrow)

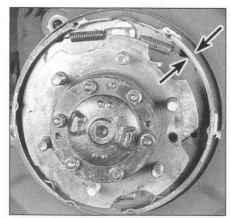

33.31 Measure the parking brake shoe lining thickness from the outer surface to the metal shoe

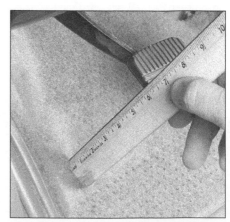

34.1 Pedal height is the distance between the pedal pad and the floor

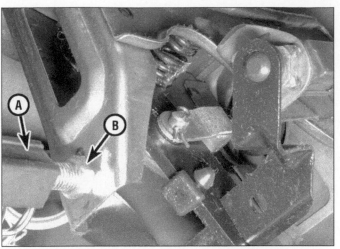

34.2 Back the stopper bolt or switch (A) out for clearance, then loosen the locknut on the pushrod (B). Turn the pushrod to adjust the pedal height

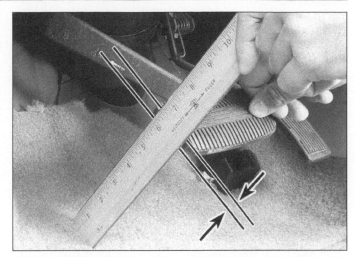

34.5 Pedal freeplay is the distance from the natural resting point of the pedal to the point at which resistance is felt

34   Clutch/brake pedal height and freeplay adjustment (every 15,000 miles or 12 months)

Pedal height

Refer to illustrations 34.1 and 34.2

1   The height of the clutch and brake pedal is the distance the pedal sits off the floor **(see illustration)**. If the pedal height is not within the specified range, it must be adjusted.

2   To adjust the clutch pedal loosen the locknut and back the stopper bolt or switch out for clearance, then loosen the locknut on the clutch pushrod. Turn the pushrod to adjust the pedal height in the middle of the specified range, then retighten the locknut **(see illustration)**.

3   Before measuring the brake pedal height make sure the pedal is in the fully returned position, then start the engine and depress the accelerator several times to activate the brake power booster. Measure the pedal height and adjust if necessary. To adjust the pedal height loosen the locknut and back the stopper bolt or switch out for clearance, then loosen the locknut on the brake pushrod. Turn the pushrod to adjust the pedal height in the middle of the specified range, then retighten the locknut.

4   Adjust the pedal switch or stopper bolt by turning it clockwise until the switch body or stopper bolt just contacts the pedal arm.

Pedal freeplay

Refer to illustration 34.5

5   The freeplay is the pedal slack, or the distance the pedal can be depressed before it begins to have any effect on the clutch or brake

system **(see illustration)**. If the pedal freeplay is not within the specified range, it must be adjusted.

6   To adjust the clutch pedal freeplay loosen the locknut on the clutch pushrod. Then back off the pushrod to adjust the pedal freeplay to the specified range and retighten the locknut.

7   Before measuring the brake pedal freeplay turn off the engine and depress the brake pedal a half dozen times. Measure the pedal freeplay and adjust if necessary. Loosen the locknut on the brake pushrod, then back off the pushrod to adjust the pedal freeplay to the specified range and retighten the locknut.

35   Fuel filter replacement (every 15,000 miles or 12 months)

Warning: *Gasoline is extremely flammable, so take extra precautions when you work on any part of the fuel system. Don't smoke or allow open flames or bare light bulbs near the work area, and don't work in a garage where a natural gas-type appliance (such as a water heater or clothes dryer) with a pilot light is present. Since gasoline is carcinogenic, wear latex gloves when there's a possibility of being exposed to fuel, and, if you spill any fuel on your skin, rinse it off immediately with soap and water. Mop up any spills immediately and do not store fuel-soaked rags where they could ignite. When you perform any kind of work on the fuel system, wear safety glasses and have a Class B Type fire extinguisher on hand.*

1   This job should be done with the engine cold (after sitting at least three hours). Place a metal container, rags or newspapers under the filter to catch spilled fuel.

Carbureted models

Refer to illustration 35.5

2   The fuel filter is located under the vehicle, near the gas tank or in the engine compartment. If the vehicle must be raised to change the filter, be sure to support it safely on jackstands!

3   To replace the filter, release the clamps and slide them down the hoses, past the fittings on the filter.

4   Carefully twist and pull on the hoses to separate them from the filter. If the hoses are in bad shape, now would be a good time to replace them with new ones. Slide off the old clamps and install new ones.

5   Pull the filter out of the clip and install the new one, then hook up the hoses and reposition the clamps. Note that the arrow on the filter must point in the direction of fuel flow (toward the carburetor) **(see illustration)**. Start the engine and check carefully for leaks at the filter hose connections.

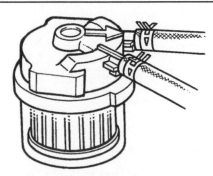

35.5 The arrow on the fuel filter indicates the outlet side (carbureted models)

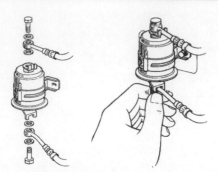

35.6 EFI fuel filter mounting details - always tighten the banjo bolts by hand before using a wrench

Fuel-injected models

Refer to illustration 35.6

Warning: *Refer to* Chapter 4 *and depressurize the fuel system before removing the filter!*

6   Loosen the banjo bolts on both ends of the fuel filter **(see illustration)** with a flare nut wrench. Disconnect both lines.
7   Remove both bracket bolts and detach the old filter and the filter support bracket.
8   Remove the filter clamp bolt and separate the old filter from the bracket. Note that the inlet and outlet lines are clearly labeled and that the flanged end of the filter faces down.
9   Install the new filter and bracket assembly and tighten the bracket bolts securely. Make sure that the new filter is installed flanged end down.
10   Using the new sealing washers - two per banjo fitting - provided by the filter manufacturer, install the inlet and outlet banjo bolts and tighten them to the specified torque.
11   The remainder of installation is the reverse of the removal.

36 Spark plug wire, distributor cap and rotor check and replacement (every 30,000 miles or 24 months)

Refer to illustrations 36.11a, 36.11b, 36.12a and 36.12b

1   The spark plug wires should be checked whenever new spark plugs are installed.
2   Begin this procedure by making a visual check of the spark plug wires while the engine is running. In a darkened garage (make sure there is adequate ventilation) start the engine and observe each plug wire. Be careful not to come into contact with any moving engine parts. If there is a break in the wire, you will see arcing or a small spark at the damaged area. If arcing is noticed, make a note to obtain new wires, then allow the engine to cool and check the distributor cap and rotor.
3   The spark plug wires should be inspected one at a time to prevent mixing up the order, which is essential for proper engine operation. Each original plug wire should be numbered to help identify its location. If the number is illegible, a piece of tape can be marked with the correct number and wrapped around the plug wire.
4   Disconnect the plug wire from the spark plug. A removal tool can be used for this purpose or you can grasp the rubber boot, twist the boot half a turn and pull the boot free. Do not pull on the wire itself.
Note: *Later models require removing a cover on the top of the valve cover to access the spark plugs.*
5   Check inside the boot for corrosion, which will look like a white crusty powder.
6   Push the wire and boot back onto the end of the spark plug. It should fit tightly onto the end of the plug. If it doesn't, remove the wire and use pliers to carefully crimp the metal connector inside the wire boot until the fit is snug.
7   Using a clean rag, wipe the entire length of the wire to remove built-up dirt and grease. Once the wire is clean, check for burns, cracks and other damage. Do not bend the wire sharply, because the

36.11a Remove the distributor cap retaining screws - pull the cap out and up to access the rotor

conductor might break.
8   Remove the rubber boot (if equipped) and disconnect the wire from the distributor. Again, pull only on the rubber boot. Check for corrosion and a tight fit. Replace the wire in the distributor.

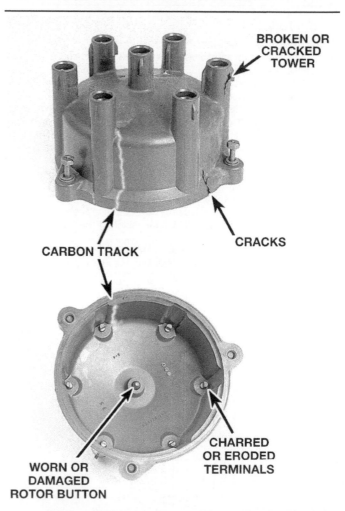

36.11b Shown here are some of the common defects to look for when inspecting the distributor cap (if in doubt about its condition, install a new one)

36.12a Pull off the rotor and inspect it thoroughly

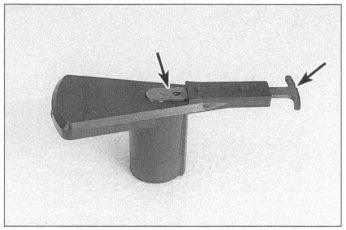

36.12b The ignition rotor should be checked for wear and corrosion as indicated here (if in doubt about its condition, buy a new one)

9   Inspect the remaining spark plug wires, making sure that each one is securely fastened at the distributor and spark plug when the check is complete.

10   If new spark plug wires are required, purchase a set for your specific engine model. Remove and replace the wires one at a time to avoid mix-ups in the firing order.

11   Detach the distributor cap by loosening the cap retaining screws. Look inside it for cracks, carbon tracks and worn, burned or loose contacts **(see illustrations)**.

12   Pull the rotor off the distributor shaft and examine it for cracks and carbon tracks **(see illustrations)**. Replace the cap and rotor if any damage or defects are noted.

13   It is common practice to install a new cap and rotor whenever new spark plug wires are installed, but if you wish to continue using the old cap, check the resistance between the spark plug wires and the cap first. If the indicated resistance is more than the maximum value listed in this Chapter's Specifications, replace the cap and/or wires.

14   When installing a new cap, remove the wires from the old cap one at a time and attach them to the new cap in the exact same location **Note:** *If an accidental mix-up occurs, refer to the firing order Specifications at the beginning of this Chapter.*

---

37   Ignition timing check and adjustment (every 30,000 miles or 24 months)

Refer to illustrations 37.2, 37.3, 37.6a and 37.6b

Note: *It is imperative that the procedures included on the tune-up or Vehicle Emissions Control Information (VECI) label be followed when adjusting the ignition timing. The label will include all information concerning preliminary steps to be performed before adjusting the timing, as well as the timing specifications. If no VECI label is found refer to the Specifications chart at the beginning of this Chapter.*

1   At the specified intervals, the ignition timing should be checked and adjusted.

2   Various procedures (depending on the model of the vehicle) must be performed to disable the distributor or computer advance mechanism before attempting to check the timing **(see illustration)**. Locate the Tune-up or VECI label under the hood and read through and perform all preliminary instructions concerning ignition timing. If no VECI label is found refer to the Specifications Section at the beginning of this Chapter.

3   Before attempting to check the timing some special tools will be needed for this procedure **(see illustration)**.

4   Check that the idle speed is as specified (Section 30).

5   Connect a timing light in accordance with the manufacturer's instructions. Generally, the light will be connected to power and ground sources and to the number one spark plug wire (refer to the cylinder location diagram in this Chapter's Specifications).

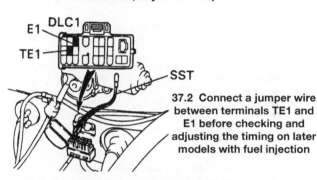

37.2 Connect a jumper wire between terminals TE1 and E1 before checking and adjusting the timing on later models with fuel injection

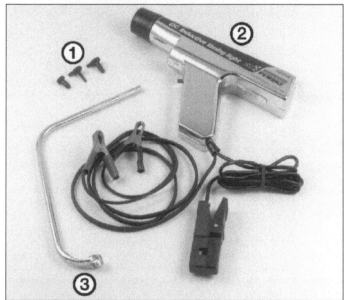

37.3 Tools needed to check and adjust the ignition timing

1   ***Vacuum plugs*** *- Vacuum hoses will, in most cases, have to be disconnected and plugged. Molded plugs in various shapes and sizes are available for this*

2   ***Inductive pick-up timing light*** *- Flashes a bright, concentrated beam of light when the number one spark plug fires. Connect the leads according to the instructions supplied with the light*

3   ***Distributor wrench*** *- On some models, the hold-down bolt for the distributor is difficult to reach and turn with conventional wrenches or sockets. A special wrench like this must be used*

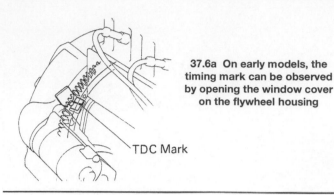

37.6a On early models, the timing mark can be observed by opening the window cover on the flywheel housing

TDC Mark

37.6b On later models, the ignition timing indicator is located at the front of the engine near the crankshaft pulley

6   Locate the timing marks on the engine **(see illustrations)**. Clean them off with solvent if necessary so you can see the numbers or marks and small grooves.
7   Use chalk or paint to mark the groove in the crankshaft pulley.
8   Mark the timing tab in accordance with the number of degrees called for on the VECI label or the tune-up label in the engine compartment
9   Make sure timing light is clear of all moving engine components, then start the engine and warm it up to normal operating temperature.
10   Aim the flashing timing light at the timing mark by the crankshaft pulley, again being careful not to come in contact with moving parts. The marks should appear to be stationary. If the marks are in alignment, the timing is correct.
11   If the notch on the crankshaft pulley is not aligned with the correct mark on the timing tab, loosen the distributor hold-down bolt and rotate the distributor until the notch is aligned with the correct timing mark.
12   Retighten the hold-down bolt and recheck the timing.
13   Turn off the engine and disconnect the timing light. Reconnect the vacuum advance hose, if removed, and any other components which were disconnected. On later models remove the ground wire from the service connector.

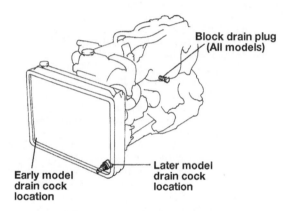

Block drain plug
(All models)

Early model
drain cock
location

Later model
drain cock
location

38.3 Coolant drain locations

38   Cooling system servicing (draining, flushing and refilling) (every 30,000 miles or 24 months)

Refer to illustrations 38.3 and 38.4
Warning: *Do not allow antifreeze to come in contact with your skin or painted surfaces of the vehicle. Rinse off spills immediately with plenty of water. Antifreeze is highly toxic if ingested. Never leave antifreeze lying around in an open container or in puddles on the floor; children and pets are attracted by it's sweet smell and may drink it. Check with local authorities about disposing of used antifreeze. Many communities have collection centers which will see that antifreeze is disposed of safely.*
1   Periodically, the cooling system should be drained, flushed and refilled to replenish the antifreeze mixture and prevent formation of rust and corrosion, which can impair the performance of the cooling system and cause engine damage. When the cooling system is serviced, all hoses and the radiator cap should be checked and replaced if necessary.
2   Apply the parking brake and block the wheels. **Warning:** *If the vehicle has just been driven, wait several hours to allow the engine to cool down before beginning this procedure.*
3   Move a large container under the radiator drain to catch the coolant. The radiator drain plug is located on the side of the lower tank of the radiator **(see illustration)**. Attach a 3/8-inch diameter hose to the drain fitting (if possible) to direct the coolant into the container, then open the drain fitting (a pair of pliers may be required to turn it).
4   Remove the radiator cap and allow the radiator to drain, then, move the container under the drivers side of the engine block **(see illustration)**. Remove the engine block drain plug and allow the coolant in the block to drain **(see illustration 38.3)**. **Note:** *Frequently,*

the coolant will not drain from the block after the plug is removed. This is due to a rust layer that has built up behind the plug. Insert a Phillips screwdriver into the hole to break the rust barrier.
5   While the coolant is draining, check the condition of the radiator hoses, heater hoses and clamps (refer to Section 10 if necessary).
6   Replace any damaged clamps or hoses.
7   Once the system is completely drained, flush the radiator with fresh water from a garden hose until it runs clear at the drain. The

38.4 Push the radiator cap downward and rotate it counterclockwise to remove

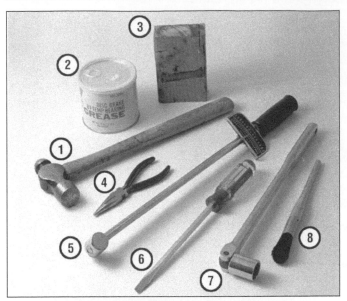

39.1 Tools and materials needed for front wheel bearing maintenance

1   **Hammer** - A common hammer will do just fine
2   **Grease** - High-temperature grease that is formulated specially for front wheel bearings should be used
3   **Wood block** - If you have a scrap piece of 2x4, it can be used to drive the new seal into the hub
4   **Needle-nose pliers** - Used to straighten and remove the cotter pin in the spindle
5   **Torque wrench** - This is very important in this procedure; if the bearing is too tight, the wheel won't turn freely - if it's too loose, the wheel will "wobble" on the spindle. Either way, it could mean extensive damage
6   **Screwdriver** - Used to remove the seal from the hub (a long screwdriver is preferred)
7   **Socket/breaker bar** - Needed to loosen the nut on the spindle if it's extremely tight
8   **Brush** - Together with some clean solvent, this will be used to remove old grease from the hub and spindle

flushing action of the water will remove sediments from the radiator but will not remove rust and scale from the engine and cooling tube surfaces.
8   These deposits can be removed with a chemical cleaner. Follow the procedure outlined in the manufacturer's instructions. If the radiator is severely corroded, damaged or leaking, it should be removed (see Chapter 3) and taken to a radiator repair shop.
9   Remove the overflow hose from the coolant reservoir and flush the reservoir with clean water, then reconnect the hose.
10   Close and tighten the radiator drain fitting. Install and tighten the block drain plug.
11   Place the heater temperature control in the maximum heat position.
12   Slowly add new coolant (a 50/50 mixture of water and antifreeze) to the radiator until it's full. Add coolant to the reservoir up to the lower mark.
13   Leave the radiator cap off and run the engine in a well-ventilated area until the thermostat opens (coolant will begin flowing through the radiator and the upper radiator hose will become hot).
14   Turn the engine off and let it cool. Add more coolant mixture to bring the level back up to the lip on the radiator filler neck.
15   Squeeze the upper radiator hose to expel air, then add more coolant mixture if necessary. Replace the radiator cap.
16   Start the engine, allow it to reach normal operating temperature and check for leaks.

39   Front hub and wheel bearing check, repack and adjustment (every 30,000 miles or 24 months)

Check and repack

Refer to illustrations 39.1, 39.7a, 39.7b, 39.7c, 39.8, 39.10, 39.14 and 39.18

1   In most cases the front wheel bearings will not need servicing until the brake shoes or pads are changed. However, the bearings should be checked whenever the front of the vehicle is raised for any reason. Several items, including a torque wrench and special grease, are required for this procedure **(see illustration)**.
2   With the vehicle securely supported on jackstands, spin each wheel and check for noise, rolling resistance and endplay.
3   Grasp the top of each tire with one hand and the bottom with the other. Move the wheel in and out on the spindle. If there's any noticeable movement, the bearings should be checked and then repacked with grease, or replaced if necessary.
4   Remove the wheel.
5   Remove the disc brake caliper (see Chapter 9) and hang it out of the way on a piece of wire.
6   Remove the front locking hub assemblies (see Chapter 8).
7   Using a small chisel and hammer, bend back the tabs on the lock washer which is located between the inner and outer locknuts. Remove the outer locknut, lock washer, inner locknut and the thrust washer **(see illustrations)**.
8   Pull the hub assembly out slightly, then push it back in. This should force the outer bearing and cup off the spindle so it can be removed **(see illustration)**.

39.7a Use a small chisel and hammer to bend back the tabs on the lock washer, then remove the outer locknut

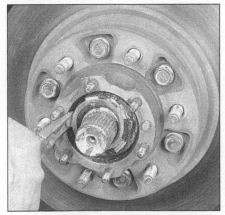
39.7b Remove the lock washer, the inner locknut . . .

39.7c . . .and the thrust washer

39.8 Pull the hub assembly out slightly, then push it back in to disengage the outer wheel bearing

39.10 Use a screwdriver or seal removal tool to pry out the grease seal

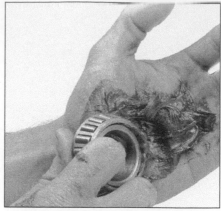

39.14 If a bearing packing tool is not available, work grease into the bearing rollers by pressing it against the palm of your hand

9   Pull the hub assembly off the spindle.

10   Use a screwdriver or a seal puller tool to pry the seal out of the rear of the hub assembly **(see illustration)**. Note how the seal is installed.

11   Remove the inner wheel bearing from the hub assembly.

12   Use solvent to remove all traces of the old grease from the bearings, hub and spindle. A small brush may prove helpful; however make sure no bristles from the brush embed themselves inside the bearing rollers. Allow the parts to air dry.

13   Carefully inspect the bearings for cracks, heat discoloration, worn rollers, etc. Check the bearing races inside the hub for wear and damage. If the bearing races are defective, drive the bearing race out of the hub using a brass drift. Drive the new race into the hub using the appropriate size bearing driver (inexpensive bearing driver sets are available at most automotive parts stores). Note that the bearings and races are replaced as matched sets; used bearings should never be installed on new races.

14   Use only high-temperature front wheel bearing grease to pack the bearings. Inexpensive bearing packing tools are available at automotive parts stores, but not entirely necessary. If one is not available, pack the grease by hand completely into the bearings, forcing it between the rollers, cone and cage from the back side **(see illustration)**.

15   Apply a thin coat of grease to the spindle at the outer bearing seat, inner bearing seat, shoulder and seal seat.

16   Place a small quantity of grease inboard of each bearing race inside the hub. Using your finger, form a dam at these points to provide extra grease availability and to keep thinned grease from flowing out of

the bearing.

17   Place the grease-packed inner bearing into the rear of the hub and put a little more grease outboard of the bearing.

18   Place a new seal over the inner bearing and tap the seal evenly into place with a hammer and block of wood until it's flush with the hub **(see illustration)**.

Adjustment

19   Carefully place the hub assembly onto the spindle and push the grease-packed outer bearing into position.

20   Install the thrust washer and the inner locknut on to the spindle. Then tighten the inner locknut to 43 ft-lbs while rotating the hub to seat the bearings.

21   Back off the inner locknut until it can be loosened by hand. Tighten the inner locknut to 48 in-lbs while rotating the hub assembly. Verify that the hub assembly rotates smoothly and no endplay is present.

22   Install the lock washer and tighten outer locknut to a 47 ft-lbs. Recheck the bearing preload, If there's any noticeable endplay readjust the inner locknut. Bend one lock washer tab towards the inner locknut and another lock washer tab towards the outer locknut.

23   The remainder of the installation is the reverse of removal. **Note:** *Lightly grease the internal components of the manual locking hub assembly with chassis grease. DO NOT pack the hubs full of grease or poor operation may occur.*

40   Automatic transmission fluid and filter change (every 30,000 miles or 24 months)

Refer to illustrations 40.5, 40.7, 40.9, 40.11 and 40.12

1   At the specified intervals, the transmission fluid should be drained and replaced. Since the fluid will remain hot long after driving, perform this procedure only after the engine has cooled down completely.

2   Before beginning work, purchase the specified transmission fluid (see *Recommended lubricants and fluids* at the front of this Chapter) and a new filter.

3   Other tools necessary for this job include a floor jack, jackstands to support the vehicle in a raised position, a drain pan capable of holding at least eight quarts, newspapers and clean rags.

4   Raise the vehicle and support it securely on jackstands.

5   Place the drain pan underneath the transmission pan and remove the drain plug. Allow the fluid to completely drain from the transmission **(see illustration),** then reinstall the drain plug.

6   Detach the transmission pan rock shield (if equipped) and remove the pan mounting bolts from the outer edges of the pan.

7   Using a rubber mallet, carefully tap on the pan to break the layer of gasket sealer between the pan and the transmission case **(see illustration)**. **Note:** *Prying between the pan and the transmission case with*

39.18 A block of wood and a hammer can be used to reinstall the new grease seal

40.5 The transmission drain plug is located on the bottom of transmission pan

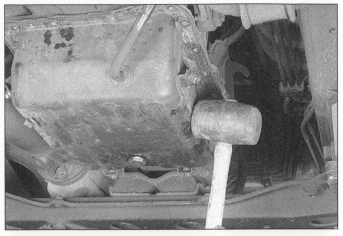

40.7 Using a rubber mallet, carefully tap on the pan to break the gasket seal between the pan and the transmission case

a screwdriver or similar tool may result in damage to the sealing surface on the transmission case.

8   Lower the pan from the vehicle and drain the remaining transmission fluid from the pan.

9   Remove the filter retaining screws from the valve body and remove the filter **(see illustration)**.

10   Thoroughly inspect the bottom of the pan, the filter and the fluid. Although normally bright red, transmission fluid may turn dark red or brown during normal use. If you find the fluid very dark colored, or if it smells burned, it usually indicates the transmission has been overheated. If you find small pieces of metal or clutch material in the pan or filter, it indicates wear or damage have occurred to the internal parts or clutches. If you have any concerns about the condition of your transmission based on what you find in the fluid, pan and filter, it's a good idea to take your vehicle to your dealer or a transmission shop for further evaluation.

11   Clean the pan with solvent and dry it with compressed air if available. Use a gasket scraper to remove any traces of old gasket material remaining on the transmission case or valve body. **Note:** *Be very careful not to gouge the delicate aluminum gasket surface on the valve body.* Install a new filter and gasket **(see illustration)**.

12   Make sure the gasket surface on the transmission pan is clean, then install a bead of RTV sealant to the pan **(see illustration)**. Reinsert the filler tube onto the dipstick tube and place the pan against the transmission case. Working around the pan, tighten each bolt a little at a time to the torque listed in this Chapter's Specifications.

13   Lower the vehicle and add approximately three quarts of the specified type of automatic transmission fluid through the filler tube

40.9 Remove the filter bolts (arrows)

(see Section 6).

14   With the transmission in Park and the parking brake set, start the engine.

15   Move the gear selector through each range and back to Park. Check the fluid level and add fluid, if necessary, until the level is within the correct range on the dipstick.

16   Check under the vehicle for leaks during the first few trips. Check the fluid level again when the transmission is hot (see Section 6).

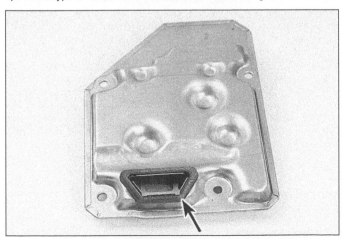

40.11 Make sure to install a new gasket (arrow) on the filter

40.12 Place a bead of RTV sealant all the way around the pan mating surface, inboard of the bolt holes

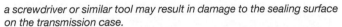

41 Manual transmission lubricant change (every 30,000 miles or 24 months)

1   Raise the vehicle and support it securely on jackstands.
2   Move a drain pan, rags, newspapers and wrenches under the transmission.
3   Remove the transmission drain plug at the bottom of the case and allow the lubricant to drain into the pan **(see illustration 17.2)**.
4   After the lubricant has drained completely, reinstall the plug and tighten it securely.
5   Remove the fill plug from the side of the transmission case. Using a hand pump, syringe or funnel, fill the transmission with the specified lubricant until it is level with the lower edge of the filler hole. Reinstall the fill plug and tighten it securely.
6   Lower the vehicle.
7   Drive the vehicle for a short distance, then check the drain and fill plugs for leakage.

42 Transfer case lubricant change (every 30,000 miles or 24 months)

1   Drive the vehicle for at least 15 minutes to warm the lubricant in the case. Perform this warm-up procedure with 4WD engaged, if possible. Use all gears, including Reverse, to ensure the lubricant is sufficiently warm to drain completely.
2   Raise the vehicle and support it securely on jackstands.
3   Remove the drain plug from the lower part of the case and allow the old lubricant to drain completely **(see illustrations 18.2)**.
4   After the lubricant has drained completely, reinstall the plug and tighten it securely
5   Remove the filler plug from the case
6   Fill the case with the specified lubricant until it is level with the lower edge of the filler hole.
7   Install the filler plug and tighten it securely.
8   Drive the vehicle for a short distance and recheck the lubricant level. In some instances a small amount of additional lubricant will have to be added.

43 Differential lubricant change (every 30,000 miles or 24 months)

Note: *The following procedure is used for the rear differential as well as the front differential.*
1   Drive the vehicle for several miles to warm up the differential oil, then raise the vehicle and support it securely on jackstands.
2   Move a drain pan, rags, newspapers and the proper tools under the vehicle.
3   With the drain pan under the differential, use a socket and ratchet to loosen the drain plug. It's the lower of the two plugs **(see illustration 19.2)**.
4   Once loosened, carefully unscrew it with your fingers until you can remove it from the case.
5   Allow all of the oil to drain into the pan, then replace the drain plug and tighten it securely.
6   Feel with your hands along the bottom of the drain pan for any metal bits that may have come out with the oil. If there are any, it's a sign of excessive wear, indicating that the internal components should be carefully inspected in the near future.
7   Remove the differential check/fill plug (see Section 19). Using a hand pump, syringe or funnel, fill the differential with the correct amount and grade of oil (see the Specifications) until the level is just at the bottom of the plug hole.
8   Reinstall the plug and tighten it securely.
9   Lower the vehicle. Check for leaks at the drain plug after the first few miles of driving.

44.6a With the No. 1 piston at TDC on the compression stroke, check and adjust the clearance of the indicated valves

44.6b Check the clearance for each valve with a feeler gauge of the specified thickness - if the clearance is correct, you should feel a slight drag on the gauge as you pull it out

44 Valve clearance check and adjustment (1FZ-FE engines)(every 60,000 miles or 48 months)

Refer to illustrations 44.6a, 44.6b, 44.7, 44.9a, 44.9b, 44.9c, 44.10 and 44.11
Note 1: *This procedure requires the use of special valve lifter tools. The tools are available from the dealer, specialty tool manufacturers and auto parts stores. It is impossible to perform this task without them.*
Note 2: *Both camshafts will have to be removed when replacing the shims on the number 6 cylinder because of clearance problems with the firewall (see Chapter 2B). Be sure to measure the valve clearance first as described in Steps 6 and 7 before removing the camshafts.*
1   Disconnect the negative cable from the battery. **Caution:** *If the stereo in your vehicle is equipped with an anti-theft system, make sure you have the correct activation code before disconnecting the battery.*
2   Disconnect the spark plug wires and remove any other components that will interfere with valve cover removal.
3   Blow out the recessed area around the spark plug openings with compressed air, if available, to remove any debris that might fall into the cylinders, then remove the spark plugs (see Section 28).
4   Remove the valve cover (refer to Chapter 2B).
5   Refer to Chapter 2 and position the number 1 piston at TDC on the compression stroke.
6   Measure the clearances of the indicated valves with feeler gauges

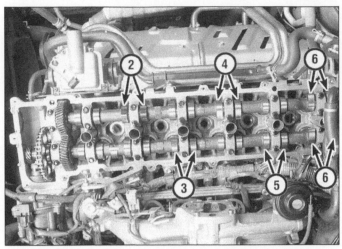

44.7 Rotate the crankshaft 360-degrees from No. 1 TDC, then check and adjust the clearance of the indicated valves

44.9a Install the valve lifter tool as shown and squeeze the handles together to depress the valve lifter, then hold the lifter down with the smaller tool so the shim can be removed

44.9b Keep pressure on the lifter with the smaller tool and remove the shim with a small screwdriver . . .

44.9c . . . a pair of tweezers or a magnet as shown here

(see illustrations). Record the measurements which are out of specification. They will be used later to determine the required replacement shims.

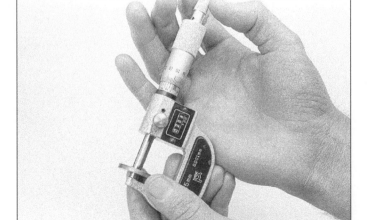

44.10 Measure the shim thickness with a micrometer

7   Turn the crankshaft one complete revolution and realign the timing marks. Measure the remaining valves (see illustration).
8   After all the valves have been measured, turn the crankshaft pulley until the camshaft lobe above the first valve which you intend to adjust is pointing upward, away from the shim.
9   Position the notch in the valve lifter toward the spark plug. Then depress the valve lifter with the special valve lifter tools (see illustration). Place the special valve lifter tool in position as shown, with the longer jaw of the tool gripping the lower edge of the cast lifter boss and the upper, shorter jaw gripping the upper edge of the lifter itself. Depress the valve lifter by squeezing the handles of the valve lifter tool together, then hold the lifter down with the smaller tool and remove the larger one. Remove the adjusting shim with a small screwdriver or a pair of tweezers (see illustrations). Note that the wire hook on the end of some valve lifter tool handles can be used to clamp both handles together to keep the lifter depressed while the shim is removed.
10   Measure the thickness of the shim with a micrometer (see illustration). To calculate the correct thickness of a replacement shim that will place the valve clearance within the specified value, use the following formula:

$$N = T + (A - V)$$

T = thickness of the old shim
A = valve clearance measured
N = thickness of the new shim
V = desired valve clearance (see this Chapters Specifications)

11   Select a shim with a thickness as close as possible to the valve clearance calculated. Shims, which are available in 17 sizes in increments of 0.0020-inch (0.050 mm), range in size from 0.0984-inch (2.500 mm) to 0.1299-inch (3.300 mm) **(see illustration)**. **Note:** *Through careful analysis of the shim sizes needed to bring the out-of-specification valve clearance within specification, it is often possible to simply move a shim that has to come out anyway to another valve lifter requiring a shim of that particular size, thereby reducing the number of new shims that must be purchased.*

12   Place the special valve lifter tool in position as shown in **illustration 44.9a**, with the longer jaw of the tool gripping the lower edge of the cast lifter boss and the upper, shorter jaw gripping the upper edge of the lifter itself, press down the valve lifter by squeezing the handles of the valve lifter tool together and install the new adjusting shim (note that the wire hook on the end of one valve lifter tool handle can be used to clamp the handles together to keep the lifter depressed while the shim is inserted. Measure the clearance with a feeler gauge to make sure that your calculations are correct.

13   Repeat this procedure until all the valves which are out of clearance have been corrected.

14   Installation of the spark plugs, valve cover, spark plug wires and boots, etc. is the reverse of removal.

New shim thickness   mm (in.)

Shim No.	Thickness	Shim No.	Thickness
1	2.500 (0.0984)	10	2.950 (0.1161)
2	2.550 (0.1004)	11	3.000 (0.1181)
3	2.600 (0.1024)	12	3.050 (0.1201)
4	2.650 (0.1043)	13	3.100 (0.1220)
5	2.700 (0.1063)	14	3.150 (0.1240)
6	2.750 (0.1083)	15	3.200 (0.1260)
7	2.800 (0.1102)	16	3.250 (0.1280)
8	2.850 (0.1122)	17	3.300 (0.1299)
9	2.900 (0.1142)		

44.11 Valve adjusting shim thickness chart

Chapter 2 Part A
Pushrod engines

Contents

	Section
Camshaft and bearings - inspection	15
Camshaft and timing gears - removal and installation	14
CHECK ENGINE light	See Chapter 6
Crankshaft front oil seal - replacement	12
Crankshaft pulley - removal and installation	11
Cylinder compression check	See Chapter 2C
Cylinder head - removal and installation	9
Drivebelt - check, adjustment and replacement	See Chapter 1
Engine mounts - check and replacement	20
Engine oil and filter change	See Chapter 1
Engine overhaul - general information	See Chapter 2C
Engine - removal and installation	See Chapter 2C
Flywheel/driveplate - removal and installation	18
General information	1
Intake and exhaust manifolds - removal and installation	8
Valve lifters - removal, inspection and installation	10
Oil cooler - removal and installation	See Chapter 3
Oil pan - removal and installation	16
Oil pump - removal and installation	17
Pushrod cover - removal and installation	5
Rear main oil seal - replacement	19
Repair operations possible with the engine in the vehicle	2
Rocker arms and pushrods - removal, inspection and installation	6
Spark plug replacement	See Chapter 1
Timing gear cover - removal and installation	13
Top Dead Center (TDC) for number one piston - locating	3
Valve clearance check and adjustment	See Chapter 1
Valve cover - removal and installation	4
Valves - servicing	See Chapter 2C
Valve springs, retainers and seals - replacement	7
Water pump - removal and installation	See Chapter 3

Specifications

2F engine

General

Displacement	257 cu. in. (4.2 liters)
Cylinder numbers (front-to-rear)	1-2-3-4-5-6
Firing order	1-5-3-6-2-4

Camshaft

Journal diameter	
No. 1	1.8880 to 1.8888 in. (47.955 to 47.975 mm)
No. 2	1.8289 to 1.8297 in. (46.455 to 46.475 mm)
No. 3	1.7699 to 1.7707 in. (44.955 to 44.975 mm)
No. 4	1.7108 to 1.7116 in. (43.455 to 43.475 mm)
Journal-to-bearing (oil) clearance	
Standard	0.0010 to 0.0030 in. (0.025 to 0.075 mm)
Limit	0.0039 inch (0.1 mm)
Journal out-of-round limit	0.0059 in. (0.15 mm)
Lobe height	
Standard	
Intake	1.5102 to 1.5142 in. (38.36 to 38.46 mm)
Exhaust	1.5059 to 1.5098 in. (38.25 to 38.35 mm)
Limit	
Intake	1.496 in. (38.0 mm)
Exhaust	1.492 in. (37.9 mm)

Camshaft (continued)

Thrust clearance
  Standard.. 0.0079 to 0.0103 in. (0.20 to 0.262 mm)
  Limit .... 0.012 in. (0.3 mm)
Timing gear backlash
  Standard.. 0.0020 to 0.0051 in. (0.05 to 0.13 mm)
  Limit... 0.008 in. (0.2 mm)

Torque specifications   **Ft-lbs** (unless otherwise indicated)

*Note: One foot-pound (ft-lb) of torque is equivalent to 12 inch-pounds (in-lbs) of torque. Torque values below approximately 15 ft-lbs
are expressed in inch-pounds, because most foot-pound torque wrenches are not accurate at these smaller values.*

Camshaft thrust plate bolts-to-block .... 108 in-lbs
Crankshaft pulley nut... 144
Cylinder head bolts.. 90
Driveplate-to-crankshaft bolts... 64
Intake/exhaust manifold bolts
  California 40
  All others 32
Flywheel-to-crankshaft bolts 60
Oil pan mounting bolts 11
Oil pump bolts 13
Rocker arm bolts
  8 mm bolts 17
  10 mm bolts 25
Valve cover-to-cylinder head bolts.. 78 in-lbs
Timing gear cover-to-block
  10 mm bolt 16
  6 mm bolt .... 60 in-lbs

**Cylinder location and distributor
rotation diagram - 2F and
3F-E engines**

*The blackened terminal shown on
the distributor cap indicates the
Number One spark plug wire position*

3F-E engine

General
Displacement... 232 cu. in. (4.0 liters)
Cylinder numbers (front-to-rear).. 1-2-3-4-5-6
Firing order 1-5-3-6-2-4

Camshaft
Journal diameter
  No. 1.. 1.8880 to 1.8888 in. (47.955 to 47.975 mm)
  No. 2.. 1.8289 to 1.8297 in. (46.455 to 46.475 mm)
  No. 3.. 1.7699 to 1.7707 in. (44.955 to 44.975 mm)
  No. 4.. 1.7108 to 1.7116 in. (43.455 to 43.475 mm)
Journal-to-bearing (oil) clearance
  Standard.. 0.0010 to 0.0030 in. (0.025 to 0.075 mm)
  Limit .... 0.0039 inch (0.10 mm)
Journal out-of-round limit... 0.0059 in. (0.15 mm)
Lobe height
  Standard
    Intake .... 1.5102 to 1.5142 in. (38.36 to 38.46 mm)
    Exhaust... 1.5059 to 1.5098 in. (38.25 to 38.35 mm)
  Limit
    Intake .... 1.496 in. (38.0 mm)
    Exhaust... 1.492 in. (37.9 mm)
Thrust clearance
  Standard.. 0.0079 to 0.0114 in. (0.200 to 0.290 mm)
  Limit... 0.0130 inch (0.33 mm)
Timing gear backlash
  Standard.. 0.0039 to 0.0072 in. (0.100 to 0.183 mm)
  Limit... 0.0098 in. (0.25 mm)

Torque specifications   **Ft-lbs** (unless otherwise indicated)

*Note: One foot-pound (ft-lb) of torque is equivalent to 12 inch-pounds (in-lbs) of torque. Torque values below approximately 15 ft-lbs
are expressed in inch-pounds, because most foot-pound torque wrenches are not accurate at these smaller values.*

Camshaft thrust plate bolts-to-block .... 108 in-lbs
Crankshaft pulley bolt.. 253
Cylinder head bolts.. 90
Cylinder head-to-brace bolts... 22
Intake manifold-to-brace bolts 22
Driveplate-to-crankshaft bolts... 64

Torque specifications (continued)

Ft-lbs (unless otherwise indicated)

Intake/exhaust manifold bolts
    17 mm bolt head (A)... 51
    14 mm bolt head (B)... 37
    14 mm nut (C)... 41
Driveplate-to-crankshaft bolts 64
Oil pan mounting bolts .... 84 in-lbs
Oil pump bolts 13
Rocker arm bolts
    12 mm bolt head 17
    14 mm bolt head 25
Valve cover-to-cylinder head bolts.. 78 in-lbs
Timing gear cover-to-block
    10 mm bolt head (A)... 43 in-lbs
    10 mm bolt head (C)... 43 in-lbs
    14 mm bolt head (B)... 18

1 General information

This Part of Chapter 2 is devoted to in-vehicle engine repair procedures for the 2F and 3F-E six-cylinder pushrod engines. Information concerning engine removal and installation, as well as engine block and cylinder head overhaul, is in Part C of this Chapter.

The following repair procedures are based on the assumption that the engine is installed in the vehicle. If the engine has been removed from the vehicle and mounted on a stand, many of the steps included in this Part of Chapter 2 will not apply.

The Specifications included in this Part of Chapter 2 apply only to the engine and procedures in this Part. The Specifications necessary for rebuilding the block and cylinder head are found in Part C.

Toyota equipped the Land Cruiser with in-line six cylinder, pushrod engines from 1980 through 1992. The 2F engine was installed from 1980 through 1987. Later models (1988 through 1992) are equipped with the 3F-E engine. Both types use pushrods and valve lifters with timing gears as part of the valve train.

The 2F engine is a carbureted 4.2 liter engine displacement, while the 3F-E engine is a 4.0 liter engine displacement with electronic fuel injection and increased power. The 3F-E engine uses a shorter stroke with the same bore and a lower deck height.

2 Repair operations possible with the engine in the vehicle

Many major repair operations can be accomplished without removing the engine from the vehicle.

Clean the engine compartment and the exterior of the engine with some type of pressure washer before any work is done. A clean engine will make the job easier and will help keep dirt out of the internal areas of the engine.

If vacuum, exhaust, oil or coolant leaks develop, indicating a need for gasket or seal replacement, the repairs can generally be made with the engine in the vehicle. The intake and exhaust manifold gaskets, oil pan gasket and cylinder head gasket are all accessible with the engine in place.

Exterior engine components such as the intake and exhaust manifolds, the oil pan (and the oil pump), the water pump, the starter motor, the alternator, the distributor and the carburetor or fuel injection components can be removed for repair with the engine in place.

Since the cylinder head can be removed without pulling the engine, valve component servicing can also be accomplished with the engine in the vehicle.

In extreme cases caused by a lack of necessary equipment, repair or replacement of piston rings, pistons, connecting rods and rod bearings is possible with the engine in the vehicle. However, this practice is not recommended because of the cleaning and preparation work that must be done to the components involved.

3.4 Use a large socket and breaker bar placed on the crankshaft pulley bolt to rotate the crankshaft

3 Top Dead Center (TDC) for number one piston - locating

Refer to illustrations 3.4, 3.6, 3.8a and 3.8b

Note 1: *The following procedure is based on the assumption that the distributor is correctly installed. If you are trying to locate TDC to install the distributor correctly, piston position must be determined by feeling for compression at the number one spark plug hole, then aligning the ignition timing marks as described in Step 8.*

Note 2: *This procedure covers the 2F and 3F-E pushrod engines, as well as the 1FZ-FE DOHC engine.*

1   Top Dead Center (TDC) is the highest point in the cylinder that each piston reaches as it travels up-and-down when the crankshaft turns. Each piston reaches TDC on the compression stroke and again on the exhaust stroke, but TDC generally refers to piston position on the compression stroke.

2   Positioning the piston(s) at TDC is an essential part of many procedures such as rocker arm removal, camshaft and timing chain/sprocket removal and distributor removal.

3   Before beginning this procedure, be sure to place the transmission in Neutral and apply the parking brake or block the rear wheels. Also, remove the spark plugs (see Chapter 1) and disable the ignition system. Detach the coil wire from the center terminal of the distributor cap and ground it on the block with a jumper wire.

4   In order to bring any piston to TDC, the crankshaft must be turned using one of the methods outlined below. When looking at the front of the engine, normal crankshaft rotation is clockwise.

    a) *The preferred method is to turn the crankshaft with a socket and ratchet attached to the bolt threaded into the front of the crankshaft (see illustration).*

3.6 Paint a mark on the distributor body directly below the number one spark plug wire terminal (arrow) on the distributor cap

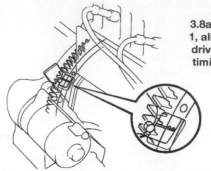

3.8a To obtain TDC number 1, align the TDC mark on the driveplate/flywheel with the timing pointer (3F-E engine shown, 2F similar)

b) *A remote starter switch, which may save some time, can also be used. Follow the instructions included with the switch. Once the piston is close to TDC, use a socket and ratchet, as described in the previous paragraph.*

c) *If an assistant is available to turn the ignition switch to the Start position in short bursts, you can get the piston close to TDC without a remote starter switch. Make sure your assistant is out of the vehicle, away from the ignition switch, then use a socket and ratchet (as described in Paragraph a) to complete the procedure.*

5   Note the position of the terminal for the number one spark plug wire on the distributor cap. If the terminal isn't marked, follow the plug wire from the number one cylinder spark plug to the cap.

6   Use a felt-tip pen or chalk to make a mark on the distributor body directly under the terminal **(see illustration)**.

7   Detach the cap from the distributor and set it aside (see Chapter 1 if necessary).

8   Turn the crankshaft (see Paragraph 3 above) until the TDC mark on the flywheel/driveplate aligns with the timing pointer (2F and 3F-E engines) or the notch in the crankshaft pulley is aligned with the 0 on the timing plate (1FZ-FE engine) **(see illustrations)**.

9   Look at the distributor rotor - it should be pointing directly at the mark you made on the distributor body. If it is, go to Step 12.

10   If the rotor is 180 degrees off, the number one piston is at TDC on the exhaust stroke. Go to Step 11.

11   To get the piston to TDC on the compression stroke, turn the crankshaft one complete turn (360 degrees) clockwise. The rotor

should now be pointing at the mark on the distributor. When the rotor is pointing at the number one spark plug wire terminal in the distributor cap and the ignition timing marks are aligned, the number one piston is at TDC on the compression stroke. **Note:** *If it is impossible to align the ignition timing marks when the rotor is pointing to the alignment mark on the number 1 cylinder, the distributor may be installed incorrectly, the timing chain or gears may be installed incorrectly or the timing chain (if equipped) has jumped off the timing sprocket.*

12   After the number one piston has been positioned at TDC on the compression stroke, TDC for any of the remaining pistons can be located by turning the crankshaft and following the firing order. Mark the remaining spark plug wire terminal locations on the distributor body just like you did for the number one terminal, then number the marks to correspond with the cylinder numbers. As you turn the crankshaft, the rotor will also turn. When it's pointing directly at one of the marks on the distributor, the piston for that particular cylinder is at TDC on the compression stroke.

4   Valve cover - removal and installation

Refer to illustration 4.4

Removal

1   Disconnect the negative cable from the battery. **Caution:** *If the radio in your vehicle is equipped with an anti-theft system, make sure you have the correct activation code before disconnecting the battery.*

2   On carbureted engines, remove the air cleaner. On fuel injected models, remove the air intake hose, the airflow sensor and the air cleaner assembly (see Chapter 4 Part B).

3   Label and then remove all hoses and/or wires necessary to provide clearance for valve cover removal.

4   Remove the valve cover retaining bolts and lift off the cover **(see illustration)**. The cover may stick. If the valve cover is stuck to the cylinder head, bump the end with a wood block and hammer to jar it loose. If that doesn't work, try to detach the cover by breaking the seal with a putty knife or razor blade. Locations for prying have been provided. **Caution:** *Don't pry at the valve cover-to-cylinder head joint or damage to the sealing surfaces may occur, leading to oil leaks after the valve cover is reinstalled.*

4.4 Remove the valve cover bolts

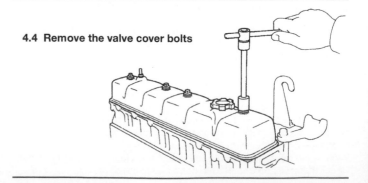

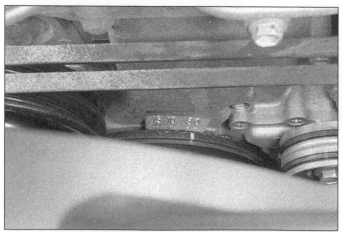

3.8b Turn the crankshaft clockwise until the notch on the crankshaft pulley aligns with the 0 on the timing cover (1FZ-FE engine shown)

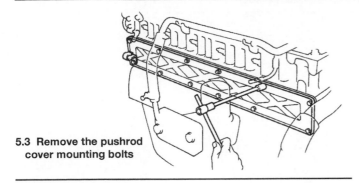

5.3 Remove the pushrod cover mounting bolts

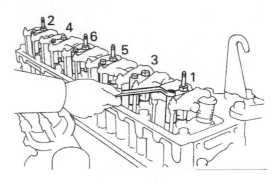

6.2 Loosen the rocker arm assembly bolts a little at a time in the sequence shown

Installation

5   Prior to installation, remove all traces of dirt, oil and old gasket material from the cover and cylinder head. Clean the mating surfaces with lacquer thinner or acetone and a clean rag.

6   Inspect the mating surface on the cover for damage and warpage. Correct or replace as necessary.

7   The rubber gasket can be reused unless it has high mileage and the rubber has hardened or cracked. If necessary, remove the rubber gasket and clean the mating surfaces. Install a new rubber gasket , pressing it evenly into the groove around the underside of the valve cover. If there's residue or oil around the mating surfaces when the valve cover is installed, oil leaks may develop. **Note:** *Be sure to install new rubber seals into the valve cover retainers.*

8   Place the valve cover on the cylinder head and install the mounting bolts. Tighten the bolts a little at a time until the torque listed in this Chapter's Specifications is reached.

9   Complete the installation procedure by reversing the removal procedure.

10   Start the engine and check for oil leaks.

5   Pushrod cover - removal and installation

Refer to illustration 5.3

1   Disconnect the negative cable at the battery. **Caution:** *If the radio in your vehicle is equipped with an anti-theft system, make sure you have the correct activation code before disconnecting the battery.*

2   Remove the oil cooler line mounting brackets and position them to the side of the engine compartment (see Chapter 3).

3   Remove the pushrod cover bolts **(see illustration)** and carefully pry the pushrod cover from the side of the block, using a thin-blade knife if necessary to break the gasket seal. Use caution not to distort the cover sealing flanges.

4   Scrape all old gasket and sealer from the pushrod cover and the block.

5   Install new gaskets on the pushrod cover, using a small amount of RTV sealer to hold them in place.

6   Installation is the reverse of the removal procedure.

6   Rocker arms and pushrods - removal, inspection and installation

Refer to illustrations 6.2, 6.3, 6.4 and 6.11

Removal

1   Detach the valve cover from the cylinder head (see Section 4).

2   Beginning at the front of the cylinder head, loosen the rocker arm bolts in the recommended sequence **(see Illustration)**. Loosen each bolt a little at t time, being careful to allow the assembly to remain level to the plane of the cylinder head.

3   Remove the rocker arm assembly as a unit **(see illustration)**. Be careful to not allow any of the components slide off the rocker shaft.

4   Remove the pushrods and store them separately to make sure they don't get mixed up during installation **(see illustration)**.

Inspection

5   Check each rocker arm for wear, cracks and other damage, especially where the pushrods and valve stems contact the rocker arm faces.

6   Make sure the hole at the pushrod end of each rocker arm is open.

7   Check each rocker arm pivot area for wear, cracks and galling. If the rocker arms are worn or damaged, replace them with new ones and use new pivots as well.

8   Inspect the pushrods for cracks and excessive wear at the ends. Roll each pushrod across a piece of plate glass to see if it's bent (if it wobbles, it's bent).

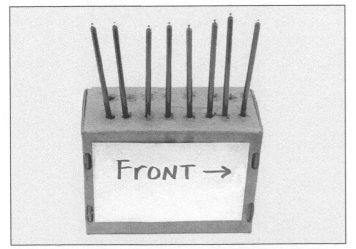

6.3 Store the pushrods in a perforated cardboard box to prevent mix-ups during installation - note the label indicating the front of the engine

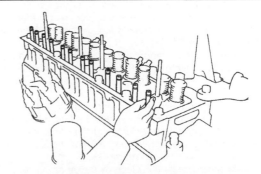

6.4 Lift the pushrods from the engine

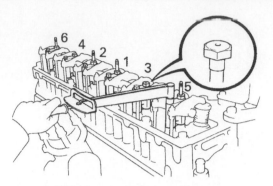

6.11 Rocker arm tightening sequence

Installation

9   Lubricate the lower ends of the pushrods with clean engine oil or moly-base grease and install them in their original locations. Make sure each pushrod seats completely in the lifter socket.

10   Apply moly-base grease to the ends of the valve stems and the upper ends of the pushrods before positioning the rocker arms and installing the capscrews.

11   Set the rocker arms in place, then install the pivots, bridges and capscrews. Apply moly-base grease to the pivots to prevent damage to the mating surfaces before engine oil pressure builds up. Tighten the bolts a little at a time in the recommended sequence to the torque listed in this Chapter's Specifications **(see illustration)**.

12   Reinstall the valve cover and run the engine. Check for oil leaks and unusual valve train noises.

7   Valve spring, retainer and seals - replacement

Refer to illustrations 7.4, 7.9, 7.16a, 7.16b, 7.16c and 7.17
Note: *Broken valve springs and defective valve stem seals can be replaced without removing the cylinder heads. Two special tools and a compressed air source are normally required to perform this operation, so read through this Section carefully and rent or buy the tools before beginning the job. If compressed air isn't available, a length of nylon rope can be used to keep the valves from falling into the cylinder during this procedure.*

1   Remove the valve cover referring to Section 4.

2   Remove the spark plug from the cylinder which has the defective component. If all of the valve stem seals are being replaced, all of the spark plugs should be removed.

3   Turn the crankshaft until the piston in the affected cylinder is at top dead center (TDC) on the compression stroke (see Section 3 for instructions). If you're replacing all of the valve stem seals, begin with cylinder number one and work on the valves for one cylinder at a time. Move from cylinder-to-cylinder following the firing order sequence (see the Specifications listed at the beginning of this Chapter).

4   Thread an adapter into the spark plug hole **(see illustration)** and connect an air hose from a compressed air source to it. Most auto parts stores can supply the air hose adapter. **Note:** *Many cylinder compression gauges utilize a screw-in fitting that may work with your air hose quick-disconnect fitting.*

5   Remove the rocker arm and pivot for the valve with the defective part and pull out the pushrod. If all of the valve stem seals are being replaced, all of the rocker arms and pushrods should be removed (see Section 5).

6   Apply compressed air to the cylinder. **Warning:** *The piston may be forced down by compressed air, causing the crankshaft to turn suddenly. If the wrench used when positioning the number one piston at TDC is still attached to the bolt in the crankshaft nose, it could cause damage or injury when the crankshaft moves.*

7   The valves should be held in place by the air pressure.

8   If you don't have access to compressed air, an alternative method can be used. Position the piston at a point just before TDC on the

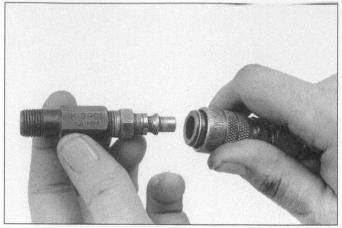

7.4 This is what the air hose adapter that threads into the spark plug looks like - they're commonly from auto part stores

compression stroke, then feed a long piece of nylon rope through the spark plug hole until it fills the combustion chamber. Be sure to leave the end of the rope hanging out of the engine so it can be removed easily. Use a large ratchet and socket to rotate the crankshaft in the normal direction of rotation until slight resistance is felt.

9   Stuff shop rags into the cylinder head holes above and below the valves to prevent parts and tools from falling into the engine, then use a valve spring compressor to compress the spring. Remove the keepers with small needle-nose pliers or a magnet **(see illustration)**. **Note:** *A couple of different types of tools are available for compressing the valve springs with the cylinder head in place. One type grips the lower spring coils and presses on the retainer as the knob is turned, while the other type, shown here, utilizes the rocker arm capscrew for leverage. Both types work very well, although the lever type is usually less expensive.*

10   Remove the spring retainer, oil shield and valve spring, then remove the guide seal. **Note:** *If air pressure fails to hold the valve in the closed position during this operation, the valve face and/or seat is probably damaged. If so, the cylinder head will have to be removed for additional repair operations.*

11   Wrap a rubber band or tape around the top of the valve stem so the valve won't fall into the combustion chamber, then release the air pressure. **Note:** *If a rope was used instead of air pressure, turn the crankshaft slightly in the direction opposite normal rotation.*

12   Inspect the valve stem for damage. Rotate the valve in the guide and check the end for eccentric movement, which would indicate that the valve is bent.

7.9 Once the spring is depressed, the keepers can be removed with a small magnet or needle-nose pliers (a magnet is preferred to prevent dropping the keepers)

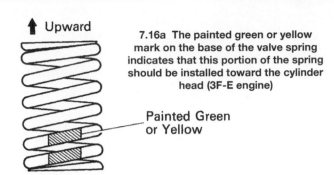

↑ Upward

7.16a The painted green or yellow mark on the base of the valve spring indicates that this portion of the spring should be installed toward the cylinder head (3F-E engine)

Painted Green or Yellow

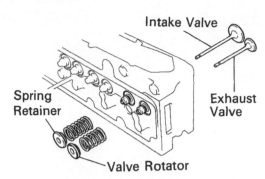

Intake Valve

Spring Retainer

Exhaust Valve

Valve Rotator

7.16b The exhaust valves are equipped with a valve rotator with a single locking groove on the valve stem. The intake valves are equipped with a spring retainer with dual locking grooves (3F-E engine)

with Spring Retainer

7.16c Valve spring retainers for the 3F-E engine

with Valve Rotator

8 Intake and exhaust manifolds - removal and installation

Refer to illustrations 8.7, 8.13, 8.17a and 8.17b
Warning: *Allow the engine to cool to room temperature before following this procedure.*

Removal

13   Move the valve up-and-down in the guide and make sure it doesn't bind. If the valve stem binds, either the valve is bent or the guide is damaged. In either case, the cylinder head will have to be removed for repair.
14   Reapply air pressure to the cylinder to retain the valve in the closed position, then remove the tape or rubber band from the valve stem. If a rope was used instead of air pressure, rotate the crankshaft in the normal direction of rotation until slight resistance is felt.
15   Lubricate the valve stem with engine oil and install a new guide seal.
16   Install the spring and shield in position over the valve **(see illustrations)**.
17   Install the valve spring retainer. Compress the valve spring and carefully position the keepers in the groove. Apply a small dab of grease to the inside of each keeper to hold it in place **(see illustration)**.
18   Remove the pressure from the spring tool and make sure the keepers are seated.
19   Disconnect the air hose and remove the adapter from the spark plug hole. If a rope was used in place of air pressure, pull it out of the cylinder.
20   Refer to Section 6 and install the rocker arms and pushrods.
21   Install the spark plugs and connect the spark plug wires.
22   Refer to Section 4 and install the valve cover.
23   Start and run the engine, then check for oil leaks and unusual sounds coming from the valve cover area.

1   Disconnect the negative cable from the battery. **Caution:** *If the radio in your vehicle is equipped with an anti-theft system, make sure you have the correct activation code before disconnecting the battery.*
2   Drain the cooling system (see Chapter 1). Disconnect any coolant hoses that connect to the intake manifold.
3   On carbureted models, remove the carburetor (see Chapter 4 Part A). On fuel injected models, remove the air intake plenum (see Chapter 4 Part B).
4   Remove the air injection pump and the air injection manifold (see Chapter 6).
5   Disconnect the bolt that retains the intake manifold-to-exhaust manifold bracket. Remove the bracket from the engine compartment.
6   On fuel injected models, disconnect the fuel line from the fuel rail (see Chapter 4B).
7   Remove the heat shields from the exhaust manifold to gain access to the mounting nuts **(see illustration)**.
8   Unbolt the power steering pump (if equipped), and set it aside without disconnecting the hoses (see Chapter 10).
9   Disconnect the throttle valve (TV) linkage, if equipped with an automatic transmission (see Chapter 7 Part B).
10   Disconnect the EGR tube from the manifolds (see Chapter 6).
11   Remove the two nuts and bolts that secure the exhaust pipe to the exhaust manifold. It may be necessary to apply penetrating oil to the threads.
12   Disconnect the oxygen sensor electrical connector (if equipped).
13   Remove the mounting bolts, nuts and spacers **(see illustration)** and detach the manifolds from the engine.

7.17 Apply a small dab of grease to the keepers before installation - it will hold them in place on the valve stem as the spring is released

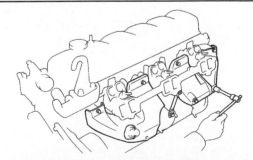

8.7 Remove the heat shield mounting bolts (3F-E engine)

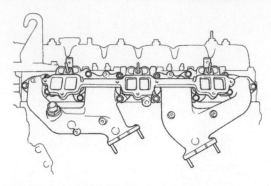

8.13 Remove the bolts retaining the manifolds to the cylinder head (3F-E engine)

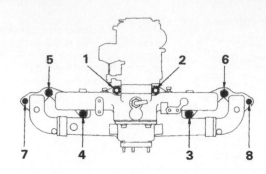

8.17a Tightening sequence on the intake/exhaust manifolds on the 2F engine

Installation

14   Thoroughly clean the mating surfaces, removing all traces of gasket material.

15   If the manifold is being replaced, transfer all necessary components to the new manifold.

16   Position the replacement gasket on the cylinder head and install the manifolds.

17   Working from the center out towards the ends, tighten the fasteners to the torque listed in this Chapter's Specifications (see illustrations).

18   Reinstall the remaining parts in the reverse order of removal.

19   Run the engine and check for leaks and proper operation.

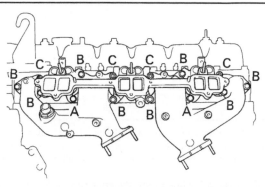

8.17b Intake and exhaust manifold bolt head designations on the 3F-E engine

| A | 17 mm bolt | B | 14 mm bolt | C | Nut |

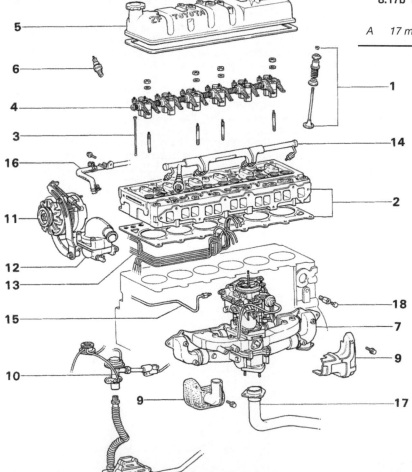

9.6a Exploded view of the cylinder head and components on the 2F engine

1   Valve and spring
2   Cylinder head gasket
3   Push rod
4   Rocker arm assembly
5   Valve cover
6   Spark plug
7   Intake/exhaust manifold assembly
8   EGR cooler
9   Insulator
10   EGR valve
11   Alternator
12   Thermostat housing
13   Vacuum pipe assembly
14   Air injection manifold
15   Fuel line
16   Oil cooler hose
17   Exhaust pipe
18   Coolant drain plug

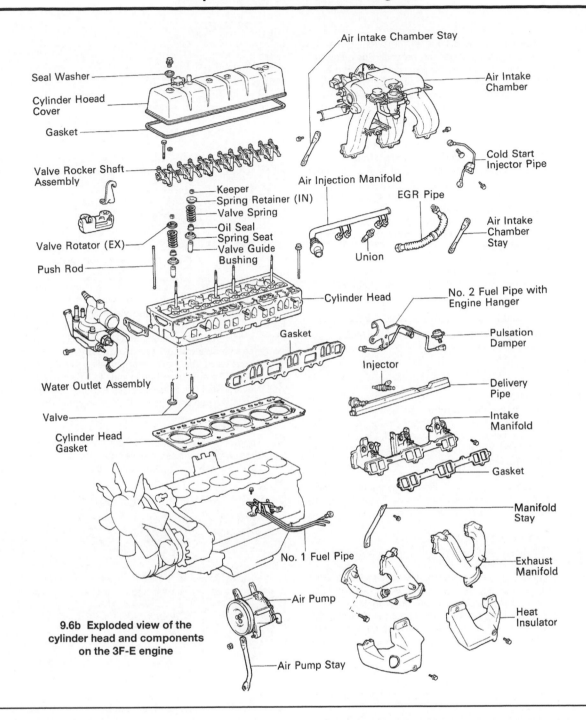

Air Intake Chamber Stay

Air Intake Chamber

Seal Washer

Cylinder Hoead Cover

Gasket

Cold Start Injector Pipe

Valve Rocker Shaft Assembly

Air Injection Manifold

EGR Pipe

Keeper
Spring Retainer (IN)
Valve Spring
Oil Seal
Spring Seat
Valve Guide Bushing

Air Intake Chamber Stay

Union

Valve Rotator (EX)

Push Rod

Cylinder Head

No. 2 Fuel Pipe with Engine Hanger

Pulsation Damper

Gasket

Injector

Water Outlet Assembly

Delivery Pipe

Intake Manifold

Valve

Cylinder Head Gasket

Gasket

Manifold Stay

No. 1 Fuel Pipe

Exhaust Manifold

Air Pump

Heat Insulator

9.6b Exploded view of the cylinder head and components on the 3F-E engine

Air Pump Stay

---

9 Cylinder head - removal and installation

Refer to illustrations 9.6a, 9.6b, 9.10, 9.13 and 9.14

Warning: *Allow the engine to cool to room temperature before following this procedure.*

Removal

1   Remove the rocker arms and pushrods (see Section 6).
2   Remove the intake and exhaust manifolds (see Section 8).
3   Remove the drivebelt(s) as described in Chapter 1.
4   Unbolt the power steering pump (if equipped) and set it aside without disconnecting the hoses.

5   Remove the air pump and bracket assembly from the engine (see Chapter 6).
6   Remove the air injection manifold from the cylinder head **(see illustration)** (see Chapter 6).
7   Drain the engine coolant (see Chapter 1), label and then remove the coolant temperature sending unit on the cylinder head (if equipped).
8   On air conditioned models, remove the air conditioner compressor without disconnecting the hoses and set the assembly off to the side (see Chapter 3).
9   Following the reverse order of the tightening sequence **(see illustration 9.14)**, loosen and remove the cylinder head bolts. Carefully lift the cylinder head off the engine. If the cylinder head is stuck to the engine block, it may be necessary to tap it with a soft-face hammer and a wood block to break the gasket seal.

Installation

10　Stuff clean shop towels into the cylinders. Thoroughly clean the gasket surfaces, removing all traces of gasket material. Run an appropriate sized tap into the bolt holes in the engine block and run a die over the bolt threads (see illustration). Ensure all bolt holes are clean and dry. **Caution:** *Because cylinder head bolts are often damaged by stretching and thread collapsing or stripping, have each cylinder head bolt examined by a qualified machinist. If in doubt about the integrity of the cylinder head bolts, replace them with new cylinder head bolts.*

11　Inspect the cylinder head for cracks and check it for warpage. Refer to Chapter 2C, for cylinder head servicing procedures.

12　These engines use a composition gasket. Install it dry (without any sealing compound).

13　Install the new gasket with the word TOP (see illustration) on the cylinder head side. Place the cylinder head on the engine.

14　Install the cylinder head bolts and tighten them in several steps and in the recommended sequence (see illustration) to the torque listed in this Chapter's Specifications.

15　Install the remaining components in the reverse order of removal.

16　Change the oil and filter (see Chapter 1).

17　Refill the cooling system and run the engine, checking for leaks and proper operation.

10　Valve lifters - removal, inspection and installation

Removal

Refer to illustrations 10.5 and 10.6

1　A defective valve lifter can cause a "ticking" noise while the engine is running. The defective lifter can be isolated with the engine idling. Place a length of hose or tubing on the valve cover near the position of each valve while listening at the other end. Or remove the valve cover and, with the engine idling, press on each rocker arm, one at a time, with a small wood block. If a valve lifter is defective, it'll be evident from the change in noise.

2　The most likely cause of a noisy valve lifter is worn lifters or damaged camshaft lobes. It is recommended that the camshaft as well as the valve lifters be replaced if the lifters are diagnosed as worn and damaged.

3　Remove the valve cover (see Section 4).

4　Remove the rocker arm assembly and both pushrods at the cylinder with the noisy valve lifter (see Section 7).

5　Remove the pushrod cover (see Section 5). Remove the valve lifters through the pushrod cover opening. A special removal tool is available (see illustration), but isn't always necessary. On newer engines without a lot of varnish buildup, valve lifters can often be removed with a magnet.

6　Store the valve lifters in a clearly labeled box to insure their reinstallation in the same lifter bore (see illustration).

Inspection

Refer to illustrations 10.8a, 10.8b and 10.8c

7　Clean the valve lifters with solvent and dry them thoroughly. Do this one lifter at a time to avoid mixing them up.

8　Check each valve lifter wall, pushrod seat and foot for scuffing, score marks and uneven wear. Each valve lifter foot (the surface that

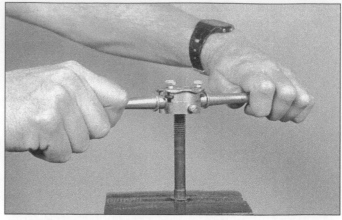

9.10　A die should be used to remove sealant and corrosion from the cylinder head bolt threads

rides on the cam lobe) must be slightly convex, although this can be difficult to determine by eye. If the base of the valve lifter is concave or rough (see illustrations), the valve lifters and camshaft must be replaced. If the lifter walls are damaged or worn (which isn't very likely), inspect the valve lifter bores in the engine block as well. If the pushrod seats (see illustration) are worn, check the pushrod ends.

9　If new valve lifters are being installed, a new camshaft must also be installed. If a new camshaft is installed, then use new valve lifters as well. Never install used lifters unless the original camshaft is used and the valve lifters can be installed in their original locations!

Installation

10　The used valve lifters must be installed in their original bores. Coat them with moly-base grease or engine assembly lube.

11　Lubricate the bearing surfaces of the valve lifter bores with engine oil.

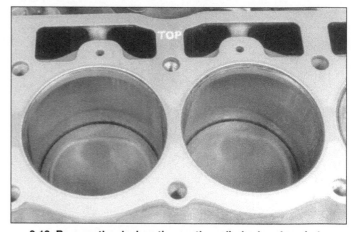

9.13　Be sure the designation on the cylinder head gasket - TOP - faces up

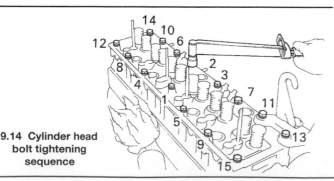

9.14　Cylinder head bolt tightening sequence

10.5　Remove the valve lifters through the pushrod cover opening on the side of the engine block

12   Install the valve lifter(s) in the lifter bore(s).
13   Install the pushrods and rocker arm assembly (see Section 5).
14   Tighten the rocker arm bolts to the torque listed in this Chapter's Specifications.
15   Install the valve cover (Section 4).

11   Crankshaft pulley - removal and installation

Refer to illustration 11.4, 11.5 and 11.6

Removal

1   Remove the cable from the negative battery terminal. **Caution:** *If the radio in your vehicle is equipped with an anti-theft system, make sure you have the correct activation code before disconnecting the battery.*
2   Remove the drivebelts (Chapter 1). Tag each belt as it's removed to simplify reinstallation. If the vehicle is equipped with a fan shroud, unscrew the mounting bolts and position the shroud out of the way.
3   Raise the vehicle and place it securely on jackstands.
4   If equipped with power steering, remove the six bolts and the power steering pulley from the crankshaft pulley **(see illustration)**.
5   To remove the crankshaft pulley bolt/nut, hold the pulley with a special tool that bolts to the crankshaft pulley **(see illustration)**. If the special tool is not available and the vehicle is equipped with a manual transmission, apply the parking brake and put the transmission in gear to prevent the crankshaft from turning, then remove the crankshaft pulley bolt/nut. If your vehicle is equipped with an automatic transmission, it may be necessary to remove the starter motor (Chapter 5) and immobilize the starter ring gear with a large screwdriver while an assistant loosens the pulley bolt/nut.
6   Using a bolt-type puller, remove the crankshaft pulley **(see illustration)**.

10.6 If you're removing more than one valve lifter, keep them in order in a clearly labeled box

Installation

7   Refer to Section 12 for the front oil seal replacement procedure.
8   Apply a thin layer of moly-base grease to the seal contact surface of the crankshaft pulley.
9   Slide the crankshaft pulley onto the crankshaft. Note that the slot in the hub must be aligned with the Woodruff key in the end of the crankshaft. Once the key is aligned with the slot, tap the pulley onto the crankshaft with a soft-face hammer. The retaining bolt/nut can also be used to press the pulley into position.
10   Tighten the crankshaft pulley-to-crankshaft bolt/nut to the torque listed in this Chapter's Specifications.

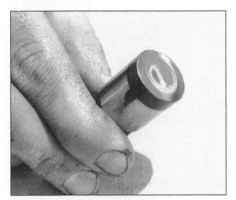

10.8a If the bottom (foot) of any valve lifter is worn concave (shown here), scratched or galled, replace the entire set with new valve lifters

10.8b The foot of each valve lifter should be slightly convex- the side of another lifter can be used as a straightedge to check it - if it appears flat, it's worn and must not be reused

10.8c If the valve lifters are pitted or rough, they shouldn't be reused

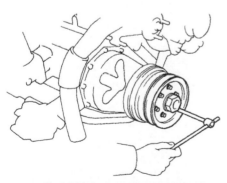

11.4 Remove the power steering pulley from the crankshaft pulley (if equipped)

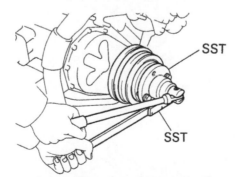

11.5 To remove the crankshaft pulley bolt/nut, hold the pulley with a special tool that bolts to the crankshaft pulley

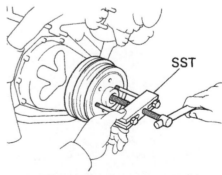

11.6 Use a bolt-type puller to remove the crankshaft pulley from the crankshaft

11   Install the power steering pulley (if equipped) on the hub and tighten the bolts to the specified torque. Use Locktite on the bolt threads.
12   Install the drivebelt(s) (Chapter 1) and replace the fan shroud (if equipped).

12   Crankshaft front oil seal - replacement

Note: *The crankshaft front oil seal can be replaced with the timing gear cover in place. However, due to the limited amount of room available, you may conclude that the procedure would be easier if the cover were removed from the engine first. If so, refer to Section 12 for the cover removal and installation procedure.*

Timing gear cover in place

Refer to illustration 12.2

1   Disconnect the negative battery cable from the battery, then remove the crankshaft pulley (see Section 10). **Caution:** *If the radio in your vehicle is equipped with an anti-theft system, make sure you have the correct activation code before disconnecting the battery.*
2   Note how the seal is installed - the new one must face the same direction! Carefully pry the oil seal out of the cover with a seal puller or a large screwdriver **(see illustration)**. Be very careful not to distort the cover or scratch the crankshaft!
3   Apply clean engine oil or multi-purpose grease to the outer edge of the new seal, then install it in the cover with the lip (open end) facing in. Drive the seal into place with a large socket or section of pipe and a hammer. Make sure the seal enters the bore squarely and stop when the front face is flush with the cover.
4   Install the crankshaft pulley (see Section 11).

Timing gear cover removed

Refer to illustrations 12.6, 12.8a and 12.8b

5   Remove the timing gear cover as described in Section 12.
6   Using a large screwdriver, pry the old seal out of the cover **(see illustration)**. Be careful not to distort the cover or scratch the wall of the seal bore. If the engine has accumulated a lot of miles, apply penetrating oil to the seal-to-cover joint and allow it to soak in before attempting to remove the seal.
7   Clean the bore to remove any old seal material and corrosion. Support the cover on a block of wood and position the new seal in the bore with the lip (open end) of the seal facing in. A small amount of oil applied to the outer edge of the new seal will make installation easier - don't overdo it!
8   Drive the seal into the bore with a large socket or section of pipe and hammer until it's completely seated **(see illustration)**. Select a socket or section of pipe that's the same outside diameter as the seal. A block of wood can be used if a socket or section of pipe isn't available **(see illustration)**.
9   Reinstall the timing gear cover.

12.2 The crankshaft front seal can be removed with a seal removal tool or a large screwdriver

13   Timing gear cover - removal and installation

Refer to illustrations 13.4 and 13.10

Removal

1   Remove the crankshaft pulley (see Section 11).
2   Remove the fan and hub assembly (see Chapter 3).
3   Remove the air conditioning compressor (if equipped) and the alternator bracket assembly from the cylinder head and set it aside.
4   Remove the timing gear cover bolts **(see illustration)**.
5   Separate the timing gear cover from the engine. Avoid damaging the sealing surfaces; do not force tools between the cover and block.

Installation

6   Thoroughly clean the cover and all sealing surfaces, removing any traces of gasket material. Drive the old oil seal out from the rear of the timing gear cover and replace it with a new one (see Section 11).
7   Apply gasket sealing compound to both sides of the new timing cover gasket and position the gasket on the engine block.
8   Position the timing gear cover on the engine block. Install the bolts finger-tight only at this time. Apply thread sealer to the two lower bolts **(see illustration)**.
9   Install the crankshaft pulley to center the timing gear cover.
10   Tighten the timing gear cover bolts to the torque listed in this Chapter's Specifications.
11   Reinstall the remaining parts in the reverse order of removal.
12   Run the engine and check for oil leaks.

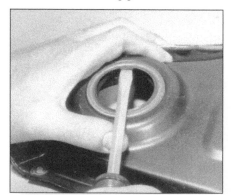

12.6 Once the timing chain cover is removed, place it on a flat surface and gently pry the old seal out with a large screwdriver

12.8a Clean the bore, then apply a small amount of oil to the outer edge of the new seal and drive it squarely into the opening with a large socket . . .

12.8b . . . or a block of wood and a hammer - don't damage the seal or the cover in the process!

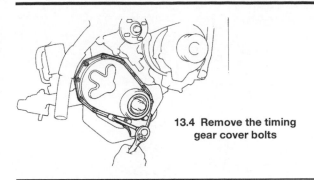

13.4 Remove the timing gear cover bolts

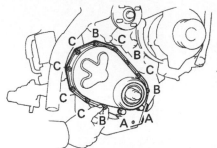

13.8 Bolt head designations for the timing cover - 3F-E engine

A   14 mm bolt head (coat bolt threads with thread sealant)

B   10 mm bolt head

C   10 mm bolt head

14   Camshaft and timing gears - removal and installation

Removal

Refer to illustrations 14.11, 14.12a, 14.12b, 14.13, 14.14, 14.15 and 14.16

1   Disconnect the cable from the negative battery terminal. **Caution:** *If the radio in your vehicle is equipped with an anti-theft system, make sure you have the correct activation code before disconnecting the battery.*

2   Set the number six piston at Top Dead Center (TDC) on the compression stroke (see Section 3). **Note:** *To locate No. six piston at TDC, set No. 1 piston at TDC, then rotate the crankshaft 360-degrees.*

3   Remove the radiator (see Chapter 3).

4   On models equipped with air conditioning, unbolt the air conditioning compressor and set it aside without disconnecting the refrigerant lines.

5   On carburetor equipped models, remove the fuel pump (see Chapter 4A).

6   Remove the distributor (see Chapter 5). **Note:** *The distributor rotor should be pointing to the No. six spark plug wire terminal. Mark the rotor and distributor base in this position before removal.*

7   Remove the valve cover (see Section 4).

8   Remove the rocker arm assembly and pushrods (see Section 5).

9   Remove the valve lifters (see Section 9).

10   Remove the timing gear cover (see Section 13).

11   Check the timing gear backlash. If the backlash is incorrect, replace both the crankshaft gear and the camshaft gear **(see illustration)**. Refer to the Specifications listed in the beginning of this Chapter.

12   Remove the thrust washer bolts and carefully slide the camshaft out of the engine block **(see illustrations)**. **Caution:** *To avoid damage to the camshaft bearings as the lobes pass over them, support the camshaft near the block as it is withdrawn.*

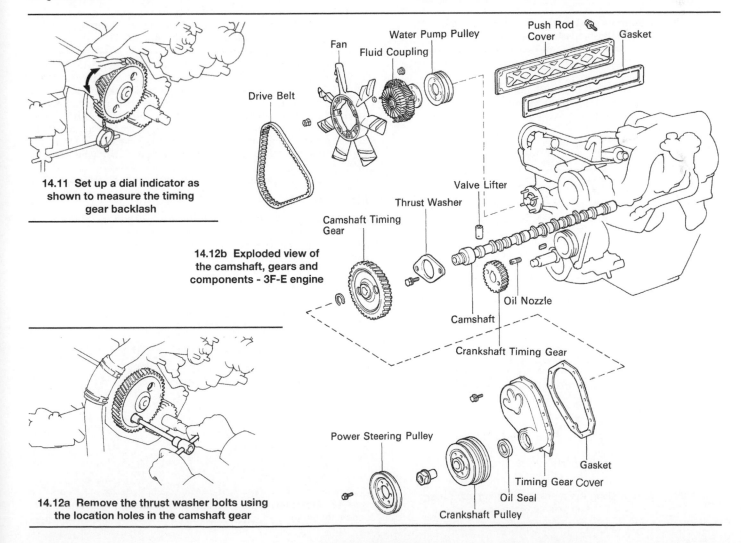

14.11 Set up a dial indicator as shown to measure the timing gear backlash

14.12b Exploded view of the camshaft, gears and components - 3F-E engine

14.12a Remove the thrust washer bolts using the location holes in the camshaft gear

Fan

Water Pump Pulley

Fluid Coupling

Push Rod Cover

Gasket

Drive Belt

Valve Lifter

Thrust Washer

Camshaft Timing Gear

Camshaft

Oil Nozzle

Crankshaft Timing Gear

Power Steering Pulley

Gasket

Timing Gear Cover

Oil Seal

Crankshaft Pulley

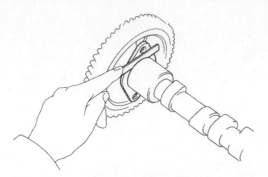

14.13 Inspect the thrust plate clearance using the
correct feeler gauge

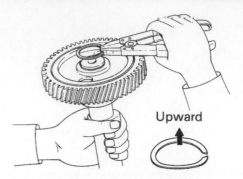

14.14 Before the camshaft timing gear can be pressed off,
the snap-ring must be removed

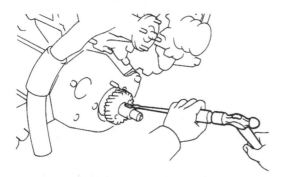

14.15 Remove the crankshaft pulley key using
a flat-bladed screwdriver

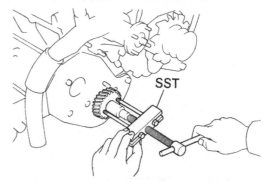

14.16 Remove the crankshaft timing gear from the
crankshaft using a bolt-type puller

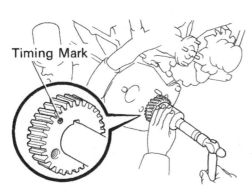

14.18a Install the crankshaft sprocket with the timing mark facing
out - align the slot with the crankshaft key and carefully
drive the gear onto the crankshaft

13   Using a feeler gauge, measure the thrust plate clearance (see
illustration). Compare your findings with this Chapter's Specifications.
If the clearance is excessive, replace the thrust plate.
14   The camshaft timing gear is pressed on the camshaft, if necessary,
remove the snap-ring (see illustration) and using a suitable press and
fixtures, press off the gear. If a press is not available, take the camshaft
to a reputable automotive machine shop and have the work performed.
15   Remove the crankshaft pulley key in the crankshaft (see illustration).
16   Remove the crankshaft gear using a puller (see illustration).

Installation

Refer to illustrations 14.18a, 14.18b, 14.19, 14.20 and 14.23
17   Clean the components and inspect for wear and damage. Timing
gear teeth that are deformed, chipped, pitted or discolored call for
replacement. Always replace the timing gears as a set. Make sure the
engine block is clean and the crankshaft does not have any burrs,
scratches or damage that will prevent the crankshaft timing gear from
sliding over the surface easily.

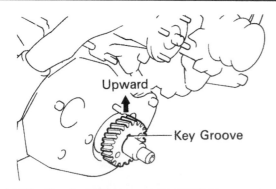

14.18b After installing the crankshaft timing gear, make
sure the key groove faces up

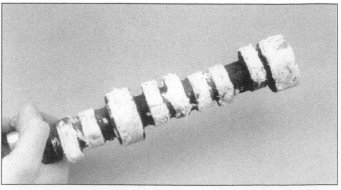

14.19 Apply moly-base grease of engine assembly lube to the
cam lobes and bearing journals before installing the camshaft

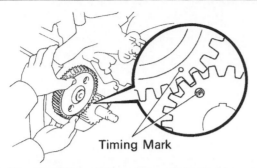

14.20 Align the timing marks on the camshaft and crankshaft gears - when the marks are aligned number six cylinder is at TDC on the compression stroke

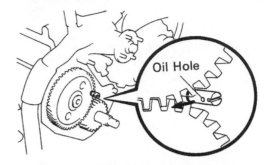

14.23 Be sure the oil hole points toward the gears at a point where the gears mesh. This is the lubrication point for the timing gears

18 Install the timing gear with the mark facing out (see illustration). Align the timing gear with the key in the crankshaft and drive the gear onto the crankshaft using a deep socket or section of pipe. Position the crankshaft with the key groove facing up (see illustration).

19 Lubricate the camshaft bearing journals and lobes with moly-base grease or engine assembly lube (see illustration). Slide the camshaft into the engine block, supporting the cam near the block. Be careful not to scrape or nick the bearings.

20 Mesh the camshaft gear with the crankshaft gear, aligning the timing marks (see illustration).

21 Install the thrust washer bolts and tighten them to the torque listed in this Chapter's Specifications.

22 Install the crankshaft pulley key.

23 If removed, install the oil nozzle and make sure the hole points toward the gears (see illustration). Note: Examine the oil pressure nozzle hole and ensure it is free of debris.

24 Install the remaining components in the reverse order of removal. Refer to the appropriate Sections for installation instructions. Note: If the original cam and valve lifters are being reinstalled, be sure to install the lifters in their original locations. If a new camshaft was installed, be sure to install new lifters, as well.

25 Add coolant and change the oil and filter (see Chapter 1).

26 Start the engine and check the ignition timing. Check for leaks and unusual noises.

15 Camshaft and bearings - inspection

Camshaft lobe lift check

Refer to illustration 15.3

1 To determine the extent of cam lobe wear, the lobe lift should be checked prior to camshaft removal. Refer to Section 4 and remove the valve cover.

2 Position the number one piston at TDC on the compression stroke (see Section 3).

3 Beginning with the number one cylinder valves, mount a dial indicator on the engine and position the plunger against the top surface of the first rocker arm. The plunger should be directly above and in line with the pushrod (see illustration).

4 Zero the dial indicator, then very slowly turn the crankshaft in the normal direction of rotation until the indicator needle stops and begins to move in the opposite direction. The point at which it stops indicates maximum cam lobe lift.

5 Record this figure for future reference, then reposition the piston at TDC on the compression stroke.

6 Move the dial indicator to the remaining number one cylinder rocker arm and repeat the check. Be sure to record the results for each valve.

7 Repeat the check for the remaining valves. Since each piston must be at TDC on the compression stroke for this procedure, work from cylinder-to-cylinder following the firing order sequence.

8 After the check is complete, compare the results to the specifications. If camshaft lobe lift is less than specified, cam lobe wear has occurred and a new camshaft should be installed.

Inspection

Refer to illustration 15.10

9 After the camshaft has been removed from the engine (see Section 14), cleaned with solvent and dried, inspect the bearing journals for uneven wear, pitting and evidence of seizure. If the journals are damaged, the bearing inserts in the block are probably damaged also. Both the camshaft and bearings will have to be replaced.

10 If the bearing journals are in good condition, measure them with a micrometer and record the measurements (see illustration). Measure each journal at several locations around its circumference. If you get different measurements at different locations, the journal is out of round.

11 Check the inside diameter of each camshaft bearing with a

15.3 When checking the camshaft lobe lift, the dial indicator plunger must be positioned directly above the pushrod

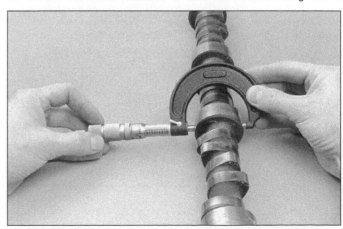

15.10 The camshaft bearing journal diameters are checked to pinpoint excessive wear and out-of-round conditions

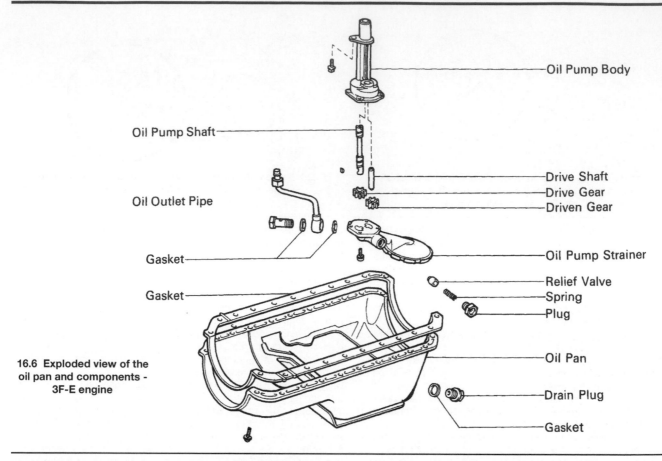

Oil Pump Shaft

Oil Outlet Pipe

Gasket

Gasket

Oil Pump Body

Drive Shaft
Drive Gear
Driven Gear

Oil Pump Strainer

Relief Valve
Spring
Plug

Oil Pan

Drain Plug

Gasket

16.6 Exploded view of the oil pan and components - 3F-E engine

telescoping gauge and measure the gauge with a micrometer. Subtract each cam journal diameter from the corresponding camshaft bearing inside diameter to obtain the bearing oil clearance. Compare the clearance for each bearing to the specifications. If it is excessive, for any of the bearings, have new bearings installed by an automotive machine shop.

12   Inspect the distributor drive gear for wear. Replace the camshaft if the gear is worn.

13   Inspect the camshaft lobes (including the fuel pump lobe on carburetor equipped models) for heat discoloration, score marks, chipped areas, pitting and uneven wear. Using a micrometer, measure the camshaft lobe height. If the lobes are in good condition and if the lobe height measurements are as specified, the camshaft can be reused.

Bearing replacement

14   Camshaft bearing replacement requires special tools and expertise that place it outside the scope of the home mechanic. Take the engine block (see Part C of this Chapter) to an automotive machine shop to ensure the job is done correctly.

16   Oil pan - removal and installation

Refer to illustrations 16.6

Removal

1   Disconnect the cable from the negative battery terminal. **Caution:** *If the radio in your vehicle is equipped with an anti-theft system, make sure you have the correct activation code before disconnecting the battery.*

2   Raise the vehicle and support it securely on jackstands.

3   Drain the engine oil and remove the oil filter (Chapter 1).

4   Disconnect the exhaust pipe at the manifold and hangers and tie the system aside.

5   Remove the starter (see Chapter 5) and the bellhousing dust cover.

6   Remove the bolts **(see illustration)** and detach the oil pan. Don't

pry between the block and pan or damage to the sealing surfaces may result and oil leaks could develop. If the pan is stuck, dislodge it with a soft-face hammer or a block of wood and a hammer.

Installation

7   Use a scraper to remove all traces of sealant from the pan and block, then clean the mating surfaces with lacquer thinner or acetone.

8   Using gasket adhesive, position new gasket on the engine.

9   Install the oil pan and tighten the mounting bolts to the torque listed in this Chapter's Specifications. Start at the center of the pan and work out toward the ends in a spiral pattern.

10   Install the bellhousing dust cover and the starter, then reconnect the exhaust pipe to the manifold and hanger brackets.

11   Lower the vehicle.

12   Install a new filter and add oil to the engine.

13   Reconnect the negative battery cable.

14   Start the engine and check for leaks.

17   Oil pump - removal and installation

Refer to illustrations 17.5

Note: *It is advisable to replace the oil pump with a new unit whenever the oil pan is removed. If low oil pressure is suspected, be sure to check the oil pressure before removing the oil pan (see Chapter 2 Part C for information concerning the oil pressure check).*

Removal

1   Remove the oil pan (see Section 16).

2   Remove the two oil pump attaching bolts from the engine block **(see illustration 16.6)**.

3   Detach the oil pump and strainer assembly from the block.

4   If the pump is defective, replace it with a new one. If the engine is being completely overhauled, install a new oil pump - don't reuse the original or attempt to rebuild it.

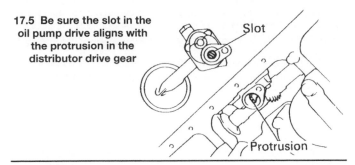

17.5 Be sure the slot in the oil pump drive aligns with the protrusion in the distributor drive gear

Slot

Protrusion

18.3 Before removing the flywheel, apply an alignment mark to ensure correct alignment on installation

Installation

5   To install the pump, turn the shaft so the gear tang mates with the slot on the lower end of the distributor drive **(see illustration)**. The oil pump should slide easily into place. If it doesn't, pull it off and turn the tang until it's aligned with the distributor drive.

6   Install the pump attaching bolts. Tighten them to the torque listed in this Chapter's Specifications.

7   Reinstall the oil pan (see Section 16).

8   Add oil, run the engine and check for leaks.

18   Flywheel/driveplate - removal and installation

Refer to illustrations 18.3 and 18.4

Removal

1   Raise the vehicle and support it securely on jackstands, then refer to Chapter 7 and remove the transmission. If it's leaking, now would be a very good time to replace the front pump seal/O-ring (automatic transmission only).

2   If equipped with a manual transmission, remove the pressure plate and clutch disc (see Chapter 8). Now is a good time to check/replace the clutch components and pilot bearing.

3   Use paint or a center-punch to make alignment marks on the flywheel/driveplate and crankshaft to ensure correct alignment during reinstallation **(see illustration)**.

4   Remove the bolts that secure the flywheel/driveplate to the crankshaft **(see illustration)**. If the crankshaft turns, hold the flywheel with a pry bar or wedge a screwdriver into the ring gear teeth to hold the flywheel.

5   Remove the flywheel/driveplate from the crankshaft. Since the flywheel is fairly heavy, be sure to support it while removing the last bolt.

Installation

6   Clean the flywheel to remove grease and oil. Inspect the surface for cracks, rivet grooves, burned areas and score marks. Light scoring can be removed with emery cloth. Check for cracked and broken ring

gear teeth or a loose ring gear. Lay the flywheel on a flat surface and use a straightedge to check for warpage.

7   Clean and inspect the mating surfaces of the flywheel/driveplate and the crankshaft. If the crankshaft rear seal is leaking, replace it before reinstalling the flywheel/driveplate.

8   Position the flywheel/driveplate against the crankshaft. Be sure to align the marks made during removal. Note that some engines have an alignment dowel or staggered bolt holes to ensure correct installation. Before installing the bolts, apply thread locking compound to the threads.

9   Wedge a screwdriver into the ring gear teeth to keep the flywheel/driveplate from turning. Using a criss-cross tightening sequence, tighten the bolts to the torque listed in this Chapter's Specifications.

10   The remainder of installation is the reverse of the removal procedure.

19   Rear main oil seal - replacement

Refer to illustration 19.5

1   The rear main bearing oil seal can be replaced without removing the oil pan or crankshaft.

2   Remove the transmission (see Chapter 7).

3   If equipped with a manual transmission, remove the pressure plate and clutch disc (see Chapter 8).

4   Remove the flywheel or driveplate (see Section 18).

5   Using a seal removal tool or a large screwdriver, carefully pry the seal out of the block **(see illustration)**. Don't scratch or nick the crankshaft in the process.

18.4 To prevent the flywheel from turning, hold a pry bar against two bolts or wedge a large screwdriver into the flywheel ring gear

19.5 Carefully pry the oil seal out with a removal tool or a screwdriver - don't nick or scratch the crankshaft or the new seal will be damaged and leaks will develop

6   Clean the bore in the block and the seal contact surface on the crankshaft. Check the crankshaft surface for scratches and nicks that could damage the new seal lip and cause oil leaks. If the crankshaft is damaged, the only alternative is a new or different crankshaft.

7   Apply a light coat of engine oil or multi-purpose grease to the outer edge of the new seal. Lubricate the seal lip with moly-base grease.

8   Carefully work the seal lip over the end of the crankshaft, then tap the new seal into place with a hammer and a large socket (if available) or a rounded punch. The seal lip must face toward the front of the engine.

9   Install the flywheel or driveplate.

10   If equipped with a manual transmission, reinstall the clutch disc and pressure plate.

11   Reinstall the transmission as described in Chapter 7.

20   Engine mounts - check and replacement

1   Engine mounts seldom require attention, but broken or deteriorated mounts should be replaced immediately or the added strain placed on the driveline components may cause damage or wear.

Check

2   During the check, the engine must be raised slightly to remove the weight from the mounts.

3   Raise the vehicle and support it securely on jackstands, then position a jack under the engine oil pan. Place a large block of wood between the jack head and the oil pan, then carefully raise the engine just enough to take the weight off the mounts. **Warning:** *DO NOT place any part of your body under the engine when it's supported only by a jack!*

4   Check the mounts to see if the rubber is cracked, hardened or separated from the metal plates. Sometimes the rubber will split right down the center.

5   Check for relative movement between the mount plates and the engine or frame (use a large screwdriver or pry bar to attempt to move the mounts). If movement is noted, lower the engine and tighten the mount fasteners.

6   Rubber preservative should be applied to the mounts to slow deterioration.

Replacement

7   Disconnect the negative battery cable from the battery, then raise the vehicle and support it securely on jackstands (if not already done). **Caution:** *If the radio in your vehicle is equipped with an anti-theft system, make sure you have the correct activation code before disconnecting the battery.*

8   Loosen the nut on the through bolt and remove the bolt and nut that secure the mount to the frame bracket.

9   Raise the engine slightly with a jack or hoist (make sure the fan doesn't hit the radiator or shroud). Remove the through bolt and nut and detach the mount.

10   Installation is the reverse of removal. Use thread locking compound on the mount bolts and be sure to tighten them securely.

Chapter 2 Part B
Dual Overhead Camshaft (DOHC) engine

Contents

	Section
Camshafts and valve lifters - removal, inspection, and installation	6
CHECK ENGINE light	See Chapter 6
Crankshaft front oil seal - replacement	9
Cylinder compression check	See Chapter 2C
Cylinder head - removal and installation	7
Drivebelt - check, adjustment and replacement	See Chapter 1
Engine mounts - check and replacement	14
Engine oil and filter change	See Chapter 1
Engine overhaul - general information	See Chapter 2C
Engine - removal and installation	See Chapter 2C
Exhaust manifold - removal and installation	5
Flywheel/driveplate - removal and installation	12
General information	1

	Section
Intake manifold - removal and installation	4
Oil cooler - removal and installation	See Chapter 3
Oil pan - removal and installation	10
Oil pump - removal, inspection and installation	11
Rear main oil seal - replacement	13
Repair operations possible with the engine in the vehicle	2
Spark plug replacement	See Chapter 1
Timing cover and chain - removal, inspection and installation	8
Top Dead Center (TDC) for number one piston - locating	See Chapter 2A
Valve clearance check and adjustment	See Chapter 1
Valve cover - removal and installation	3
Valves - servicing	See Chapter 2C
Water pump - removal and installation	See Chapter 3

Specifications

General

Engine designation	1FZ-FE
Displacement	273 cu. in. (4.5 liters)
Cylinder numbers (drivebelt end-to-transmission end)	1-2-3-4-5-6
Firing order	1-5-3-6-2-4

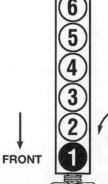

The blackened terminal shown on the distributor cap indicates the Number One spark plug wire position

Cylinder location and distributor rotation - 1FZ-FE engines

FRONT

92056-1-SPECS1FZFE HAYNES

Camshaft

Journal diameter	1.0614 to 1.0620 inches (26.959 to 26.975 mm)
Journal-to-bearing (oil) clearance	
Standard	0.0010 to 0.0024 inch (0.025 to 0.062 mm)
Limit	0.0039 inch (0.10 mm)
Journal out-of-round limit	0.0024 inch (0.06 mm)
Lobe height	
Standard	1.9925 to 1.9965 inches (50.61 to 50.71 mm)
Limit	1.9886 inches (50.51 mm)
Thrust clearance (endplay)	
Standard	0.0012 to 0.0031 inch (0.030 to 0.080 mm)
Limit	0.0039 inch (0.10 mm)
Camshaft gear backlash	
Standard	0.0008 to 0.0079 inch (0.020 to 0.200 mm)
Limit	0.0188 inch (0.30 mm)
Camshaft gear spring free length	0.717 to 0.740 inch (18.2 to 18.8 mm)

Oil pump

Rotor-to-body clearance	
Standard	0.0039 to 0.0067 inch (0.100 to 0.170 mm)
Service limit	0.0118 inch (0.30 mm)
Rotor tip clearance	
Standard	0.0012 to 0.0063 inch (0.030 to 0.160 mm)
Service limit	0.0098 inch (0.25 mm)
Rotor-to-cover clearance	
Standard	0.0012 to 0.0035 inch (0.030 to 0.090 mm)
Service limit	0.0059 inch (0.15 mm)

Torque specifications

Ft-lbs (unless otherwise indicated)

Note: One foot-pound (ft-lb) of torque is equivalent to 12 inch-pounds (in-lbs) of torque. Torque values below approximately 15 ft-lbs are expressed in inch-pounds, because most foot-pound torque wrenches are not accurate at these smaller values.

Air intake plenum	15
Intake manifold bolts	15
Exhaust manifold nuts	29
Exhaust pipe-to-exhaust manifold	46
Heat insulator-to-exhaust manifold	14
Air pipe (PAIR system)	
Bolt	14
Nut	15
Crankshaft pulley-to-crankshaft bolt	304
Flywheel bolts (manual transmission)	58
Driveplate bolts (automatic transmission)	47
Heater inlet pipe and hose bolt	15
Idler pulley bolts	27
Cylinder head bolts	
Step 1	29
Step 2	Tighten an additional 90-degrees
Step 3	Tighten an additional 90-degrees
Cylinder head-to-timing cover bolts	15
Camshaft bearing cap bolts	12
Camshaft sprocket bolt	54
Oil pump bolts	12
Oil pick-up/strainer nuts/bolts	14
Oil jet	14
Oil pan bolts	
Oil pan number 1 (upper)	
14 mm bolt head	32
12 mm bolt head	14
Oil pan number 1-to-timing chain cover	14
Oil pan number 2 (lower)	
Bolt head	69 in-lbs
Nut	78 in-lbs
Oil level sensor-to-block mounting bolts	48 in-lbs
Oil relief valve	36
Rear crankshaft oil seal retainer bolts	15
Timing cover bolts and nuts	15
Timing chain tensioner nuts	15

3.5 Remove the bolts around the perimeter of the valve cover

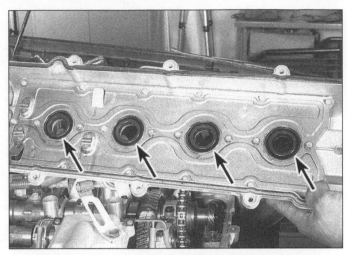

3.6 Make sure the spark plug tube seals (arrows) are in place before replacing the valve cover

1   General information

This Part of Chapter 2 is devoted to in-vehicle repair procedures for the DOHC inline six-cylinder engine. All information concerning engine removal and installation and engine block and cylinder head overhaul can be found in Part C of this Chapter.

The following repair procedures are based on the assumption that the engine is installed in the vehicle. If the engine has been removed from the vehicle and mounted on a stand, many of the steps outlined in this Part of Chapter 2 will not apply.

The Specifications included in this Part of Chapter 2 apply only to the procedures contained in this Part. Part C of Chapter 2 contains the Specifications necessary for cylinder head and engine block rebuilding.

The DOHC inline six-cylinder engine is installed in 1993 and later models and is designated 1FZ-FE. The engine displaces 4.5 liters and is considerably more powerful than previous models.

Each cylinder is equipped with two intake valves and two exhaust valves. The left camshaft (driver's side) operates the intake valves while the right camshaft (passenger's side) operates the exhaust valves. The engine is equipped with an electronically controlled multi-port fuel injection system.

2   Repair operations possible with the engine in the vehicle

Many major repair operations can be accomplished without removing the engine from the vehicle.

Clean the engine compartment and the exterior of the engine with some type of degreaser before any work is done. It will make the job easier and help keep dirt out of the internal areas of the engine.

Depending on the components involved, it may be helpful to remove the hood to improve access to the engine as repairs are performed (refer to Chapter 11 if necessary). Cover the fenders to prevent damage to the paint. Special pads are available, but an old bedspread or blanket will also work.

If vacuum, exhaust, oil or coolant leaks develop, indicating a need for gasket or seal replacement, the repairs can generally be made with the engine in the vehicle. The intake and exhaust manifold gaskets, oil pan gasket, crankshaft oil seals and cylinder head gasket are all accessible with the engine in place.

Exterior engine components, such as the intake and exhaust manifolds, the oil pan, the oil pump, the water pump, the starter motor, the alternator, the distributor and the fuel system components can be removed for repair with the engine in place.

Since the cylinder head can be removed without pulling the

engine, camshaft and valve component servicing can also be accomplished with the engine in the vehicle. Replacement of the timing chain and pulleys is also possible with the engine in the vehicle.

In extreme cases caused by a lack of necessary equipment, repair or replacement of piston rings, pistons, connecting rods and rod bearings is possible with the engine in the vehicle. However, this practice is not recommended because of the cleaning and preparation work that must be done to the components involved.

3   Valve cover - removal and installation

Removal

Refer to illustration 3.5

1   Disconnect the negative cable from the battery. **Caution:** *If the stereo in your vehicle is equipped with an anti-theft system, make sure you have the correct activation code before disconnecting the battery.*

2   Remove the air intake duct. Detach the PCV hoses from the valve cover.

3   Remove the spark plug wire covers. Remove the spark plug wires from the spark plugs, handling them by the boots, not pulling on the wires.

4   Disconnect the accelerator and cruise control cables from the throttle body and position them out of the way. Remove the number one engine hanger located on the front section of the intake manifold.

5   Remove the valve cover mounting bolts, then detach the valve cover and gasket from the cylinder head **(see illustration)**. If the valve cover is stuck to the cylinder head, bump the end with a wood block and a hammer to jar it loose. If that doesn't work, try to slip a flexible putty knife between the cylinder head and valve cover to break the seal. **Caution:** *Don't pry at the valve cover-to-cylinder head joint or damage to the sealing surfaces may occur, leading to oil leaks after the valve cover is reinstalled.*

Installation

Refer to illustrations 3.6 and 3.7

6   The mating surfaces of the cylinder head and valve cover must be clean when the valve cover is installed. The rubber sealing gasket can be reused unless it has high mileage and the rubber has hardened or cracked. If necessary, pull out the rubber seal and clean the mating surfaces with lacquer thinner or acetone. Install a new rubber gasket, pressing it evenly into the groove around the underside of the valve cover. If there's residue or oil on the mating surfaces when the valve cover is installed, oil leaks may develop. **Note:** *Make sure that the spark plug tube gaskets are in place on the underside of the valve cover before reinstalling it* **(see illustration)**.

3.7 Apply sealant (arrow) to the rubber plugs before installing the valve cover

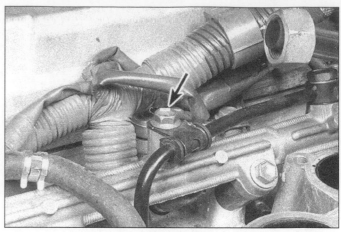

4.7a Remove the ground strap bolt from the cylinder head

7   Apply RTV sealant to the semi-circular rubber plugs and install them onto the cylinder head **(see illustration)**.

8   Install the valve cover and bolts.

9   Tighten the bolts to the torque listed in this Chapter's Specifications in three or four equal steps.

10   Reinstall the remaining components, run the engine and check for oil leaks.

4   Intake manifold - removal and installation

Removal

Refer to illustrations 4.7a and 4.7b

Note: *If the intake manifold is to be unbolted only for removal of the cylinder head, then the intake manifold can simply be unbolted from the cylinder head and pushed away from the cylinder head, without disconnecting any hoses, wires or linkage. The following procedure is for complete removal of the manifold from the vehicle.*

1   Disconnect the negative cable from the battery. **Caution:** *If the stereo in your vehicle is equipped with an anti-theft system, make sure you have the correct activation code before disconnecting the battery.*

2   Refer to Chapter 4B and relieve the fuel system pressure.

3   Label and detach the PCV and vacuum hoses connected to the intake manifold, including those the brake booster and the air conditioning idle-up actuator.

4   Remove the throttle body and air intake plenum (see Chapter 4B).

5   Remove the fuel rail and injectors as an assembly from the intake manifold (see Chapter 4B).

6   Disconnect the vacuum hoses from the EGR valve and vacuum modulator (if equipped), label them correctly and unbolt the EGR pipe from the intake manifold (see Chapter 4B). Set the EGR assembly aside.

7   Remove the ground strap and mounting nuts/bolts, then detach the manifold from the engine **(see illustrations)**. **Note:** *From under the vehicle, it will be necessary to unbolt the wiring harness, the lower intake manifold bolts, and the two nuts securing the pair of steel lines to the underside of the intake manifold.*

Installation

8   Clean the mating surfaces of the intake manifold and the cylinder head mounting surface with lacquer thinner or acetone. If the gasket shows signs of leaking, have the manifold checked for warpage at an automotive machine shop and resurfaced if necessary.

9   Install a new gasket, then position the manifold on the cylinder head and install the nuts/bolts.

10   Tighten the nuts/bolts in three equal steps to the torque listed in this Chapter's Specifications. Work from the center out towards the ends to avoid warping the manifold.

11   Install the remaining parts in the reverse order of removal.

12   Before starting the engine, check the throttle linkage for smooth operation.

13   Run the engine and check for coolant and vacuum leaks.

14   Road test the vehicle and check for proper operation of all accessories, including the cruise control system, if equipped.

5   Exhaust manifold - removal and installation

Warning: *The engine must be completely cool before beginning this procedure.*

Removal

Refer to illustrations 5.3 and 5.5

1   Disconnect the negative cable from the battery. **Caution:** *If the stereo in your vehicle is equipped with an anti-theft system, make sure you have the correct activation code before disconnecting the battery.*

2   If the engine is equipped with the Pulse Air Injection (PAIR) system, remove the pipes and valve assembly from the cylinder head and exhaust manifold area (see Chapter 6).

3   Remove the upper heat insulator from the manifold **(see illustration)**.

4   Apply penetrating oil to the exhaust manifold mounting nuts, and the nuts retaining the exhaust pipe to the manifold. After the nuts have soaked, remove the nuts retaining the exhaust pipe to the manifold (see Chapter 4B).

5   Remove the nuts and detach the manifold and gasket **(see illustration)**.

4.7b Remove the intake manifold bolts/nuts (arrows) and remove the intake manifold

5.3 Remove the bolts that retain the heat insulator
to the exhaust manifold (arrows)

6.2 With the No. 1 piston at TDC on the compression stroke, the
two dots (2) on the camshaft drive and driven gears face out
and while the inner marks (1) are aligned together

Installation

6   Use a scraper to remove all traces of old gasket material and carbon deposits from the manifold and cylinder head mating surfaces. If the gasket was leaking, have the manifold checked for warpage at an automotive machine shop and resurfaced if necessary.

7   Position a new gasket over the cylinder head studs. **Note:** *The marks on the gasket should face out (away from the cylinder head). Consult the parts representative concerning the position of the gasket*

5.5 Remove the nuts (arrows) and remove the exhaust manifold

if in doubt.

8   Install the manifold and thread the mounting nuts into place.

9   Working from the center out, tighten the nuts to the torque listed in this Chapter's Specifications in three or four equal steps.

10   Reinstall the remaining parts in the reverse order of removal.

11   Run the engine and check for exhaust leaks.

6   Camshafts and valve lifters - removal, inspection and installation

Note: *Before beginning this procedure, obtain two 6 x 1.0 mm bolts 16 to 20 mm long. They will be referred to as service bolts in the text.*

Removal

Refer to illustrations 6.2, 6.3, 6.5a, 6.5b, 6.6a, 6.6b, 6.6c, 6.8, 6.9, 6.10, 6.13, 6.17a, 6.17b, 6.18, 6.19a and 6.19b

1   Remove the valve cover as described in Section 3.

2   Refer to Chapter 2A and place the engine on TDC for number 1 cylinder. Make sure the alignment marks (dots) of the camshaft drive and driven gears are in straight line in relation to the cylinder head surface **(see illustration)**. If not, double-check the TDC mark on the crankshaft pulley and rotate the engine 360-degrees to obtain the correct setting.

3   Measure the camshaft thrust clearance (endplay) with a dial indicator **(see illustration)**. If the clearance is greater than the service limit, replace the camshaft and/or the cylinder head.

4   Remove the distributor (see Chapter 5).

5   Remove the timing chain tensioner **(see illustrations)**.

6.3 Mount a dial indicator as shown to
measure camshaft endplay - zero the
dial and pry the camshaft forward
and back to read the endplay

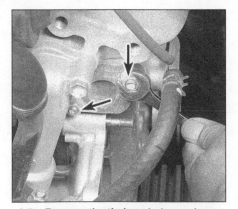

6.5a Remove the timing chain tensioner
cover bolts (arrows) . . .

6.5b . . . and withdraw the tensioner from
the timing cover

6.6a Apply a paint mark on the timing chain link aligned with the camshaft sprocket timing mark (arrows)

6.6b Use a wrench placed on the hex provided to hold the camshaft while removing the camshaft sprocket bolt

6   Apply a paint mark on the timing chain link aligned with the timing mark on the camshaft sprocket. Hold the intake camshaft with a wrench and remove the camshaft sprocket bolt. Remove the distributor drive gear. Carefully remove the camshaft sprocket and chain from the camshaft and rest it on the chain guides **(see illustrations)**.

Exhaust camshaft

7   First, remove the exhaust camshaft. **Caution:** *The camshafts must be removed from the cylinder head in sequence and kept horizontal to the plane of the cylinder head when loosened to avoid damaging the portion of the cylinder head that regulates the thrust clearance of the camshafts.*

8   Rotate the camshaft until the service bolt hole of the camshaft sub gear is facing up **(see illustration)**. Secure the sub gear to the main gear with a service bolt (6-mm thread diameter, 1-mm thread pitch and 16 to 20 mm long).

9   Set the timing marks (two dots) approximately 35 degrees from level by turning the intake camshaft **(see illustration)**.

10   Lightly push the camshafts toward the rear of the engine compartment and loosen the number 1 bearing cap bolts by alternately loosening the right and left bolts uniformly **(see illustration)**.

11   Loosen and remove the numbers 2, 3, 5 and 7 bearing cap bolts alternately loosening the right and left bolts uniformly. Do not remove the numbers 4 and 6 cap bolts.

12   Carefully remove the numbers 4 and 6 bearing cap bolts watching to make sure the camshaft rises evenly in a horizontal plane to the cylinder head. If the camshaft does not come up evenly, tighten the numbers 4 and 6 bearing caps back down, install numbers 1, 2, 3, 5 and 7 and reverse the entire procedure. Start once again and make sure the camshaft stays level.

Intake camshaft

13   Next, remove the intake camshaft. Place the timing mark (two dots) of the camshaft drive gear at an approximately 25-degree angle by rotating the intake camshaft **(see illustration)**.

14   Lightly push the camshaft toward the front of the engine and loosen the number 1 bearing cap bolts by alternately loosening the right and left bolts uniformly **(see illustration 6.10)**.

15   Loosen and remove the numbers 3, 4, 6 and 7 bearing cap bolts by alternately loosening the right and left bolts uniformly. Do not remove the numbers 2 and 5 cap bolts.

16   Carefully remove the numbers 2 and 5 bearing cap bolts, watching to make sure the camshaft rises evenly in a horizontal plane to the

6.6c Remove the camshaft sprocket and chain from the intake camshaft - lower the sprocket onto the chain guides

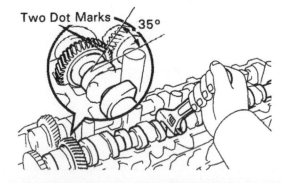

6.9 Rotate the intake camshaft until the alignment marks (two dots) are 35 degrees from level

6.8 Install a service bolt through the sub-gear and thread it into the main gear (arrow)

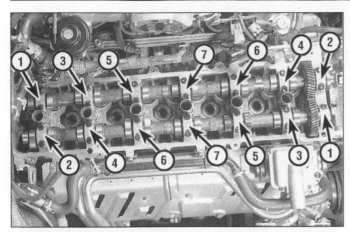

6.10 Removal sequence for the exhaust and intake camshafts. These numbers do not designate the bearing cap numbers, but callout the proper order for removal

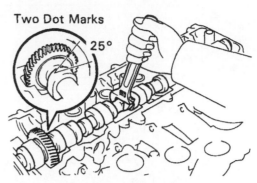

6.13 Rotate the intake camshaft 25-degrees from the horizontal or level position

cylinder head. If the camshaft does not come up evenly, tighten the numbers 2 and 5 bearing caps back down, install numbers 1, 3, 4, 6 and 7 caps and reverse the entire procedure. Start once again and make sure the camshaft stays level. Remove the intake camshaft.

17 Clean the oil from the valve lifter shims, mark them with a felt-tip marker and remove the lifters, keeping the shims with their respective lifters (see illustration). Store the camshaft bearing caps, lifters and shims so they can be reinstalled without mix-ups (see illustration).

18 Position the intake camshaft in a vise, clamping it on the hex portion. Using a two-pin spanner, rotate the sub-gear clockwise and remove the service bolt from the threaded hole, then allow the sub-gear to rotate back until all tension is relieved (see illustration).

19 Remove the sub-gear snap-ring. The wave washer, sub-gear and camshaft gear spring can now be removed from the camshaft (see illustrations).

6.17a Mark the lifters/shims (I for intake, E for exhaust, and number their location) and remove them with a magnetic retrieval tool

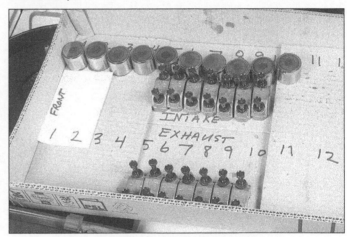

6.17b Mark up a cardboard box to store the lifters/shims and bearing caps

6.18 With the hex portion of the intake camshaft clamped in a vise, use a spanner to relieve the tension on the service bolt, remove the bolt, then release the tension on the sub-gear

6.19a Remove the snap-ring with a pair of snap-ring pliers

6.19b Remove the wave washer (1), the camshaft sub-gear (2) and the gear spring (3)

Inspection

Refer to illustrations 6.20, 6.21, 6.22, 6.23, 6.24a, 6.24b and 6.25

20   Measure the free length (distance between the ends) of the camshaft gear spring **(see illustration)** and compare it to this Chapter's Specifications. If not as specified, replace the spring.

21   Inspect each lifter for scuffing and score marks **(see illustration)**.

22   Visually examine the cam lobes and bearing journals for score marks, pitting, galling and evidence of overheating (blue, discolored areas). Look for flaking away of the hardened surface layer of each lobe. Using a micrometer, measure the height of each camshaft lobe **(see illustration)**. Compare your measurements with this Chapter's

Specifications. If the height for any one lobe is less than the specified minimum, replace the camshaft.

23   Using a micrometer, measure the diameter of each journal at several points **(see illustration)**. Compare your measurements with this Chapter's Specifications. If the diameter of any one journal is less than specified, replace the camshaft.

24   Check the oil clearance for each camshaft journal as follows:

a) *Clean the bearing caps and the camshaft journals with lacquer thinner or acetone.*

b) Carefully lay the camshaft(s) in place in the cylinder head. Don't install the lifters or intake camshaft sub-gear and don't use any lubrication.

c) *Lay a strip of Plastigage on each journal.*

d) *Install the bearing caps with the arrows pointing toward the front (timing chain end) of the engine* **(see illustration)**.

e) *Tighten the bolts to the torque listed in this Chapter's Specifications in 1/4-turn increments.* **Note:** *Don't turn the camshaft while the Plastigage is in place.*

f) *Remove the bolts and detach the caps.*

g) *Compare the width of the crushed Plastigage (at its widest point) to the scale on the Plastigage envelope* **(see illustration)**.

h) *If the clearance is greater than specified, replace the camshaft and/or cylinder head.*

i) *Scrape off the Plastigage with your fingernail or the edge of a credit card - don't scratch or nick the journals or bearing caps.*

25   With the caps reinstalled temporarily, use a dial indicator to measure the backlash between the two camshaft gears. Hold one camshaft from turning (using a wrench on the hex portion) while measuring the movement in the other camshaft gear, and compare the results to Specifications **(see illustration)**. If the backlash is beyond Specifications, replace both camshafts.

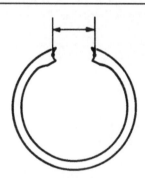

6.20 Measure the distance between the ends of the camshaft gear spring

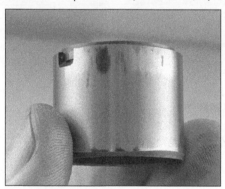

6.21 Inspect each lifter for wear and scuffing

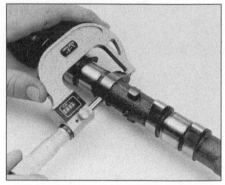

6.22 Measure the lobe heights on each camshaft - if any lobe height is less than the specified allowable minimum, replace that camshaft

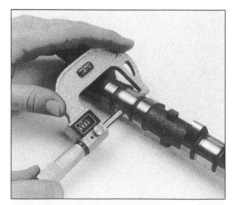

6.23 Measure each journal diameter with a micrometer (if any journal measures less than the specified limit, replace the camshaft)

6.24a The camshaft bearing caps are numbered with an arrow facing the front of the engine

6.24b Compare the width of the crushed Plastigage to the scale on the envelope to determine the oil clearance

6.25 Position a dial indicator as shown here to measure gear backlash - hold one camshaft steady with a wrench while moving the other camshaft with another wrench

Installation

Refer to illustrations 6.28, 6.29, 6.37 and 6.39

Intake camshaft

26   Apply moly-base grease or engine assembly lube to the lifters, then install them in their original locations. Make sure the valve adjustment shims are on place on the lifters.

27   Apply moly-base grease or engine assembly lube to the camshaft lobes and bearing journals. Also, apply lubrication to the thrust portion of the intake camshaft.

28   Position the intake camshaft in the cylinder head with the number 1 and 4 cam lobes facing down and the alignment mark at approximately 25-degrees from level **(see illustration)**.

29   Lightly push the camshaft toward the front of the engine and install the number 2 and 5 bearing caps in the proper location **(see illustration)**. Make sure the arrows are pointing toward the front of the engine.

30   Temporarily tighten the number 2 and 5 bearing caps alternately until snug.

31   Install the number 3, 4, 6, and 7 bearing caps in their proper location and temporarily tighten the bolts uniformly and in sequence.

32   Install the number 1 bearing cap and temporarily tighten the bolts uniformly.

33   After the intake camshaft is secured, tighten all the intake bearing cap bolts to the torque listed in this Chapter's Specifications in several stages.

Exhaust camshaft

34   Reassemble the exhaust camshaft sub-gear. Install the camshaft gear spring, sub-gear and wave washer. Secure them with the snap-ring **(see illustrations 6.19a and 6.19b)**.

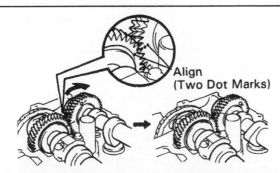

6.37 Align the camshaft gears as shown here using the two dot alignment marks

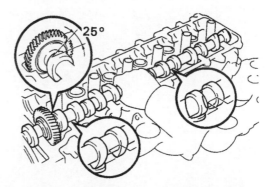

6.28 Gently place the intake camshaft into position with the number 1 and 4 lobes facing down

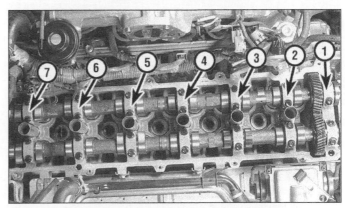

6.29 Intake camshaft bearing cap designations

35   Apply moly-base grease or engine assembly lube to the lifters, then install them in their original locations. Make sure the valve adjustment shims are in place on the lifters.

36   Apply moly-base grease or engine assembly lube to the camshaft lobes and bearing journals.

37   Align the intake camshaft gear with the exhaust camshaft gear by matching up the two dot alignment marks on the gears **(see illustration)**. This is exactly opposite or 180 degrees from the location of the marks for removal. This is a non-interference engine and the camshafts must be tightened at this position and later rotated back into the TDC position to allow the thrust surface area to mate evenly without the hindrance of the spring pressure upon certain lobes. Follow the instructions carefully.

38   Roll the exhaust camshaft down into position. Turn the exhaust camshaft back and forth a little until the exhaust camshaft sits in the bearings evenly.

39   Install the number 4 and 6 bearing caps in the proper location **(see illustration)** the arrows pointing toward the timing chain end of the engine.

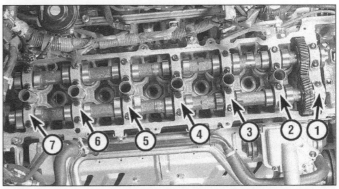

6.39 Exhaust camshaft bearing cap designations

7.11 Remove the coolant hoses and vacuum lines from the cylinder head (arrows)

7.13 Cylinder head bolt LOOSENING sequence

40   Temporarily tighten the number 4 and 6 bearing caps alternately until snug.

41   Install the number 2, 3, 5, and 7 bearing caps in their proper location and temporarily tighten them being careful to tighten them uniformly and in sequence.

42   Install the number 1 bearing cap and temporarily tighten the bolts uniformly.

43   Tighten all the exhaust bearing cap bolts to the torque listed in this Chapter's Specifications in several stages.

44   Rotate the camshafts until the service bolt is up and remove the service bolt. Make sure the camshafts rotate smoothly.

45   Position the camshafts with the timing marks aligned at TDC **(see illustration 6.2)**. Make sure the crankshaft pulley is still aligned at TDC.

46   Install the camshaft sprocket and chain. Make sure the alignment marks made in Step 6 are aligned. Install the bolt and tighten it to the torque listed in this Chapter's Specifications.

47   Install the timing chain tensioner with a new gasket. Press the tensioner in until it's seated on the timing cover, install the nuts and tighten them to the torque listed in this Chapter's Specifications.

48   The remainder of installation is the reverse of the removal procedure. Check and adjust the valve clearance, if necessary (see Chapter 1).

7   Cylinder head - removal and installation

Note: *The engine must be completely cool before beginning this procedure.*

Removal

Refer to illustrations 7.11, 7.13 and 7.14

1   Disconnect the negative cable from the battery. **Caution:** *If the stereo in your vehicle is equipped with an anti-theft system, make sure you have the correct activation code before disconnecting the battery.*

2   Drain the coolant from the engine block and radiator (see Chapter 1).

3   Drain the engine oil and remove the oil filter (see Chapter 1).

4   Remove the throttle body, air intake plenum, fuel injectors and fuel rail (see Chapter 4B).

5   Remove the intake manifold (see Section 4).

6   Remove the exhaust manifold (see Section 5). **Note:** *It is possible to leave the intake and exhaust manifolds attached to the cylinder head, to be removed along with the cylinder head for disassembly on the bench.*

7   Remove the alternator and distributor (see Chapter 5).

8   Remove the camshaft sprocket from the intake camshaft and rest it on the timing chain guides (see Section 6).

9   Remove the camshafts and valve lifters (see Section 6).

10   Unbolt the upper bracket of the power steering pump and set the pump aside without disconnecting the hoses.

11   Label and remove any remaining items, such as coolant fittings,

7.14 Carefully lift the cylinder head from the engine compartment

tubes, cables and hoses **(see illustration)**. Disconnect the wiring harness connectors from the various sensors.

12   Refer to Chapter 3 and detach the water necks from each end of the cylinder head.

13   Remove the two cylinder head-to-timing cover bolts. Using an 8 mm hex-head socket bit and a breaker bar, loosen the cylinder head bolts in 1/4-turn increments until they can be removed by hand. Loosen the cylinder head bolts using the proper sequence to avoid warping or cracking the cylinder head **(see illustration)**. **Caution:** *Because cylinder head bolts often damaged by stretching and thread collapsing or stripping, have each head bolt examined by a qualified machinist. If in doubt about the integrity of the cylinder head bolts, replace them with new head bolts.*

14   Lift the cylinder head off the engine block. If it's stuck, very carefully pry up at the transmission end, beyond the gasket surface **(see illustration)**.

15   Remove all external components from the cylinder head to allow for thorough cleaning and inspection. See Chapter 2, Part C, for cylinder head servicing procedures.

Installation

Refer to illustrations 7.17, 7.22 and 7.25

16   The mating surfaces of the cylinder head and block must be perfectly clean when the cylinder head is installed.

17   Use a gasket scraper to remove all traces of carbon and old gasket material **(see illustration)**, then clean the mating surfaces with lacquer thinner or acetone. If there's oil on the mating surfaces when the cylinder head is installed, the gasket may not seal correctly and leaks could develop. When working on the block, stuff the cylinders with clean shop rags to keep out debris. Use a vacuum cleaner to remove material that falls into the cylinders.

7.17 Remove all traces of old gasket material - the cylinder head and block mating surfaces must be perfectly clean to ensure a good gasket seal

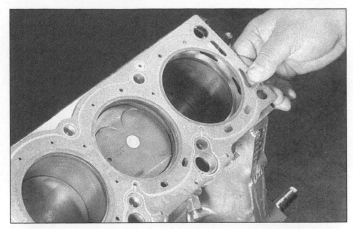

7.22 Place the new cylinder head gasket over the dowels in the block, noting the markings for UP on the gasket

18   Check the block and cylinder head mating surfaces for nicks, deep scratches and other damage. If damage is slight, it can be removed with a file; if it's excessive, machining may be the only alternative.

19   Use a tap of the correct size to chase the threads in the cylinder head bolt holes, then clean the holes with compressed air - make sure that nothing remains in the holes. **Warning:** *Wear eye protection when using compressed air!*

20   Mount each bolt in a vise and run a die down the threads to remove corrosion and restore the threads. Dirt, corrosion, sealant and damaged threads will affect torque readings.

21   Install the components that were removed from the cylinder head.

22   Position the new gasket over the dowel pins in the block **(see illustration)**.

23   Carefully set the cylinder head on the block without disturbing the gasket.

24   Before installing the cylinder head bolts, apply a small amount of clean engine oil to the threads and under the bolt heads.

25   Install the bolts in their original locations and tighten them finger tight. Install the shorter bolts along the intake side of the cylinder head and the longer bolts along the exhaust side. Following the recommended sequence, tighten the bolts in three steps to the torque listed in this Chapter's Specifications **(see illustration)**. Steps 2 and 3 of the tightening sequence each require the bolts to be tightened an additional 90-degrees. If you don't have an angle-torque attachment for your torque wrench, simply apply a paint mark at one edge of each cylinder head bolt and tighten the bolt until that mark is 90-degrees from where you started (Step 2). After Step 3, the marks will be 180-degrees from where they started. Install the two cylinder head-to-timing cover bolts and tighten them to the torque listed in this Chapter's Specifications

26   The remaining installation steps are the reverse of removal, refer to the appropriate Sections for component installation.

27   Check and adjust the valves as necessary (see Chapter 1).

28   Refill the cooling system, install a new oil filter and add oil to the engine (see Chapter 1).

29   Run the engine and check for leaks. Adjust the ignition timing (see Chapter 5) and road test the vehicle.

8   Timing cover and chain - removal, inspection and installation

Removal

Refer to illustrations 8.7, 8.8, 8.14a, 8.14b, 8.15a and 8.15b

1   Disconnect the negative cable from the battery. **Caution:** *If the stereo in your vehicle is equipped with an anti-theft system, make sure you have the correct activation code before disconnecting the battery.*

2   Block the rear wheels and apply the parking brake.

7.25 Cylinder head bolt TIGHTENING sequence

3   Raise the front of the vehicle and support it securely on jackstands.

4   If equipped, remove the protective cover under the front of the engine compartment.

5   Remove the coolant expansion tank (see Chapter 3).

6   Remove the spark plugs and drivebelts (see Chapter 1).

7   Detach the air conditioning compressor, without disconnecting the refrigerant lines, and position the compressor aside. Remove the compressor mounting bracket **(see illustration)**.

8.7 Remove the bolts (arrows) that retain the air conditioning compressor bracket assembly to the timing cover and the engine block

8.8 Remove the bolts (arrows) that retain the adjustment bracket to the timing cover. Remove the idler pulley by loosening the center bolt

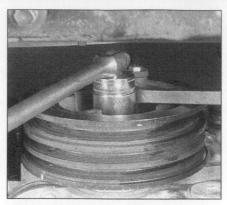

8.14a Using a long breaker bar, remove the crankshaft pulley bolt

8.14b Remove the crankshaft pulley with a puller that bolts to the threaded holes in the pulley

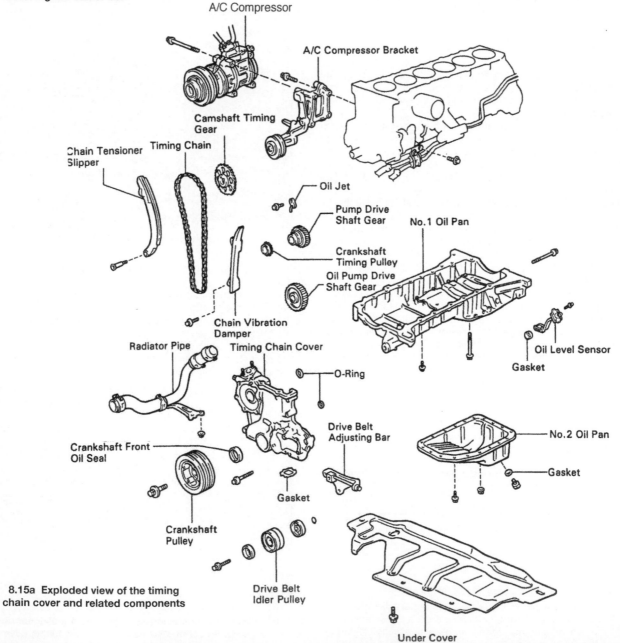

A/C Compressor

A/C Compressor Bracket

Camshaft Timing Gear

Chain Tensioner Slipper

Timing Chain

Oil Jet

Pump Drive Shaft Gear

No.1 Oil Pan

Crankshaft Timing Pulley

Oil Pump Drive Shaft Gear

Oil Level Sensor

Gasket

Chain Vibration Damper

Radiator Pipe

Timing Chain Cover

O-Ring

Crankshaft Front Oil Seal

Drive Belt Adjusting Bar

No.2 Oil Pan

Gasket

Gasket

Crankshaft Pulley

8.15a Exploded view of the timing chain cover and related components

Drive Belt Idler Pulley

Under Cover

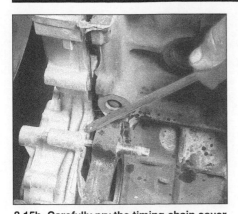

8.15b Carefully pry the timing chain cover off the front of the engine block with a dull screwdriver

8.18a The tensioner plunger must move freely in the bore of the housing

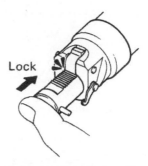

8.18b Release the ratchet pawl and make sure the plunger locks in place

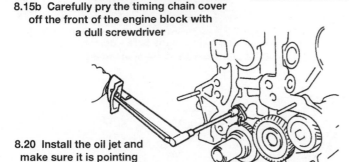

8.20 Install the oil jet and make sure it is pointing toward the timing gear

8   Remove the alternator, the alternator bracket assembly, the belt idler pulley and the adjustment bracket from the engine **(see illustrations)**.
9   Unbolt the cruise control actuator (if equipped) and set it aside.
10   Detach the lower radiator hose from the inlet and remove the two nuts from the bracket. Remove the water pump (see Chapter 3).
11   Position the number one piston at TDC on the compression stroke (see Chapter 2A).
12   Remove the cylinder head (see Section 7).
13   Remove the oil pan (see Section 10).
14   Use a special crankshaft pulley holding tool, or thread a bolt into the pulley and use a long prybar to hold the pulley while loosening the crankshaft pulley bolt **(see illustration)**. The crankshaft pulley should slide off the crankshaft. If necessary, use a bolt-type puller to pull it off **(see illustration)**.
15   Remove the timing chain cover bolts and nuts **(see illustration)**. Carefully pry the timing chain cover with a dull screwdriver being careful not to nick or gouge the aluminum **(see illustration)**.
16   Remove the camshaft sprocket and timing chain.

Inspection

Refer to illustrations 8.18a and 8.18b

17   Inspect the individual sprocket teeth and keyways for wear and damage. Check the chain for cracked plates, pitted or worn rollers, Check the chain guides for wear or damage. Replace any worn or excessively worn or defective parts with new ones. **Caution:** *If excessive plastic material is missing from the chain guides, the oil pan should be removed and cleaned of all debris. Check the oil pick-up tube and screen. Replace the assembly if it is clogged.*
18   Check the timing chain tensioner for proper operation:
   a) *Check that the plunger moves smoothly when the ratchet pawl is raised with your finger* **(see illustration)**.
   b) *Release the ratchet pawl and make sure the plunger is locked in place by the ratchet pawl and does not move when pushed with your finger* **(see illustration)**.
19   Remove the oil jet and inspect the orifice for restrictions.

Installation

Refer to illustrations 8.20, 8.21a, 8.21b and 8.23

20   Remove all dirt, oil and grease from the timing chain area at the front of the engine and the timing chain cover. Clean the sealing surface of the cover and install two new O-rings. Install the oil jet and make sure it is directed toward the timing gear **(see illustration)**.
21   Install the timing chain over the camshaft sprocket aligning the timing mark on the sprocket with the bright link on the chain. Place the timing chain/camshaft sprocket assembly into position on the engine, looping the chain around the crankshaft gear and aligning the other bright link with the mark on the crankshaft sprocket. Tie the timing chain guides together to retain the chain in position **(see illustrations)**.
22   Apply a 3 mm bead of RTV sealant to the timing cover sealing surface. Make sure the O-rings are in place and install the timing chain cover onto the engine block, engaging the splined shaft of the oil pump

8.21a Align the mark on the camshaft sprocket with the bright link on the chain - use a tie strap, rubber band or a piece of cord to secure the timing chain in position

8.21b Align the mark on the crankshaft sprocket with the other bright link - note the crankshaft keyway should be pointing straight down

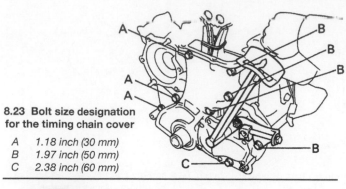

8.23 Bolt size designation for the timing chain cover

A	1.18 inch (30 mm)
B	1.97 inch (50 mm)
C	2.38 inch (60 mm)

9.2b The seal may be removed more easily by carefully cutting the lip as indicated

Cut Position

9.2a Wrap tape around the screwdriver tip and carefully work the crankshaft front oil seal out of the bore - do not nick or scratch the crankshaft in the process

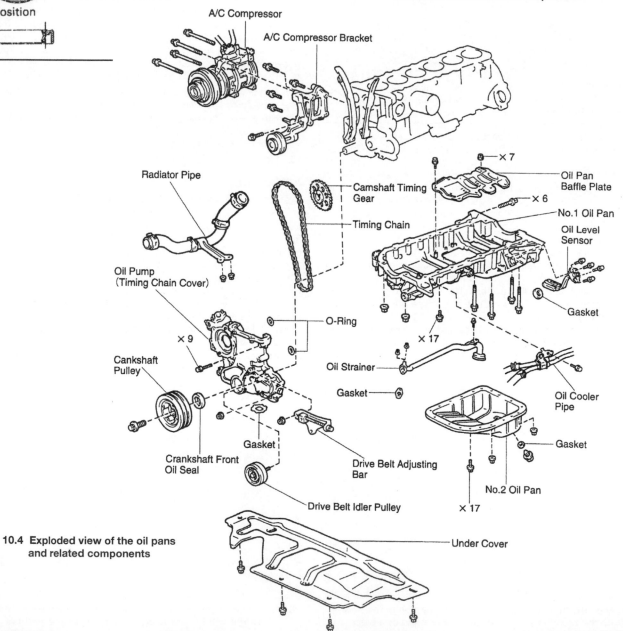

A/C Compressor

A/C Compressor Bracket

Radiator Pipe

Camshaft Timing Gear

× 7

Oil Pan Baffle Plate

Timing Chain

× 6

No.1 Oil Pan

Oil Level Sensor

Oil Pump (Timing Chain Cover)

O-Ring

Gasket

× 9

× 17

Crankshaft Pulley

Oil Strainer

Gasket

Oil Cooler Pipe

Gasket

Crankshaft Front Oil Seal

Gasket

Drive Belt Adjusting Bar

No.2 Oil Pan

Drive Belt Idler Pulley

× 17

10.4 Exploded view of the oil pans and related components

Under Cover

drive rotor with the oil pump drive gear. **Note:** *Installation must be completed within 5 minutes of applying the sealer.*
23   Install the timing cover bolts and tighten them to the torque listed in this Chapter's Specifications **(see illustration)**. Remove the cord securing the timing chain and guides.
24   Reinstall the remaining parts in the reverse order of removal, referring to the appropriate sections for component installation. **Caution:** *DO NOT start the engine until you're absolutely certain that the timing chain is installed correctly. Serious and costly engine damage could occur if the chain is installed incorrectly.*
25   Run the engine and check for proper operation.

9   Crankshaft front oil seal - replacement

Refer to illustrations 9.2a and 9.2b
1   Remove the drivebelts (see Chapter 1). Remove the crankshaft pulley (see Section 8).
2   Note how far the seal is recessed in the bore, then carefully pry it out of the timing chain cover with a screwdriver or seal removal tool **(see illustration)**. Don't scratch the cover bore or damage the crankshaft in the process (if the crankshaft is damaged, the new seal will end up leaking). **Note:** *The seal may be easier to remove if the old seal lip is cut with a sharp utility knife first* **(see illustration)**.
3   Clean the bore in the cover and coat the outer edge of the new seal with engine oil or multi-purpose grease. Apply moly-base grease to the seal lip.
4   Using a socket with an outside diameter slightly smaller than the outside diameter of the seal, carefully drive the new seal into place with a seal driver or large socket. Make sure it's installed squarely and driven in to the same depth as the original. If a socket isn't available, a short section of large diameter pipe will also work. Check the seal after installation to make sure the spring didn't pop out of place.
5   Reinstall the crankshaft pulley and drivebelts.
6   Run the engine and check for oil leaks at the front seal.

10   Oil pan - removal and installation

Removal

Refer to illustrations 10.4, 10.8, 10.9a, 10.9b, 10.10a, 10.10b, 10.11, 10.12, 10.13a and 10.13b
1   Disconnect the negative cable from the battery. **Caution:** *If the stereo in your vehicle is equipped with an anti-theft system, make sure you have the correct activation code before disconnecting the battery.*
2   Set the parking brake and block the rear wheels.
3   Raise the front of the vehicle and support it securely on jackstands.
4   If equipped, remove the engine protector plate under the engine **(see illustration)**.
5   Drain the engine oil and remove the oil filter (see Chapter 1). Remove the oil dipstick.

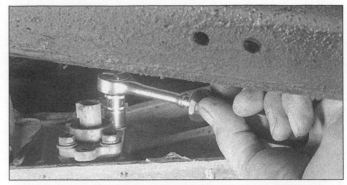

10.8 Remove the oil level sensor from the side of the engine block

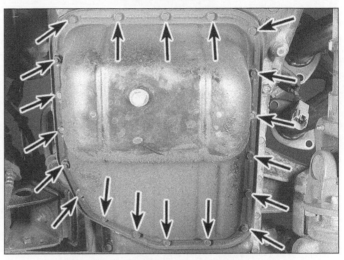

10.9a Remove the bolts and nuts around the perimeter of oil pan number 1

6   Remove the front stabilizer bar (see Chapter 10).
7   Remove the two nuts retaining the front exhaust pipe to the exhaust manifold, then the two bolts/nuts connecting the pipe to the rear of the exhaust system (see Chapter 4). Remove the exhaust assembly.
8   Remove the oil level sensor **(see illustration)**. Inspect the oil level sensor (see Step 18).
9   Remove the bolts around the perimeter of the number 2 oil pan **(see illustration)**, then carefully pry between the steel and aluminum sections to loosen the oil pan **(see illustration)**. Be careful not to gouge the softer aluminum or distort the flange of the steel pan.
10   Remove the bolts around the perimeter of the number 1 oil pan **(see illustrations)**.

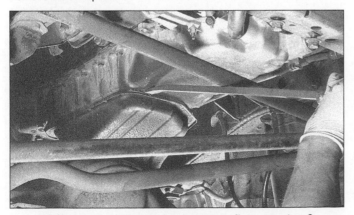

10.9b Insert a large pry bar between oil pan number 2 and the flange on oil pan number 1

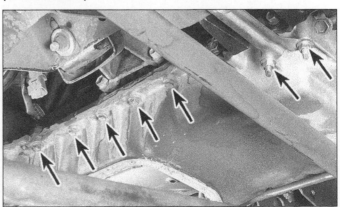

10.10a Remove the mounting bolts on the number 2 oil pan

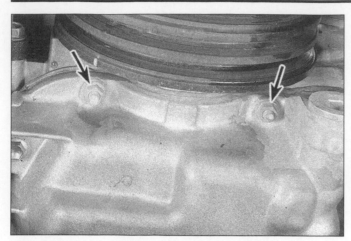

10.10b Don't forget the two forward mounting nuts directly under the front pulley

10.11 Break the gasket seal with a putty knife or flat-bladed screwdriver

11   Carefully pry the number 1 oil pan away from the engine block **(see illustration)**.
12   Angle the oil pan down and away from the front stabilizer bar to make clearance **(see illustration)**.
13   Unbolt the pick-up tube/oil strainer assembly and remove it for cleaning **(see illustrations)**.

Installation

Refer to illustration 10.18

14   Use a scraper to remove all traces of old gasket material and sealant from the block and oil pan. Clean the mating surfaces with lacquer thinner or acetone.
15   Make sure the threaded bolt holes in the block are clean.
16   Check the oil pan flange for distortion, particularly around the bolt holes. If necessary, place the oil pan on a wood block and use a hammer to flatten and restore the gasket surface.
17   Inspect the oil pump pick-up tube assembly for cracks and a blocked strainer. Install the oil pick-up tube.
18   Inspect the oil level sensor assembly for cracks or damaged electrical connections. Replace the gasket **(see illustration)**.
19   Install a new O-ring into the lower section of the timing chain cover. This seals the oil pump galley to the pick-up tube.
20   Apply a 3 mm wide bead of RTV sealant to the oil pan flange on the number 1 oil pan first. **Note:** *Installation must be completed within 5 minutes of applying the sealer.*
21   Carefully position the oil pan on the engine block and install the bolts. Working from the center out. Tighten the bolts to the torque listed in this Chapter's Specifications in three or four steps.

22   Apply a 3 mm wide bead of RTV sealant to the oil pan flange on the number 2 oil pan next. **Note:** *Installation must be completed within 5 minutes of applying the sealer.*
23   Carefully position the number 2 oil pan on the engine block and install the bolts. Working from the center out. Tighten the bolts to the torque listed in this Chapter's Specifications in three or four steps.

10.12 Remove the oil pan from the engine block by angling the assembly down and away from the front suspension

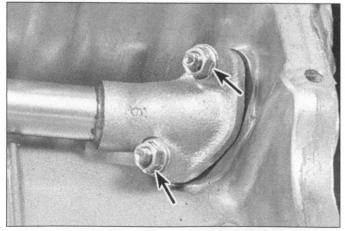

10.13a Remove the oil pick-up tube mounting nuts (arrows) from the number 1 oil pan

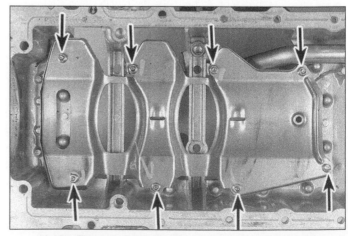

10.13b Remove the baffle plate mounting nuts (arrows)

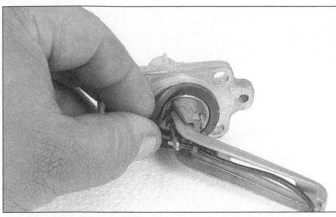

10.18 Remove the gasket from the oil level sensor
assembly and replace it

11.4 Remove the screws and separate the oil pump
assembly from the front cover

24   The remainder of installation is the reverse of removal. Be sure to add oil and install a new oil filter. Use new gasket/seals on the front exhaust pipe.
25   Run the engine and check for oil pressure and leaks.

11   Oil pump - removal, inspection and installation

Removal

Refer to illustrations 11.4 and 11.6
1   Remove the drivebelts and the idler pulley from the front of the engine.
2   Remove the oil pan (see Section 10).
3   Remove the crankshaft pulley and timing chain cover (see Section 7).
4   Remove the seven screws and detach the oil pump body from the timing cover **(see illustration)**.
5   Use a scraper to remove all traces of sealant and old gasket material from the pump body and timing cover, then clean the mating surfaces with lacquer thinner or acetone.
6   Lift out the drive and driven rotors **(see illustration)**.
7   Remove the oil pressure relief valve snap-ring retainer, spring and piston located on the side of the cover. **Warning:** *The spring is tightly compressed - be careful and wear eye protection.*

Inspection

Refer to illustrations 11.10a, 11.10b and 11.10c
8   Clean all components with solvent, inspect them for wear and damage.
9   Check the oil pressure relief valve piston sliding surface and valve spring. If either the spring or the valve is damaged, they must be replaced as a set.

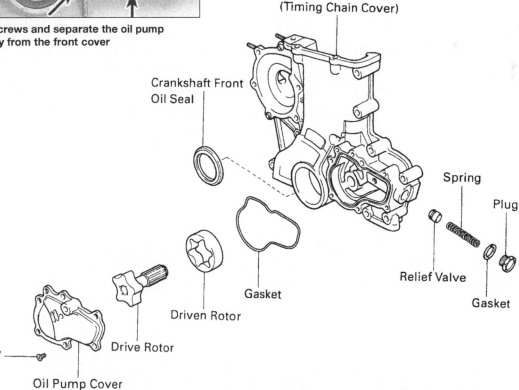

11.6 Exploded view of the oil pump assembly

**11.10a Measure the driven rotor-to-body clearance
with a feeler gauge**

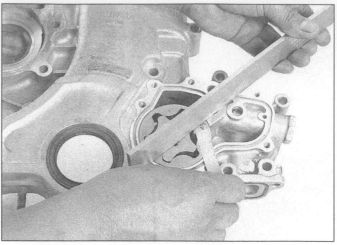

**11.10b Using a straightedge and feeler gauge, measure
the rotor-to-cover clearance**

10   Check the driven rotor-to-body clearance, rotor-to-cover clearance and drive rotor tip clearance with a feeler gauge **(see illustrations)** and compare the results to this Chapter's Specifications. If any clearance is excessive, replace the rotors as a set. If necessary, replace the oil pump body (timing cover).

Installation

Refer to illustration 11.15

11   Lubricate the drive and driven rotors with clean engine oil and place them in the pump body.
12   Pack the pump cavity with petroleum jelly and attach the pump cover, tighten the screws to the torque listed in this Chapter's Specifications.
13   Lubricate the oil pressure relief valve piston with clean engine oil and reinstall the valve components in the pump body.
14   Place a new gasket on the engine block (the dowel pins should hold it in place).
15   Install a new oil passage gasket in the timing cover **(see illustration)**.
16   Install the timing cover onto the engine block, engaging the splined shaft of the oil pump drive rotor with the oil pump drive gear (see Section 8).
17   Reinstall the remaining parts in the reverse order of removal,

referring to the appropriate Sections for component installation.
18   Add oil, start the engine and check for oil pressure and leaks.

---

12   Flywheel/driveplate - removal and installation

Removal

Refer to illustration 12.9

1   Raise the vehicle and support it securely on jackstands, then refer to Chapter 7 and remove the transmission.
2   If equipped, remove the pressure plate and clutch disc (see Chapter 8) (manual transmission equipped vehicles).
3   Use a center punch or paint to make alignment marks on the flywheel/driveplate and crankshaft to ensure correct alignment during reinstallation.
4   Remove the bolts securing the flywheel/driveplate to the crankshaft. If the crankshaft turns, wedge a screwdriver in the ring gear teeth to hold the flywheel/driveplate.
5   Remove the flywheel/driveplate from the crankshaft. Since the flywheel is fairly heavy, be sure to support it while removing the last bolt. Automatic transmission equipped vehicles have spacers on both sides of the driveplate. Keep them with the driveplate. **Warning:** *The ring-gear teeth may be sharp, wear gloves to protect your hands.*

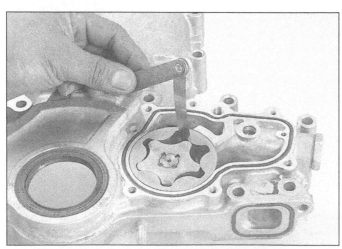

**11.10c Measure the rotor tip clearance with a feeler gauge -
install the rotors with the marks facing out,
against the cover**

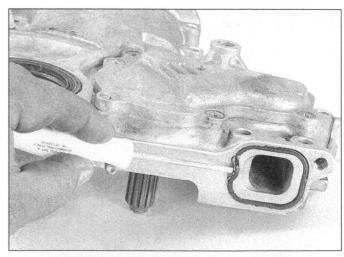

**11.15 Install a new O-ring into the bottom of the
timing chain cover**

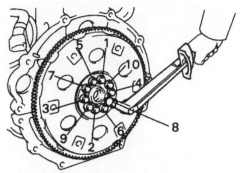

12.9 Follow the tightening sequence when installing the flywheel/driveplate bolts

Installation

6   Clean the flywheel/driveplate to remove grease and oil. Inspect the surface for cracks, rivet grooves, burned areas and score marks. Light scoring can be removed with emery cloth. Check for cracked or broken ring gear teeth. Lay the flywheel/driveplate on a flat surface and use a straightedge to check for warpage.
7   Clean and inspect the mating surfaces of the flywheel/driveplate and the crankshaft. If the crankshaft rear seal is leaking, replace it before reinstalling the flywheel/driveplate (see Section 13).
8   Position the flywheel/driveplate against the crankshaft. Be sure to align the marks made during removal. Note that some engines have an alignment dowel or staggered bolt holes to ensure correct installation. Before installing the bolts, apply thread-locking compound to the threads.
9   Wedge a screwdriver in the ring gear teeth to keep the flywheel/driveplate from turning and tighten the bolts to the torque listed in this Chapter's Specifications. Follow a criss-cross pattern and work up to the final torque in three or four steps (see illustration).
10   The remainder of installation is the reverse of the removal procedure.

13 Rear main oil seal - replacement

Refer to illustrations 13.2, 13.5 and 13.6
1   Remove the transmission (see Chapter 7). Remove the rear end plate.
2   The seal can be replaced without removing the oil pan or seal retainer. However, this method is not recommended because the lip of the seal is quite stiff and it's possible to cock the seal in the retainer bore or damage it during installation. If you want to take the chance, pry out the old seal with a screwdriver (see illustration). Apply multi-purpose grease to the crankshaft seal journal and the lip of the new seal and carefully press the new seal into place. The lip is stiff so carefully work it onto the seal journal of the crankshaft with a smooth object like the end of an extension as you tap the seal into place. Don't rush it

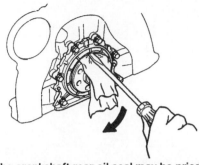

13.2 The crankshaft rear oil seal may be pried out with a screwdriver, lubricate the crankshaft journal and the lip of the new seal with moly-base grease and press the new seal into place - the seal lip is very stiff and can be easily damaged during installation if you're not careful

or you may damage the seal.
3   The following method is recommended but requires removing the seal retainer and resealing the rear of the oil pan (see Section 10).
4   After removing the two rearmost oil pan-to-seal retainer bolts, break the seal between the rear of the oil pan and the bottom of the seal retainer with a putty knife. Remove the retainer-to-engine block bolts, detach the seal retainer and remove all the old gasket material. remove the sealant from the top of the oil pan flange. **Note:** *Cover the open area of the oil pan with clean rags to keep debris out while bracing the pan flange.*
5   Position the seal and retainer assembly between two wood blocks on a workbench and drive the old seal out from the back side with a screwdriver (see illustration).
6   Drive the new seal into the retainer with a wood block (see illustration) or a section of pipe slightly smaller in diameter than the outside diameter of the seal.
7   Lubricate the crankshaft seal journal and the lip of the new seal with multi-purpose grease. Position a new gasket on the engine block. Apply a bead of RTV sealant to the exposed portion of oil pan flange and particularly at the pan-to-block mating surface.
8   Slowly and carefully push the seal and retainer onto the crankshaft. The seal lip is stiff, so work it onto the crankshaft with a smooth object such as the end of an extension as you push the retainer against the block.
9   Install and tighten the retainer bolts to the torque listed in this Chapter's Specifications.
10   The remainder of installation is the reverse of removal.

13.5 After removing the retainer assembly from the engine block, support it between two wooden blocks and drive out the old seal with a screwdriver and hammer

13.6 Drive the new seal into the retainer with a wood block or a section of pipe - make sure that you don't cock the seal in the retainer bore

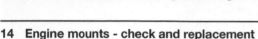

14.4a Location of the right side engine mount

14.4b Location of the left side engine mount

14   Engine mounts - check and replacement

1   Engine mounts seldom require attention, but broken or deteriorated mounts should be replaced immediately or the added strain placed on the driveline components may cause damage or wear.

Check

Refer to illustrations 14.4a and 14.4b

2   During the check, the engine must be raised slightly to remove the weight from the mounts.

3   Raise the vehicle and support it securely on jackstands, then position a jack under the engine oil pan. Place a large wood block between the jack head and the oil pan, then carefully raise the engine just enough to take the weight off the mounts. Do not position the wood block under the drain plug. **Warning:** *DO NOT place any part of your body under the engine when it's supported only by a jack!*

4   Check the mounts **(see illustrations)** to see if the rubber is cracked, hardened or separated from the metal plates. Sometimes the rubber will split down the center.

5   Check for relative movement between the mount plates and the engine or frame (use a large screwdriver or pry bar to attempt to move

the mounts). If movement is noted, lower the engine and tighten the mount fasteners.

6   Rubber preservative should be applied to the mounts to slow deterioration.

Replacement

7   Disconnect the negative battery cable from the battery, then raise the vehicle and support it securely on jackstands (if not already done). Support the engine as described in Step 3. **Caution:** *If the stereo in your vehicle is equipped with an anti-theft system, make sure you have the correct activation code before disconnecting the battery.*

8   To remove the right engine mount, remove the nuts from underneath, one bolt from above, and the upper section will separate from the engine bracket.

9   Remove the mount-to-chassis nuts and detach the mount.

10   To remove the left engine mount, remove the mounting bolt and nut, separate the upper and lower mount.

11   Remove the bolts from the engine block and chassis retaining the insulator

12   Installation is the reverse of removal. Use thread locking compound on the mount bolts/nuts and be sure to tighten them securely.

13   See Chapter 7 for transmission mount replacement.

Chapter 2 Part C
General engine overhaul procedures

Contents

	Section
CHECK ENGINE light .... See Chapter 6	
Crankshaft - inspection..	19
Crankshaft - installation and main bearing oil clearance check	23
Crankshaft - removal..	14
Cylinder compression check...	4
Cylinder head - cleaning and inspection....................................	10
Cylinder head - disassembly..	9
Cylinder head - reassembly	12
Cylinder honing..	17
Engine block - cleaning..	15
Engine block - inspection..	16
Engine overhaul - disassembly sequence....................................	8
Engine overhaul - general information	2
Engine overhaul - reassembly sequence	21
Engine rebuilding alternatives	7

	Section
Engine - removal and installation	6
Engine removal - methods and precautions	5
General information..	1
Initial start-up and break-in after overhaul...............................	26
Main and connecting rod bearings - inspection	20
Oil cooler - removal and installation .... See Chapter 3	
Pistons/connecting rods - inspection	18
Pistons/connecting rods - installation and rod bearing oil clearance check..	25
Piston/connecting rods - removal.......................................	13
Piston rings - installation..	22
Rear main oil seal installation...	24
Vacuum gauge diagnostic checks	3
Valves - servicing	11

Specifications

2F engine

General

Cylinder compression pressure...	150 psi
Maximum variation between cylinders	30 psi
Oil pressure	
At idle (600 rpm)...	4.3 psi
Above 1600 rpm...	37 to 75 psi

Engine block

Cylinder bore diameter (standard)...	3.7008 to 3.7027 inches (94.00 to 94.05 mm)
Maximum allowable taper and out-of-round.........................	0.008 inch (0.2 mm)
Warpage limit..	0.006 inch (0.15 mm)

Cylinder head and valves

Cylinder head warpage limit .... 0.006 inch (0.15 mm)
Valve margin limit
    Intake.. 0.031 inch (0.80 mm)
    Exhaust .... 0.039 inch (1.00 mm)
Valve stem diameter
    Intake.. 0.3138 to 0.3144 inch (7.970 to 7.985 mm)
    Exhaust .... 0.3134 to 0.3140 inch (7.960 to 7.975 mm)
Valve stem-to-guide clearance
    Standard
        Intake .... 0.0012 to 0.0024 inch (0.030 to 0.060 mm)
        Exhaust .... 0.0016 to 0.0028 inch (0.040 to 0.070 mm)
    Limit
        Intake .... 0.0039 inch (0.10 mm)
        Exhaust .... 0.0047 inch (0.12 mm)
Valve spring free length .... 2.028 inches (51.5 mm)
Valve spring installed height... 1.693 inches (43.0 mm)

Crankshaft and connecting rods

Connecting rod journal
    Diameter.. 2.1252 to 2.1260 inches (53.98 to 54.00 mm)
    Bearing oil clearance .... 0.0008 to 0.0024 inch (0.02 to 0.06 mm)
Connecting rod side clearance (endplay)
    Standard.. 0.0043 to 0.0091 inch (0.11 to 0.23 mm)
    Limit.. 0.012 inch (0.30 mm)
Main bearing journal
    Diameter
        No. 1 .... 2.6367 to 2.6376 inches (66.972 to 66.996 mm)
        No. 2 .... 2.6957 to 2.6967 inches (68.472 to 68.496 mm)
        No. 3 .... 2.7548 to 2.7557 inches (69.972 to 69.996 mm)
        No. 4 .... 2.8139 to 2.8148 inches (71.472 to 71.496 mm)
    Bearing oil clearance.. 0.0008 to 0.0017 inches (0.020 to 0.044 mm)
Crankshaft endplay
    Standard.. 0.0024 to 0.0063 inch (0.06 to 0.016 mm)
    Limit.. 0.012 inch (0.30 mm)
Maximum taper and out-of-round .... 0.004 inch (0.1 mm)

Pistons and rings

Piston diameter (standard) .... 3.6996 to 3.7016 inches (93.97 to 94.02 mm)
Piston-to-bore clearance... 0.0012 to 0.0020 inch (0.03 to 0.05 mm)
Piston ring end gap
    Top compression ring .... 0.0079 to 0.0150 inch (0.20 to 0.38 mm)
    Second compression ring .... 0.0079 to 0.0150 inch (0.20 to 0.38 mm)
Piston side clearance
    Top compression ring .... 0.0012 to 0.0024 inch (0.03 to 0.06 mm)
    Second compression ring .... 0.0008 to 0.0024 inch (0.02 to 0.06 mm)

Torque specifications *

Ft-lbs

Connecting rod cap nuts.. 55
Main bearing cap bolts
    No. 1, 2 and 3.. 108
    No. 4... 94

* **Note:** *Refer to Part A for additional torque specifications.*

3F-E engines

General

Cylinder compression pressure... 150 psi
Maximum variation between cylinders .... 30 psi
Oil pressure
    At idle (600 rpm) .... 4.3 psi
    Above 1600 rpm... 37 to 75 psi

Engine block

Maximum warpage .... 0.006 inch (0.15 mm)
Cylinder bore diameter (standard)... 3.7008 to 3.7020 inches (94.000 to 94.030 mm)
Maximum taper and out-of-round .... 0.001 inch (0.025 mm)

Cylinder head and valves

Warpage limit	0.006 inch (0.15 mm)
Valve margin limit	
Intake	0.039 inch (1.0 mm)
Exhaust	0.047 inch (1.2 mm)
Valve stem diameter	
Intake	0.3138 to 0.3144 inch (7.970 to 7.985 mm)
Exhaust	0.3134 to 0.3140 inch (7.960 to 7.975 mm)
Valve stem-to-guide clearance	
Standard	
Intake	0.0010 to 0.0024 inch (0.025 to 0.060 mm)
Exhaust	0.0014 to 0.0028 inch (0.035 to 0.070 mm)
Limit	
Intake	0.0039 inch (0.10 mm)
Exhaust	0.0047 inch (0.12 mm)
Valve spring free length	2.028 inches (51.5 mm)
Valve spring installed height	1.693 inches (43.0 mm)

Crankshaft and connecting rods

Connecting rod journal	
Diameter	2.0861 to 2.0866 inches (52.988 to 53.000 mm)
Bearing oil clearance	0.0008 to 0.0020 inch (0.020 to 0.050 mm)
Connecting rod side clearance (endplay)	
Standard	0.0063 to 0.0118 inch (0.160 to 0.300 mm)
Limit	0.0156 inch (0.40 mm)
Main bearing journal	
Diameter	
No. 1	2.6367 to 2.6376 inches (66.972 to 66.996 mm)
No. 2	2.6957 to 2.6967 inches (68.472 to 68.496 mm)
No. 3	2.7548 to 2.7557 inches (69.972 to 69.996 mm)
No. 4	2.8139 to 2.8148 inches (71.472 to 71.496 mm)
Bearing oil clearance	0.0008 to 0.0017 inch (0.020 to 0.044 mm)
Crankshaft endplay	
Standard	0.0006 to 0.0080 inch (0.015 to 0.204 mm)
Limit	0.0118 inch (0.30 mm)
Maximum taper and out-of-round	0.0008 inch (0.02 mm)

Pistons and rings

Piston diameter (standard)	3.6992 to 3.7004 inches (93.960 to 93.990 mm)
Piston-to-bore clearance	0.0011 to 0.0019 inch (0.027 to 0.047 mm)
Piston ring end gap	
Top compression ring	0.0079 to 0.0150 inch (0.200 to 0.420 mm)
Second compression ring	0.0197 to 0.0283 inch (0.500 to 0.720 mm)
Oil ring	0.0079 to 0.0323 inch (0.200 to 0.820 mm)
Piston side clearance	
Top compression ring	0.0012 to 0.0028 inch (0.030 to 0.070 mm)
Second compression ring	0.0020 to 0.0035 inch (0.050 to 0.090 mm)

Torque specifications *

	Ft-lbs
Connecting rod cap nuts	43
Main bearing cap bolts	
19 mm bolt head	99
17 mm bolt head	85

* **Note:** *Refer to Part A for additional torque specifications.*

1FZ-FE engines

General

Cylinder compression pressure	171 psi
Maximum variation between cylinders	30 psi
Oil pressure	
At idle (600 rpm)	4.3 psi
Above 1600 rpm	37 to 75

Engine block

Maximum warpage	0.0020 inch (0.05 mm)
Cylinder bore diameter (standard)	
Mark 1	3.9370 to 3.9374 inches (100.000 to 100.010 mm)
Mark 2	3.9374 to 3.9378 inches (100.010 to 100.020 mm)
Mark 3	3.9378 to 3.9382 inches (100.020 to 100.030 mm)
Maximum taper and out-of-round	0.001 inch (0.025 mm)

Cylinder head and valves

Warpage limit..	0.006 inch (0.15 mm)
Valve margin limit	
Intake..	0.039 inch (1.0 mm)
Exhaust	0.039 inch (1.0 mm)
Valve stem diameter	
Intake..	0.2744 to 0.2750 inch (6.970 to 6.985 mm)
Exhaust	0.2742 to 0.2748 inch (6.965 to 6.980 mm)
Valve stem-to-guide clearance	
Standard	
Intake	0.0010 to 0.0024 inch (0.025 to 0.060 mm)
Exhaust	0.0012 to 0.0026 inch (0.030 to 0.065 mm)
Limit	
Intake	0.0031 inch (0.08 mm)
Exhaust	0.0039 inch (0.10 mm)
Valve spring free length	1.7299 to 1.7740 inches (43.94 to 45.06 mm)
Valve spring installed height	1.437 inches (36.5 mm)

Crankshaft and connecting rods

Connecting rod journal	
Diameter..	2.2434 to 2.2441 inches (56.982 to 57.000 mm)
Bearing oil clearance..	0.0013 to 0.0020 inch (0.032 to 0.050 mm)
Connecting rod side clearance (endplay)	
Standard..	0.0063 to 0.0103 inch (0.160 to 0.262 mm)
Limit	0.0143 inch (0.362 mm)
Main bearing journal	
Diameter..	2.7158 to 2.7165 inches (68.982 to 69.000 mm)
Bearing oil clearance..	0.0017 to 0.0024 inch (0.042 to 0.060 mm)
Crankshaft endplay	
Standard..	0.0008 to 0.0087 inch (0.020 to 0.022 mm)
Limit	0.0118 inch (0.30 mm)
Maximum taper and out-of-round	0.0008 inch (0.02 mm)

Pistons and rings

Piston diameter (standard)	
Mark 1	3.9350 to 3.9354 inches (99.950 to 99.960 mm)
Mark 2	3.9354 to 3.9358 inches (99.960 to 99.970 mm)
Mark 3	3.9358 to 3.9362 inches (99.970 to 99.980 mm)
Piston-to-bore clearance...	0.0016 to 0.0024 inch (0.040 to 0.060 mm)
Piston ring end gap	
Top compression ring	0.0118 to 0.0205 inch (0.300 to 0.520 mm)
Second compression ring	0.0177 to 0.0264 inch (0.450 to 0.670 mm)
Oil ring	0.0059 to 0.0205 inch (0.150 to 0.520 mm)
Piston side clearance	
Top compression ring	0.0016 to 0.0031 inch (0.040 to 0.080 mm)
Second compression ring	0.0012 to 0.0028 inch (0.030 to 0.070 mm)

Torque specifications *

	Ft-lbs
Crankshaft oil nozzles..	18
Main bearing cap bolts	
Step 1...	54
Step 2...	Tighten an additional 90-degrees
Connecting rod cap nuts	
Step 1...	35
Step 2...	Tighten an additional 90-degrees

* Note: *Refer to Part B for additional torque specifications.*

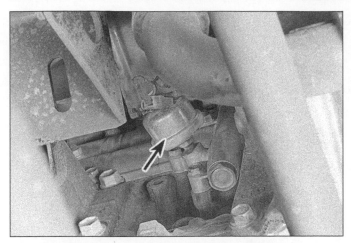

2.4a Remove the oil pressure sending unit (arrow) . . .

2.4b . . . and connect an oil pressure test gauge to
check the oil pressure

1   General information

Included in this portion of Chapter 2 are the general overhaul procedures for the cylinder head(s) and internal engine components.

The information ranges from advice concerning preparation for an overhaul and the purchase of replacement parts to detailed, step-by-step procedures covering removal and installation of internal engine components and the inspection of parts.

The following Sections have been written based on the assumption that the engine has been removed from the vehicle. For information concerning in-vehicle engine repair, as well as removal and installation of the external components necessary for the overhaul, see Part A or B of this Chapter and Section 8 of this Part.

The Specifications included in this Part are only those necessary for the inspection and overhaul procedures which follow. Refer to Parts A and B for additional Specifications.

2   Engine overhaul - general information

Refer to illustrations 2.4a and 2.4b

It's not always easy to determine when, or if, an engine should be completely overhauled, as a number of factors must be considered.

High mileage is not necessarily an indication that an overhaul is needed, while low mileage doesn't preclude the need for an overhaul. Frequency of servicing is probably the most important consideration. An engine that's had regular and frequent oil and filter changes, as well as other required maintenance, will most likely give many thousands of miles of reliable service. Conversely, a neglected engine may require an overhaul very early in its life.

Excessive oil consumption is an indication that piston rings, valve seals and/or valve guides are in need of attention. Make sure that oil leaks aren't responsible before deciding that the rings and/or guides are bad. Perform a cylinder compression check to determine the extent of the work required (see Section 4).

Check the oil pressure with a gauge installed in place of the oil pressure sending unit **(see illustrations)** and compare it to the Specifications. If it's extremely low, the bearings and/or oil pump are probably worn out.

Loss of power, rough running, knocking or metallic engine noises, excessive valve train noise and high fuel consumption rates may also point to the need for an overhaul, especially if they're all present at the same time. If a complete tune-up doesn't remedy the situation, major mechanical work is the only solution.

An engine overhaul involves restoring the internal parts to the specifications of a new engine. During an overhaul, the piston rings are replaced and the cylinder walls are reconditioned (rebored and/or honed). If a rebore is done by an automotive machine shop, new oversize pistons will also be installed. The main bearings, connecting rod bearings and camshaft bearings are generally replaced with new ones and, if necessary, the crankshaft may be reground to restore the journals. Generally, the valves are serviced as well, since they're usually in less than-perfect condition at this point. While the engine is being overhauled, other components, such as the distributor, starter and alternator, can be rebuilt as well. The end result should be a like new engine that will give many thousands of trouble free miles. **Note:** *Critical cooling system components such as the hoses, drivebelts, thermostat and water pump MUST be replaced with new parts when an engine is overhauled. The radiator should be checked carefully to ensure that it isn't clogged or leaking (see Chapter 3). Also, we don't recommend overhauling the oil pump - always install a new one when an engine is rebuilt.*

Before beginning the engine overhaul, read through the entire procedure to familiarize yourself with the scope and requirements of the job. Overhauling an engine isn't difficult if you have the right equipment and follow the instructions carefully, but it is time consuming. Plan on the vehicle being tied up for a minimum of two weeks, especially if parts must be taken to an automotive machine shop for repair or reconditioning. Check on availability of parts and make sure that any necessary special tools and equipment are obtained in advance. Most work can be done with typical hand tools, although a number of precision measuring tools are required for inspecting parts to determine if they must be replaced. Often an automotive machine shop will handle the inspection of parts and offer advice concerning reconditioning and replacement. **Note:** *Always wait until the engine has been completely disassembled and all components, especially the engine block, have been inspected before deciding what service and repair operations must be performed by an automotive machine shop. Since the block's condition will be the major factor to consider when determining whether to overhaul the original engine or buy a rebuilt one, never purchase parts or have machine work done on other components until the block has been thoroughly inspected. As a general rule, time is the primary cost of an overhaul, so it doesn't pay to install worn or substandard parts.*

As a final note, to ensure maximum life and minimum trouble from a rebuilt engine, everything must be assembled with care in a spotlessly clean environment.

3   Vacuum gauge diagnostic checks

Refer to illustration 3.4

A vacuum gauge provides valuable information about what is going on in the engine at a low-cost. You can check for worn rings or cylinder walls, leaking head or intake manifold gaskets, incorrect carburetor adjustments, restricted exhaust, stuck or burned valves, weak valve springs, improper ignition or valve timing and ignition problems.

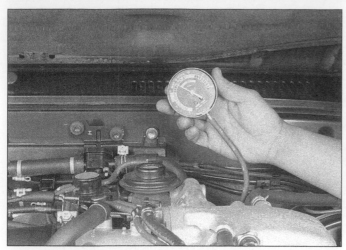

3.4 **Use a vacuum gauge attached to a manifold vacuum port**

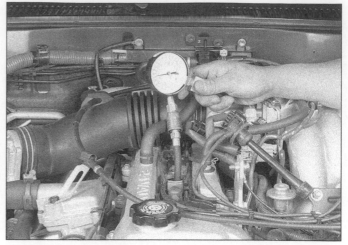

4.6 **A compression gauge with a threaded fitting for the spark plug hole is preferred over the type that requires hand pressure to maintain the seal**

Unfortunately, vacuum gauge readings are easy to misinterpret, so they should be used in conjunction with other tests to confirm the diagnosis.

Both the absolute readings and the rate of needle movement are important for accurate interpretation. Most gauges measure vacuum in inches of mercury (in-Hg). The following references to vacuum assume the diagnosis is being performed at sea level. As elevation increases (or atmospheric pressure decreases), the reading will decrease. For every 1,000 foot increase in elevation above approximately 2,000 feet, the gauge readings will decrease about one inch of mercury.

Connect the vacuum gauge directly to intake manifold vacuum, not to ported (above the throttle plate) vacuum **(see illustration)**. Be sure no hoses are left disconnected during the test or false readings will result.

Before you begin the test, allow the engine to warm up completely. Block the wheels and set the parking brake. With the transmission in neutral (or Park, on automatics), start the engine and allow it to run at normal idle speed. **Warning:** *Carefully inspect the fan blades for cracks or damage before starting the engine. Keep your hands and the vacuum tester clear of the fan and do not stand in front of the vehicle or in line with the fan when the engine is running.*

Read the vacuum gauge; an average, healthy engine should normally produce between 17 and 22 in-Hg of vacuum with a fairly steady needle.

Refer to the following vacuum gauge readings and what they indicate about the engines condition:

1   A low steady reading usually indicates a leaking gasket between the intake manifold and carburetor or throttle body, a leaky vacuum hose, late ignition timing or incorrect camshaft timing. Check ignition timing with a timing light and eliminate all other possible causes, utilizing the tests provided in this Chapter before you remove the timing belt cover to check the timing marks.

2   If the reading is three to eight inches below normal and it fluctuates at that low reading, suspect an intake manifold gasket leak at an intake port or a faulty injector.

3   If the needle has regular drops of about two to four inches at a steady rate the valves are probably leaking. Perform a compression or leak-down test to confirm this.

4   An irregular drop or down-flick of the needle can be caused by a sticking valve or an ignition misfire. Perform a compression or leak-down test and read the spark plugs.

5   A rapid vibration of about four in-Hg vibration at idle combined with exhaust smoke indicates worn valve guides. Perform a leak-down test to confirm this. If the rapid vibration occurs with an increase in engine speed, check for a leaking intake manifold gasket or head gasket, weak valve springs, burned valves or ignition misfire.

6   A slight fluctuation, say one inch up and down, may mean ignition problems. Check all the usual tune-up items and, if necessary, run the

engine on an ignition analyzer.

7   If there is a large fluctuation, perform a compression or leak-down test to look for a weak or dead cylinder or a blown head gasket.

8   If the needle moves slowly through a wide range, check for a clogged PCV system, incorrect idle fuel mixture, throttle body or intake manifold gasket leaks.

9   Check for a slow return after revving the engine by quickly snapping the throttle open until the engine reaches about 2,500 rpm and let it shut. Normally the reading should drop to near zero, rise above normal idle reading (about 5 in-Hg over) and then return to the previous idle reading. If the vacuum returns slowly and doesn't peak when the throttle is snapped shut, the rings may be worn. If there is a long delay, look for a restricted exhaust system (often the muffler or catalytic converter). An easy way to check this is to temporarily disconnect the exhaust ahead of the suspected part and redo the test.

4   Cylinder compression check

Refer to illustration 4.6

1   A compression check will tell you what mechanical condition the upper end (pistons, rings, valves, head gaskets) of your engine is in. Specifically, it can tell you if the compression is down due to leakage caused by worn piston rings, defective valves and seats or a blown head gasket. **Note:** *The engine must be at normal operating temperature and the battery must be fully charged for this check. Also, if the engine is equipped with a carburetor, the choke valve must be all the way open to get an accurate compression reading (if the engine's warm, the choke should be open).*

2   Begin by cleaning the area around the spark plugs before you remove them (compressed air should be used, if available, otherwise a small brush or even a bicycle tire pump will work). The idea is to prevent dirt from getting into the cylinders as the compression check is being done.

3   Remove all of the spark plugs from the engine (see Chapter 1).

4   Block the throttle wide open.

5   Detach the coil wire from the center of the distributor cap and ground it on the engine block. Use a jumper wire with alligator clips on each end to ensure a good ground. On fuel-injected models, the fuel pump circuit should also be disabled (see Chapter 4B).

6   Install the compression gauge in the number one spark plug hole **(see illustration)**.

7   Crank the engine over at least seven compression strokes and watch the gauge. The compression should build up quickly in a healthy engine. Low compression on the first stroke, followed by gradually increasing pressure on successive strokes, indicates worn piston

rings. A low compression reading on the first stroke, which doesn't build up during successive strokes, indicates leaking valves or a blown head gasket (a cracked head could also be the cause). Deposits on the undersides of the valve heads can also cause low compression. Record the highest gauge reading obtained.

8   Repeat the procedure for the remaining cylinders and compare the results to the Specifications.

9   Add some engine oil (about three squirts from a plunger-type oil can) to each cylinder, through the spark plug hole, and repeat the test.

10   If the compression increases after the oil is added, the piston rings are definitely worn. If the compression doesn't increase significantly, the leakage is occurring at the valves or head gasket. Leakage past the valves may be caused by burned valve seats and/or faces or warped, cracked or bent valves.

11   If two adjacent cylinders have equally low compression, there's a strong possibility that the head gasket between them is blown. The appearance of coolant in the combustion chambers or the crankcase would verify this condition.

12   If one cylinder is about 20-percent lower than the others, and the engine has a slightly rough idle, a worn exhaust lobe on the camshaft could be the cause.

13   If the compression is unusually high, the combustion chambers are probably coated with carbon deposits. If that's the case, the cylinder head should be removed and decarbonized.

14   If compression is way down or varies greatly between cylinders, it would be a good idea to have a leak-down test performed by an automotive repair shop. This test will pinpoint exactly where the leakage is occurring and how severe it is.

5   Engine - removal methods and precautions

If you've decided that an engine must be removed for overhaul or major repair work, several preliminary steps should be taken.

Locating a suitable place to work is extremely important. Adequate work space, along with storage space for the vehicle, will be needed. If a shop or garage isn't available, at the very least a flat, level, clean work surface made of concrete or asphalt is required.

Cleaning the engine compartment and engine before beginning the removal procedure will help keep your tools and your hands clean.

engine hoist or A-frame will also be necessary. Make sure the equipment is rated in excess of the combined weight of the engine and accessories. Safety is of primary importance, considering the potential hazards involved in lifting the engine out of the vehicle.

If the engine is being removed by a novice, a helper should be available. Advice and aid from someone more experienced would also be helpful. There are many instances when one person cannot simultaneously perform all of the operations required when lifting the engine out of the vehicle.

Plan the operation ahead of time. Arrange for or obtain all of the tools and equipment you'll need prior to beginning the job. Some of the equipment necessary to perform engine removal and installation safely and with relative ease are (in addition to an engine hoist) a heavy duty floor jack, complete sets of wrenches and sockets as described in the front of this manual, wooden blocks and plenty of rags and cleaning solvent for mopping up spilled oil, coolant and gasoline. If the hoist must be rented, make sure that you arrange for it in advance and perform all of the operations possible without it beforehand. This will save you money and time.

Plan for the vehicle to be out of use for quite a while. A machine shop will be required to perform some of the work which the do-it-yourselfer can't accomplish without special equipment. These shops often have a busy schedule, so it would be a good idea to consult them before removing the engine in order to accurately estimate the amount of time required to rebuild or repair components that may need work.

Always be extremely careful when removing and installing the engine. Serious injury can result from careless actions. Plan ahead, take your time and a job of this nature, although major, can be accomplished successfully.

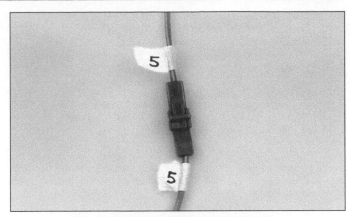

6.5 Label both ends of each wire before unplugging the connector

6   Engine - removal and installation

Warning 1: *Gasoline is extremely flammable, so take extra precautions when you work on any part of the fuel system. Don't smoke or allow open flames or bare light bulbs near the work area, and don't work in a garage where a natural gas-type appliance (such as a water heater or a clothes dryer) with a pilot light is present. Since gasoline is carcinogenic, wear latex gloves when there's a possibility of being exposed to fuel, and, if you spill any fuel on your skin, rinse it off immediately with soap and water. Mop up any spills immediately and do not store fuel-soaked rags where they could ignite. On fuel injected models, the fuel system is under constant pressure, so, if any fuel lines are to be disconnected, the fuel pressure in the system must be relieved first. When you perform any kind of work on the fuel system, wear safety glasses and have a Class B type fire extinguisher on hand.*

Warning 2: *The air conditioning system is under high pressure. DO NOT loosen any fittings or remove any components until after the system has been discharged. Air conditioning refrigerant should be properly discharged into an EPA-approved container at a dealer service department or an automotive air conditioning repair facility. Always wear eye protection when disconnecting air conditioning system fittings.*

Removal

Refer to illustration 6.5

1   Refer to Chapter 4B and relieve the fuel system pressure (fuel-injected models only), then disconnect the negative cable from the battery.

2   Cover the fenders and cowl (see Chapter 11). Special pads are available to protect the fenders, but an old bedspread or blanket will also work.

3   Remove the air cleaner assembly (see Chapter 4A or 4B).

4   Drain the cooling system (see Chapter 1).

5   Label the vacuum lines, emissions system hoses, wiring connectors, ground strap and fuel lines, to ensure correct reinstallation **(see illustration)**, then detach them. If there's any possibility of confusion, make a sketch of the engine compartment and clearly label the lines, hoses and wires.

6   Label and detach all coolant hoses from the engine.

7   Remove the cooling fan, shroud and radiator (see Chapter 3).

8   Remove the drivebelt(s) (see Chapter 1).

9   Disconnect the fuel lines running from the engine to the chassis (see Chapter 4). Plug or cap all open fittings/lines.

10   Disconnect the throttle linkage (and TV linkage/cruise control cable, if equipped) from the engine (see Chapter 4A).

11   On power steering equipped vehicles, unbolt the power steering pump (see Chapter 10). Leave the lines/hoses attached and make sure the pump is kept in an upright position in the engine compartment (use wire or rope to restrain it out of the way).

12   On air conditioned vehicles, unbolt the compressor (see Chapter 3) and set it aside. Do not disconnect the hoses.

13   Drain the engine oil (see Chapter 1) and remove the oil filter.
14   Remove the starter motor (see Chapter 5).
15   Remove the alternator (see Chapter 5).
16   Unbolt the exhaust system from the engine (see Chapter 4).
17   If you're working on a vehicle with an automatic transmission, refer to Chapter 7, Part B and remove the torque converter-to-drive-plate bolts.
18   Support the transmission with a jack. Position a block of wood between the jack and transmission to prevent damage to the transmission. Special transmission jacks with safety chains are available - use one if possible.
19   Attach an engine sling or a length of chain to the lifting brackets on the engine.
20   Roll the hoist into position and connect the sling to it. Take up the slack in the sling or chain, but don't lift the engine. **Warning:** *DO NOT place any part of your body under the engine when it's supported only by a hoist or other lifting device.*
21   Remove the transmission-to-engine block bolts.
22   Remove the engine mount-to-frame bolts.
23   Recheck to be sure nothing is still connecting the engine to the transmission or vehicle. Disconnect anything still remaining.
24   Raise the engine slightly. Carefully work it forward to separate it from the transmission. If you're working on a vehicle with an automatic transmission, be sure the torque converter stays in the transmission (clamp a pair of vise-grips to the housing to keep the converter from sliding out). If you're working on a vehicle with a manual transmission, the input shaft must be completely disengaged from the clutch. Slowly raise the engine out of the engine compartment. Check carefully to make sure nothing is hanging up.
25   Remove the flywheel/driveplate and mount the engine on an engine stand.

Installation

26   Check the engine and transmission mounts. If they're worn or damaged, replace them.
27   If you're working on a vehicle with a manual transmission, install the clutch and pressure plate (see Chapter 8). Now is a good time to install a new clutch.
28   Carefully lower the engine into the engine compartment - make sure the engine mounts line up.
29   If you're working on a vehicle with an automatic transmission, guide the torque converter into the crankshaft following the procedure outlined in Chapter 7, Part B.
30   If you're working on a vehicle with a manual transmission, apply a dab of high-temperature grease to the input shaft and guide it into the crankshaft pilot bearing until the bellhousing is flush with the engine block.
31   Install the transmission-to-engine bolts and tighten them securely. **Caution:** *DO NOT use the bolts to force the transmission and engine together!*
32   Reinstall the remaining components in the reverse order of removal.
33   Add coolant, oil, power steering and transmission fluid as needed.
34   Run the engine and check for leaks and proper operation of all accessories, then install the hood and test drive the vehicle.

7   Engine rebuilding alternatives

The do-it-yourselfer is faced with a number of options when performing an engine overhaul. The decision to replace the engine block, piston/connecting rod assemblies and crankshaft depends on a number of factors, with the number one consideration being the condition of the block. Other considerations are cost, access to machine shop facilities, parts availability, time required to complete the project and the extent of prior mechanical experience on the part of the do-it-yourselfer.

Some of the rebuilding alternatives include:

Individual parts - If the inspection procedures reveal that the engine block and most engine components are in reusable condition, purchasing individual parts may be the most economical alternative. The block, crankshaft and piston/connecting rod assemblies should all

be inspected carefully. Even if the block shows little wear, the cylinder bores should be surface honed.

Short block - A short block consists of an engine block with a crankshaft, camshaft and piston/connecting rod assemblies already installed. All new bearings are incorporated and all clearances will be correct. The existing valve train components, cylinder head and external parts can be bolted to the short block with little or no machine shop work necessary.

Long block - A long block consists of a short block plus an oil pump, oil pan, cylinder head, valve cover and valve train components, timing sprockets and chain or gears and timing cover. All components are installed with new bearings, seals and gaskets incorporated throughout. The installation of manifolds and external parts is all that's necessary.

Give careful thought to which alternative is best for you and discuss the situation with local automotive machine shops, auto parts dealers and experienced rebuilders before ordering or purchasing replacement parts.

8   Engine overhaul - disassembly sequence

Refer to illustrations 8.5a and 8.5b

1   It's much easier to disassemble and work on the engine if it's mounted on a portable engine stand. A stand can often be rented quite cheaply from an equipment rental yard. Before the engine is mounted on a stand, the flywheel/driveplate should be removed from the engine.
2   If a stand isn't available, it's possible to disassemble the engine with it blocked up on the floor. Be extra careful not to tip or drop the engine when working without a stand.
3   If you're going to obtain a rebuilt engine, all external components must come off first, to be transferred to the replacement engine, just as they will if you're doing a complete engine overhaul yourself. These include:

> *Alternator and brackets*
> *Emissions control components*
> *Distributor, spark plug wires and spark plugs*
> *Thermostat and housing cover*
> *Water pump*
> *Fuel injection components or carburetor*
> *Intake/exhaust manifolds*
> *Oil filter*
> *Engine mounts*
> *Clutch and flywheel/driveplate*
> *Engine rear plate*

Note: *When removing the external components from the engine, pay close attention to details that may be helpful or important during installation. Note the installed position of gaskets, seals, spacers, pins, brackets, washers, bolts and other small items.*

4   If you're obtaining a short block, which consists of the engine block, crankshaft, pistons and connecting rods all assembled, then the cylinder head(s), oil pan and oil pump will have to be removed as well. See *Engine rebuilding alternatives* for additional information regarding the different possibilities to be considered.
5   If you're planning a complete overhaul, the engine must be disassembled and the internal components **(see illustrations)** removed in the following order:

> *Valve cover*
> *Intake and exhaust manifolds*
> *Rocker arms and pushrods (2F and 3F-E engines)*
> *Camshafts (1FZ-FE engines)*
> *Cylinder head*
> *Valve lifters (2F and 3F-E engines)*
> *Timing cover and timing gears (2F and 3F-E engines)*
> *Timing chain and sprockets (1FZ-FE engines)*
> *Camshaft (2F and 3F-E engines)*
> *Oil pan*
> *Oil pump*
> *Piston/connecting rod assemblies*
> *Crankshaft and main bearings*

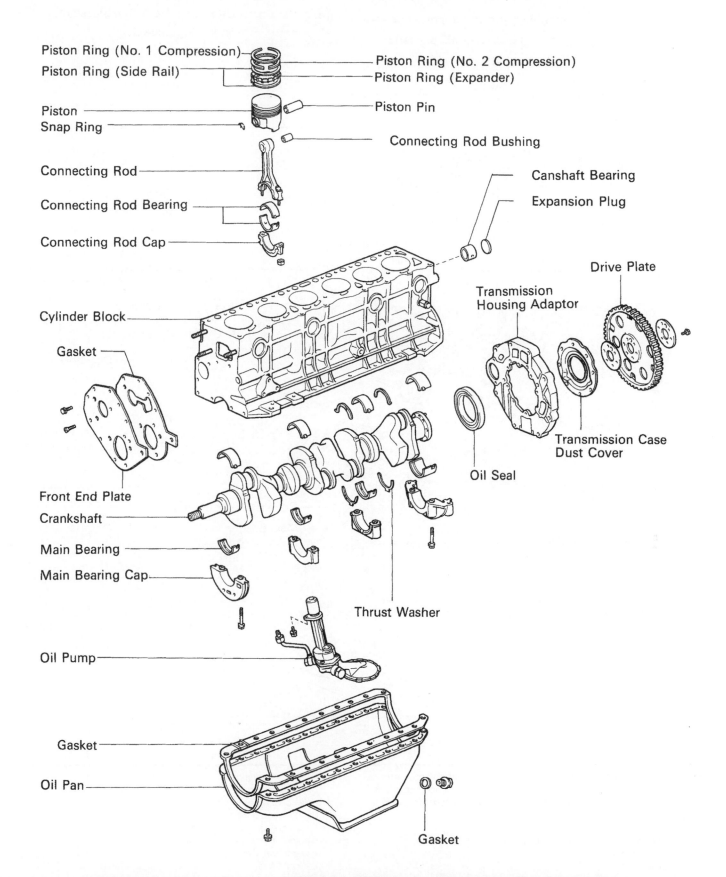

8.5a Internal engine components - exploded view (3F-E inline six-cylinder engine, 2F engine similar)

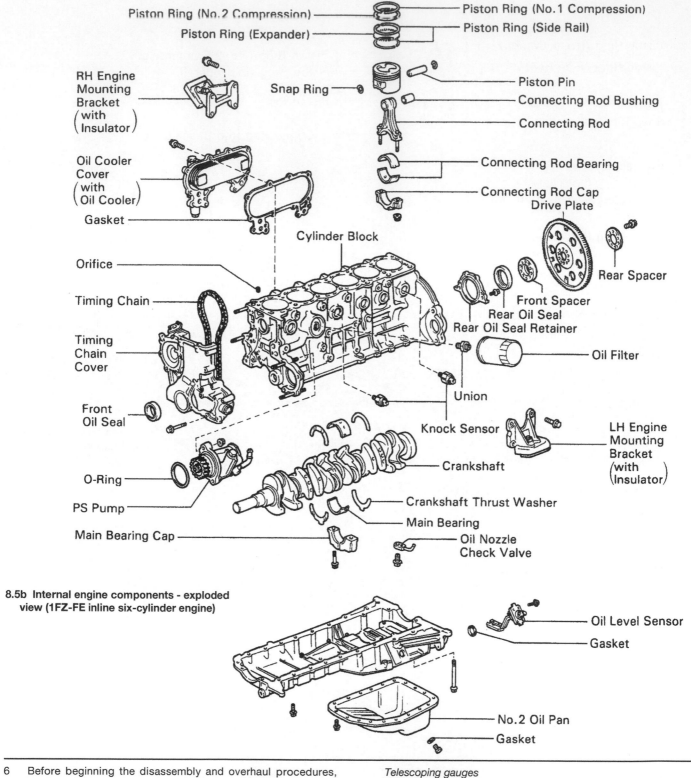

8.5b Internal engine components - exploded view (1FZ-FE inline six-cylinder engine)

6   Before beginning the disassembly and overhaul procedures, make sure the following items are available. Also, refer to *Engine overhaul - reassembly sequence* for a list of tools and materials needed for engine reassembly.

     Common hand tools
     Small cardboard boxes or plastic bags for storing parts
     Gasket scraper
     Ridge reamer
     Vibration damper puller
     Micrometers

     Telescoping gauges
     Dial indicator set
     Valve spring compressor
     Cylinder surfacing hone
     Piston ring groove cleaning tool
     Electric drill motor
     Tap and die set
     Wire brushes
     Oil gallery brushes
     Cleaning solvent

9.2 A small plastic bag, with an appropriate label, can be used to store the valve train components so they can be kept together and reinstalled in the original location

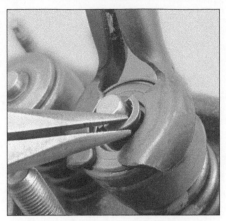

9.3 Use a valve spring compressor to compress the spring, then remove the keepers from the valve stem

9.4 If the valve won't pull through the guide, deburr the edge of the stem end and the area around the top of the keeper groove with a file or whetstone

9 Cylinder head - disassembly

Refer to illustrations 9.2, 9.3 and 9.4
Note: *New and rebuilt cylinder heads are commonly available for most engines at dealerships and auto parts stores. Due to the fact that some specialized tools are necessary for the disassembly and inspection procedures, and replacement parts may not be readily available, it may be more practical and economical for the home mechanic to purchase a replacement head rather than taking the time to disassemble, inspect and recondition the original.*

1   Cylinder head disassembly involves removal of the intake and exhaust valves and related components. If they're still in place, remove the rocker arm bolts or nuts, pivot and rocker arms from the cylinder head studs. Label the parts or store them separately so they can be reinstalled in their original locations.
2   Before the valves are removed, arrange to label and store them, along with their related components, so they can be kept separate and reinstalled in the same valve guides they were removed from **(see illustration)**.
3   Compress the springs on the first valve with a spring compressor and remove the keepers **(see illustration)**. Carefully release the valve spring compressor and remove the retainer, the spring and the spring seat (if used).
4   Pull the valve out of the head, then remove the oil seal from the guide. If the valve binds in the guide (won't pull through), push it back into the head and deburr the area around the keeper groove with a fine file or whetstone **(see illustration)**.
5   Repeat the procedure for the remaining valves. Remember to keep all the parts for each valve together so they can be reinstalled in the same locations.
6   Once the valves and related components have been removed and stored in an organized manner, the head should be thoroughly cleaned and inspected. If a complete engine overhaul is being done, finish the engine disassembly procedures before beginning the cylinder head cleaning and inspection process.

10 Cylinder head - cleaning and inspection

1   Thorough cleaning of the cylinder head and related valve train components, followed by a detailed inspection, will enable you to decide how much valve service work must be done during the engine overhaul. **Note:** *If the engine was severely overheated, the cylinder head is probably warped* (see Step 12).

Cleaning

2   Scrape all traces of old gasket material and sealing compound off the head gasket, intake manifold and exhaust manifold sealing surfaces. Be very careful not to gouge the cylinder head. Special gasket removal solvents that soften gaskets and make removal much easier are available at auto parts stores.
3   Remove all built up scale from the coolant passages.
4   Run a stiff wire brush through the various holes to remove deposits that may have formed in them.
5   Run an appropriate size tap into each of the threaded holes to remove corrosion and thread sealant that may be present. If compressed air is available, use it to clear the holes of debris produced by this operation. **Warning:** *Wear eye protection when using compressed air!*
6   Clean the rocker arm pivot bolt or stud threads with a wire brush.
7   Clean the cylinder head with solvent and dry it thoroughly. Compressed air will speed the drying process and ensure that all holes and recessed areas are clean. **Note:** *Decarbonizing chemicals are available and may prove very useful when cleaning cylinder heads and valve train components. They are very caustic and should be used with caution. Be sure to follow the instructions on the container.*
8   Clean the rocker arms, pivot balls or fulcrums, nuts or bolts and pushrods with solvent and dry them thoroughly (don't mix them up during the cleaning process). Compressed air will speed the drying process and can be used to clean out the oil passages.
9   Clean all the valve springs, spring seats, keepers and retainers (or rotators) with solvent and dry them thoroughly. Do the components from one valve at a time to avoid mixing up the parts.
10   Scrape off any heavy deposits that may have formed on the valves, then use a motorized wire brush to remove deposits from the valve heads and stems. Again, make sure the valves don't get mixed up.

Inspection

Note: *Be sure to perform all of the following inspection procedures before concluding that machine shop work is required. Make a list of the items that need attention.*

Cylinder head
Refer to illustrations 10.12 and 10.14
11   Inspect the head very carefully for cracks, evidence of coolant leakage and other damage. If cracks are found, check with an automotive machine shop concerning repair. If repair isn't possible, a new cylinder head should be obtained.

10.12 Check the cylinder head gasket surface for warpage by trying to slip a feeler gauge under the straightedge (see this Chapter's Specifications for the maximum warpage allowed and use a feeler gauge of that thickness)

10.14 A dial indicator can be used to determine the valve stem-to-guide clearance (move the valve stem as indicted by the arrows)

12   Using a straightedge and feeler gauge, check the head gasket mating surface for warpage (see illustration). If the warpage exceeds the specified limit, it can be resurfaced at an automotive machine shop.

13   Examine the valve seats in each of the combustion chambers. If they're pitted, cracked or burned, the head will require valve service that's beyond the scope of the home mechanic.

14   Check the valve stem-to-guide clearance by measuring the lateral movement of the valve stem with a dial indicator attached securely to the head (see illustration). The valve must be in the guide and approximately 1/16-inch off the seat. The total valve stem movement indicated by the gauge needle must be divided by two to obtain the actual clearance. After this is done, if there's still some doubt regarding the condition of the valve guides they should be checked by an automotive machine shop (the cost should be minimal).

Valves

Refer to illustrations 10.15a, 10.15b and 10.16

15   Carefully inspect each valve for uneven wear, deformation, cracks, pits and burned areas (see illustrations). Check the valve stem for scuffing and galling and the neck for cracks. Rotate the valve and check for any obvious indication that it's bent. Look for pits and excessive wear on the end of the stem. The presence of any of these

conditions indicates the need for valve service by an automotive machine shop.

16   Measure the margin width on each valve (see illustration). Any valve with a margin narrower than specified will have to be replaced with a new one.

Valve components

Refer to illustrations 10.17 and 10.18

17   Check each valve spring for wear (on the ends) and pits. Measure the free length and compare it to the Specifications (see illustration). Any springs that are shorter than specified have sagged and should not be reused. The tension of all springs should be checked with a special fixture before deciding that they're suitable for use in a rebuilt engine (take the springs to an automotive machine shop for this check).

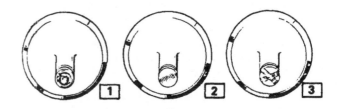

10.15b Valve stem tip wear patterns

1   *Property tip pattern (rotator functioning properly)*
2   *No rotation pattern (replace rotator and check rotation)*
3   *Partial rotation pattern (replace rotator and check rotation)*

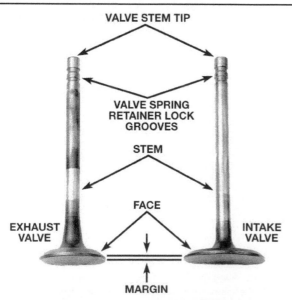

VALVE STEM TIP

VALVE SPRING RETAINER LOCK GROOVES

STEM

FACE

EXHAUST VALVE

INTAKE VALVE

MARGIN

10.15a Check for valve wear at the points shown here

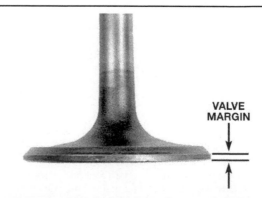

VALVE MARGIN

10.16 The margin width on each valve must be as specified (if no margin exists, the valve cannot be reused)

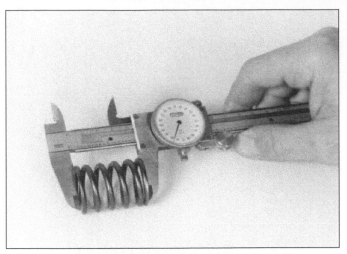

10.17 Measure the free length of each valve spring with a dial or vernier caliper

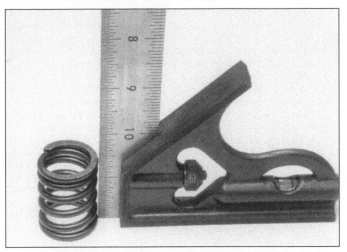

10.18 Check each valve spring for squareness

18   Stand each spring on a flat surface and check it for squareness **(see illustration)**. If any of the springs are distorted or sagged, replace all of them with new parts.

19   Check the spring retainers and keepers for obvious wear and cracks. Any questionable parts should be replaced with new ones, as extensive damage will occur if they fail during engine operation.

Rocker arm components

20   Check the rocker arm faces (the areas that contact the pushrod ends and valve stems) for pits, wear, galling, score marks and rough spots. Check the rocker arm pivot contact areas and pivot balls or fulcrums as well. Look for cracks in each rocker arm and nut or bolt.

21   Inspect the pushrod ends for scuffing and excessive wear. Roll each pushrod on a flat surface, like a piece of plate glass, to determine if it's bent.

22   Check the rocker arm bolt holes or studs in the cylinder heads for damaged threads and secure installation.

23   Any damaged or excessively worn parts must be replaced with new ones.

24   If the inspection process indicates that the valve components are in generally poor condition and worn beyond the limits specified, which is usually the case in an engine that's being overhauled, reassemble the valves in the cylinder head and refer to Section 11 for valve servicing recommendations.

11 Valves - servicing

1   Because of the complex nature of the job and the special tools and equipment needed, servicing of the valves, the valve seats and the valve guides, commonly known as a valve job, should be done by a professional.

2   The home mechanic can remove and disassemble the head, do the initial cleaning and inspection, then reassemble and deliver it to a dealer service department or an automotive machine shop for the actual service work. Doing the inspection will enable you to see what condition the head and valvetrain components are in and will ensure that you know what work and new parts are required when dealing with an automotive machine shop.

3   The dealer service department, or automotive machine shop, will remove the valves and springs, recondition or replace the valves and valve seats, recondition the valve guides, check and replace the valve springs, spring retainers or rotators and keepers (as necessary), replace the valve seals with new ones, reassemble the valve components and make sure the installed spring height is correct. The cylinder head gasket surface will also be resurfaced if it's warped.

4   After the valve job has been performed by a professional, the head will be in like-new condition. When the head is returned, be sure

to clean it again before installation on the engine to remove any metal particles and abrasive grit that may still be present from the valve service or head resurfacing operations. Use compressed air, if available, to blow out all the oil holes and passages.

12 Cylinder head - reassembly

Refer to illustrations 12.3, 12.5a, 12.5b, 12.6 and 12.8

1   Regardless of whether or not a head was sent to an automotive repair shop for valve servicing, make sure it's clean before beginning reassembly.

2   If a head was sent out for valve servicing, the valves and related components will already be in place. Begin the reassembly procedure with Step 8.

3   Install new seals on each of the intake valve guides. Using a hammer and a deep socket or seal installation tool, gently tap each seal into place until it's completely seated on the guide **(see illustration)**. Don't twist or cock the seals during installation or they won't seal properly on the valve stems. The umbrella-type seals (if used) are installed over the valves after the valves are in place.

4   Beginning at one end of the head, lubricate and install the first valve. Apply moly-base grease or clean engine oil to the valve stem.

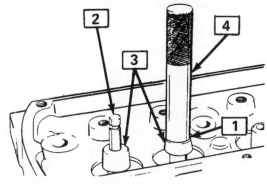

12.3 Make sure the new valve stem seals are seated against the tops of the valve guides

1   Valve seated in tool
2   End of valve stem - be sure to deburr this area before installing the seal
3   Seal
4   Valve seal installation tool (if you don't have this tool, a deep socket also will work)

5   Place the spring seat or shim(s) over the valve guide and set the valve springs, shield and retainer in place **(see illustrations)**. **Caution:** *The valve springs on 3F-E engines must be installed using the correct color code. Refer to Chapter 2A for the procedure and illustrations.*

6   Compress the springs with a valve spring compressor. Position the keepers in the upper groove, then slowly release the compressor and make sure the keepers seat properly. Add a small dab of grease to each keeper to hold it in place if necessary **(see illustration)**.

7   Repeat the procedure for the remaining valves. Be sure to return the components to their original locations - don't mix them up!

8   Check the installed valve spring height with a ruler graduated in 1/32-inch increments or a dial caliper. If the head was sent out for service work, the installed height should be correct (but don't automatically assume that it is). The measurement is taken from the top of each spring seat or shim(s) to the top of the oil shield (or the bottom of the retainer, the two points are the same) **(see illustration)**. If the height is greater than specified, shims can be added under the springs to correct it. **Caution:** *Don't, under any circumstances, shim the springs to the point where the installed height is less than specified.*

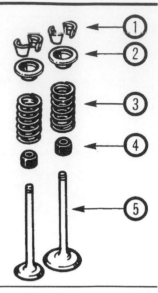

12.5a Valves and related components - exploded view

1   *Keepers*
2   *Retainers*
3   *Springs*
4   *Umbrella-type oil seals*
5   *Valves*

13   Pistons/connecting rods - removal

Refer to illustrations 13.1, 13.3 and 13.6

Note: *Prior to removing the piston/connecting rod assemblies, remove the cylinder head, the oil pan and the oil pump by referring to the appropriate Sections in Chapter 2.*

1   Use your fingernail to feel if a ridge has formed at the upper limit of ring travel (about 1/4-inch down from the top of each cylinder). If carbon deposits or cylinder wear have produced ridges, they must be completely removed with a special tool **(see illustration)**. Follow the manufacturer's instructions provided with the tool. Failure to remove the ridges before attempting to remove the piston/connecting rod assemblies may result in piston breakage.

2   After the cylinder ridges have been removed, turn the engine upside-down so the crankshaft is facing up.

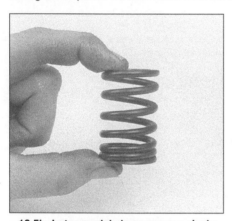

12.5b Later models have progressively wound valve springs - install them with the closely wound coils toward the cylinder head

12.6 Apply a small dab of grease to each keeper as shown here before installation - it'll hold them in place on the valve stem as the spring is released

12.8 Be sure to check the valve spring installed height (the distance from the top of the seat/shims to the top of the shield or the bottom of the retainer)

13.1 A ridge reamer is required to remove the ridge from the top of each cylinder - do this before removing the pistons!

13.3 Check the connecting rod side clearance with a feeler gauge as shown

13.6 To prevent damage to the crankshaft journals and cylinder walls, slip sections of rubber or plastic hose over the rod bolts before removing the pistons

14.1 Checking crankshaft endplay with a dial indicator

14.3 Checking crankshaft endplay with a feeler gauge

14.4a Use a center punch or number stamping dies to mark the main bearing caps to ensure installation in their original locations on the block (make the punch marks near one of the bolt heads)

3   Before the connecting rods are removed, check the endplay with feeler gauges. Slide them between the first connecting rod and the crankshaft throw until the play is removed (see illustration). The end-play is equal to the thickness of the feeler gauge(s). If the end-play exceeds the service limit, new connecting rods will be required. If new rods (or a new crankshaft) are installed, the endplay may fall under the specified minimum (if it does, the rods will have to be machined to restore it - consult an automotive machine shop for advice if necessary). Repeat the procedure for the remaining connecting rods.

4   Check the connecting rods and caps for identification marks. If they aren't plainly marked, use a small center-punch to make the appropriate number of indentations on each rod and cap (1, 2, 3, etc., depending on the engine type and cylinder they're associated with).

5   Loosen each of the connecting rod cap nuts 1/2-turn at a time until they can be removed by hand. Remove the number one connecting rod cap and bearing insert. Don't drop the bearing insert out of the cap.

6   Slip a short length of plastic or rubber hose over each connecting rod cap bolt to protect the crankshaft journal and cylinder wall as the piston is removed (see illustration).

7   Remove the bearing insert and push the connecting rod/piston assembly out through the top of the engine. Use a wooden hammer handle to push on the upper bearing surface in the connecting rod. If resistance is felt, double-check to make sure that all of the ridge was removed from the cylinder.

8   Repeat the procedure for the remaining cylinders.

9   After removal, reassemble the connecting rod caps and bearing inserts in their respective connecting rods and install the cap nuts finger tight. Leaving the old bearing inserts in place until reassembly will help prevent the connecting rod bearing surfaces from being accidentally nicked or gouged.

10   Don't separate the pistons from the connecting rods (see Section 18 for additional information).

14 Crankshaft - removal

Refer to illustrations 14.1, 14.3, 14.4a and 14.4b
Note: *The crankshaft can be removed only after the engine has been removed from the vehicle. It's assumed that the flywheel or driveplate, vibration damper, timing chain, oil pan, oil pump and piston/connecting rod assemblies have already been removed.*

1   Before the crankshaft is removed, check the endplay. Mount a dial indicator with the stem in line with the crankshaft and just touching one of the crank throws (see illustration).

2   Push the crankshaft all the way to the rear and zero the dial indicator. Next, pry the crankshaft to the front as far as possible and check the reading on the dial indicator. The distance that it moves is the

endplay. If it's greater than specified, check the crankshaft thrust surfaces for wear. If no wear is evident, new main bearings should correct the endplay.

3   If a dial indicator isn't available, feeler gauges can be used. Gently pry or push the crankshaft all the way to the front of the engine. Slip feeler gauges between the crankshaft and the front face of the thrust main bearing to determine the clearance (see illustration).

4   Check the main bearing caps to see if they're marked to indicate their locations. They should be numbered consecutively from the front of the engine to the rear. If they aren't, mark them with number stamping dies or a center-punch (see illustration). Main bearing caps generally have a cast-in arrow, which points to the front of the engine (see illustration). Loosen the main bearing cap bolts 1/4-turn at a time each, until they can be removed by hand. Note if any stud bolts are used and make sure they're returned to their original locations when the crankshaft is reinstalled.

5   Gently tap the caps with a soft-face hammer, then separate them from the engine block. If necessary, use the bolts as levers to remove the caps. Try not to drop the bearing inserts if they come out with the caps.

6   Carefully lift the crankshaft out of the engine. It may be a good idea to have an assistant available, since the crankshaft is quite heavy. With the bearing inserts in place in the engine block and main bearing caps, return the caps to their respective locations on the engine block and tighten the bolts finger tight.

14.4b The arrow on the main cap indicates the direction to the front of the engine while the stamped numbers indicate the specific journal

15.4a A hammer and large punch can be used to knock the core plugs sideways in their bores

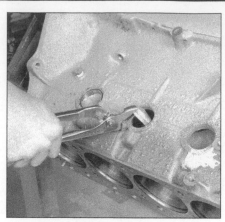

15.4b Pull the core plugs from the block with pliers

15.8 All bolt holes in the block - particularly the main bearing cap and head bolt holes - should be cleaned and restored with a tap (be sure to remove debris from the holes after this is done)

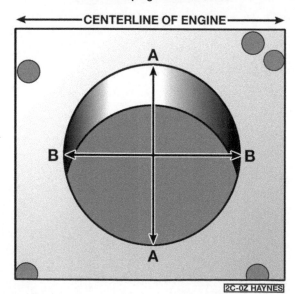

15.10 A large socket on an extension can be used to drive the new core plugs into the bores

16.4a Measure the diameter of each cylinder at a right angle to the engine centerline (A), and parallel to engine centerline (B) - out-of-round is the distance between A and B; taper is the difference between A and B at the top of the cylinder and A and B at the bottom of the cylinder

15   Engine block - cleaning

Refer to illustrations 15.4a, 15.4b, 15.8 and 15.10
Caution: *The core plugs (also known as freeze or soft plugs) may be difficult or impossible to retrieve if they're driven into the block coolant passages.*

1   Remove the main bearing caps and separate the bearing inserts from the caps and the engine block. Tag the bearings, indicating which cylinder they were removed from and whether they were in the cap or the block, then set them aside.

2   Using a gasket scraper, remove all traces of gasket material from the engine block. Be very careful not to nick or gouge the gasket sealing surfaces.

3   Remove all of the threaded oil gallery plugs from the block. The plugs are usually very tight - they may have to be drilled out and the holes retapped. Use new plugs when the engine is reassembled.

4   Remove the core plugs from the engine block. To do this, knock one side of the core plug into the engine block with a hammer and punch, then grasp the core plug with a large pair of pliers and pull it out **(see illustrations)**.

5   If the engine is extremely dirty it should be taken to an automotive machine shop to be steam cleaned or hot tanked.

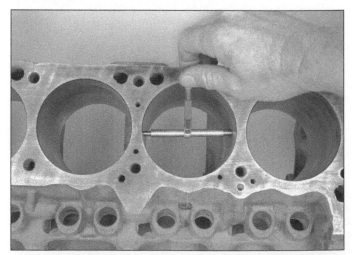

16.4b The ability to "feel" when the telescoping gauge is at the correct point will be developed over time, so work slowly and repeat the check until you're satisfied the bore measurement is accurate

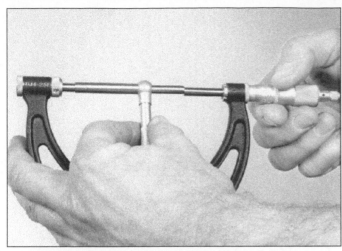

16.4c **The gauge is then measured with a micrometer to determine the bore size**

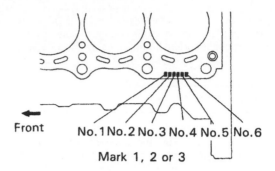

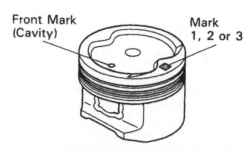

16.7 **Refer to the piston crown for the standard piston diameter Mark and the engine block for the correct Mark number and then cross-reference the number to the specifications. 1FZ-FE engine blocks can be equipped with three different standard size pistons and bore diameters**

6   After the block is returned, clean all oil holes and oil galleries one more time. Brushes specifically designed for this purpose are available at most auto parts stores. Flush the passages with warm water until the water runs clear, dry the block thoroughly and wipe all machined surfaces with a light, rust preventive oil. If you have access to compressed air, use it to speed the drying process and to blow out all the oil holes and galleries. **Warning:** *Wear eye protection when using compressed air!*

7   If the block isn't extremely dirty or sludged up, you can do an adequate cleaning job with hot soapy water and a stiff brush. Take plenty of time and do a thorough job. Regardless of the cleaning method used, be sure to clean all oil holes and galleries very thoroughly, dry the block completely and coat all machined surfaces with light oil.

8   The threaded holes in the block must be clean to ensure accurate torque readings during reassembly. Run the proper size tap into each of the holes to remove rust, corrosion, thread sealant or sludge and restore damaged threads **(see illustration)**. If possible, use compressed air to clear the holes of debris produced by this operation. Now is a good time to clean the threads on the head bolts and the main bearing cap bolts as well.

9   Reinstall the main bearing caps and tighten the bolts finger tight.

10   After coating the sealing surfaces of the new core plugs with Permatex no. 2 sealant, install them in the engine block **(see illustration)**. Make sure they're driven in straight and seated properly or leakage could result. Special tools are available for this purpose, but a large socket, with an outside diameter that will just slip into the core plug, a 1/2-inch drive extension and a hammer will work just as well.

11   Apply non-hardening sealant (such as Permatex no. 2 or Teflon pipe sealant) to the new oil gallery plugs and thread them into the holes in the block. Make sure they're tightened securely.

12   If the engine isn't going to be reassembled right away, cover it with a large plastic trash bag to keep it clean.

16   Engine block - inspection

Refer to illustrations 16.4a, 16.4b, 16.4c and 16.7

1   Before the block is inspected, it should be cleaned as described in Section 15.

2   Visually check the block for cracks, rust and corrosion. Look for stripped threads in the threaded holes. It's also a good idea to have the block checked for hidden cracks by an automotive machine shop that has the special equipment to do this type of work. If defects are found, have the block repaired, if possible, or replaced.

3   Check the cylinder bores for scuffing and scoring.

4   Check the cylinder bore for taper and out-of-round conditions as

follows **(see illustrations)**:

a) *Measure the diameter of each cylinder at the top (just under the ridge area), center and bottom of the cylinder bore, parallel to the crankshaft axis.*

b) *Next measure each cylinder's diameter at the same three locations perpendicular to the crankshaft axis.*

5   The taper of the cylinder is the difference between the bore diameter at the top of the cylinder and the diameter at the bottom. The out-of-round specification of the cylinder bore is the difference between the parallel and perpendicular readings. Compare your results to those listed in this Chapter's Specifications.

6   If the required precision measuring tools aren't available, the piston-to-bore clearance can be obtained, though not quite as accurately, using feeler gauge stock. Feeler gauge stock comes in 12-inch lengths and various thickness and is generally available at auto parts stores.

7   To check the clearance, select a feeler gauge and slip it into the cylinder along with the matching piston. The piston must be positioned exactly as it normally would be. The feeler gauge must be between the piston and cylinder on one of the thrust faces (90 degrees to the piston pin bore). 1FZ-FE engine block are equipped with three different piston bore sizes. Refer to the mark (Mark 1, 2 or 3) on the piston crown to find the correct diameter **(see illustration)**.

8   The piston should slip through the cylinder (with the feeler gauge in place) with moderate pressure.

9   If it falls through or slides through easily, the clearance is excessive and a new piston will be required. If the piston binds at the lower end of the cylinder and is loose toward the top, the cylinder is tapered. If tight spots are encountered as the piston/feeler gauge is rotated in the cylinder, the cylinder is out-of-round.

10   Repeat the procedure for the remaining pistons and cylinders.

11   If the cylinder walls are badly scuffed or scored, or if they're out-of-round or tapered beyond the limits given in the specifications, have the engine block rebored and honed at an automotive machine shop. If a rebore is done, oversize pistons and rings will be required.

12   If the cylinders are in reasonably good condition and not worn to the outside of the limits, and if the piston-to-cylinder clearances can be maintained properly, then they don't have to be rebored. Honing is all that's necessary (see Section 17).

13   Many different engines are produced with oversize or undersize components, such as oversize cylinder bores, undersize main bearing and connecting rod journals, or oversize camshaft bearing bores.
14   These engines are identified by a letter code stamped on the engine block.

17 Cylinder honing

Refer to illustrations 17.3a and 17.3b

1   Prior to engine reassembly, the cylinder bores must be honed so the new piston rings will seat correctly and provide the best possible combustion chamber seal. **Note:** *If you don't have the tools or don't want to tackle the honing operation, most automotive machine shops will do it for a reasonable fee.*
2   Before honing the cylinders, install the main bearing caps and tighten the bolts to the torque listed in this Chapter's Specifications.
3   Two types of cylinder hones are commonly available - the flex hone or "bottle brush" type and the more traditional surfacing hone with spring-loaded stones. Both will do the job, but for the less experienced mechanic the "bottle brush" hone will probably be easier to use. You'll also need some kerosene or honing oil, rags and an electric drill motor. Proceed as follows:

   a) *Mount the hone in the drill motor, compress the stones (if applicable) and slip it into the first cylinder* **(see illustration)**. *Be sure to wear safety goggles or a face shield!*
   b) *Lubricate the cylinder with plenty of honing oil or kerosene, turn on the drill and move the hone up-and-down in the cylinder at a pace that will produce a fine crosshatch pattern on the cylinder walls. Ideally, the crosshatch lines should intersect at approximately a 60-degree angle* **(see illustration)**. *Be sure to use plenty of lubricant and don't take off any more material than is absolutely necessary to produce the desired finish.* **Note:** *Piston ring manufacturers may specify a smaller crosshatch angle than the traditional 60-degrees - read and follow any instructions included with the new rings.*
   c) *Don't withdraw the hone from the cylinder while it's running. Instead, shut off the drill and continue moving the hone up-and-down in the cylinder until it comes to a complete stop, then compress the stones and withdraw the hone. If you're using a "bottle brush" type hone, stop the drill motor, then turn the chuck in the normal direction of rotation while withdrawing the hone from the cylinder.*
   d) *Wipe the oil out of the cylinder and repeat the procedure for the remaining cylinders.*

4   After the honing job is complete, chamfer the top edges of the cylinder bores with a small file so the rings won't catch when the pistons are installed. Be very careful not to nick the cylinder with the file.
5   The entire engine block must be washed again very thoroughly with warm, soapy water to remove all traces of the abrasive grit produced during the honing operation. **Note:** *The bores can be considered clean when a lint-free white cloth - dampened with clean engine oil - used to wipe them out doesn't pick up any more honing residue, which will show up as gray areas on the cloth. Be sure to run a brush through all oil holes and galleries and flush them with running water.*
6   After rinsing, dry the block and apply a coat of light rust preventive oil to all machined surfaces. Wrap the block in a plastic trash bag to keep it clean and set it aside until reassembly.

18 Pistons/connecting rods - inspection

Refer to illustrations 18.4a, 18.4b, 18.10 and 18.11

1   Before the inspection process can be carried out, the piston/connecting rod assemblies must be cleaned and the original piston rings removed from the pistons. **Note:** *Always use new piston rings when the engine is reassembled.*
2   Using a piston ring installation tool, carefully remove the rings

17.3a A "bottle brush" hone will produce better results if you've never honed cylinders before

from the pistons. Be careful not to nick or gouge the pistons in the process.
3   Scrape all traces of carbon from the top of the piston. A hand-held wire brush or a piece of fine emery cloth can be used once the majority of the deposits have been scraped away. Do not, under any circumstances, use a wire brush mounted in a drill motor to remove deposits from the pistons. The piston material is soft and may be eroded away by the wire brush.
4   Use a piston ring groove cleaning tool to remove carbon deposits from the ring grooves. If a tool isn't available, a piece broken off the old ring will do the job. Be very careful to remove only the carbon deposits - don't remove any metal and do not nick or scratch the sides of the ring grooves **(see illustrations)**.
5   Once the deposits have been removed, clean the piston/rod assemblies with solvent and dry them with compressed air (if available). Make sure the oil return holes in the back sides of the ring grooves are clear.
6   If the pistons and cylinder walls aren't damaged or worn excessively, and if the engine block is not rebored, new pistons won't be necessary. Normal piston wear appears as even vertical wear on the

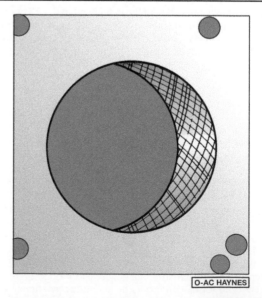

17.3b The cylinder hone should leave a smooth, crosshatch pattern with the lines intersecting at approximately a 60-degree angle

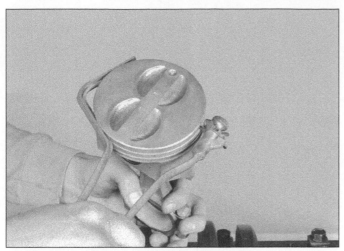

18.4a The piston ring grooves can be cleaned with a special tool, as shown here, . . .

18.4b . . . or a section of a broken ring

piston thrust surfaces and slight looseness of the top ring in its groove. New piston rings, however, should always be used when an engine is rebuilt.

7   Carefully inspect each piston for cracks around the skirt, at the pin bosses and at the ring lands.

8   Look for scoring and scuffing on the thrust faces of the skirt, holes in the piston crown and burned areas at the edge of the crown. If the skirt is scored or scuffed, the engine may have been suffering from overheating and/or abnormal combustion, which caused excessively high operating temperatures. The cooling and lubrication systems should be checked thoroughly. A hole in the piston crown is an indication that abnormal combustion (preignition) was occurring. Burned areas at the edge of the piston crown are usually evidence of spark knock (detonation). If any of the above problems exist, the causes must be corrected or the damage will occur again. The causes may include intake air leaks, incorrect fuel/air mixture, incorrect ignition timing and EGR system malfunctions.

9   Corrosion of the piston, in the form of small pits, indicates that coolant is leaking into the combustion chamber and/or the crankcase. Again, the cause must be corrected or the problem may persist in the rebuilt engine.

10   Measure the piston ring side clearance by laying a new piston ring in each ring groove and slipping a feeler gauge in beside it **(see illustration)**. Check the clearance at three or four locations around each groove. Be sure to use the correct ring for each groove - they are different. If the side clearance is greater than specified, new pistons will

have to be used.

11   Check the piston-to-bore clearance by measuring the bore (see Section 16) and the piston diameter. Make sure the pistons and bores are correctly matched. Measure the piston across the skirt, at a 90 degree angle to and in line with the piston pin **(see illustration)**. Subtract the piston diameter from the bore diameter to obtain the clearance. If it's greater than specified, the block will have to be rebored and new pistons and rings installed.

12   Check the piston-to-rod clearance by twisting the piston and rod in opposite directions. Any noticeable play indicates excessive wear, which must be corrected. The piston/connecting rod assemblies should be taken to an automotive machine shop to have the pistons and rods resized and new pins installed.

13   If the pistons must be removed from the connecting rods for any reason, they should be taken to an automotive machine shop. While they are there have the connecting rods checked for bend and twist, since automotive machine shops have special equipment for this purpose. **Note:** *Unless new pistons and/or connecting rods must be installed, do not disassemble the pistons and connecting rods.*

14   Check the connecting rods for cracks and other damage. Temporarily remove the rod caps, lift out the old bearing inserts, wipe the rod and cap bearing surfaces clean and inspect them for nicks, gouges and scratches. After checking the rods, replace the old bearings, slip the caps into place and tighten the nuts finger tight. **Note:** *If the engine is being rebuilt because of a connecting rod knock, be sure to install new rods.*

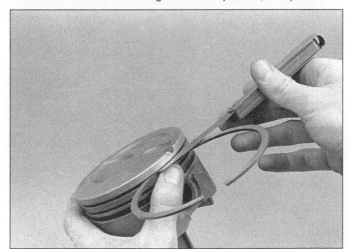

18.10 Check the ring side clearance with a feeler gauge at several points around the groove

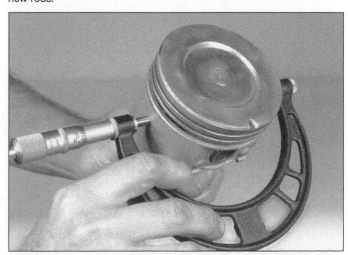

18.11 Measure the piston diameter at a 90-degree angle to the piston pin and in line with it

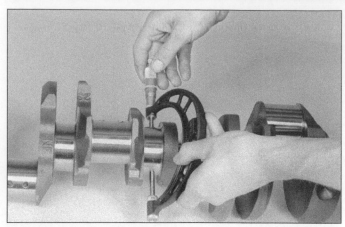

19.6 Measure the diameter of each crankshaft journal at several points to detect taper and out-of-round conditions

19   Crankshaft - inspection

Refer to illustration 19.6

1   Remove all burrs from the crankshaft oil holes with a stone, file or scraper.

2   Clean the crankshaft with solvent and dry it with compressed air (if available). Be sure to clean the oil holes with a stiff brush and flush them with solvent.

3   Check the main and connecting rod bearing journals for uneven wear, scoring, pits and cracks.

4   Rub a penny across each journal several times. If a journal picks up copper from the penny, it's too rough and must be reground.

5   Check the rest of the crankshaft for cracks and other damage. It should be magnafluxed to reveal hidden cracks - an automotive machine shop will handle the procedure.

6   Using a micrometer, measure the diameter of the main and connecting rod journals and compare the results to the specifications **(see illustration)**. By measuring the diameter at a number of points around each journal's circumference, you'll be able to determine whether or not the journal is out-of-round. Take the measurement at each end of the journal, near the crank throws, to determine if the journal is tapered.

7   If the crankshaft journals are damaged, tapered, out-of-round or worn beyond the limits given in the specifications, have the crankshaft reground by an automotive machine shop. Be sure to use the correct size bearing inserts if the crankshaft is reconditioned.

8   Check the oil seal journals at each end of the crankshaft for wear and damage. If the seal has worn a groove in the journal, or if it's nicked or scratched, the new seal may leak when the engine is reassembled. In some cases, an automotive machine shop may be able to repair the journal by pressing on a thin sleeve. If repair isn't feasible, a new or different crankshaft should be installed.

9   Refer to Section 20 and examine the main and rod bearing inserts.

20   Main and connecting rod bearings - inspection

Inspection

Refer to illustration 20.1

1   Even though the main and connecting rod bearings should be replaced with new ones during the engine overhaul, the old bearings should be retained for close examination, as they may reveal valuable information about the condition of the engine **(see illustration)**.

2   Bearing failure occurs because of lack of lubrication, the presence of dirt or other foreign particles, overloading the engine and corrosion. Regardless of the cause of bearing failure, it must be corrected before

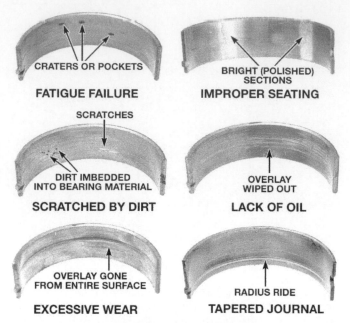

CRATERS OR POCKETS
FATIGUE FAILURE

BRIGHT (POLISHED) SECTIONS
IMPROPER SEATING

SCRATCHES
DIRT IMBEDDED INTO BEARING MATERIAL
SCRATCHED BY DIRT

OVERLAY WIPED OUT
LACK OF OIL

OVERLAY GONE FROM ENTIRE SURFACE
EXCESSIVE WEAR

RADIUS RIDE
TAPERED JOURNAL

20.1 Typical bearing failures

the engine is reassembled to prevent it from happening again.

3   When examining the bearings, remove them from the engine block, the main bearing caps, the connecting rods and the rod caps and lay them out on a clean surface in the same general position as their location in the engine. This will enable you to match any bearing problems with the corresponding crankshaft journal.

4   Dirt and other foreign particles get into the engine in a variety of ways. It may be left in the engine during assembly, or it may pass through filters or the PCV system. It may get into the oil, and from there into the bearings. Metal chips from machining operations and normal engine wear are often present. Abrasives are sometimes left in engine components after reconditioning, especially when parts are not thoroughly cleaned using the proper cleaning methods. Whatever the source, these foreign objects often end up embedded in the soft bearing material and are easily recognized. Large particles will not embed in the bearing and will score or gouge the bearing and journal. The best prevention for this cause of bearing failure is to clean all parts thoroughly and keep everything spotlessly clean during engine assembly. Frequent and regular engine oil and filter changes are also recommended.

5   Lack of lubrication (or lubrication breakdown) has a number of interrelated causes. Excessive heat (which thins the oil), overloading (which squeezes the oil from the bearing face) and oil leakage or throw off (from excessive bearing clearances, worn oil pump or high engine speeds) all contribute to lubrication breakdown. Blocked oil passages, which usually are the result of misaligned oil holes in a bearing shell, will also oil starve a bearing and destroy it. When lack of lubrication is the cause of bearing failure, the bearing material is wiped or extruded from the steel backing of the bearing. Temperatures may increase to the point where the steel backing turns blue from overheating.

6   Driving habits can have a definite effect on bearing life. Full throttle, low speed operation (lugging the engine) puts very high loads on bearings, which tends to squeeze out the oil film. These loads cause the bearings to flex, which produces fine cracks in the bearing face (fatigue failure). Eventually the bearing material will loosen in pieces and tear away from the steel backing. Short trip driving leads to corrosion of bearings because insufficient engine heat is produced to drive off the condensed water and corrosive gases. These products collect in the engine oil, forming acid and sludge. As the oil is carried to the engine bearings, the acid attacks and corrodes the bearing material.

7   Incorrect bearing installation during engine assembly will lead to bearing failure as well. Tight fitting bearings leave insufficient bearing

Standard bearing thickness (at center wall):
Mark "A"   1.484 – 1.488 mm
              (0.0584 –0.0586 in.)
Mark "B"   1.488 – 1.492 mm
              (0.0586 – 0.0587 in.)
Mark "C"   1.492 – 1.496 mm
              (0.0587 – 0.0589 in.)

20.8a Connecting rod standard bearing selection chart -
3F–E engine

oil clearance and will result in oil starvation. Dirt or foreign particles trapped behind a bearing insert result in high spots on the bearing which lead to failure.

Bearing selection

Refer to illustrations 20.8a, 20.8b, 20.8c. 20.8d, 20.8e and 20.8f

8   If the oil clearances are incorrect (see Section 23) or you are replacing the original bearings, refer to the accompanying charts to select the correct new standard bearings **(see illustrations)**.
9   If the crankshaft has been reground, new undersize bearings must be installed. Disregard steps 11 through 13 and consult the automotive machine shop that reground the crankshaft. They will provide

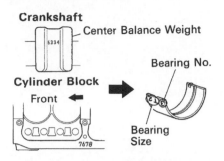

	Number marked								
Crankshaft	3			4			5		
Cylinder block	6	7	8	6	7	8	6	7	8
Bearing	T3	T4	T5	T2	T3	T4	T1	T2	T3

Example: Crankshaft "5", Cylinder Block "7"
              = Bearing "T2"

20.8b Crankshaft standard bearing selection chart - 3F-E engine

or help you select the correct size bearings.
10   Regardless of how the bearing sizes are determined, use the oil clearance measured with Plastigage (see Section 23) as a guide to ensure the proper size bearings are installed.
11   Engines are assembled at the factory with various sizes of color-coded bearing inserts as listed in illustrations 20.8a through 20.8f. The color code appears on the edge of the bearing insert.
12   The journal size codes are generally painted on the adjacent cheek toward the flanged end (rear) of the crankshaft, except for the rear main bearing journal, which is marked on the crankshaft rear flange.

Crankshaft journal diameter:
Mark "3"   No.1   66.972 – 66.980 mm
                     (2.6367 – 2.6370 in.)
              No.2   68.472 – 68.480 mm
                     (2.6957 – 2.6961 in.)
              No.3   69.972 – 69.980 mm
                     (2.7548 – 2.7551 in.)
              No.4   71.472 – 71.480 mm
                     (2.8139 – 2.8142 in.)

Mark "4"   No.1   66.980 – 66.988 mm
                     (2.6370 – 2.6373 in.)
              No.2   68.480 – 68.488 mm
                     (2.6961 – 2.6964 in.)
              No.3   69.980 – 69.988 mm
                     (2.7551 – 2.7554 in.)
              No.4   71.480 – 71.488 mm
                     (2.8142 – 2.8145 in.)

Mark "5"   No.1   66.988 – 66.996 mm
                     (2.6373 – 2.6376 in.)
              No.2   68.488 – 68.496 mm
                     (2.6964 – 2.6967 in.)
              No.3   69.988 – 69.996 mm
                     (2.7554 – 2.7557 in.)
              No.4   71.488 – 71.496 mm
                     (2.8145 – 2.8148 in.)

Cylinder block main journal bore diameter:
Mark "6"   No.1   72.010 – 72.018 mm
                     (2.8350 – 2.8353 in.)
              No.2   73.510 – 73.518 mm
                     (2.8941 – 2.8944 in.)
              No.3   75.010 – 75.018 mm
                     (2.9531 – 2.9535 in.)
              No.4   76.510 – 76.518 mm
                     (3.0122 – 3.0125 in.)

Mark "7"   No.1   72.018 – 72.026 mm
                     (2.8353 – 2.8357 in.)
              No.2   73.518 – 73.526 mm
                     (2.8944 – 2.8947 in.)
              No.3   75.018 – 75.026 mm
                     (2.9535 – 2.9538 in.)
              No.4   76.518 – 76.526 mm
                     (3.0125 – 3.0128 in.)

Mark "8"   No.1   72.026 – 72.034 mm
                     (2.8357 – 2.8360 in.)
              No.2   73.526 – 73.534 mm
                     (2.8947 – 2.8950 in.)
              No.3   75.026 – 75.034 mm
                     (2.9538 – 2.9541 in.)
              No.4   76.526 – 76.534 mm
                     (3.0128 – 3.0131 in.)

20.8c Crankshaft journal and cylinder block main journal bore diameters - 3F-E engines

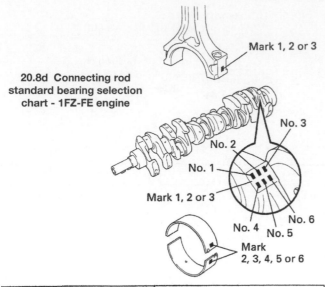

20.8d Connecting rod standard bearing selection chart - 1FZ-FE engine

Mark 1, 2 or 3

No. 3
No. 2
No. 1
Mark 1, 2 or 3
No. 4   No. 5
No. 6
Mark 2, 3, 4, 5 or 6

	Number marked								
Connecting rod	1			2			3		
Crankshaft	1	2	3	1	2	3	1	2	3
Use bearing	2	3	4	3	4	5	4	5	6

EXAMPLE: Connecting rod ''3'' + Crankshaft ''1''
= Total number 4 (Use bearing ''4'')

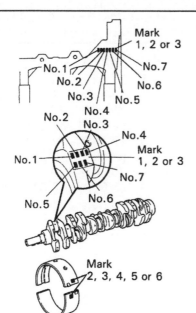

Mark 1, 2 or 3
No.1   No.7
No.2   No.6
No.3   No.5
No.4
No.2
No.3
No.4
No.1   Mark 1, 2 or 3
No.7
No.5   No.6
Mark 2, 3, 4, 5 or 6

	Number marked								
Cylinder block	1			2			3		
Crankshaft	1	2	3	1	2	3	1	2	3
Use bearing	2	3	4	3	4	5	4	5	6

EXAMPLE: Cylinder block ''2'' + Crankshaft ''1''
= Total number 3 (Use bearing ''3'')

20.8e Crankshaft standard bearing selection chart - 1FZ-FE engine

13   To obtain a select fit, upper and lower bearing inserts of different sizes may be used as a pair. For example, a standard insert is sometimes used in combination with a 0.001-inch undersize insert to reduce clearance by 0.0005-inch. **Caution:** *Never use a pair of bearing inserts with a greater size difference than 0.001-inch. When replacing inserts, the odd sized inserts must be either all on the top or all on the bottom.*

21   Engine overhaul - reassembly sequence

1   Before beginning engine reassembly, make sure you have all the necessary new parts, gaskets and seals as well as the following items on hand:

Common hand tools
3/8-inch and 1/2-inch drive torque wrenches
Piston ring installation tool
Piston ring compressor
Vibration damper installation tool
Short lengths of rubber or plastic hose
to fit over connecting rod bolts
Plastigage
Feeler gauges
A fine-tooth file
New engine oil
Engine assembly lube or moly-base grease
Gasket sealant
Thread locking compound

2   In order to save time and avoid problems, engine reassembly must be done in the following general order:

New camshaft bearings (must be done by automotive machine shop)
Piston rings

Reference:
Cylinder block main journal bore diameter:
   Mark "1"
      74.026 — 74.032 mm (2.9144 — 2.9146 in.)
   Mark "2"
      74.032 — 74.038 mm (2.9146 — 2.9149 in.)
   Mark "3"
      74.038 — 74.044 mm (2.9149 — 2.9151 in.)
Crankshaft journal diameter:
   Mark "1"
      68.994 — 69.000 mm (2.7163 — 2.7165 in.)
   Mark "2"
      68.988 — 68.994 mm (2.7161 — 2.7163 in.)
   Mark "3"
      68.982 — 68.988 mm (2.7158 — 2.7161 in.)
Standard sized bearing center wall thickness:
   Mark "2"
      2.489 — 2.492 mm (0.0980 — 0.0981 in.)
   Mark "3"
      2.492 — 2.495 mm (0.0981 — 0.0982 in.)
   Mark "4"
      2.495 — 2.498 mm (0.0982 — 0.0983 in.)
   Mark "5"
      2.498 — 2.501 mm (0.0983 — 0.0985 in.)
   Mark "6"
      2.501 — 2.504 mm (0.0985 — 0.0986 in.)

20.8f Crankshaft journal and cylinder block main journal bore diameters - 1FZ-FE engines

22.3 When checking piston ring end gap, the ring must be square in the cylinder bore (this is done by pushing the ring down with the top of a piston as shown)

Crankshaft and main bearings
Piston/connecting rod assemblies
Oil pump
Camshaft and lifters (2F and 3F-E engines)
Oil pan
Timing chain and sprockets (1FZ-FE engines)
Cylinder head, pushrods and rocker arms
Timing cover
Intake and exhaust manifolds
Valve cover
Engine rear plate
Flywheel/driveplate

22 Piston rings - installation

Refer to illustrations 22.3, 22.4, 22.5, 22.9a, 22.9b and 22.12

1 Before installing the new piston rings, the ring end gaps must be checked. It's assumed that the piston ring side clearance has been checked and verified correct (see Section 18).
2 Lay out the piston/connecting rod assemblies and the new ring sets so the ring sets will be matched with the same piston and cylinder during the end gap measurement and engine assembly.
3 Insert the top (number one) ring into the first cylinder and square it up with the cylinder walls by pushing it in with the top of the piston **(see illustration)**. The ring should be near the bottom of the cylinder, at the lower limit of ring travel.
4 To measure the end gap, slip feeler gauges between the ends of the ring until a gauge equal to the gap width is found **(see illustration)**.

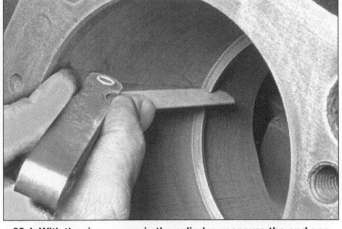

22.4 With the ring square in the cylinder, measure the end gap with a feeler gauge

The feeler gauge should slide between the ring ends with a slight amount of drag. Compare the measurement to the Specifications. If the gap is larger or smaller than specified, double-check to make sure you have the correct rings before proceeding.
5 If the gap is too small, it must be enlarged or the ring ends may come in contact with each other during engine operation, which can cause serious damage to the engine. The end gap can be increased by filing the ring ends very carefully with a fine file. Mount the file in a vise equipped with soft jaws, slip the ring over the file with the ends contacting the file face and slowly move the ring to remove material from the ends. When performing this operation, file only from the outside in **(see illustration)**.
6 Excess end gap isn't critical unless it's greater than 0.040-inch. Again, double-check to make sure you have the correct rings for your engine.
7 Repeat the procedure for each ring that will be installed in the first cylinder and for each ring in the remaining cylinders. Remember to keep rings, pistons and cylinders matched.
8 Once the ring end gaps have been checked/corrected, the rings can be installed on the pistons.
9 The oil control ring (lowest one on the piston) is usually installed first. It's normally composed of three separate components. Slip the spacer/expander into the groove **(see illustration)**. If an anti-rotation tang is used, make sure it's inserted into the drilled hole in the ring groove. Next, install the lower side rail. Don't use a piston ring installation tool on the oil ring side rails, as they may be damaged. Instead, place one end of the side rail into the groove between the spacer/expander and the ring land, hold it firmly in place and slide a finger around the piston while pushing the rail into the groove **(see illustration)**. Next, install the upper side rail in the same manner.

22.5 If the end gap is too small, clamp a file in a vise and file the ring ends (from the outside in only) to enlarge the gap slightly

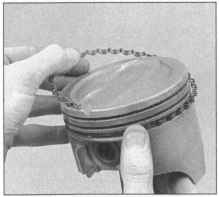

22.9a Installing the spacer/expander in the oil control ring groove

22.9b DO NOT use a piston ring installation tool when installing the oil ring side rails

22.12 Installing the compression rings with a ring expander - the mark (arrow) must face up

23.11 Lay the Plastigage strips (arrow) on the main bearing journals, parallel to the crankshaft centerline

10   After the three oil ring components have been installed, check to make sure that both the upper and lower side rails can be turned smoothly in the ring groove.

11   The number two (middle) ring is installed next. It's usually stamped with a mark which must face up, toward the top of the piston. **Note:** *Always follow the instructions printed on the ring package or box - different manufacturers may require different approaches. Do not mix up the top and middle rings, as they have different cross sections. Make sure the ring designations face up toward the piston crown.*

12   Use a piston ring installation tool and make sure the identification mark is facing the top of the piston, then slip the ring into the middle groove on the piston **(see illustration)**. Don't expand the ring any more than necessary to slide it over the piston.

13   Install the number one (top) ring in the same manner. Make sure the mark is facing up. Be careful not to confuse the number one and number two rings.

14   Repeat the procedure for the remaining pistons and rings.

23   Crankshaft - installation and main bearing oil clearance check

1   Crankshaft installation is the first step in engine reassembly. It's assumed at this point that the engine block and crankshaft have been cleaned, inspected and repaired or reconditioned.

2   Position the engine with the bottom facing up.

3   Remove the main bearing cap bolts and lift out the caps. Lay them out in the proper order to ensure correct installation.

4   If they're still in place, remove the original bearing inserts from the block and the main bearing caps. Wipe the bearing surfaces of the block and caps with a clean, lint-free cloth. They must be kept spotlessly clean.

Main bearing oil clearance check

Refer to illustrations 23.11 and 23.15

5   Clean the back sides of the new main bearing inserts and lay one in each main bearing saddle in the block. If one of the bearing inserts from each set has a large groove in it, make sure the grooved insert is installed in the block. Lay the other bearing from each set in the corresponding main bearing cap. Make sure the tab on the bearing insert fits into the recess in the block or cap. **Caution:** *The oil holes in the block must line up with the oil holes in the bearing insert. Do not hammer the bearing into place and don't nick or gouge the bearing faces. No lubrication should be used at this time.*

6   The thrust washers must be installed in the proper recess in the engine block. On 2F and 3F-E engines, it is number three (counting from the front of the engine); on 1FZ-FE engines it's number four **(see illustrations 8.5a and 8.5b)**.

7   Clean the faces of the bearings in the block and the crankshaft

main bearing journals with a clean, lint-free cloth.

8   Check or clean the oil holes in the crankshaft, as any dirt here can go only one way - straight through the new bearings.

9   Once you're certain the crankshaft is clean, carefully lay it in position in the main bearings.

10   Before the crankshaft can be permanently installed, the main bearing oil clearance must be checked.

11   Cut several pieces of the appropriate size Plastigage (they must be slightly shorter than the width of the main bearings) and place one piece on each crankshaft main bearing journal, parallel with the journal axis **(see illustration)**.

12   Clean the faces of the bearings in the caps and install the caps in their respective positions (don't mix them up) with the arrows pointing toward the front of the engine. Don't disturb the Plastigage.

13   Starting with the center main and working out toward the ends, tighten the main bearing cap bolts, in three steps, to the torque listed in this Chapter's Specifications. Don't rotate the crankshaft at any time during this operation.

14   Remove the bolts and carefully lift off the main bearing caps. Keep them in order. Don't disturb the Plastigage or rotate the crankshaft. If any of the main bearing caps are difficult to remove, tap them gently from side-to-side with a soft-face hammer to loosen them.

15   Compare the width of the crushed Plastigage on each journal to the scale printed on the Plastigage envelope to obtain the main bearing oil clearance **(see illustration)**. Check the Specifications to make sure it's correct.

23.15 Compare the width of the crushed Plastigage to the scale on the envelope to determine the main bearing oil clearance (always take the measurement at the widest point of the Plastigage); be sure to use the correct scale - standard and metric ones are included

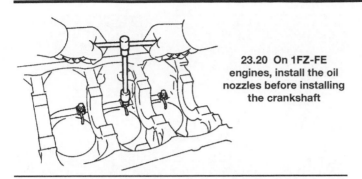

23.20 On 1FZ-FE engines, install the oil nozzles before installing the crankshaft

24.3 Tap around the circumference of the oil seal with a hammer and a punch to seat it squarely in the bore

16   If the clearance is not as specified, the bearing inserts may be the wrong size (which means different ones will be required). Before deciding that different inserts are needed, make sure that no dirt or oil was between the bearing inserts and the caps or block when the clearance was measured. If the Plastigage was wider at one end than the other, the journal may be tapered (refer to Section 19).

17   Carefully scrape all traces of the Plastigage material off the main bearing journals and/or the bearing faces. Use your fingernail or the edge of a credit card - don't nick or scratch the bearing faces.

18   Carefully lift the crankshaft out of the engine.

19   Clean the bearing faces in the block, then apply a thin, uniform layer of moly-base grease or engine assembly lube to each of the bearing surfaces. Be sure to coat the thrust faces as well as the journal face of the thrust bearing. On inline six-cylinder engines, install the rear main oil seal sections in the block and rear main bearing cap (see Section 23).

Final crankshaft installation

Refer to illustration 23.20

20   On 1FZ-FE engines, install the oil nozzles and check valves into the engine block **(see illustration)**.

21   Make sure the crankshaft journals are clean, then lay the crankshaft back in place in the block. Clean the faces of the bearings in the caps, then apply lubricant to them.

22   Install the caps in their respective positions with the arrows pointing toward the front of the engine.

23   Install the bolts.

24   Tighten all except the thrust bearing cap bolts to the specified torque (work from the center out and approach the final torque in three steps).

25   Tighten the thrust bearing cap bolts to 10-to-12 ft-lbs.

26   Tap the ends of the crankshaft forward and backward with a lead or brass hammer to align the main bearing and crankshaft thrust surfaces.

27   Retighten all main bearing cap bolts to the specified torque, starting with the center main and working out toward the ends.

28   On manual transmission equipped models, install a new pilot bearing in the end of the crankshaft (see Chapter 8).

29   Rotate the crankshaft a number of times by hand to check for any obvious binding.

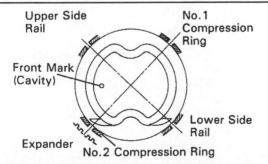

25.5 Position the piston ring gaps as shown before installing the piston/connecting rod assemblies in the engine. The dot must point towards the front of the engine

30   Finally, check the crankshaft endplay with a feeler gauge or a dial indicator as described in Section 14. The endplay should be correct if the crankshaft thrust washers aren't worn or damaged and new bearings have been installed.

31   If you are working on an engine with a one-piece rear main oil seal, refer to Section 24 and install the new seal.

24   Rear main oil seal installation

Refer to illustration 24.3

1   Clean the bore in the block/cap and the seal contact surface on the crankshaft. Check the crankshaft surface for scratches and nicks that could damage the new seal lip and cause oil leaks. If the crankshaft is damaged, the only alternative is a new or different crankshaft.

2   Apply a light coat of engine oil or multi-purpose grease to the outer edge of the new seal. Lubricate the seal lip with moly-base grease or engine assembly lube.

3   Carefully work the seal lip over the end of the crankshaft and tap the seal in with a hammer and punch until it's seated in the bore **(see illustration)**.

25   Pistons/connecting rods - installation and rod bearing oil clearance check

1   Before installing the piston/connecting rod assemblies, the cylinder walls must be perfectly clean, the top edge of each cylinder must be chamfered, and the crankshaft must be in place.

2   Remove the cap from the end of the number one connecting rod (refer to the marks made during removal). Remove the original bearing inserts and wipe the bearing surfaces of the connecting rod and cap with a clean, lint-free cloth. They must be kept spotlessly clean.

Connecting rod bearing oil clearance check

Refer to illustrations 25.5, 25.11, 25.13 and 25.17

3   Clean the back side of the new upper bearing insert, then lay it in place in the connecting rod. Make sure the tab on the bearing fits into the recess in the rod. Don't hammer the bearing insert into place and be very careful not to nick or gouge the bearing face. Don't lubricate the bearing at this time.

4   Clean the back side of the other bearing insert and install it in the rod cap. Again, make sure the tab on the bearing fits into the recess in the cap, and don't apply any lubricant. It's critically important that the mating surfaces of the bearing and connecting rod are perfectly clean and oil free when they're assembled.

5   Position the piston ring gaps at intervals around the piston **(see illustration)**.

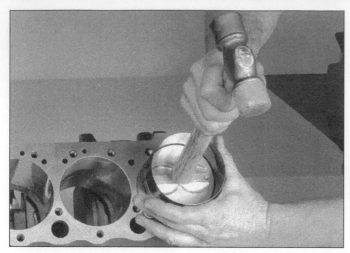

25.11 Drive the piston gently into the cylinder bore with the end of a wooden or plastic hammer handle

25.13 Lay the Plastigage strips on each rod bearing journal, parallel to the crankshaft centerline

6   Slip a section of plastic or rubber hose over each connecting rod cap bolt.

7   Lubricate the piston and rings with clean engine oil and attach a piston ring compressor to the piston. Leave the skirt protruding about 1/4-inch to guide the piston into the cylinder. The rings must be compressed until they're flush with the piston.

8   Rotate the crankshaft until the number one connecting rod journal is at BDC (bottom dead center) and apply a coat of engine oil to the cylinder walls.

9   With the arrow or notch on top of the piston facing the front of the engine **(see illustration 25.5)**, gently insert the piston/connecting rod assembly into the number one cylinder bore and rest the bottom edge of the ring compressor on the engine block.

10   Tap the top edge of the ring compressor to make sure it's contacting the block around its entire circumference.

11   Gently tap on the top of the piston with the end of a wooden hammer handle **(see illustration)** while guiding the end of the connecting rod into place on the crankshaft journal. The piston rings may try to pop out of the ring compressor just before entering the cylinder bore, so keep some down pressure on the ring compressor. Work slowly, and if any resistance is felt as the piston enters the cylinder, stop immediately. Find out what's hanging up and fix it before proceeding. Do not, for any reason, force the piston into the cylinder - you might break a ring and/or the piston.

12   Once the piston/connecting rod assembly is installed, the connecting rod bearing oil clearance must be checked before the rod cap is permanently bolted in place.

13   Cut a piece of the appropriate size Plastigage slightly shorter than the width of the connecting rod bearing and lay it in place on the number one connecting rod journal, parallel with the journal axis **(see illustration)**.

14   Clean the connecting rod cap bearing face, remove the protective hoses from the connecting rod bolts and install the rod cap. Make sure the mating mark on the cap is on the same side as the mark on the connecting rod.

15   Install the nuts and tighten them to the torque listed in this Chapter's Specifications, working up to it in three steps. **Note:** *Use a thinwall socket to avoid erroneous torque readings that can result if the socket is wedged between the rod cap and nut. If the socket tends to wedge itself between the nut and the cap, lift up on it slightly until it no longer contacts the cap. Do not rotate the crankshaft at any time during this operation.*

16   Remove the nuts and detach the rod cap, being very careful not to disturb the Plastigage.

17   Compare the width of the crushed Plastigage to the scale printed on the Plastigage envelope to obtain the oil clearance **(see illustration)**. Compare it to the Specifications to make sure the clearance is correct.

18   If the clearance is not as specified, the bearing inserts may be the wrong size (which means different ones will be required). Before deciding that different inserts are needed, make sure that no dirt or oil was between the bearing inserts and the connecting rod or cap when the clearance was measured. Also, recheck the journal diameter. If the Plastigage was wider at one end than the other, the journal may be tapered (refer to Section 19).

Final connecting rod installation

Refer to illustration 25.21

19   Carefully scrape all traces of the Plastigage material off the rod journal and/or bearing face. Be very careful not to scratch the bearing - use your fingernail or the edge of a credit card.

20   Make sure the bearing faces are perfectly clean, then apply a uniform layer of clean moly-base grease or engine assembly lube to both of them. You'll have to push the piston into the cylinder to expose the face of the bearing insert in the connecting rod - be sure to slip the protective hoses over the rod bolts first.

21   Slide the connecting rod back into place on the journal, remove the protective hoses from the rod cap bolts, install the rod cap and tighten the nuts to the torque listed in this Chapter's Specifications. Again, work up to the torque in three steps **(see illustration)**.

22   Repeat the entire procedure for the remaining pistons/connecting rods.

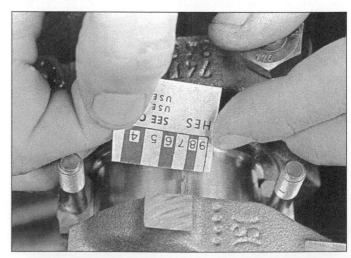

25.17 Measuring the width of the crushed Plastigage to determine the rod bearing oil clearance (be sure to use the correct scale - standard and metric ones are included)

25.21 Be sure the numbers (arrow) on the connecting rod match before tightening the cap nuts

23   The important points to remember are:
a) *Keep the back sides of the bearing inserts and the insides of the connecting rods and caps perfectly clean when assembling them.*
b) *Make sure you have the correct piston/rod assembly for each cylinder.*
c) *The notch or mark on the piston must face the front of the engine.*
d) *Lubricate the cylinder walls with clean oil.*
e) *Lubricate the bearing faces when installing the rod caps after the oil clearance has been checked.*

24   After all the piston/connecting rod assemblies have been properly installed, rotate the crankshaft a number of times by hand to check for any obvious binding.
25   As a final step, the connecting rod endplay must be checked. Refer to Section 14 for this procedure.
26   Compare the measured endplay to the Specifications to make sure it's correct. If it was correct before disassembly and the original crankshaft and rods were reinstalled, it should still be right. If new rods or a new crankshaft were installed, the endplay may be inadequate. If so, the rods will have to be removed and taken to an automotive machine shop for resizing.

26 Initial start-up and break-in after overhaul

Warning: *Have a fire extinguisher handy when starting the engine for the first time.*
1   Once the engine has been installed in the vehicle, double-check the engine oil and coolant levels.
2   With the spark plugs out of the engine and the ignition system disabled (see Section 3), crank the engine until oil pressure registers on the gauge or the light goes out.
3   Install the spark plugs, hook up the plug wires and restore the ignition system functions (see Section 3).
4   Start the engine. It may take a few moments for the fuel system to build up pressure, but the engine should start without a great deal of effort. **Note:** *If backfiring occurs through the carburetor or throttle body, recheck the valve timing and ignition timing.*
5   After the engine starts, it should be allowed to warm up to normal operating temperature. While the engine is warming up, make a thorough check for fuel, oil and coolant leaks.
6   Shut the engine off and recheck the engine oil and coolant levels.
7   Drive the vehicle to an area with minimum traffic, accelerate from 30 to 50 mph, then allow the vehicle to slow to 30 mph with the throttle closed. Repeat the procedure 10 or 12 times. This will load the piston rings and cause them to seat properly against the cylinder walls. Check again for oil and coolant leaks.
8   Drive the vehicle gently for the first 500 miles (no sustained high speeds) and keep a constant check on the oil level. It is not unusual for an engine to use oil during the break-in period.
9   At approximately 500 to 600 miles, change the oil and filter.
10   For the next few hundred miles, drive the vehicle normally. Do not pamper it or abuse it.
11   After 2000 miles, change the oil and filter again and consider the engine broken in.

Notes

Chapter 3
Cooling, heating and air conditioning systems

Contents

	Section
Air conditioning and heating system - check and maintenance	13
Air conditioning compressor - removal and installation	15
Air conditioning condenser - removal and installation	16
Air conditioning evaporator and expansion valve - removal and installation	17
Air conditioning receiver/drier - removal and installation	14
Antifreeze - general information	2
Blower unit - check and replacement	10
Coolant level check	See Chapter 1
Coolant temperature sending unit - check and replacement	9
Cooling system check	See Chapter 1
Cooling system servicing (draining, flushing and refilling)	See Chapter 1

	Section
Drivebelt check, adjustment and replacement	See Chapter 1
Engine cooling fan/clutch assembly - check, removal and installation	4
General information	1
Heater and air conditioning control assembly - check, removal and installation	12
Heater core - removal and installation	11
Oil cooler - removal and installation	6
Radiator and coolant reservoir - removal and installation	5
Thermostat - check and replacement	3
Underhood hose check and replacement	See Chapter 1
Water pump - check	7
Water pump - removal and installation	8

Specifications

General

Radiator cap pressure rating	10.7 to 14.9 psi
Thermostat rating	183 to 203-degrees F
Refrigerant type	
1993	R-12
1994 and later	R-134a
Refrigerant capacity	
Models without rear air conditioning	28 to 32 ounces
Models with rear air conditioning	53 ounces

Torque specifications

Ft-lbs (unless otherwise indicated)

Note: *One foot-pound (ft-lb) of torque is equivalent to 12 inch-pounds (in-lbs) of torque. Torque values below approximately 15 ft-lbs are expressed in inch-pounds, because most foot-pound torque wrenches are not accurate at these smaller values.*

Thermostat housing bolts	
1992 and earlier	156 in-lbs
1993 and later	15
Water pump-to-block bolts	
1992 and earlier	27
1993 and later	15

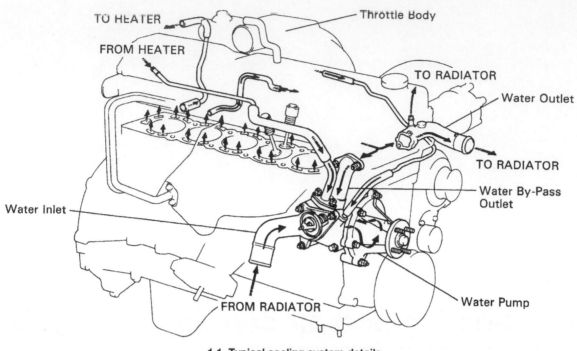

1.1 Typical cooling system details

1   General information

Engine cooling system

Refer to illustration 1.1

All vehicles covered by this manual employ a pressurized engine cooling system with thermostatically-controlled coolant circulation **(see illustration)**. An impeller type water pump mounted on the front of the block pumps coolant through the engine. The coolant flows around each cylinder and toward the rear of the engine. Cast-in coolant passages direct coolant around the intake and exhaust ports, near the spark plug areas and in proximity to the exhaust valve guides.

A wax-pellet type thermostat is located in the thermostat housing at the radiator end of the engine. During warm up, the closed thermostat prevents coolant from circulating through the radiator. When the engine reaches normal operating temperature, the thermostat opens and allows hot coolant to travel through the radiator, where it is cooled before returning to the engine.

The cooling system is sealed by a pressure-type radiator cap. This raises the boiling point of the coolant, and the higher boiling point of the coolant increases the cooling efficiency of the radiator. If the system pressure exceeds the cap pressure-relief value, the excess pressure in the system forces the spring-loaded valve inside the cap off its seat and allows the coolant to escape through the overflow tube into a coolant reservoir. When the system cools, the excess coolant is automatically drawn from the reservoir back into the radiator.

The coolant reservoir serves as both the point at which fresh coolant is added to the cooling system to maintain the proper fluid level and as a holding tank for overheated coolant.

This type of cooling system is known as a closed design because coolant that escapes past the pressure cap is saved and reused.

Heating system

The heating system consists of a blower fan and heater core located within the heater box under the dashboard, the inlet and outlet hoses connecting the heater core to the engine cooling system and the heater/air conditioning control head on the dashboard. Hot engine coolant is circulated through the heater core. When the heater mode is activated, a flap door opens to expose the heater box to the passenger compartment. A fan switch on the control head activates the blower motor, which forces air through the core, heating the air.

Air conditioning system

The air conditioning system consists of a condenser mounted in front of the radiator, an evaporator mounted adjacent to the heater core, a compressor mounted on the engine, a filter-drier which contains a high pressure relief valve and the plumbing connecting all of the above.

A blower fan forces the warmer air of the passenger compartment through the evaporator core (sort of a radiator-in-reverse), transferring the heat from the air to the refrigerant. The liquid refrigerant boils off into low pressure vapor, taking the heat with it when it leaves the evaporator. The compressor keeps refrigerant circulating through the system, pumping the warmed coolant through the condenser where it is cooled and then circulated back to the evaporator.

2   Antifreeze - general information

Refer to illustration 2.4

Warning: *Do not allow antifreeze to come in contact with your skin or painted surfaces of the vehicle. Rinse off spills immediately with plenty of water. Antifreeze is highly toxic if ingested. Never leave antifreeze lying around in an open container or in puddles on the floor; children and pets are attracted by it's sweet smell and may drink it. Check with local authorities about disposing of used antifreeze. Many communities have collection centers which will see that antifreeze is disposed of safely. Never dump used antifreeze on the ground or into drains.*
Note: *Non-toxic antifreeze is now manufactured and available at local auto parts stores, but even these types should be disposed of properly.*

The cooling system should be filled with a water/ethylene-glycol based antifreeze solution, which will prevent freezing down to at least -20 degrees F, or lower if local climate requires it. It also provides protection against corrosion and increases the coolant boiling point.

The cooling system should be drained, flushed and refilled every 30,000 miles or every two years (see Chapter 1). The use of antifreeze solutions for periods of longer than two years is likely to cause damage and encourage the formation of rust and scale in the system. If your

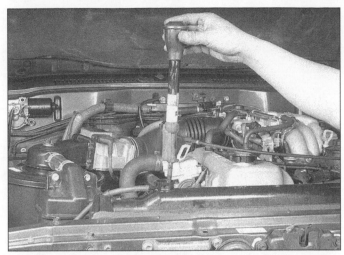

2.4 An inexpensive hydrometer can be used to test the condition of your coolant

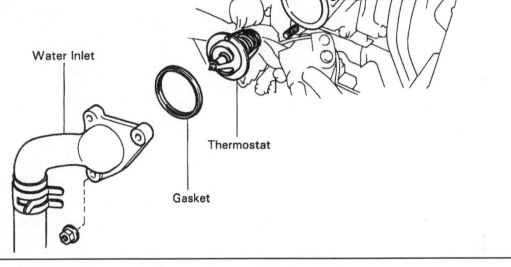

3.8a Thermostat housing on the 3F-E engine

3.8b Thermostat housing on the 1FZ-FE engine

Water Inlet

Thermostat

Gasket

tap water is "hard", i.e. contains a lot of dissolved minerals, use distilled water with the antifreeze.

Before adding antifreeze to the system, check all hose connections, because antifreeze tends to leak through very minute openings. Engines do not normally consume coolant. Therefore, if the level goes down, find the cause and correct it.

The exact mixture of antifreeze-to-water you should use depends on the relative weather conditions. The mixture should contain at least 50-percent antifreeze, but should never contain more than 70-percent antifreeze. Consult the mixture ratio chart on the antifreeze container before adding coolant. Hydrometers are available at most auto parts stores to test the ratio of antifreeze to water **(see illustration)**. Use antifreeze which meets the vehicle manufacturer's specifications.

3   Thermostat - check and replacement

Warning: *Do not attempt to remove the radiator cap, coolant or thermostat until the engine has cooled completely.*

Check

1   Before assuming the thermostat is responsible for a cooling system problem, check the coolant level (Chapter 1), drivebelt tension (Chapter 1) and temperature gauge (or light) operation.
2   If the engine takes a long time to warm up (as indicated by the

temperature gauge or heater operation), the thermostat is probably stuck open. Replace the thermostat with a new one.
3   If the engine runs hot, use your hand to check the temperature of the lower radiator hose. If the hose is not hot, but the engine is, the thermostat is probably stuck in the closed position, preventing the coolant inside the engine from traveling through the radiator. Replace the thermostat. **Caution:** *On models equipped with fuel injection, do not drive the vehicle without a thermostat. The computer may stay in open loop and emissions and fuel economy will suffer.*
4   If the lower radiator hose is hot, it means that the coolant is flowing and the thermostat is open. Consult the *Troubleshooting* section at the front of this manual for further diagnosis.

Replacement

Refer to illustrations 3.8a, 3.8b, 3.9 and 3.10
5   Disconnect the negative cable from the battery. **Caution:** *If the stereo in your vehicle is equipped with an anti-theft system, make sure you have the correct activation code before disconnecting the battery.*
6   Drain the cooling system (see Chapter 1).
7   If equipped, disconnect the vacuum hoses to the BVSV on the thermostat housing.
8   Detach the housing from the engine **(see illustrations)**. Be prepared for some coolant to spill as the gasket seal is broken. The radiator hose can be left attached to the housing, unless the housing itself is to be replaced.

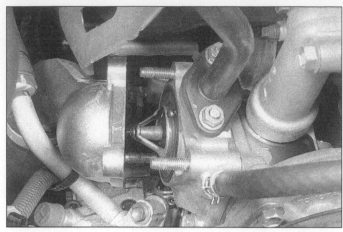

3.9 The thermostat is installed with the spring end towards the cylinder head

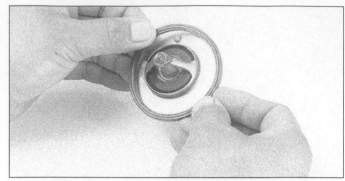

3.10 On 1993 and later models, the thermostat gasket, which is actually a grooved sealing ring, fits around the edge of the thermostat

9   Remove the thermostat, noting the direction in which it was installed in the housing, and thoroughly clean the sealing surfaces **(see illustration)**.
10   Install a new gasket onto the thermostat or housing **(see illustration)**. Make sure it is evenly fitted all the way around.
11   Install the thermostat and housing, positioning the jiggle pin, if equipped, at the highest point.
12   Tighten the housing fasteners to the torque listed in this Chapter's Specifications and reinstall the remaining components in the reverse order of removal.
13   Refill the cooling system, run the engine and check for leaks and proper operation.

4   Engine cooling fan/clutch assembly - check, removal and installation

Check

1   Check the fan carefully for cracks, especially around the base of each blade. If any cracks are evident, replace the fan.
2   Check the clutch for fluid leakage. If it's leaking, replace the clutch.
3   Try to move the fan blades in a front-to-rear direction. If they wobble, replace the fan clutch.
4   Spin the fan by hand. It should resist spinning (more resistance when the engine is hot). If it spins freely, or won't spin at all, replace the fan clutch.

4.5 Remove the fan shroud mounting bolts (arrows) and lift the assembly from the engine

Removal

Refer to illustrations 4.5, 4.6 and 4.7

5   Remove the fan shroud mounting bolts and separate the shroud from the radiator **(see illustration)**. Remove the drivebelt(s) (see Chapter 1).
6   Remove the fan assembly mounting screws **(see illustration)** and separate the fan from the water pump.
7   To separate the fan blades from the viscous clutch hub, remove the mounting nuts **(see illustration)**.
8   Installation is the reverse of removal.

4.6 Unscrew the four nuts (arrows) that retain the fan to the water pump, then detach the fan/clutch assembly

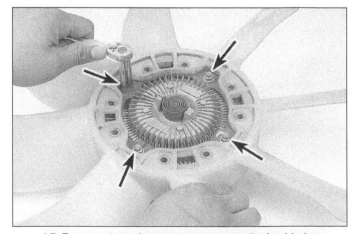

4.7 Remove these four nuts to separate the fan blades from the clutch

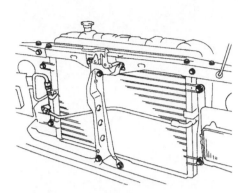

5.7 Remove the bolts from the grille center support

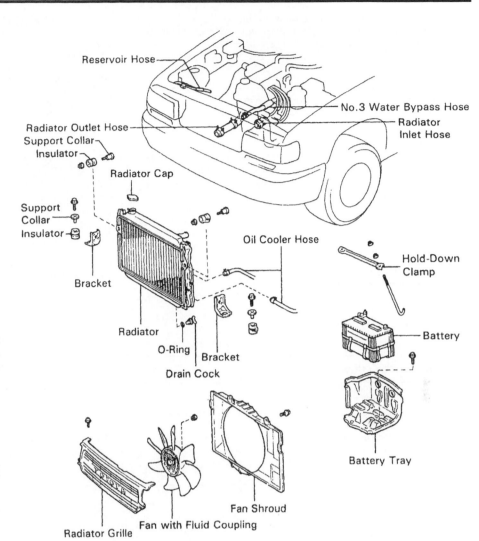

5.8a Remove the bolts (arrows) that mount the radiator to the side support braces

5.8b Radiator mounting details

5   Radiator and coolant reservoir - removal and installation

Warning: *Do not start this procedure until the engine is completely cool.*

Radiator

Refer to illustrations 5.7, 5.8a and 5.8b

1   Disconnect the negative battery cable. **Caution:** *If the stereo in your vehicle is equipped with an anti-theft system, make sure you have the correct activation code before disconnecting the battery.*
2   Drain the coolant into a container (see Chapter 1).
3   Remove both the upper and lower radiator hoses.
4   Disconnect the reservoir hose from the radiator filler neck.
5   Remove the cooling fan (see Section 4).
6   If equipped with an automatic transmission, disconnect the cooler lines from the radiator. Place a drip pan to catch the fluid and cap the fittings.
7   Remove the grille center support **(see illustration)**.
8   Remove the two upper radiator mounting brackets **(see illustration)**. Remove all the bolts that retain the radiator to the grille frame **(see illustration)**.

9   Lift out the radiator. Be aware of dripping fluids and the sharp fins.
10   With the radiator removed, it can be inspected for leaks, damage and internal blockage. If in need of repairs, have a professional radiator shop or dealer service department perform the work as special techniques are required.
11   Bugs and dirt can be cleaned from the radiator with compressed air and a soft brush. Don't bend the cooling fins as this is done. **Warning:** *Wear eye protection when using compressed air.*
12   Installation is the reverse of the removal procedure. Be sure the rubber mounts are in place on the bottom of the radiator.
13   After installation, fill the cooling system with the proper mixture of antifreeze and water. Refer to Chapter 1 if necessary.
14   Start the engine and check for leaks. Allow the engine to reach normal operating temperature, indicated by both radiator hoses becoming hot. Recheck the coolant level and add more if required.
15   On automatic transmission equipped models, check and add fluid as needed.

Coolant reservoir

Refer to illustration 5.16

16   On most models, the coolant reservoir simply pulls up and out of

the bracket next to the battery **(see illustration)**.

17   Pour the coolant into a container. Wash out and inspect the reservoir for cracks and chafing. Replace it if damaged.

18   Installation is the reverse of removal.

6   Oil cooler - removal and installation

Refer to illustrations 6.9a and 6.9b

1   Disconnect the negative battery cable. **Caution:** *If the stereo in your vehicle is equipped with an anti-theft system, make sure you have the correct activation code before disconnecting the battery.*

2   Drain the coolant into a container (see Chapter 1).

3   Remove the air cleaner assembly (see Chapter 4A or 4B).

4   Remove the front exhaust pipe assembly from the vehicle.

5   If equipped with the PAIR emission system, remove the air pipe (see Chapter 6).

6   On 1993 and later models, remove the exhaust manifold (see Chapter 2B).

7   Remove the oil pressure switch (see Chapter 2C).

8   Remove the oil cooler hoses, if equipped.

9   Remove the oil cooler mounting bolts and nuts **(see illustrations)**.

10   Installation is the reverse of removal.

7   Water pump - check

Refer to illustration 7.3

1   A failure in the water pump can cause serious engine damage due to overheating.

2   With the engine running and warmed to normal operating temperature, squeeze the upper radiator hose. If the water pump is working properly, a pressure surge should be felt as the hose is released.

5.16 Pull the coolant reservoir hose assembly (left arrow), then pull the reservoir bottle out of its bracket (right arrow)

Warning: *Keep hands away from fan blades!*

3   Water pumps are equipped with weep or vent holes **(see illustration)**. If a failure occurs in the pump seal, coolant will leak from this hole. In most cases it will be necessary to use a flashlight to find the hole on the water pump by looking through the space behind the pulley just below the water pump shaft.

4   If the water pump shaft bearings fail there may be a howling sound at the front of the engine while it is running. Bearing wear can be felt if the water pump pulley is rocked up and down. Do not mistake drivebelt slippage, which causes a squealing sound, for water pump failure. Spray automotive drivebelt dressing on the belts to eliminate the belt as a possible cause of the noise.

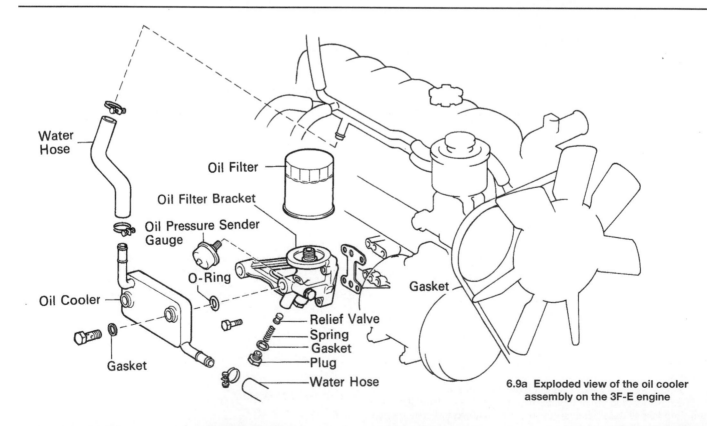

6.9a Exploded view of the oil cooler assembly on the 3F-E engine

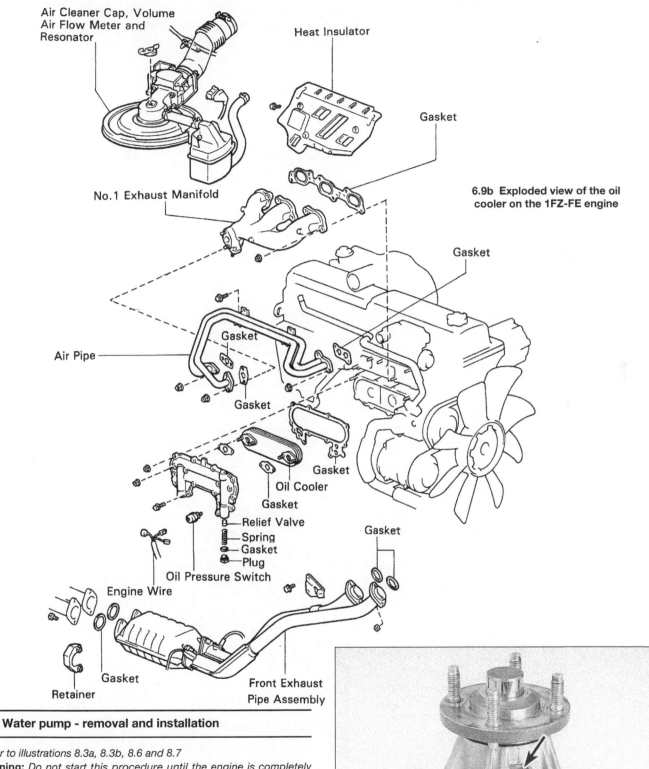

Air Cleaner Cap, Volume
Air Flow Meter and
Resonator

Heat Insulator

Gasket

No.1 Exhaust Manifold

**6.9b Exploded view of the oil
cooler on the 1FZ-FE engine**

Gasket

Gasket

Air Pipe

Gasket

Gasket

Gasket

Oil Cooler

Gasket

Relief Valve
Spring
Gasket
Plug

Gasket

Oil Pressure Switch

Engine Wire

Gasket

Retainer

Front Exhaust
Pipe Assembly

8   Water pump - removal and installation

Refer to illustrations 8.3a, 8.3b, 8.6 and 8.7

Warning: *Do not start this procedure until the engine is completely
cool.*

1   Disconnect the negative battery cable and drain the cooling system (see Chapter 1). **Caution:** *If the stereo in your vehicle is equipped
with an anti-theft system, make sure you have the correct activation
code before disconnecting the battery.*

2   Refer to Chapter 1 and remove the power steering belt, alterna-
tor/water pump belt and air conditioning belt. On 1992 and earlier
models, remove the power steering idler pulley and bracket and the
alternator brace.

7.3 Location of the weep hole on the water pump (arrow)

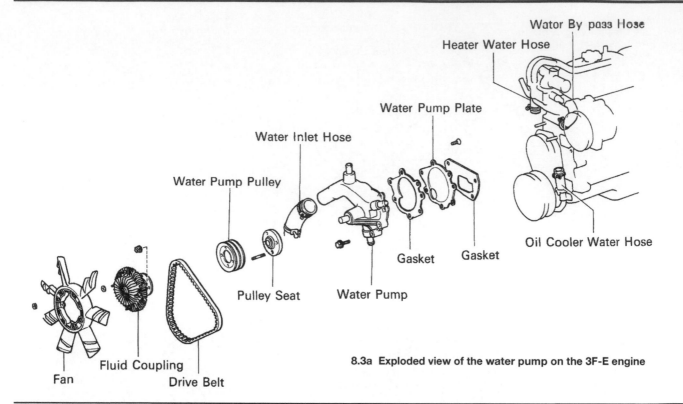

Water By pass Hose

Heater Water Hose

Water Pump Plate

Water Inlet Hose

Water Pump Pulley

Gasket   Gasket

Oil Cooler Water Hose

Pulley Seat   Water Pump

Fluid Coupling

Fan   Drive Belt

8.3a Exploded view of the water pump on the 3F-E engine

3   Remove the fan shroud from the radiator (see Section 4) **(see accompanying illustrations and illustration 4.5)**.
4   Remove the fan assembly from the water pump (see Section 4).
5   Remove the water pump pulley from the water pump **(see illus-**trations 8.3a and 8.3b)**. Disconnect the radiator inlet hose and the bypass hose.
6   Remove the bolts retaining the water pump to the engine block and remove the water pump **(see illustration)**.

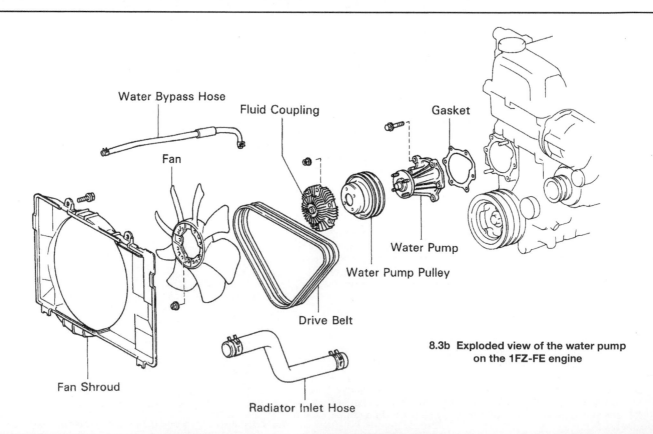

Water Bypass Hose   Fluid Coupling   Gasket

Fan

Water Pump

Water Pump Pulley

Drive Belt

Fan Shroud

Radiator Inlet Hose

8.3b Exploded view of the water pump on the 1FZ-FE engine

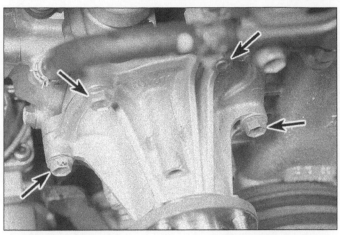

8.6 Remove the water pump assembly-to-block bolts (arrows)

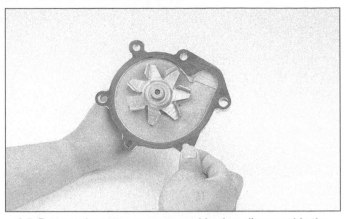

8.7 Remove the water pump assembly, then disassemble the pump and replace the gaskets

7   Disassemble the water pump and replace the gaskets. Thoroughly clean all sealing surfaces **(see illustration)**. Assemble the water pump assembly.

8   Installation is the reverse of removal. Tighten the water pump bolts to the torque listed in this Chapter's Specifications.

9   Refill the cooling system (see Chapter 1), run the engine and check for leaks.

9   Coolant temperature sending unit - check and replacement

Warning: *Do not start this procedure until the engine is completely cool.*

Check

Refer to illustrations 9.3a and 9.3b

1   If the coolant temperature gauge is inoperative, check the fuses first (see Chapter 12).

2   If the temperature gauge indicates excessive temperature after running awhile, see the *Troubleshooting* section in the front of the manual.

3   If the temperature gauge indicates HOT as soon as the engine is started cold, disconnect the wire at the coolant temperature sender **(see illustrations)**. If the gauge reading drops, replace the sending unit. If the reading remains high, the wire to the gauge may be shorted to ground or the gauge is faulty. **Note:** *The coolant temperature sending unit on the 3F-E engines is located near the thermostat housing. Refer to the component location diagrams at the beginning of* Chapter 4B.

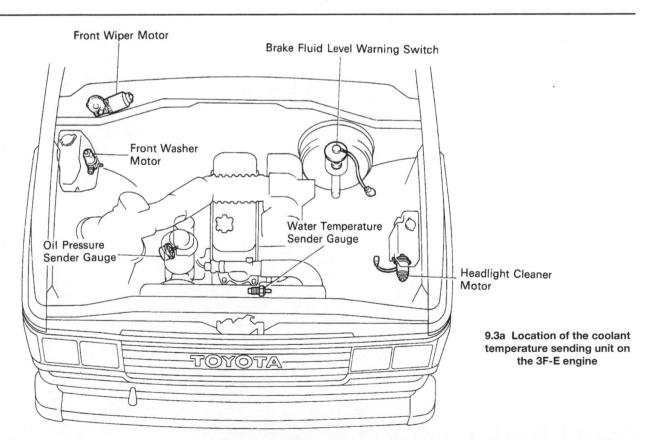

Front Wiper Motor

Brake Fluid Level Warning Switch

Front Washer Motor

Water Temperature Sender Gauge

Oil Pressure Sender Gauge

Headlight Cleaner Motor

TOYOTA

9.3a Location of the coolant temperature sending unit on the 3F-E engine

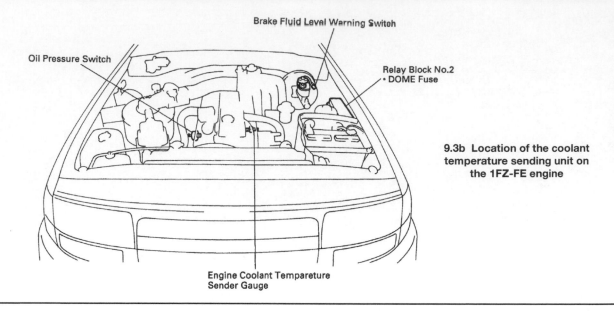

9.3b Location of the coolant temperature sending unit on the 1FZ-FE engine

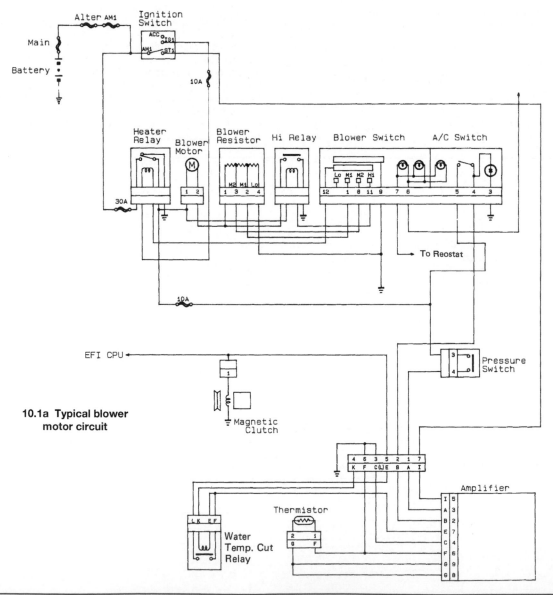

10.1a Typical blower motor circuit

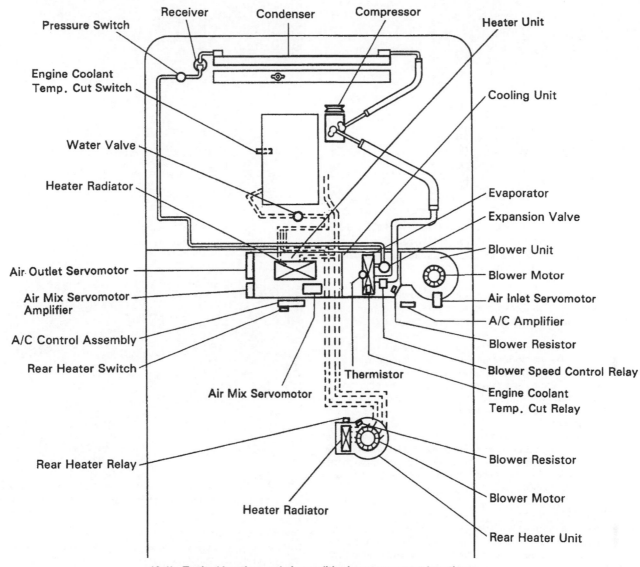

10.1b Typical heating and air conditioning component locations

---

4   If the coolant temperature gauge fails to show any indication after the engine has been warmed up, (approximately 10 minutes) and the fuses checked out OK, shut off the engine. Disconnect the wire at the sending unit and, using a jumper wire, connect the wire to a clean ground on the engine. Briefly turn on the ignition without starting the engine. If the gauge now indicates Hot, replace the sending unit.

5   If the gauge fails to respond, the circuit may be open or the gauge may be faulty - see Chapter 12 for additional information.

Replacement

6   Drain the coolant (see Chapter 1).
7   Disconnect the wiring connector from the sending unit.
8   Using a deep socket or a wrench, remove the sending unit.
9   Install the new unit and tighten it securely. Do not use thread sealant as it may electrically insulate the sending unit.
10   Reconnect the wiring connector, refill the cooling system and check for coolant leakage and proper gauge function.

---

10   Blower unit - check and replacement

Warning: *Some models covered by this manual are equipped with*

airbags. The airbag is armed and can deploy (inflate) anytime the battery is connected. To prevent accidental deployment (and possible injury), turn the ignition key to LOCK and disconnect the negative battery cable whenever working near airbag components. After the battery is disconnected, wait at least two minutes before beginning work (the system has a back-up capacitor that must fully discharge). For more information see Chapter 12.

Caution: *If the stereo in your vehicle is equipped with an anti-theft system, make sure you have the correct activation code before disconnecting the battery.*

Check

Refer to illustrations 10.1a, 10.1b and 10.3

1   The blower unit is located in the cooling assembly under the dash and in the passenger compartment on rear mounted A/C units. If the blower doesn't work, check the fuse and all connections in the circuit for looseness and corrosion **(see illustrations)**. Make sure the battery is fully charged.

2   Locate the blower unit electrical connector.

3   If the blower motor does not operate, disconnect the electrical connector at the blower motor, turn the ignition key On (engine not

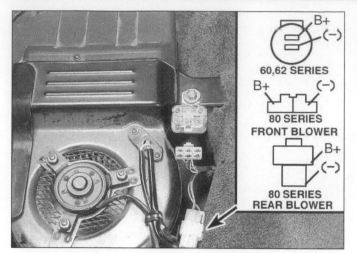

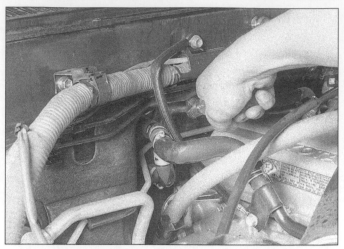

10.3 Unplug the electrical connector (arrow) and check for battery voltage to the blower motor with the ignition key ON (engine not running)

11.3 Disconnect the heater hoses at the firewall

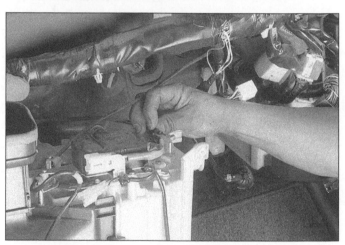

11.9 Disconnect the cables from the heater assembly

running) and check for battery voltage with the blower speed switch ON **(see illustration)**. If battery voltage is not present, there is a problem in the ignition feed circuit or in the circuit from the blower switch to the motor.

4   If battery voltage is present, reconnect the terminal to the blower motor and backprobe the correct terminal with a jumper wire connected to ground. If the motor still does not operate, the motor is probably faulty. Apply fused power and ground connections to the blower motor terminals, if the motor does not operate, replace the motor. Refer to the wiring schematics at the end of Chapter 12 for the correct wire color designations.

5   If the motor is good, but doesn't operate the blower resistor or the heater/air conditioning control switch is probably faulty. Remove the control assembly (see Section 12) and check for continuity through the switch in each position. Remove the resistor and check for continuity across each of the terminals.

Replacement

6   If the blower motor must be replaced, remove the bracket retaining the wiring connector to the blower, then remove the three mounting screws and lower the blower assembly from the housing. The fan can be removed and reused on the new blower motor.

7   Installation is the reverse of removal. Check for proper operation.

11   Heater core - removal and installation

Refer to illustrations 11.3, 11.9, 11.10 and 11.11

Warning: *These models are equipped with airbags. The airbag is armed and can deploy (inflate) anytime the battery is connected. To prevent accidental deployment (and possible injury), turn the ignition key to LOCK and disconnect the negative battery cable whenever working near airbag components. After the battery is disconnected, wait at least two minutes before beginning work (the system has a back-up capacitor that must fully discharge). For more information see Chapter 12.*

Note 1: *The following procedure details removing the heater core assembly by itself, but it is much easier to remove the heater core housing if the evaporator (cooling) assembly is removed first, although this necessitates having the refrigerant discharged and recovered before work begins. See Section 17 for evaporator removal, and do not discharge the refrigerant without reading the **Warning** in Section 12.*

Note 2: *The factory recommends removal of the entire instrument panel and lowering the steering column to remove the heater core. This involves disconnecting numerous electrical connectors and there is the potential for breakage of delicate plastic tabs on various components.*

This is a difficult job for the average home mechanic.

Note 3: *It is possible to remove the heater core assembly without removing the instrument panel on early 60 series vehicles.*

1   Disconnect the negative cable from the battery. **Caution:** *If the stereo in your vehicle is equipped with an anti-theft system, make sure you have the correct activation code before disconnecting the battery.*

2   Wait until the engine is completely cool, then drain the cooling system (see Chapter 1).

3   Working in the engine compartment, disconnect the heater hoses at the firewall **(see illustration)**. Push the rubber seal around the hoses toward the inside of the vehicle, releasing it from the sheetmetal. Plug the heater core tubes to prevent leakage when it is removed.

4   Refer to Chapters 11 and 12 and remove the center console, glove compartment, glove compartment liner, ashtray, radio and center dash bezels and instrument panel.

5   Refer to Section 12 of this Chapter to remove the heater/air conditioning controls.

6   Refer to Chapter 6 and remove the ECM without disconnecting the connectors. Set the ECM aside to allow room under the heater core housing.

7   Remove the heater unit braces.

8   Unbolt and remove the center ventilation duct.

9   Remove the heater duct/control door that is directly above and in front of the heater unit **(see illustration)**.

10   Loosen the nuts on the studs that retain the heater tubes to the housing **(see illustration)**.

11   Pull the heater unit out from behind the dash **(see illustration)**. Keep plenty of towels or rags on the carpeting to catch any coolant

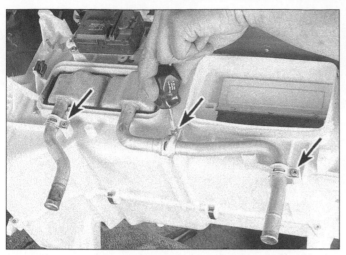

11.10 Remove the screws (arrows) from the heater tube clamps

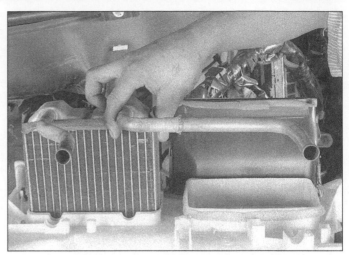

11.11 Lift the heater core from the heater assembly

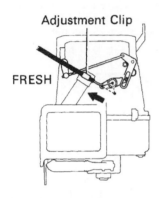

12.9 To adjust the FRESH door cable, move the arm up and tighten the clamp

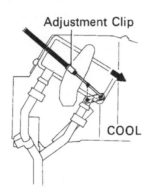

12.10 To adjust the COOL control cable, push the lever away from the cable clamp and tighten the clamp

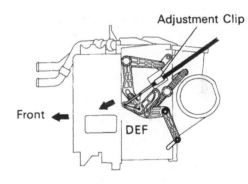

12.11 Adjust the DEF control cable with the lever pulled toward the front, then tighten the clamp

that may drip. **Caution:** *Work slowly and carefully to avoid breaking any plastic components during removal. The assembly must be pulled out (toward the rear of the vehicle) far enough for the plastic tabs to clear anything that might inhibit removal.*

12   Installation is the reverse order of removal.

13   Refill the cooling system, reconnect the battery and run the engine. Check for leaks and proper system operation.

12   Heater and air conditioning control assembly - check, removal and installation

Warning: *Some models covered by this manual are equipped with airbags. The airbag is armed and can deploy (inflate) anytime the battery is connected. To prevent accidental deployment (and possible injury), turn the ignition key to LOCK and disconnect the negative battery cable whenever working near airbag components. After the battery is disconnected, wait at least two minutes before beginning work (the system has a back-up capacitor that must fully discharge). For more information see Chapter 12.*

Note: *Early models are equipped with cable actuated heating and A/C controls. Later models are equipped with electronic actuating controls for the heating and A/C system blend doors and components.*

Removal and installation

1   Disconnect the negative cable from the battery. **Caution:** *If the*

stereo in your vehicle is equipped with an anti-theft system, make sure you have the correct activation code before disconnecting the battery.

2   Remove the center cluster trim panels (see Chapter 11).

3   Pull off the control knobs.

4   Remove the mounting screws located on the front of the control assembly (see Chapter 12).

5   Pull the control out slightly. If equipped with cables, twist the flags on the control cable mounts and remove the cables from the control. Disconnect the electrical connectors.

6   Installation is the reverse of the removal procedure.

7   Run the engine and check for proper functioning of the heater (and air conditioning, if equipped).

Cable adjustment

Refer to illustrations 12.9, 12.10 and 12.11

8   With the cables attached at the control end and the control assembly installed in the dash, adjust the cables at their ends. The controls should be set to: RECIRC, COOL, and DEF.

9   To adjust the air inlet damper control cable set the damper lever to FRESH, install the cable and clamp it in place **(see illustration)**.

10   To adjust the air mix control cable, set the air mix damper to COOL, install the cable and lock the clamp while applying slight pressure on the outer cable **(see illustration)**.

11   To adjust the mode damper control cable, set the mode damper to the DEF mode, hook the cable end on and tighten the clamp **(see illustration)**.

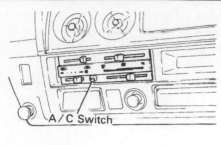

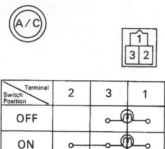

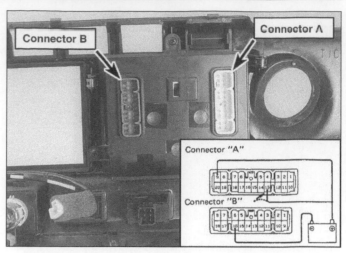

Switch Position / Terminal	2	3	1
OFF		○—⊗—○	
ON	○—○—⊗		

12.12a Remove the A/C switch and check the continuity (60, 62 series)

12.12b Connect a jumper wire from the positive (+) battery terminal to connector A, terminal number 9 and a jumper wire from negative (-) battery terminal to terminal 16. Push each blower speed selection IN and make sure the indicator light illuminates. Connect the positive lead to terminal 13 and make sure all the indicator lights dim (80 series)

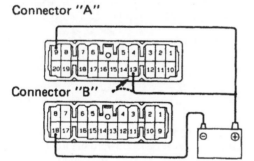

12.12c Connect a jumper wire from the positive (+) battery terminal to connector A, terminal number 9 and a jumper wire from negative (-) battery terminal to terminal 18. Push the A/C button IN and make sure the indicator light illuminates. Connect the positive lead to terminal 13 and make sure all the indicator lights dim (80 series)

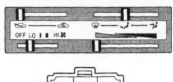

Terminals / Condition	1	2	3	4	5	6
OFF						
I		○—	—○—	—○		
II	○—		—○—	—○		
III			○—	—○—		—○
IV (Hi)				○—	—○—	—○

12.13a Check the continuity of the heater blower switch with the heater switch in the various positions (62 series)

Electrical checks

Refer to illustrations 12.12a, 12.12b, 12.12c, 12.13a and 12.13b

12   Check the air conditioning switch following the accompanying tables **(see illustrations)**.

13   Check the heater control switch following the accompanying tables **(see illustrations)**. If the switch fails any of the tests, replace the control assembly.

13 Air conditioning and heating system - check and maintenance

Air conditioning system

Refer to illustration 13.1

Warning: *The air conditioning system is under high pressure. Do not loosen any hose fittings or remove any components until the system has been discharged. Air conditioning refrigerant should be properly discharged into an EPA-approved recovery/recycling unit by a dealer service department or an automotive air conditioning repair facility.*

Always wear eye protection when disconnecting air conditioning system fittings.

1   The following maintenance checks should be performed on a regular basis to ensure that the air conditioner continues to operate at peak efficiency **(see illustration)**.

a) *Inspect the condition of the compressor drivebelt. If it is worn or deteriorated, replace it (see Chapter 1).*

b) *Check the drivebelt tension and, if necessary, adjust it (see Chapter 1).*

c) *Inspect the system hoses. Look for cracks, bubbles, hardening and deterioration. Inspect the hoses and all fittings for oil bubbles or seepage. If there is any evidence of wear, damage or leakage, replace the hose(s).*

d) *Inspect the condenser fins for leaves, bugs and any other foreign material that may have embedded itself in the fins. Use a "fin comb" or compressed air to remove debris from the condenser.*

e) *Make sure the system has the correct refrigerant charge.*

2   It's a good idea to operate the system for about ten minutes at least once a month. This is particularly important during the winter months because long term non-use can cause hardening, and subse-

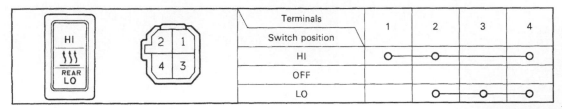

Terminals Switch position	1	2	3	4
HI	O——————O		O
OFF				
LO		O——————O		O

12.13b Check the continuity of the rear heater blower switch with the rear heater switch in the various positions (62 series)

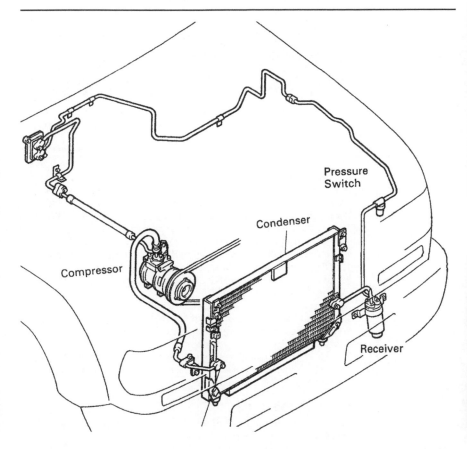

13.1 Basic components of the air conditioning system

13.10 Check the temperature of the output air in the center register with a thermometer - it should be approximately 35 to 40-degrees F below the ambient air temperature

13.11 The sight glass (arrow) is located in the line to the receiver/drier on 80 series vehicles (on 60 and 62 series vehicles it's located on the top of the receiver/drier)

quent failure, of the seals.

3   Leaks in the air conditioning system are best spotted when the system is brought up to operating temperature and pressure, by running the engine with the air conditioning ON for five minutes. Shut the engine off and inspect the air conditioning hoses and connections. Traces of oil usually indicate refrigerant leaks.

4   Because of the complexity of the air conditioning system and the special equipment required to effectively work on it, accurate troubleshooting of the system should be left to a professional technician.

5   If the air conditioning system doesn't operate at all, check the fuse panel and the air conditioning relay, located in the fuse/relay box in the engine compartment. Refer to Sections 10 and 11 for electrical checks of heating/air conditioning system components.

6   The most common cause of poor cooling is simply a low system refrigerant charge. If a noticeable drop in cool air output occurs, the following quick check will help you determine if the refrigerant level is low. For more information on air conditioning systems, refer to the *Haynes Automotive Heating and Air Conditioning Manual.*

Checking the refrigerant charge
Refer to illustrations 13.9, 13.10 and 13.11

7   Warm the engine up to normal operating temperature.

8   Place the air conditioning temperature selector at the coldest setting and put the blower at the highest setting. Open the doors (to make sure the air conditioning system doesn't cycle off as soon as it cools the passenger compartment).

9   With the compressor engaged, the clutch will make an audible click and the center of the clutch will rotate. After the system reaches operating temperature, feel the two pipes connected to the compressor.

10   There should be a noticeable difference in temperature between the two pipes. If not, the system probably needs a charge. Insert a thermometer in the center air distribution duct while operating the air conditioning system **(see illustration)** - the temperature of the output air should be 35 to 40-degrees F below the ambient air temperature (down to approximately 40-degrees F). If the ambient (outside) air temperature is very high, say 110 degrees F, the duct air temperature may be as high as 60-degrees F, but generally the air conditioning is 35 to 40-degrees F cooler than the ambient air. If the air isn't as cold as it used to be, the system probably needs a charge. Further inspection or testing of the system is beyond the scope of the home mechanic and should be left to a professional.

11   Inspect the sight glass If the refrigerant looks foamy when running, it's low **(see illustration)**. When ambient temperatures are very

13.12 A basic charging kit is available at most auto parts stores - it must say R-134a and so must the cans of refrigerant you buy

13.15 Add R-134a refrigerant to the low-side port only - the procedure will go faster if you wrap the can with a warm, wet towel to prevent icing

hot, bubbles may show in the sight glass even with the proper amount of refrigerant. With the proper amount of refrigerant, when the air conditioning is turned off, the sight glass should show refrigerant that foams, then clears.

Adding refrigerant

Refer to illustrations 13.12 and 13.15

Caution: *Refrigerant has changed from the use of R-12 (through 1993 models), to the "environmentally friendly" R-134a used in 1994 and later models. The two refrigerants are NOT compatible. Even after purging and evacuating an R-12 system, there is enough residual oil and refrigerant in the hoses and components that simply filling the system with R-134a cannot be done. Special fittings and manifold gauge sets are used on the different refrigerant types so that an accidental hook-up of the two systems cannot be made. When replacing entire components, additional refrigerant oil should be added equal to the amount that is removed with the component being replaced. Refrigerant oils, just like refrigerant R-12 vs. R-134a, are not compatible. Be sure to read the can before adding any oil to the system, to make sure it is compatible with the type of system being repaired.*

Note: *Because of Federal regulations, R-12 refrigerant is not available for home-mechanic use, however, cans of R-134 refrigerant are commonly available in auto parts stores. Models with R-12 systems will have to be serviced at a dealership or air conditioning shop.*

12   Buy an automotive charging kit at an auto parts store. A charging kit includes a 14-ounce can of R-134a refrigerant, a tap valve and a short section of hose that can be attached between the tap valve and the system low side service valve **(see illustration)**. Because one can of refrigerant may not be sufficient to bring the system charge up to the proper level, it's a good idea to buy a couple of additional cans. Try to find at least one can that contains red refrigerant dye. If the system is leaking, the red dye will leak out with the refrigerant and help you pinpoint the location of the leak.

13   Connect the charging kit by following the manufacturer's instructions.

14   Back off the valve handle on the charging kit and screw the kit onto the refrigerant can, making sure first that the O-ring or rubber seal inside the threaded portion of the kit is in place. **Warning:** *Wear protective eye wear when dealing with pressurized refrigerant cans.*

15   Remove the dust cap from the low-side charging port and attach the quick-connect fitting on the kit hose **(see illustration)**. **Warning:** *DO NOT hook the charging kit hose to the system high side! The fittings on the charging kit are designed to fit only on the low side of the system.*

16   Warm the engine to normal operating temperature and turn on the air conditioner. Keep the charging kit hose away from the fan and other moving parts.

17   Turn the valve handle on the kit until the stem pierces the can, then back the handle out to release the refrigerant. You should be able to hear the rush of gas. Add refrigerant to the low side of the system until both the outlet and the evaporator inlet pipe feel about the same temperature. Allow stabilization time between each addition. **Warning:** *Never add more than two cans of refrigerant to the system. The can may tend to frost up, slowing the procedure. Wrap a shop towel wet with hot water around the bottom of the can to keep it from frosting.*

18   If you have an accurate thermometer, you can place it in the center air conditioning duct inside the vehicle to monitor the air temperature. A charged system that is working properly, should output air down to approximately 40-degrees F.

19   When the can is empty, turn the valve handle to the closed position and release the connection from the low-side port. Replace the dust cap.

20   Remove the charging kit from the can and store the kit for future use with the piercing valve in the UP position, to prevent inadvertently piercing the can on the next use.

Heating systems

21   If the air coming out of the heater vents isn't hot, the problem could stem from any of the following causes:

a) *The thermostat is stuck open, preventing the engine coolant from warming up enough to carry heat to the heater core. Replace the thermostat (see Section 3).*

b) *A heater hose is blocked, preventing the flow of coolant through the heater core. Feel both heater hoses at the firewall. They should be hot. If one of them is cold, there is an obstruction in one of the hoses or in the heater core, or the heater control valve is shut. Detach the hoses and back flush the heater core with a water hose. If the heater core is clear but circulation is impeded, remove the two hoses and flush them out with a water hose.*

c) *If flushing fails to remove the blockage from the heater core, the core must be replaced. (see Section 11).*

22   If the blower motor speed does not correspond to the setting selected on the blower switch, the problem could be a bad fuse, circuit, switch, blower motor resistor or motor (see Sections 9 and 10).

23   If there isn't any air coming out of the vents:

a) *Turn the ignition ON and activate the fan control. Place your ear at the heating/air conditioning register (vent) and listen. Most motors are audible. Can you hear the motor running?*

b) *If you can't (and have already verified that the blower switch and the blower motor resistor are good), the blower motor itself is probably bad (see Section 9).*

24   If the carpet under the heater core is damp, or if antifreeze vapor or steam is coming through the vents, the heater core is leaking. Remove it (see Section 11) and install a new unit (most radiator shops will not repair a leaking heater core).

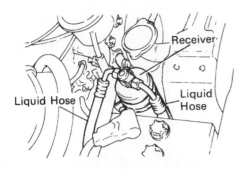

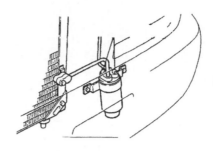

14.3 After the system has been discharged, detach the two refrigerant lines from the top of the receiver/drier and cap them

60,62 SERIES

80 SERIES

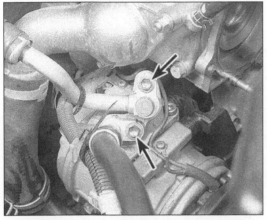

15.4 Disconnect the electrical connector at the compressor, then unbolt the flanges (arrows) and detach the refrigerant lines from the compressor

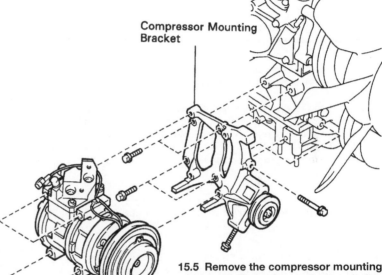

Compressor Mounting Bracket

15.5 Remove the compressor mounting bolts (arrows indicate the two bottom bolts, two more are near the top of the compressor) and remove the compressor

25 Inspect the drain hose from the heater/air conditioning assembly, make sure it is not clogged.

14 Air conditioning receiver/drier - removal and installation

Refer to illustration 14.3

Warning: *The air conditioning system is under high pressure. Do not loosen any hose fittings or remove any components until the system has been discharged. Air conditioning refrigerant should be properly discharged into an EPA-approved recovery/recycling unit by a dealer service department or an automotive air conditioning repair facility. Always wear eye protection when disconnecting air conditioning system fittings.*

1 Have the refrigerant discharged and recovered by an air conditioning technician.
2 Refer to Chapter 11 and remove the grille.
3 Disconnect the refrigerant lines from the receiver/drier and cap the open fittings to prevent entry of moisture **(see illustration)**.
4 Loosen the pinch bolt (or bracket mounting bolts, whichever is easier to get a wrench on) and remove the receiver/drier.
5 Installation is the reverse of removal.
6 Have the system evacuated, charged and leak tested by the shop that discharged it. If the receiver was replaced, have them add new

refrigeration oil to the compressor, about 0.7 ounces for R-12 systems, or 0.34 ounces for R-134a systems. Use only the refrigerant oil compatible with the refrigerant of your system (R-12 vs. R-134a).

15 Air conditioning compressor - removal and installation

Refer to illustrations 15.4 and 15.5

Warning: *The air conditioning system is under high pressure. Do not loosen any hose fittings or remove any components until the system has been discharged. Air conditioning refrigerant should be properly discharged into an EPA-approved recovery/recycling unit by a dealer service department or an automotive air conditioning repair facility. Always wear eye protection when disconnecting air conditioning system fittings.*

1 Have the refrigerant discharged by an automotive air conditioning technician.
2 Disconnect the negative cable from the battery. **Caution:** *If the stereo in your vehicle is equipped with an anti-theft system, make sure you have the correct activation code before disconnecting the battery.*
3 Remove the drivebelt from the compressor (see Chapter 1).
4 Detach the electrical connector and disconnect the refrigerant lines **(see illustration)**.
5 Unbolt the compressor and lift it from the vehicle **(see illustration)**.

16.3 Disconnect the refrigerant line on the right side by unbolting this flange (arrow)

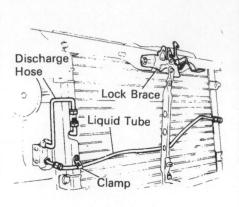

16.4 Remove the condenser mounting bolts - there is one bolt at each end (right side shown)

17.2 Disconnecting the air conditioning lines at the firewall

6   If a new or rebuilt compressor is being installed, follow the directions supplied with the compressor regarding the proper level of oil prior to installation.

7   Installation is the reverse of removal. Replace any O-rings with new ones specifically made for the type of refrigerant in your system and lubricate them with refrigerant oil, also designed specifically for your system (R-12 vs. R-134a).

8   Have the system evacuated, recharged and leak tested by the shop that discharged it.

16 Air conditioning condenser - removal and installation

Refer to illustrations 16.3 and 16.4

Warning: *The air conditioning system is under high pressure. Do not loosen any hose fittings or remove any components until the system has been discharged. Air conditioning refrigerant should be properly discharged into an EPA-approved recovery/recycling unit by a dealer service department or an automotive air conditioning repair facility. Always wear eye protection when disconnecting air conditioning system fittings.*

1   Have the refrigerant discharged by an air conditioning technician.

2   Remove the radiator as described in Section 5.

3   Remove the front grille, the hood lock brace and center support brace for access (see Chapter 11). Disconnect the condenser inlet and outlet fittings **(see illustration)**. Cap the open fittings immediately to keep moisture and contamination out of the system. **Note:** *The left-*

side fitting is connected to the receiver/drier (see Section 14).

4   Remove the condenser mounting bolts, pull the condenser back and lift it out **(see illustration)**.

5   Install the condenser, brackets and bolts, making sure the rubber cushions fit on the mounting points properly.

6   Reconnect the refrigerant lines, using new O-rings where needed. If a new condenser has been installed, add approximately 1.4 to 1.7 ounces (50 cc) of new refrigerant oil of the correct type (R-12 vs. R-134a).

7   Reinstall the remaining parts in the reverse order of removal.

8   Have the system evacuated, charged and leak tested by the shop that discharged it.

17 Air conditioning evaporator and expansion valve - removal and installation

Refer to illustrations 17.2, 17.4a, 17.4b, 17.5a and 17.5b

Warning: *The air conditioning system is under high pressure. Do not loosen any hose fittings or remove any components until the system has been discharged. Air conditioning refrigerant should be properly discharged into an EPA-approved recovery/recycling unit by a dealer service department or an automotive air conditioning repair facility. Always wear eye protection when disconnecting air conditioning system fittings.*

1   Have the air conditioning system discharged (see the Warning above). Disconnect the negative cable from the battery. **Caution:** *If the*

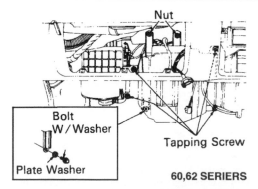

60,62 SERIERS

17.4a Remove the mounting bolts (arrows) at the evaporator housing (62 series)

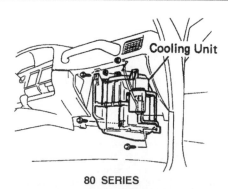

80 SERIES

17.4b Location of the mounting bolts for the evaporator housing on the 80 series

stereo in your vehicle is equipped with an anti-theft system, make sure you have the correct activation code before disconnecting the battery. Remove the glove compartment assembly (see Chapter 11).

2   Disconnect the air conditioning lines at the firewall, use a back-up wrench so as not to damage the fittings **(see illustration)**. Cap the open fittings after disassembly to prevent the entry of air or dirt.

3   On 1FZ-FE engines, remove the ECM (see Chapter 6).

4   Remove nuts and screws retaining the evaporator unit to the firewall and pull the unit out of the vehicle **(see illustrations)**.

5   With the evaporator unit on the bench, remove the screws and clips and separate the top and bottom halves of the case and pull out the evaporator **(see illustrations)**.

6   Pull the thermistor sensor probe from the evaporator core and unbolt the expansion valve and the two short refrigerant lines.

7   The evaporator core can be cleaned with a "fin comb" and blown off with compressed air. **Warning:** *Be sure to wear eye protection when using compressed air.*

8   If the evaporator core is replaced with a new unit, add 1.4 ounces of new refrigerant oil of the correct type (R-12 vs. R-134a) to the system.

9   The remainder of the installation is the reverse of the removal process. Be sure to use new O-rings, and new gaskets on the expansion valve.

10   Have the system evacuated, charged and leak tested by the shop that discharged it.

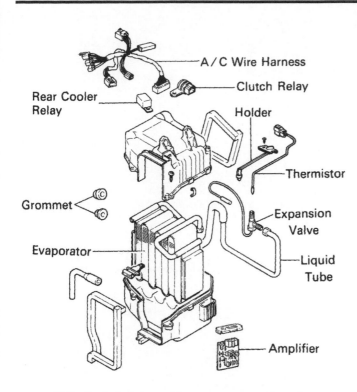

17.5a Exploded view of the evaporator and housing on the 62 series

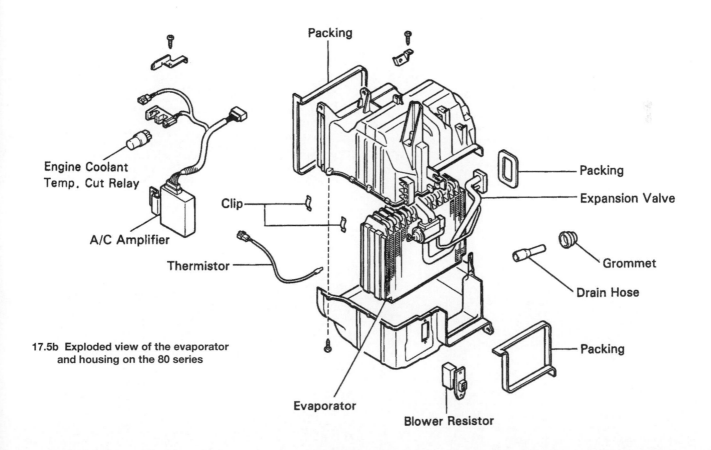

17.5b Exploded view of the evaporator and housing on the 80 series

Notes

Chapter 4 Part A
Fuel and exhaust systems - carbureted engines

Contents

	Section
Accelerator cable - removal, installation and adjustment	9
Air cleaner housing - removal and installation	8
Air filter replacement	See Chapter 1
Carburetor adjustments	11
Carburetor - diagnosis and overhaul	10
Carburetor - removal and installation	12
Catalytic converter	See Chapter 6
Exhaust system check	See Chapter 1

	Section
Exhaust system - servicing and general information	13
Fuel pump - removal and installation	6
Fuel level sending unit - check and replacement	7
Fuel lines and fittings - inspection and replacement	3
Fuel pump/fuel pressure check	2
Fuel tank cleaning and repair - general information	5
Fuel tank - removal and installation	4
General information	1

1 General information

Warning: *Gasoline is extremely flammable, so take extra precautions when you work on any part of the fuel system. Don't smoke or allow open flames or bare light bulbs near the work area, and don't work in a garage where a natural gas-type appliance (such as a water heater or a clothes dryer) with a pilot light is present. Since gasoline is carcinogenic, wear latex gloves when there's a possibility of being exposed to fuel, and, if you spill any fuel on your skin, rinse it off immediately with soap and water. Mop up any spills immediately and do not store fuel-soaked rags where they could ignite. When you perform any kind of work on the fuel system, wear safety glasses and have a Class B type fire extinguisher on hand.*

The fuel system on carbureted models consists of a fuel tank mounted in various locations on the chassis, a mechanically operated fuel pump and a carburetor. A combination of metal and rubber fuel hoses is used to connect these components.

The carburetor is of the two-barrel downdraft type.

2 Fuel pump/fuel pressure check

Warning: *Gasoline is extremely flammable, so take extra precautions when you work on any part of the fuel system. See **Warning** in Section 1.*

Note: *It is a good idea to check the fuel pump and lines for any obvious damage or fuel leakage. Also check all hoses from the tank to the pump, particularly the suction hoses at the fuel tank and pump which, if they have are cracked, may not allow fuel to the fuel pump. It is also possible for the fuel pump diaphragm to rupture internally and leak fuel into the crankcase. If you have excess fuel consumption or fuel smell, and you can't find any external leaks, check the condition of the engine oil for any signs of fuel mixing with the engine oil; the oil level will usually be abnormally high and the oil will be thinned out and have a fuel smell.*

1   Disconnect the fuel line from the carburetor and install a T-fitting. Connect a fuel pressure gauge to the T-fitting with a section of fuel hose and connect the fuel line to the carburetor.

2   Start the engine and allow it to idle. The pressure on the gauge should be 2-1/2 to 8 psi. It should remain constant and return to zero slowly when the engine is shut off. **Note:** *If the engine will not start, crank the engine until you get a reading on the gauge.*

3   An instant pressure drop indicates a faulty outlet valve and the fuel pump must be replaced.

4   If the pressure is too high, check the air vent to see if it is plugged before replacing the pump.

5   If the pressure is too low, be sure the fuel hoses and lines are in good shape and not plugged, then replace the pump.

3   Fuel lines and fittings - inspection and replacement

Warning: *Gasoline is extremely flammable, so take extra precautions when you work on any part of the fuel system. See* **Warning** *in Section 1.*

Inspection

1   Once in a while, you will have to raise the vehicle to service or replace some component (an exhaust pipe hanger, for example). Whenever you work under the vehicle, always inspect the fuel lines and fittings for possible damage or deterioration.

2   Check all hoses and pipes for cracks, kinks, deformation or obstructions.

3   Make sure all hose and pipe clips attach their associated hoses or pipes securely to the underside of the vehicle.

4   Verify all hose clamps attaching rubber hoses to metal fuel lines or pipes are snug enough to assure a tight fit between the hoses and pipes.

Replacement

5   If you must replace any damaged sections, use hoses approved for use in fuel systems or pipes made from steel only (it's best to use an original-type pipe from a dealer that's already flared and pre-bent). Do not install substitutes constructed from inferior or inappropriate material, as this could result in a fuel leak and a fire.

6   Always, before detaching or disassembling any part of the fuel line system, note the routing of all hoses and pipes and the orientation of all clamps and clips to assure that replacement sections are installed in exactly the same manner.

7   Before detaching any part of the fuel system, be sure to relieve the fuel tank pressure by removing the fuel filler cap. Also, disconnect the cable from the negative terminal of the battery. **Caution:** *If the stereo in your vehicle is equipped with an anti-theft system, make sure you have the correct activation code before disconnecting the battery.*

8   Always use new hose clamps after loosening or removing them.

9   While you're under the vehicle, it's a good idea to check the following related components:

a) *Check the condition of the fuel filter - make sure that it's not clogged or damaged (see Chapter 1).*

b) *Inspect the evaporative emission control (EVAP) system. Verify that all hoses are attached and in good condition (see Chapter 6).*

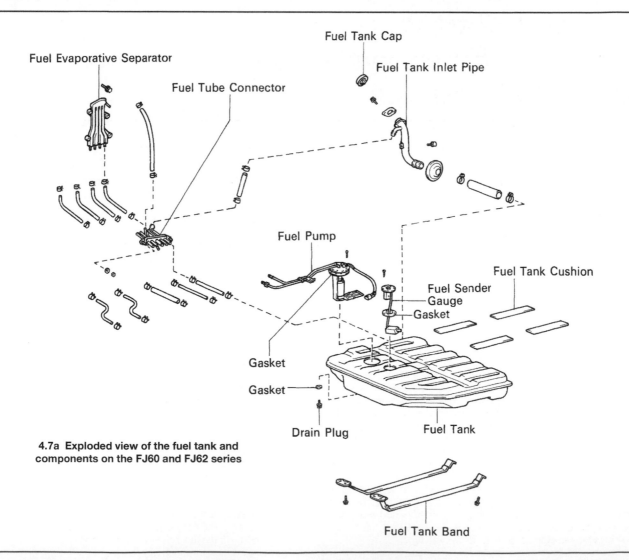

4.7a Exploded view of the fuel tank and components on the FJ60 and FJ62 series

4   Fuel tank - removal and installation

Refer to illustrations 4.7a and 4.7b

Warning: *Gasoline is extremely flammable, so take extra precautions when you work on any part of the fuel system. See* **Warning** *in Section 1.*

Note: *The following procedure is much easier to perform if the fuel tank is empty. Some tanks have a drain plug for this purpose. If the tank does not have a drain plug, drain the fuel into an approved fuel container using a commercially available siphoning kit (NEVER start the siphoning action by mouth) or wait until the fuel tank is nearly empty, if possible.*

1   Remove the fuel tank filler cap to relieve fuel tank pressure.
2   Detach the cable from the negative terminal of the battery. **Caution:** *If the stereo in your vehicle is equipped with an anti-theft system,* make sure you have the correct activation code before disconnecting the battery.
3   If the tank still has fuel in it, you can drain it at the fuel filler hose after raising the vehicle. If the tank has a drain plug, remove it and allow the fuel to collect in an approved gasoline container.
4   Raise the vehicle and place it securely on jackstands.
5   Remove the screws from the top of the fuel filler neck and disconnect the fuel filler neck from the body of the vehicle.
6   Disconnect the fuel lines and the vapor return line. **Note:** *The fuel feed and return lines and the vapor return line are three different diameters, so reattachment is simplified. If you have any doubts, however, clearly label the three lines and the fittings. Be sure to plug the hoses to prevent leakage and contamination of the fuel system.*
7   Loosen the hose clamp(s) and detach the fuel filler neck from the tank **(see illustrations)**. If there is still fuel in the tank, siphon it out

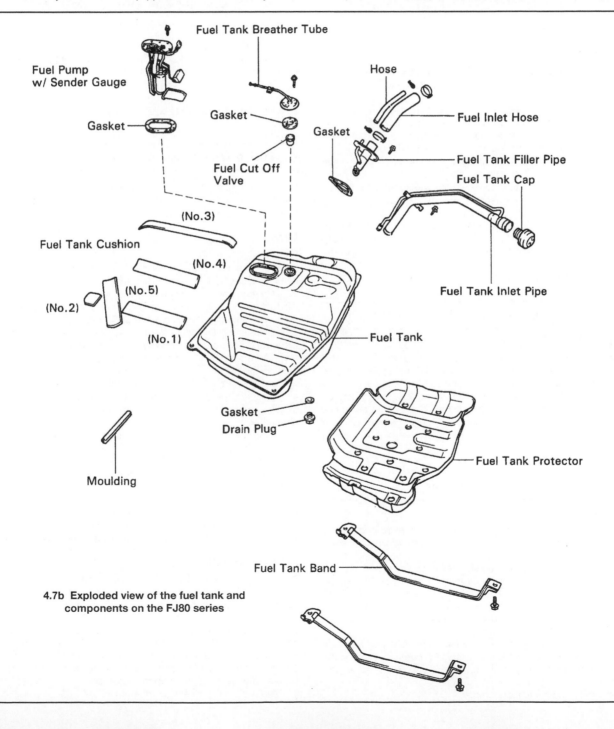

4.7b Exploded view of the fuel tank and components on the FJ80 series

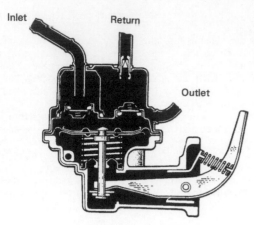

6.3 Cross-sectional schematic of the fuel pump

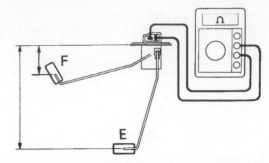

7.6 Connect the ohmmeter to the terminals on the fuel level sending unit electrical connector and check the resistance of the sending unit at the empty and full positions

from the fuel feed port. Remember - NEVER start the siphoning action by mouth! Use a siphoning kit, which can be purchased at most auto parts stores.
8   If the fuel tank is equipped with a protective shield, remove the nuts and detach it from the underside of the chassis.
9   Support the fuel tank with a floor jack. Position a piece of wood between the jack head and the fuel tank to protect the tank.
10   Disconnect both fuel tank retaining straps and pivot them down until they are hanging out of the way.
11   Lower the tank enough to disconnect the wires and ground strap from the fuel pump/fuel gauge sending unit, if you have not already done so. Remove the tank from the vehicle.
12   Some models are equipped with an auxiliary fuel tank. Removal of these tanks is similar to the main fuel tank, but instead of being retained by straps, they are retained by bolts around the flange of the tank.
13   Installation is the reverse of removal.

---

5   Fuel tank cleaning and repair - general information

1   All repairs to the fuel tank or filler neck should be carried out by a professional who has experience in this critical and potentially danger-ous work. Even after cleaning and flushing of the fuel system, explo-sive fumes can remain and ignite during repair of the tank.
2   If the fuel tank is removed from the vehicle, it should not be placed in an area where sparks or open flames could ignite the fumes coming out of the tank. Be especially careful inside garages where a natural gas-type appliance is located, because the pilot light could cause an explosion.

---

6   Fuel pump - removal and installation

Refer to illustration 6.3
Warning: *Gasoline is extremely flammable, so take extra precautions when you work on any part of the fuel system. See* **Warning** *in Sec-tion 1.*
1   The fuel pump is located on the side of the engine block.
2   Disconnect the cable from the negative terminal of the battery.
Caution: *If the stereo in your vehicle is equipped with an anti-theft sys-tem, make sure you have the correct activation code before discon-necting the battery.* Place rags under the fuel pump to catch any gaso-line which may be spilled during removal.
3   Carefully unscrew the fuel line clamps from the fuel lines and slide them from the pump. Remove the fuel lines by carefully twisting the rubber hoses from the fuel pump inlet, outlet and return ports **(see illustration)**.
4   Unbolt and remove the fuel pump.

5   Before installation, coat both sides of the gasket surface with RTV sealant, position the gasket and fuel pump against the block and install the bolts, tightening them securely.
6   Attach the lines to the pump. Use new hose clamps on the fuel hose.
7   Run the engine and check for leaks.

---

7   Fuel level sending unit - check and replacement

Warning: *Gasoline is extremely flammable, so take extra precautions when you work on any part of the fuel system. See* **Warning** *in Sec-tion 1.*

Check

Refer to illustration 7.6
1   Before performing any tests on the fuel level sending unit, deter-mine the actual fuel level in the fuel tank.
2   On early models, access the fuel level sending unit harness near the fuel tank. On later models, remove the middle seat and the fuel pump/sending unit access cover (see Chapter 4B).
3   Disconnect the cable from the negative terminal of the battery.
Caution: *If the stereo in your vehicle is equipped with an anti-theft sys-tem, make sure you have the correct activation code before discon-necting the battery.* Disconnect the fuel level sending unit electrical connector located on top of the fuel tank.
4   Position the ohmmeter probes onto the fuel level sending unit electrical connector and check for resistance. Use the 200 scale on the ohmmeter.
5   With the fuel tank completely full, the resistance of the sending unit should be approximately 17.0 ohms. With the tank empty, the resistance should be approximately 120 ohms.
6   If the readings are incorrect, replace the sending unit. **Note:** *The test can also be performed with the fuel level sending unit removed from the fuel tank. Using an ohmmeter, check the resistance of the sending unit with the swing arm completely down (tank empty) and with the arm up (tank full). The resistance should change steadily from 120 ohms (E) to approximately 17.0 ohms (F)* **(see illustration)**.

Replacement

7   On early models, it will be necessary to remove the fuel tank from the vehicle. On later 80 series models, remove the middle seat to gain access to the fuel pump/fuel level sending unit assembly (see Chap-ter 4B).
8   Carefully angle the sending unit out of the opening without dam-aging the fuel level float located at the bottom of the assembly.
9   Disconnect the electrical connectors from the sending unit.
10   Remove the screw from the side of the sending unit bracket and separate the sending unit from the assembly.
11   Installation is the reverse of removal.

8 Air cleaner housing - removal and installation

1   Remove the air filter from the air cleaner housing (see Chapter 1).
2   Disconnect any vacuum hoses or electrical connectors that would interfere with air cleaner removal and mark them with pieces of numbered tape for reassembly purposes.
3   Lift the air cleaner housing from the engine compartment.
4   Installation is the reverse of removal.

9 Accelerator cable - removal, installation and adjustment

Removal

1   Remove the accelerator cable from the throttle linkage by prying the cable end off the shaft using a small screwdriver.
2   Loosen the accelerator cable locknut and remove the cable from the bracket assembly.
3   Detach the screws and the clips retaining the lower instrument trim panel on the driver's side and remove the trim piece.
4   Pull the cable end out and then up from the accelerator pedal recess.
5   To disconnect the cable at the firewall, remove the cable retainer bracket bolts and push the cable assembly through the firewall from inside the passenger compartment.

Installation

6   Installation is the reverse of removal. **Note:** *To prevent possible interference, flexible components (hoses, wires, etc.) must not be routed within two inches of moving parts, unless routing is controlled.*
7   Operate the accelerator pedal and check for any binding condition by completely opening and closing the throttle.
8   If necessary, at the engine compartment side of the firewall, apply sealant around the accelerator cable to prevent water from entering the passenger compartment.

Adjustment

9   There is no specific adjustment on these type of accelerator cables but it will be necessary to adjust the kickdown cable for the automatic transmission (see Chapter 7B).

10 Carburetor - diagnosis and overhaul

Warning: *Gasoline is extremely flammable, so take extra precautions when you work on any part of the fuel system. See* **Warning** *in Section 1.*

Diagnosis

1   A thorough road test and check of carburetor adjustments should be done before any major carburetor service. Specifications for some adjustments are listed on the *Vehicle Emissions Control Information* (VECI) label found in the engine compartment.
2   Carburetor problems usually show up as flooding, hard starting, stalling, severe backfiring and poor acceleration. A carburetor that's leaking fuel and/or covered with wet looking deposits definitely needs attention.
3   Some performance complaints directed at the carburetor are actually a result of loose, out-of-adjustment or malfunctioning engine or electrical components. Others develop when vacuum hoses leak, are disconnected or are incorrectly routed. The proper approach to analyzing carburetor problems should include the following items:

a) Inspect all vacuum hoses and actuators for leaks and correct installation (see Chapters 1 and 6).
b) Tighten the intake manifold and carburetor mounting nuts/bolts evenly and securely.
c) Perform a compression test and vacuum test (see Chapter 2C).
d) Clean or replace the spark plugs as necessary (see Chapter 1).
e) Check the spark plug wires (see Chapter 1).
f) Inspect the ignition primary wires.
g) Check the ignition timing (follow the instructions printed on the Emissions Control Information label).
h) Check the fuel pump pressure/volume (see Section 2).
i) Check the heat control valve in the air cleaner for proper operation (see Chapter 1).
j) Check/replace the air filter element (see Chapter 1).
k) Check the PCV system (see Chapter 6).
l) Check/replace the fuel filter (see Chapter 1). Also, the strainer in the tank could be restricted.
m) Check for a plugged exhaust system.
n) Check EGR valve operation (see Chapter 6).
o) Check the choke - it should be completely open at normal engine operating temperature (see Chapter 1).
p) Check for fuel leaks and kinked or dented fuel lines (see Chapters 1 and 4)
q) Check accelerator pump operation with the engine off (remove the air cleaner cover and operate the throttle as you look into the carburetor throat - you should see a stream of gasoline enter the carburetor).
r) Check for incorrect fuel or bad gasoline.
s) Check the valve clearances (if applicable) and camshaft lobe lift (see Chapters 1 and 2).
t) Have a dealer service department or other repair shop check the electronic engine and carburetor controls.

4   Diagnosing carburetor problems may require that the engine be started and run with the air cleaner off. While running the engine without the air cleaner, backfires are possible. This situation is likely to occur if the carburetor is malfunctioning, but just the removal of the air cleaner can lean the fuel/air mixture enough to produce an engine backfire. **Warning:** *Do not position any part of your body, especially your face, directly over the carburetor during inspection and servicing procedures. Wear eye protection!*

Overhaul

Refer to illustration 10.5

5   Once it's determined that the carburetor needs an overhaul, several options are available. If you're going to attempt to overhaul the carburetor yourself, first obtain a good-quality carburetor rebuild kit (which will include all necessary gaskets, internal parts, instructions and a parts list). You'll also need some special solvent and a means of blowing out the internal passages of the carburetor with air **(see illustration on following page)**.
6   An alternative is to obtain a new or rebuilt carburetor. They are readily available from dealers and auto parts stores. Make absolutely sure the exchange carburetor is identical to the original. A tag is usually attached to the top of the carburetor or a number is stamped on the float bowl. It will help determine the exact type of carburetor you have. When obtaining a rebuilt carburetor or a rebuild kit, make sure the kit or carburetor matches your application exactly. Seemingly insignificant differences can make a large difference in engine performance.
7   If you choose to overhaul your own carburetor, allow enough time to disassemble it carefully, soak the necessary parts in the cleaning solvent (usually for at least one-half day or according to the instructions listed on the carburetor cleaner) and reassemble it, which will usually take much longer than disassembly. When disassembling the carburetor, match each part with the illustration in the carburetor kit and lay the parts out in order on a clean work surface. Overhauls by inexperienced mechanics can result in an engine which runs poorly or not at all. To avoid this, use care and patience when disassembling the carburetor so you can reassemble it correctly.
8   Because carburetor designs are constantly modified by the manufacturer in order to meet increasingly stringent emissions regulations, it isn't feasible to include a step-by-step overhaul of each type. You'll receive a detailed, well illustrated set of instructions with the carburetor overhaul kit.

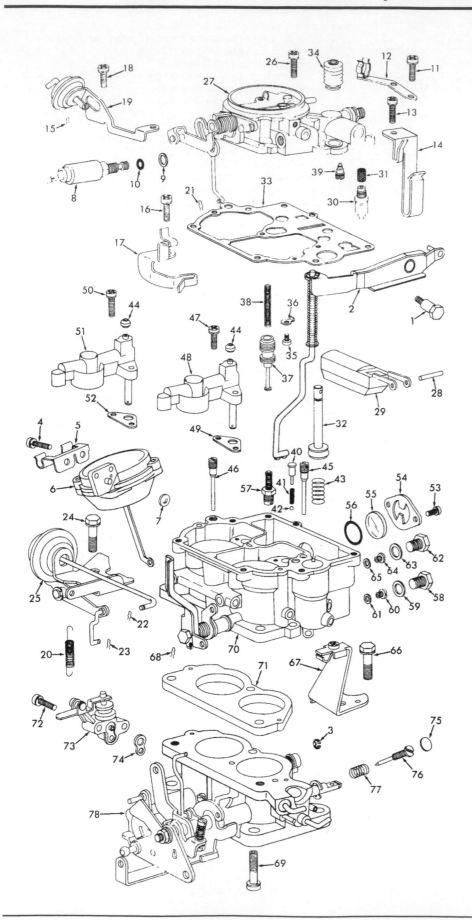

10.5 Exploded view of the carburetor

1. SCREW - PUMP LEVER
2. PUMP LEVER & ROD ASSY.
3. E-CLIP - SEC. DIAPHRAGM LINK
4. SCREW & LOCKWASHER (2) - SEC. DIAPHRAGM ASSY. & HOSE BRACKET
5. BRACKET - HOSE
6. SECONDARY DIAPHRAGM ASSY.
7. GASKET - SEC. DIAPHRAGM ASSY.
8. SOLENOID ASSY. - IDLE SHUT OFF
9. GASKET - SOLENOID
10. O-RING - SOLENOID
11. SCREW & LOCKWASHER (2) - WIRE SOCKET BRACKET
12. BRACKET - WIRE SOCKET
13. SCREW & LOCKWASHER - HOSE & WIRE BRACKET
14. BRACKET - HOSE & WIRE
15. RETAINER - CHOKE PULL OFF ROD
16. SCREW & LOCKWASHER - WIRE BRACKET & CHOKE PULL OFF ASSY.
17. BRACKET - WIRE
18. SCREW & LOCKWASHER - CHOKE PULL OFF BRACKET
19. CHOKE PULL OFF DIAPHRAGM ASSY.
20. SPRING - SEC. THROTTLE RETURN
21. RETAINER - FAST IDLE ROD
22. RETAINER - THROTTLE POSITIONER ROD
23. RETAINER - LINK
24. SCREW & LOCKWASHER - THROTTLE POSITIONERS ASSY.
25. THROTTLE POSITIONER, BRACKET & LEVER ASSY.
26. SCREW & LOCKWASHER (2) - BOWL COVER
27. BOWL COVER ASSY.
28. PIN - FLOAT
29. FLOAT & LEVER ASSY.
30. NEEDLE & SEAT ASSY.
31. SCREEN - NEEDLE SEAT
32. PUMP PLUNGER ASSY.
33. GASKET - BOWL COVER
34. BOOT - PUMP
35. SCREW & LOCKWASHER - LOCK
36. LOCK - POWER PISTON
37. POWER PISTON
38. SPRING - POWER PISTON
39. JET - PRI. SLOW AIR --
40. STOPPER - PUMP CHECK SPRING
41. SPRING - PUMP CHECK BALL
42. BALL - PUMP DISC. CHECK
43. SPRING - PUMP RETURN
44. SEAL (2) - VENTURI CLUSTER
45. JET - PRI. SLOW --
46. JET - SEC. SLOW --
47. SCREW & LOCKWASHER (2) - PRI. VENTURI
48. PRI. VENTURI ASSY.
49. GASKET - VENTURI
50. SCREW & LOCKWASHER (2) - SEC. VENTURI
51. SEC. VENTURI ASSY.
52. GASKET - VENTURI
53. SCREW (2) - WINDOW RETAINER
54. RETAINER - WINDOW
55. WINDOW - FUEL BOWL
56. O-RING - FUEL BOWL WINDOW
57. POWER VALVE ASSY.
58. PLUG - PRI. MAIN JET
59. GASKET - PLUG
60. JET - PRI. MAIN (BRASS) --
61. GASKET - PRI. MAIN JET
62. PLUG - SEC. MAIN JET
63. GASKET - PLUG
64. JET - SEC. MAIN (CROME) --
65. GASKET - SEC. MAIN JET
66. SCREW & LOCKWASHER - BRACKET & THROTTLE BODY
67. BRACKET - CHOKE CABLE
68. RETAINER - UNLOADER LINK
69. SCREW & LOCKWASHER - THROTTLE BODY
70. FLOAT BOWL ASSY.
71. INSULATOR BLOCK & GASKETS THROTTLE BODY
72. SCREW & LOCKWASHER (2) - SEC. IDLE SHUT OFF
73. SEC. IDLE SHUT OFF VALVE ASSY.
74. GASKET - SEC. IDLE VALVE ASSY.
75. PLUG - IDLE NEEDLE
76. NEEDLE - IDLE ADJUSTING
77. SPRING - IDLE NEEDLE
78. THROTTLE BODY ASSY.

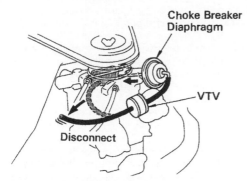

11.1 Disconnect the vacuum hose from the carburetor

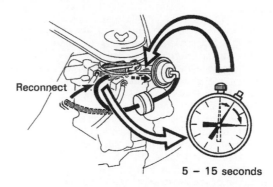

11.3 With the engine warmed up, make sure the choke breaker linkage is pulled back into the diaphragm when the hose is reconnected

11   Carburetor adjustments

Note: *These carburetor adjustments are strictly in-vehicle adjustments. During overhaul, refer to the instructions included in the overhaul kit for complete procedures and any additional adjustments that are required.*

Choke breaker

Refer to illustrations 11.1 and 11.3

1   Start the engine and allow it to idle. Disconnect the vacuum hose between the carburetor and the VTV at the carburetor side **(see illustration)**.
2   Observe that the choke breaker linkage returns quickly to recoiled position by spring pressure
3   Reconnect the hose and confirm that the choke breaker linkage is pulled back within 5 to 15 seconds of reconnecting **(see illustration)**.

Choke opener

Refer to illustrations 11.5 and 11.7

4   With the coolant temperature below 41-degrees F (5 degrees C), start the engine and disconnect the hose from the choke opener diaphragm and reconnect it.
5   Make sure the choke linkage remains steady and does not move **(see illustration)**.
6   With the engine at normal operating temperature, disconnect the hose from the choke opener and observe that the choke linkage returns.
7   Reconnect the hose and make sure the choke is pulled back by the choke opener **(see illustration)**.

Choke knob and cable (models with manual choke)

8   Remove the air cleaner assembly and pull the choke knob out all the way and make sure the choke valve is fully closed (horizontal position).

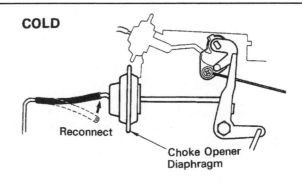

11.5 Disconnect the hose from the choke opener diaphragm

9   Make sure the choke valve is fully open (vertical position) when the choke knob is pushed in all the way.

Fast idle speed adjustment

Refer to illustration 11.14

10   With the engine completely warmed to operating temperature, turn the ignition key OFF and remove the air cleaner assembly
11   Pull the choke knob out completely.
12   Disconnect the vacuum hoses from the distributor and plug the hose ends.
13   Disconnect the vacuum hoses from port S of the VCV for the EVAP and EGR valve and plug both ends.
14   Install a tachometer and adjust the fast idle speed to 1,800 rpm **(see illustration)**.
15   Push the choke knob in all the way and confirm that the idle speed returns to normal (see Chapter 1).

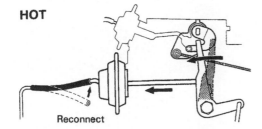

11.7 With the engine at operating temperature, make sure the choke linkage is pulled back into the diaphragm when the hose is reconnected

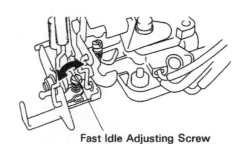

11.14 Location of the fast idle adjusting screw

Idle mixture

Note: Be sure to check the ignition timing, the distributor cap and rotor, the condition of the spark plugs and spark plug wires and, if necessary, perform a complete tune-up before attempting to make the following carburetor adjustment. Often, problems with the mixture are due to a vacuum leak (see Troubleshooting at the front of this manual and the vacuum gauge checks in Chapter 2C).

16   After the engine has reached normal operating temperature, set the parking brake and block the wheels, then remove the air cleaner and turn the idle speed adjusting screw out as far as possible without the engine running rough, since the throttle plate(s) must be as nearly closed as possible when adjusting the idle mixture.

17   Turn the idle mixture screw clockwise until the idle speed drops a noticeable amount **(see illustration 10.5, item no. 76)**. Now slowly turn the mixture screw out until the maximum idle speed is reached, but no further. **Note:** If it was necessary to turn the idle speed screw in (to keep the engine running) before setting the idle mixture, adjust the idle speed, then perform the mixture adjustment procedure again. This will ensure that the idle mixture is set with the throttle plates fully closed, so the engine is drawing the fuel mixture only from the idle circuit.

12 Carburetor - removal and installation

Warning: Gasoline is extremely flammable, so take extra precautions when you work on any part of the fuel system. See **Warning** in Section 1.

Removal

1   Remove the fuel filler cap to relieve fuel tank pressure. Disconnect the cable from the negative terminal of the battery. **Caution:** If the stereo in your vehicle is equipped with an anti-theft system, make sure you have the correct activation code before disconnecting the battery.

2   Remove the air cleaner housing (see Section 8).

3   Disconnect the accelerator cable from the throttle lever (see Section 9).

4   If the vehicle is equipped with an automatic transmission, disconnect the kickdown cable or linkage from the throttle lever.

5   Clearly label all vacuum hoses and fittings, then disconnect the hoses.

6   Disconnect the fuel line from the carburetor.

7   Label the wires and terminals, then unplug all the electrical connectors.

8   Remove the mounting fasteners and detach the carburetor from the intake manifold. Remove the carburetor mounting gasket. Stuff a rag into the intake manifold openings to prevent debris from entering.

Installation

9   Use a gasket scraper to remove all traces of gasket material and sealant from the intake manifold (and the carburetor, if it's being reinstalled), then remove the shop rag from the manifold openings. Clean the mating surfaces with lacquer thinner or acetone.

10   Place a new gasket on the intake manifold.

11   Position the carburetor on the gasket and install the mounting fasteners.

12   To prevent carburetor distortion or damage, tighten the fasteners to approximately 16 ft-lbs in a criss-cross pattern, 1/4-turn at a time.

13   The remaining installation steps are the reverse of removal.

14   Check and, if necessary, adjust the idle speed (see Chapter 1).

15   If the vehicle is equipped with an automatic transmission, refer to Chapter 7B for the kickdown cable or linkage adjustment procedure.

16   Start the engine and check carefully for fuel leaks.

13 Exhaust system - servicing and general information

Warning: Inspection and repair of exhaust system components should be done only after enough time has elapsed after driving the vehicle to allow the system components to cool completely. Also, when working under the vehicle, make sure it is securely supported on jackstands.

1   The exhaust system consists of the exhaust manifold(s), the catalytic converter(s), the muffler, the tailpipe and all connecting pipes, brackets, hangers and clamps. The exhaust system is attached to the body with mounting brackets and rubber hangers. If any of the parts are improperly installed, excessive noise and vibration will be transmitted to the body.

2   Conduct regular inspections of the exhaust system to keep it safe and quiet. Look for any damaged or bent parts, open seams, holes, loose connections, excessive corrosion or other defects which could allow exhaust fumes to enter the vehicle. Deteriorated exhaust system components should not be repaired; they should be replaced with new parts.

3   If the exhaust system components are extremely corroded or rusted together, welding equipment will probably be required to remove them. The convenient way to accomplish this is to have a muffler repair shop remove the corroded sections with a cutting torch. If, however, you want to save money by doing it yourself (and you don't have a welding outfit with a cutting torch), simply cut off the old components with a hacksaw. If you have compressed air, special pneumatic cutting chisels can also be used. If you do decide to tackle the job at home, be sure to wear safety goggles to protect your eyes from metal chips and work gloves to protect your hands.

4   Here are some simple guidelines to follow when repairing the exhaust system:

a) Work from the back to the front when removing exhaust system components.

b) Apply penetrating oil to the exhaust system component fasteners to make them easier to remove.

c) Use new gaskets, hangers and clamps when installing exhaust system components.

d) Apply anti-seize compound to the threads of all exhaust system fasteners during reassembly.

e) Be sure to allow sufficient clearance between newly installed parts and all points on the underbody to avoid overheating the floor pan and possibly damaging the interior carpet and insulation. Pay particularly close attention to the catalytic converter and heat shield.

Chapter 4 Part B
Fuel and exhaust systems -
fuel-injected engines

Contents

	Section			Section
Accelerator cable - removal, installation and adjustment	8		Fuel level sending unit - check and replacement	6
Air cleaner assembly - removal and installation	7		Fuel lines and fittings - inspection and replacement	4
Catalytic converter	See Chapter 6		Fuel pressure relief	2
CHECK ENGINE light	See Chapter 6		Fuel pump/fuel pressure - check	3
Electronic Fuel Injection (EFI) system - check	10		Fuel pump - removal and installation	5
Electronic Fuel Injection (EFI) system - component check			Fuel system check	See Chapter 1
and replacement	11		Fuel tank cap gasket replacement	See Chapter 1
Electronic Fuel Injection (EFI) system - general information	9		Fuel tank cleaning and repair - general information	See Chapter 4A
Exhaust manifold - removal and installation	See Chapter 2A, 2B		Fuel tank - removal and installation	See Chapter 4A
Exhaust system check	See Chapter 1		General information	1
Exhaust system servicing - general information	See Chapter 4A		Intake manifold - removal and installation	See Chapter 2A, 2B
Fuel filter replacement	See Chapter 1		Underhood hose check and replacement	See Chapter 1

Specifications

Fuel system

Fuel pressure
Ignition ON, engine not running	38 to 46 psi

Engine idling
Vacuum sensing hose detached	38 to 44 psi
Vacuum sensing hose attached	31 to 37 psi
Fuel system hold pressure	21 psi
Fuel injector resistance	13.4 to 14.2 ohms

Idle Speed Control (ISC) valve resistance
(1988 through 1992 models)	10 to 30 ohms

Idle Air Control (IAC) valve resistance (1993 and later models)
Cold (below 122-degrees F)	15 to 25 ohms
Hot (above 122-degrees F)	20 to 30 ohms
Cold start injector resistance	2 to 4 ohms

Idle Speed .... See Chapter 1

Torque specifications

	Ft-lbs
Throttle body mounting bolts	14
Fuel rail mounting bolts	14
Fuel line banjo fitting at fuel pump	22
Air intake plenum bolts/nuts	15

1 General information

The fuel system consists of a fuel tank, an electric fuel pump (located in the fuel tank), an EFI main relay, fuel injectors, a fuel pressure regulator, an air cleaner assembly and a throttle body unit. The fuel system on 1988 through 1992 models equipped with the 3F-E engine is an Electronic Fuel Injection (EFI) system. 1993 and later models equipped with the 1FZ-FE engine are equipped with an updated sequential EFI system. There are some slight variations in the fuel pressure controls and the air intake detection systems on later models. Refer to Chapter 4A for Land Cruiser models equipped with carbureted fuel systems installed on the 2F engine.

Electronic Fuel Injection (EFI) system

Electronic fuel injection uses timed impulses to inject the fuel directly into the intake port of each cylinder. The injectors are controlled by the Electronic Control Module (ECM). The ECM monitors various engine parameters and delivers the exact amount of fuel, in the correct sequence, into the intake ports. The throttle body serves only to control the amount of air passing into the system. Because each cylinder is equipped with an injector mounted immediately adjacent to the intake valve, much better control of the air/fuel mixture ratio is possible.

EFI models are equipped with a Vacuum Switching Valve (VSV) for fuel pressure control. This valve switches vacuum to the fuel pressure regulator during warm start-up if the computer detects high

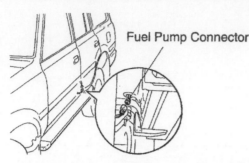

2.3 To depressurize the fuel system, unplug the fuel pump electrical connector, start the engine and allow it to stall (later model shown - on earlier models without this connector, remove the circuit opening relay and crank the engine)

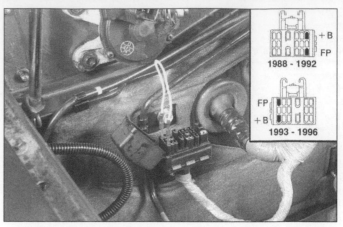

3.2 Bridge terminals FP and +B on the test connector using a jumper wire or paper clip

coolant temperatures. This allows for less fuel pressure and a leaner air/fuel mixture. Early models are equipped with the High Temperature Line-Up Pressure system that uses a temperature switch to regulate the VSV directly. Later models use the computer to regulate the VSV.

Fuel pump and lines

Fuel is circulated from the fuel tank to the fuel injection system, and back to the fuel tank, through a pair of metal lines running along the underside of the vehicle. An electric fuel pump is attached to the fuel sending unit inside the fuel tank. A vapor return system routes all vapors and hot fuel back to the fuel tank through a separate return line.

The fuel pump will operate as long as the engine is cranking or running and the ECM is receiving ignition reference pulses from the electronic ignition system (see Chapter 5). If there are no reference pulses, the fuel pump will shut off after 2 or 3 seconds.

Exhaust system

The exhaust system includes an exhaust manifold fitted with an exhaust oxygen sensor, a catalytic converter, an exhaust pipe, and a muffler.

The catalytic converter is an emission control device added to the exhaust system to reduce pollutants. A single-bed converter is used in combination with a three-way (reduction) catalyst. Refer to Chapter 6 for more information regarding the catalytic converter.

2   Fuel pressure relief

Refer to illustration 2.3
Warning: *Gasoline is extremely flammable, so take extra precautions when you work on any part of the fuel system. Don't smoke or allow open flames or bare light bulbs near the work area, and don't work in a garage where a natural gas-type appliance (such as a water heater or a clothes*

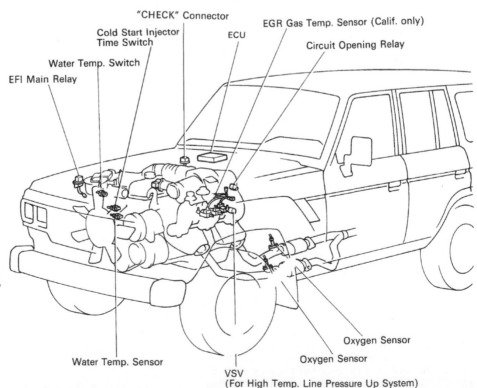

3.6a Location of the EFI main relay, the circuit opening relay and other fuel system components on 1988 through 1990 models

dryer) with a pilot light is present. Since gasoline is carcinogenic, wear latex gloves when there's a possibility of being exposed to fuel, and, if you spill any fuel on your skin, rinse it off immediately with soap and water. Mop up any spills immediately and do not store fuel-soaked rags where they could ignite. The fuel system is under constant pressure, so, if any fuel lines are to be disconnected, the fuel pressure in the system must be relieved first. When you perform any kind of work on the fuel system, wear safety glasses and have a Class B type fire extinguisher on hand.

1   Before servicing any fuel system component, you must relieve the fuel pressure to minimize the risk of fire or personal injury.

2   Remove the fuel filler cap - this will relieve any pressure built up in the tank.

3   Unplug the fuel pump electrical connector **(see illustration)**. On later models this connector is easily accessible through the left rear wheel opening. On 1991 through 1994 models, you can unplug the electrical connector at the fuel pump after removing the rear seats, floor mat and the access cover (see Section 5). On 1988 through 1990 models, remove the circuit opening relay and crank the engine over (if you can't find a fuel pump connector near the fuel tank).

4   Start the engine and wait for the engine to stall, then turn the ignition key to Off.

5   The fuel system is now depressurized. **Note:** *Place a rag around the fuel line before removing any hose clamp or fitting to absorb any fuel that spills out.*

6   Before working on the fuel system, disconnect the cable from the negative terminal of the battery. **Caution:** *If the stereo in your vehicle is equipped with an anti-theft system, make sure you have the correct activation code before disconnecting the battery.*

---

3   Fuel pump/fuel pressure - check

Warning: *Gasoline is extremely flammable, so take extra precautions when you work on any part of the fuel system. Don't smoke or*

allow open flames or bare light bulbs near the work area, and don't work in a garage where a natural gas-type appliance (such as a water heater or a clothes dryer) with a pilot light is present. Since gasoline is carcinogenic, wear latex gloves when there's a possibility of being exposed to fuel, and, if you spill any fuel on your skin, rinse it off immediately with soap and water. Mop up any spills immediately and do not store fuel-soaked rags where they could ignite. The fuel system is under constant pressure, so, if any fuel lines are to be disconnected, the fuel pressure in the system must be relieved first. When you perform any kind of work on the fuel system, wear safety glasses and have a Class B type fire extinguisher on hand.

Fuel pump operation check

Refer to illustrations 3.2, 3.6a, 3.6b, 3.6c, 3.6d, 3.6e and 3.6f

1   Turn ON the ignition switch (but do not start the engine).

2   Bridge terminals +B and FP of the test connector with a jumper wire **(see illustration)**. **Note:** *1988 through 1994 models use a Test Connector with a different configuration and terminal arrangement than later models. Look carefully at the terminal guide under the Test Connector cover for the proper terminal designations. On 1995 and later models, it is not possible to power the fuel pump using the test connector. On these models, start the engine to obtain fuel pressure readings.*

3   The fuel pump is now activated. Listen for fuel pump noise from the fuel tank (under the rear seat) and verify that there is pressure in the hose from the fuel filter.

4   Remove the jumper wire. Close the cap on the test connector.

5   Turn the ignition switch OFF.

6   If the fuel pump did not operate, inspect the following electrical components: the EFI 15-amp fuse and the ignition switch 30-amp fuse (AM2) (see Chapter 12) and/or the EFI main relay and the circuit opening relay, as described later in this section, battery voltage to the fuel pump, and the wiring and electrical connectors **(see illustrations)**.

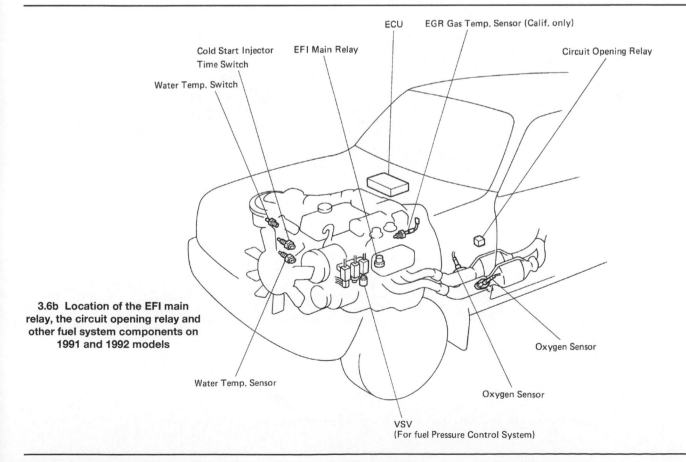

3.6b Location of the EFI main relay, the circuit opening relay and other fuel system components on 1991 and 1992 models

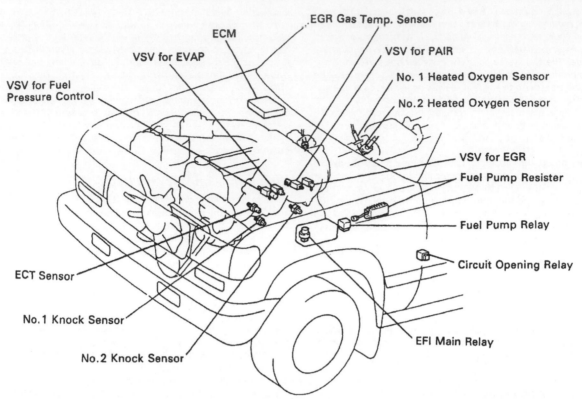

3.6c Location of the EFI main relay, the circuit opening relay, fuel pump relay and other fuel system components on 1993 and later models

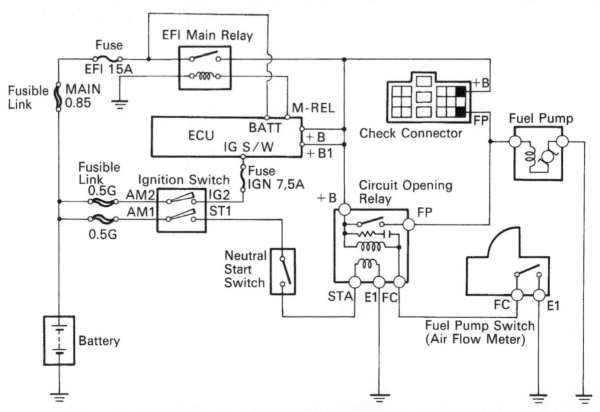

3.6d Typical electrical schematic of the fuel pump circuit (1988 through 1992 models)

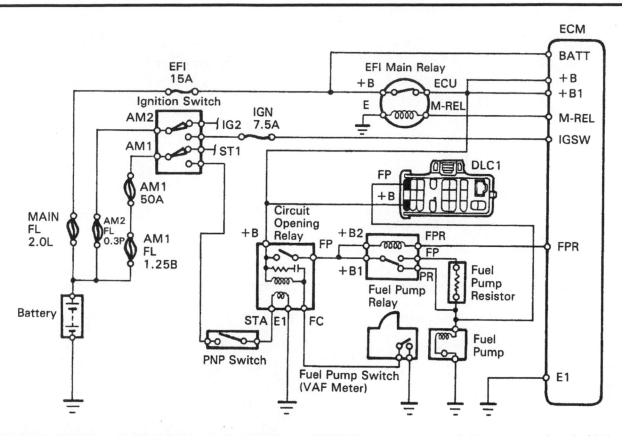

3.6e Typical electrical schematic of the fuel pump circuit (1993 and 1994) (1995 and later models similar but not equipped with the DLC1)

3.6f Check for battery voltage directly at the fuel pump electrical connector on the top of the fuel tank

Note: *1993 and later models are equipped with a fuel pump relay located near the Main Relay. On these models, check for battery voltage to the fuel pump relay and the operation of the relay.*

Fuel pressure check
Refer to illustrations 3.7, 3.11a and 3.11b

7   A fuel pressure gauge, capable of measuring a minimum of 50-psi, equipped with a banjo fitting on the end of the hose is required for the following procedure (available at an automotive parts store). There are a couple of alternatives if you are unable to obtain the special fuel pressure gauge set:

a) *Obtain a 12 mm banjo fitting that will adapt to your fuel pressure gauge hose with a hose clamp.*

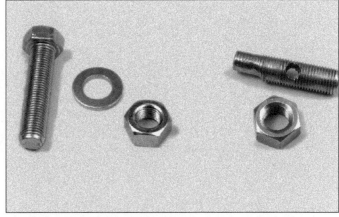

3.7 Cut the head off the bolt (12 mm diameter/1.25 thread pitch) and drill a hole directly through the bolt, lengthwise. Grind the end to accept the pressure gauge hose and drill another hole to allow system pressure to flow. It is important that this hole is drilled in the correct location. The easiest method is to use the banjo bolt that was removed from the fuel filter and place it directly next to the tool for the correct alignment of this passage for fuel flow

b) *If you can't find the correct size banjo fitting, obtain a 12 mm bolt with 1.25 mm thread pitch, cut the head off and drill a hole through the center and one through the side, perpendicular to the lengthwise hole. Add a locknut with the same thread pitch and seal the threads with Teflon tape* **(see illustration).**

8   Remove the fuel tank cap to relieve any pressure that has built up in the tank.
9   Verify that the battery voltage is 12 volts or more (see Chapter 5).
10   Relieve the fuel pressure (see Section 2).

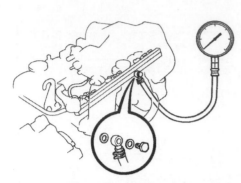

3.11a On 1988 through 1992 models, remove the fuel line fitting at the cold start injector pipe and install the fuel pressure gauge using the homemade tool or a special banjo fitting

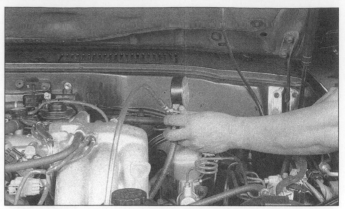

3.11b On 1993 and later models, connect the fuel pressure gauge to the fuel filter inlet fitting

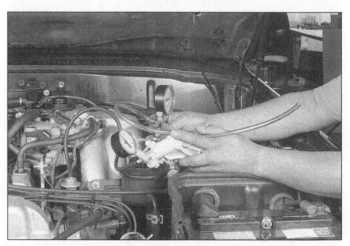

3.20 First check the fuel pressure without vacuum applied to the fuel pressure regulator and then with vacuum applied - fuel pressure should DECREASE as vacuum INCREASES

11   There are two methods possible for testing the fuel pressure (**see illustrations**):

a) *If you have the special banjo fitting required to attach the fuel pressure gauge, remove the fuel rail bolt at the end of the fuel rail and install the special tool into the fuel rail.*

b) *If you have constructed the drilled-out bolt and locknut tool, detach one side of the fuel rail and install the fuel pressure gauge at this point.*

12   To attach the fuel pressure gauge:

a) *If you are using the special tool setup, use the special banjo fitting on the fuel rail to attach the fuel pressure gauge to the fuel rail. Be sure to use crush washers on both sides of the banjo fitting.*

b) *If you are using a drilled-out 12 mm bolt, working in the engine compartment, remove the fuel line fitting and install the home made tool, tighten the locknut and attach the fuel pressure gauge hose with a hose clamp.*

13   Wipe off any gasoline that has leaked out of the fuel rail (or filter).

14   Place the transmission in Neutral (manual) or Park (automatic) and apply the parking brake.

15   On 1988 through 1994 models, bridge terminals +B and FP of the test connector (**see illustration 3.2**). Note: *On 1995 and later models, it is not possible to power the fuel pump using the test connector. On these models, start the engine to obtain fuel pressure readings.*

16   Turn the ignition to ON (engine not running). Measure the fuel pressure and compare it to the fuel pressure listed in this Chapter's Specifications.

a) *If the pressure is high, check for a restricted fuel return line. If the line is clear, replace the pressure regulator.*

b) *If the pressure is low, pinch the fuel return line. If the pressure goes up, replace the fuel pressure regulator. If the pressure does not increase, check the fuel feed line, the fuel pump and the fuel filter.*

17   Remove the jumper wire from the service electrical connector or fuel pump connector.

18   Start the engine.

a) *Measure the fuel pressure at idle and compare your reading to the fuel pressure listed in this Chapter's Specifications.*

b) *If the pressure is not as specified, check the vacuum sensing hose and fuel pressure regulator (see Steps 20 through 24).*

19   Stop the engine and verify that the fuel pressure remains at 21 psi or more for five minutes after the engine is turned off. If the pressure bleeds down, the fuel pressure regulator, the fuel pump or a fuel injector may be leaking.

Fuel pressure regulator check

Refer to illustrations 3.20 and 3.21

20   Disconnect and plug the vacuum hose from the fuel pressure regulator and connect a hand-held vacuum pump to the regulator. Start the engine and read the fuel pressure gauge without vacuum applied

to the fuel pressure regulator. Apply vacuum to the regulator and check the fuel pressure again (**see illustration**). The fuel pressure should decrease as vacuum increases. Compare your readings with the values listed in this Chapter's Specifications.

21   Reconnect the vacuum hose to the regulator and check the fuel pressure at idle, comparing your reading with the value listed in this Chapter's Specifications. Disconnect the hose and watch the gauge - the pressure should rise to the maximum specified pressure as soon as the hose is disconnected. If the pressure at idle was too high (with the hose disconnected), connect a vacuum gauge to the hose and check for vacuum (**see illustration**). If there is no reading on the gauge, check the hose, the air intake plenum and intake manifold for a vacuum leak.

22   EFI models are equipped with a Vacuum Switching Valve (VSV) for fuel pressure control. This valve switches vacuum to the fuel pressure regulator during warm start-up if the computer detects high coolant temperatures. This allows for less fuel pressure and a leaner air/fuel mixture. Early models are equipped with the High Temperature Line-Up Pressure system that uses a temperature switch to regulate the VSV directly. Later models use the computer to regulate the VSV. Have the VSV checked by a dealer service department.

23   If the fuel pressure is LOW, pinch the fuel return line shut and watch the gauge. If the pressure doesn't rise, the fuel pump is defective or there is a restriction in the fuel feed line. If the pressure rises sharply, replace the fuel pressure regulator (see Section 11).

24   If the indicated fuel pressure is too high, relieve the fuel pressure (see Section 2), disconnect the fuel return line and blow through it to check for blockage. If there is no blockage, replace the fuel pressure

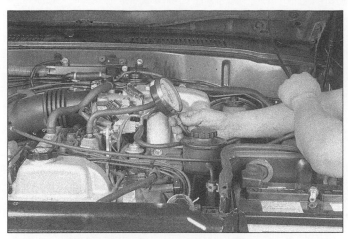

3.21 Connect a vacuum gauge to the vacuum line leading to the fuel pressure regulator and check for vacuum

3.31 Locate and remove the EFI relay and check for battery voltage to the relay with the ignition key ON (1994 1FZ-FE engine shown)

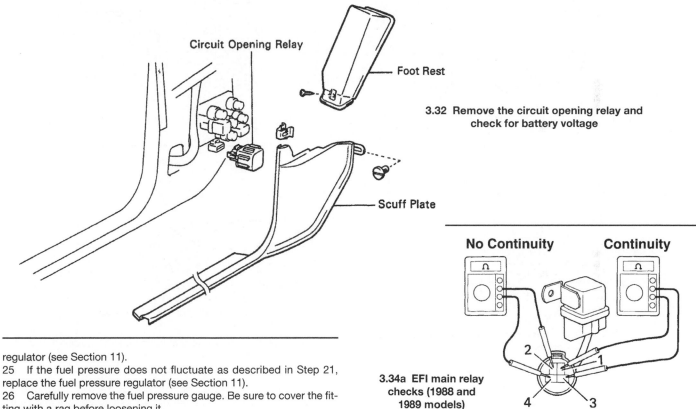

Circuit Opening Relay

Foot Rest

3.32 Remove the circuit opening relay and check for battery voltage

Scuff Plate

No Continuity   Continuity

3.34a EFI main relay checks (1988 and 1989 models)

regulator (see Section 11).

25   If the fuel pressure does not fluctuate as described in Step 21, replace the fuel pressure regulator (see Section 11).

26   Carefully remove the fuel pressure gauge. Be sure to cover the fitting with a rag before loosening it.

27   Using new sealing washers, reattach the fuel line and banjo fitting to the fuel rail.

28   Wipe up any spilled gasoline.

29   Start the engine and check for leaks.

EFI main relay, fuel pump relay and circuit opening relay checks

Voltage checks

Refer to illustrations 3.31 and 3.32

30   There are two relays involved in the fuel pump circuit. First, test for battery voltage to the EFI main relay and then the circuit opening relay. Later models are also equipped with a fuel pump relay **(see illustrations 3.6a, 3.6b and 3.6c)**.

31   Remove the EFI main relay from the electrical connector and with

the ignition key ON (engine not running), check for battery voltage **(see illustration)**.

32   If battery voltage is present, insert the relay back into the connector and check for battery voltage at the circuit opening relay **(see illustration)**. The circuit opening relay is located behind the driver's side kick panel.

33   If battery voltage is present at the relay connectors, check the relays.

EFI main relay

Refer to illustrations 3.34a, 3.34b, 3.35a and 3.35b

34   Using an ohmmeter, check for continuity across terminals 1 and 3 **(see illustrations)**. Check that there is no continuity across terminals 2 and 4.

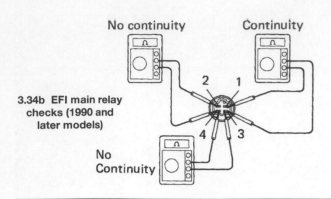

3.34b EFI main relay checks (1990 and later models)

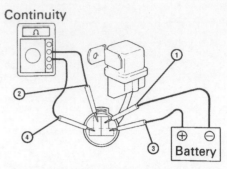

3.35a Apply battery voltage to terminals 1 and 3 and check for continuity between terminals 2 and 4 (1988 and 1989 models)

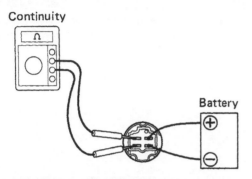

3.35b Apply battery voltage to terminals 1 and 3 and check for continuity between terminals 2 and 4 (1990 and later models)

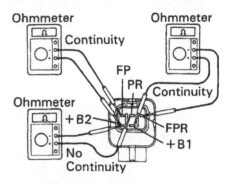

3.36 Fuel pump relay checks (1993 through 1995 models)

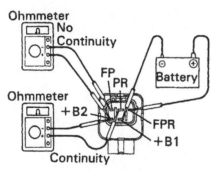

3.37 Apply battery voltage and check for continuity across the designated terminals

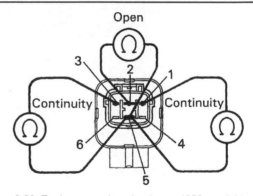

3.38 Fuel pump relay checks on 1996 models

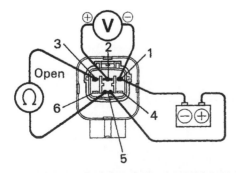

3.39 Apply battery voltage to terminals 1 and 5 and check for continuity between terminals 5 and 3 (1996 models)

35   Connect a 12 volt battery to terminals 1 and 3 (see illustrations). Using an ohmmeter, check for continuity across terminals 2 and 4. Continuity should exist. If the test results are incorrect, replace the relay.

Fuel pump relay (1993 through 1995 models)

Refer to illustrations 3.36 and 3.37

36   Using an ohmmeter, check for continuity across terminals +B1 and FPR (see illustration) and +B2 and FP. Check that there is no continuity across terminals +B2 and PR.

37   Connect a 12 volt battery to terminals +B1 and FPR (see illustration). Using an ohmmeter, check for continuity across terminals +B2 and PR. Continuity should exist. If the test results are incorrect, replace the relay.

Fuel pump relay (1996 models)

Refer to illustrations 3.38 and 3.39

38   Using an ohmmeter, check for continuity across terminals 3 and 5 (see illustration). Check that there is continuity across terminals 1

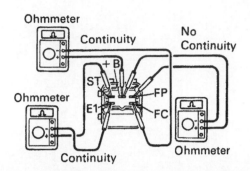

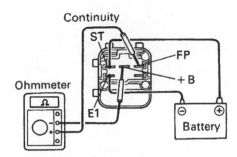

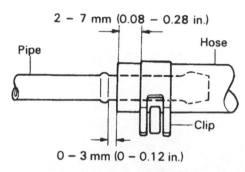

3.40 Circuit opening relay checks on 1988 through 1994 models

3.41 Apply battery voltage to terminals E1 and ST1 and check for continuity between terminals B+ and FP (1988 through 1994 models)

3.43 Check the resistance of the fuel pump resistor

4.6 When attaching a section of rubber hose to a metal fuel line, be sure to overlap the hose as shown, secure it to the line with a new hose clamp of the proper type

and 5 is 80 ohms.

39   Connect a 12 volt battery to terminals 1 and 5 **(see illustration)**. There should be battery voltage between terminals 1 and 2. Also, using an ohmmeter, check for continuity across terminals 3 and 5. Continuity should exist. If the test results are incorrect, replace the relay.

Circuit opening relay

1988 through 1994 models

Refer to illustrations 3.40 and 3.41

40   Using an ohmmeter, check for continuity across terminals ST and E1 **(see illustration)**. Also, check for continuity across +B and FC. Check that there is no continuity across terminals +B and FP.

41   Connect a 12 volt battery to terminals ST and E1 **(see illustration)**. Using an ohmmeter, check for continuity across terminals +B and FP. Continuity should exist. If the test results are incorrect, replace the relay.

1995 and later models

42   The circuit opening relay on these models is checked using the same tests as for the EFI main relay. Refer to Steps 34 and 35 and **illustrations 3.34b and 3.35b**.

Fuel pump resistor

Refer to illustration 3.43

43   Using an ohmmeter, check the resistance across the terminals. It should be 0.73 ohms at ambient temperature **(see illustration)**.

44   If there is no continuity, replace the resistor.

---

4   Fuel lines and fittings - inspection and replacement

Warning: *Gasoline is extremely flammable, so take extra precautions when you work on any part of the fuel system. Don't smoke or allow open flames or bare light bulbs near the work area, and don't work in a garage where a natural gas-type appliance (such as a water heater or a clothes dryer) with a pilot light is present. Since gasoline is carcinogenic, wear latex gloves when there's a possibility of being exposed to fuel, and, if you spill any fuel on your skin, rinse it off immediately with soap and water. Mop up any spills immediately and do not store fuel-soaked rags where they could ignite. The fuel system is under constant pressure, so, if any fuel lines are to be disconnected, the fuel pressure in the system must be relieved first. When you perform any kind of work on the fuel system, wear safety glasses and have a Class B type fire extinguisher on hand.*

Inspection

1   Once in a while, you will have to raise the vehicle to service or replace some component (an exhaust pipe hanger, for example). Whenever you work under the vehicle, always inspect fuel lines and all fittings and connections for damage or deterioration.

2   Check all hoses and pipes for cracks, kinks, deformation or obstructions.

3   Make sure all hoses and pipe clips attach their associated hoses or pipes securely to the underside of the vehicle.

4   Verify all hose clamps attaching rubber hoses to metal fuel lines or pipes are snug enough to assure a tight fit between the hoses and pipes.

Replacement

Refer to illustration 4.6

5   If you must replace any damaged sections, use original equipment replacement hoses or pipes constructed from exactly the same material as the section you are replacing. Do not install substitutes constructed from inferior or inappropriate material or you could cause a fuel leak or a fire.

6   Always, before detaching or disassembling any part of the fuel line system, note the routing of all hoses and pipes and the orientation of all clamps and clips to assure that replacement sections are installed in exactly the same manner. When attaching hoses to metal

5.4 Remove the fuel pump/fuel level sending unit access cover screws (arrows) and lift the cover off

5.5a Disconnect the fuel pump electrical connector

5.5b Disconnect the fuel tank vapor line

5.5c On models so equipped, remove the banjo fitting and separate the fuel line. Retrieve the washers - new ones should be used on reassembly

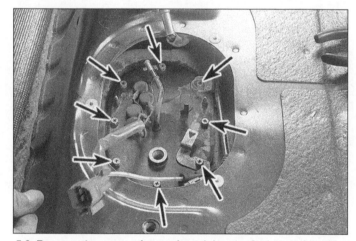

5.6 Remove the screws (arrows) retaining the fuel pump/sending unit to the fuel tank

lines, overlap them as shown **(see illustration)**.

7   Before detaching any part of the fuel system, be sure to relieve the fuel line and tank pressure (see Section 2). Cover the fitting being disconnected with a rag to absorb any fuel that may spray out.

---

5   Fuel pump - removal and installation

Warning: *Gasoline is extremely flammable, so take extra precautions when you work on any part of the fuel system. Don't smoke or allow open flames or bare light bulbs near the work area, and don't work in a garage where a natural gas-type appliance (such as a water heater or a clothes dryer) with a pilot light is present. Since gasoline is carcinogenic, wear latex gloves when there's a possibility of being exposed to fuel, and, if you spill any fuel on your skin, rinse it off immediately with soap and water. Mop up any spills immediately and do not store fuel-soaked rags where they could ignite. The fuel system is under constant pressure, so, if any fuel lines are to be disconnected, the fuel pressure in the system must be relieved first. When you perform any kind of work on the fuel system, wear safety glasses and have a Class B type fire extinguisher on hand.*
Note: *1988 through 1990 models are not equipped with an access hole for fuel pump removal. Remove the fuel tank to gain access to the fuel pump. On 1991 and later models, remove the middle seat to uncover the fuel pump access cover for fuel pump testing and replacement procedures.*

Removal

Refer to illustrations 5.4, 5.5a, 5.5b, 5.5c, 5.6, 5.7, 5.9, 5.10 and 5.13

1   Relieve the fuel system pressure (see Section 2)..
2   Disconnect the cable from the negative terminal of the battery.
Caution: *If the stereo in your vehicle is equipped with an anti-theft system, make sure you have the correct activation code before disconnecting the battery.*
3   On 1988 through 1990 models, remove the fuel tank from the vehicle (see Chapter 4A).
4   On 1991 and later models, remove the middle seat from inside the passenger compartment (see Chapter 11). Remove the fuel pump/sending unit access cover **(see illustration)**.
5   Disconnect the electrical connector. Disconnect the fuel lines **(see illustrations)**.
6   Remove the fuel pump/sending unit retaining bolts **(see illustration)**.
7   Carefully withdraw the fuel pump/fuel level sending unit assembly from the fuel tank **(see illustration)**.
8   Pry the lower end of the fuel pump loose from the bracket.
9   Remove the rubber cushion from the lower end of the fuel pump **(see illustration)**.
10   Remove the clip securing the inlet screen to the pump **(see illustration)**.
11   Remove the screen and inspect it for contamination. If it is dirty, replace it.
12   If you are only replacing the fuel pump inlet screen, install the new screen, the clip and the rubber cushion, push the lower end of the

5.7 Lift the fuel pump/sending unit assembly from the fuel tank at an angle so as not to damage the inlet screen or float arm

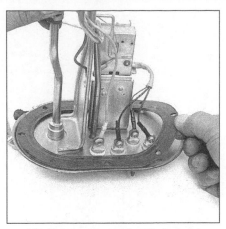

5.9 Remove the rubber O-ring and replace it with a new part to prevent vapor leakage

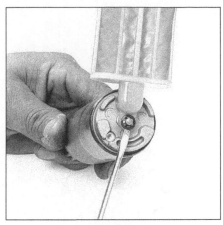

5.10 Remove the clip that retains the pump screen to the bottom of the fuel pump

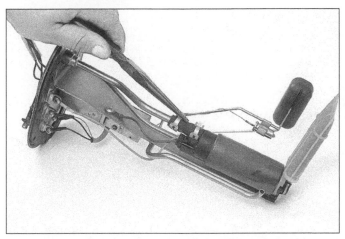

5.13 Remove the clamp from the fuel line and separate the fuel line from the fuel pump

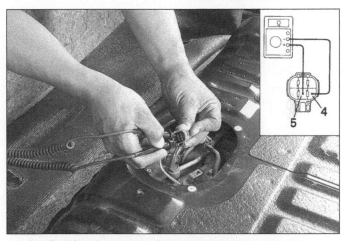

6.4 Position the probes of the ohmmeter onto the correct terminals and observe the fuel level sending unit resistance

pump back into the bracket and install the pump/sending unit assembly in the fuel tank.

13   If you are replacing the fuel pump, remove the hose clamp at the upper end of the pump and disconnect the pump from the hose (see illustration).

14   Disconnect the wires from the pump terminals and remove the pump.

Installation

15   Installation is the reverse of removal. On models so equipped, be sure to use new sealing washers on either side of the banjo fitting, and tighten the fitting bolt to the torque listed in this Chapter's Specifications.

---

6   Fuel level sending unit - check and replacement

Warning: *Gasoline is extremely flammable, so take extra precautions when you work on any part of the fuel system. Don't smoke or allow open flames or bare light bulbs near the work area, and don't work in a garage where a natural gas-type appliance (such as a water heater or a clothes dryer) with a pilot light is present. Since gasoline is carcinogenic, wear latex gloves when there's a possibility of being exposed to fuel, and, if you spill any fuel on your skin, rinse it off immediately with soap and water. Mop up any spills immediately and do not store fuel-soaked rags where they could ignite. The fuel system is under constant*

pressure, so, if any fuel lines are to be disconnected, the fuel pressure in the system must be relieved first. When you perform any kind of work on the fuel system, wear safety glasses and have a Class B type fire extinguisher on hand.

Note: *1988 through 1990 models are not equipped with an access hole for fuel level sending unit removal. Remove the fuel tank to gain access to the fuel level sending unit. On 1991 through 1996 models, remove the middle seat to uncover the fuel pump access cover for fuel pump/fuel level sending unit testing and replacement procedures.*

Check

Refer to illustrations 6.4 and 6.8

1   Before performing the tests on the fuel level sending unit, determine the actual fuel level in the fuel tank.

2   On 1988 through 1990 models, locate the harness connector near the fuel tank from below to check resistance of the sending unit. On 1991 and later models, remove the rear seat and the fuel pump/sending unit access cover (see illustration 5.4).

3   Disconnect the fuel level sending unit electrical connector.

4   Position the ohmmeter probes on the connector terminals and check for resistance (see illustration). Use the 200-ohm scale on the ohmmeter.

5   With the fuel tank completely full, the resistance should be approximately 17 ohms on 1988 through 1990 models and 3.0 ohms on 1991 and later models.

6   When the tank is nearly empty, the resistance of the sending unit

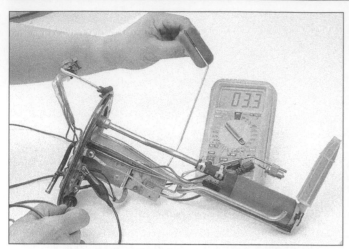

6.8 An accurate check of the sending unit can be made by removing it from the fuel tank and observing the resistance with the float down (empty) and then up (full)

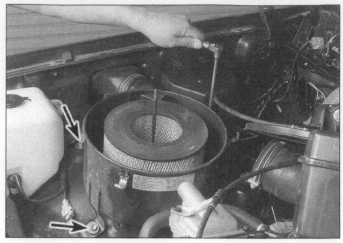

7.3 Remove the bolts (arrows) that retain the air cleaner housing to the engine compartment

should be approximately 120 ohms on 1988 through 1990 models and 110 ohms on 1991 and later models.

7   If the readings are incorrect, replace the sending unit.

8   The test can also be performed with the fuel level sending unit removed from the fuel tank. Using an ohmmeter, check the resistance of the sending unit with the float arm completely down (tank empty) and with the arm up (tank full) **(see illustration)**. The resistance should change steadily from empty to full.

Replacement

9   Remove the fuel pump/fuel level sending unit assembly from the fuel tank (see Section 5).

10   Carefully angle the sending unit out of the opening without damaging the fuel level float located at the bottom of the assembly **(see illustration 5.7)**.

11   Disconnect the electrical connectors from the sending unit.

12   Remove the screw from the side of the sending unit bracket and separate the sending unit from the assembly.

13   Installation is the reverse of removal.

7   Air cleaner assembly - removal and installation

Refer to illustration 7.3

1   Detach the clips and remove the air filter and the filter element

(see Chapter 1).

2   Disconnect the air intake hose from the assembly.

3   Remove the three bolts and remove the air cleaner assembly from the engine compartment **(see illustration)**.

4   Installation is the reverse of removal.

8   Accelerator cable - removal, installation and adjustment

Refer to illustrations 8.2, 8.3 and 8.4

Removal

1   Detach the cable from the negative terminal of the battery. **Caution:** *If the stereo in your vehicle is equipped with an anti-theft system, make sure you have the correct activation code before disconnecting the battery*.

2   Loosen the locknut on the threaded portion of the throttle cable at the throttle body **(see illustration)**.

3   Rotate the throttle lever and slip the throttle cable end out of the slot in the lever **(see illustration)**.

4   Detach the throttle cable from the accelerator pedal **(see illustration)**. Remove the two bolts securing the cable casing to the firewall.

5   From inside the engine compartment, pull the cable through the firewall.

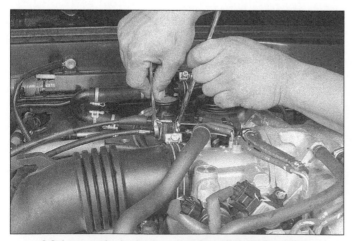

8.2 Loosen the locknut on the threaded portion of the accelerator cable

8.3 Rotate the throttle lever and remove the cable end from the slot

Installation and adjustment

6   Installation is the reverse of removal. Make sure the cable casing grommet seats properly in the firewall.
7   To adjust the cable, fully depress the accelerator pedal and check that the throttle is fully opened.
8   If not fully opened, loosen the locknuts, depress accelerator pedal and adjust the cable until the throttle is fully open.
9   Tighten the locknuts and recheck the adjustment. Make sure the throttle closes fully when the pedal is released.

9   Electronic Fuel Injection (EFI) system - general information

Refer to illustrations 9.1a and 9.1b
1   These models are equipped with an Electronic Fuel Injection (EFI) system. The EFI system is composed of three basic subsystems: the fuel delivery system, the air induction system and the electronic control system **(see illustrations)**.

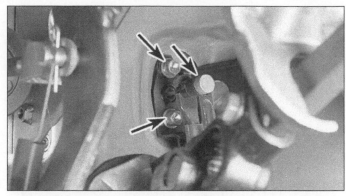

8.4 Separate the cable from the accelerator pedal and slide the cable end (center arrow) out of the slot in the housing - to free the cable casing, remove the two bolts (upper and lower arrows)

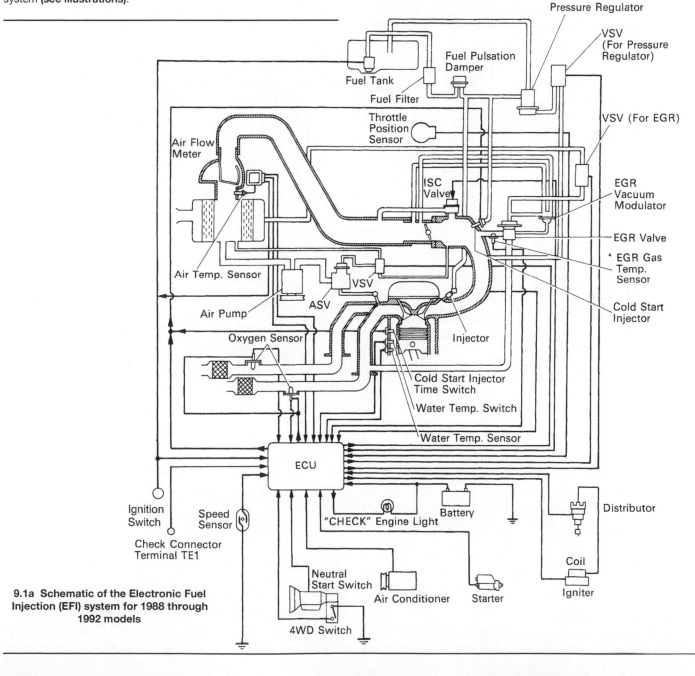

9.1a Schematic of the Electronic Fuel Injection (EFI) system for 1988 through 1992 models

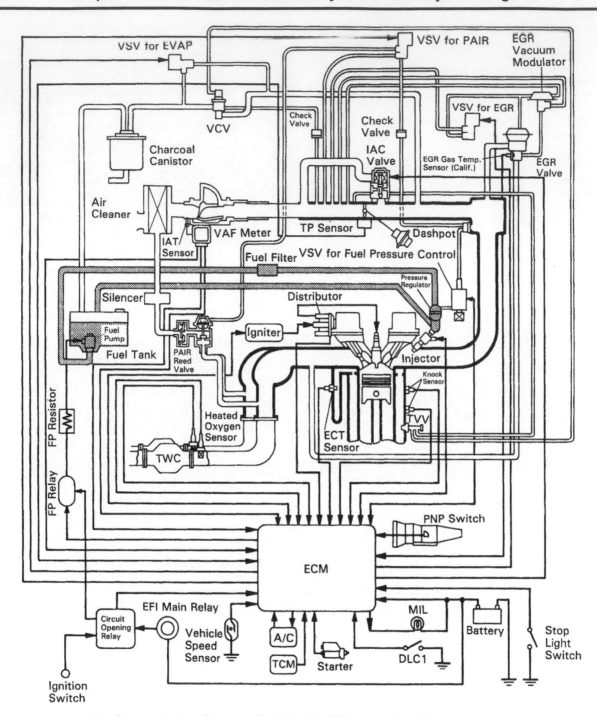

9.1b Schematic of the Electronic Fuel Injection (EFI) system for 1993 and later models

Fuel system

2   An electric fuel pump located inside the fuel tank supplies fuel under constant pressure to the fuel rail, which distributes fuel evenly to all injectors. From the fuel rail, fuel is injected into the intake ports, just above the intake valves, by fuel injectors. The amount of fuel supplied by the injectors is precisely controlled by an Electronic Control Module (ECM). A pressure regulator controls system pressure in relation to intake manifold vacuum. A fuel filter between the fuel pump and the fuel rail filters fuel to protect the components of the system.

Air induction system

3   The air induction system consists of an air filter housing, the throttle body and the duct connecting the two. An Intake Air Temperature

(IAT) sensor monitors the temperature of the incoming air. This information helps the ECM determine the amount of fuel to be injected by the injectors. The throttle plate inside the throttle body is controlled by the driver. As the throttle plate opens, the speed of the incoming air increases, which lowers the temperature of the air. The IAT sends this information to the ECM and the ECM signals the injectors to increase the amount of fuel delivered to the intake ports.

Electronic control system

4   The Computer Control System controls the EFI and other systems by means of an Electronic Control Module (ECM), which employs a microcomputer. The ECM receives signals from a number of information sensors which monitor such variables as intake air temperature,

10.6 With the engine off, use aerosol carburetor cleaner (make sure it is safe for use with catalytic converters and oxygen sensors) and a rag to clean the throttle body - open the throttle plate so you can clean behind it

throttle angle, coolant temperature, engine rpm, vehicle speed and exhaust oxygen content. These signals help the ECM determine the injection duration necessary for the optimum air/fuel ratio. Some of these sensors and their corresponding ECM-controlled relays are not contained within EFI components, but are located throughout the engine compartment. For further information regarding the ECM and its relationship to the engine electrical and ignition system, see Chapter 6.

10   Electronic Fuel Injection (EFI) system - check

Refer to illustrations 10.6, 10.7, 10.8 and 10.9

1   Check the ground wire connections for tightness. Check all wiring and electrical connectors that are related to the system. Loose electrical connectors and poor grounds can cause many problems that resemble more serious malfunctions.

2   Check to see that the battery is fully charged, as the control unit and sensors depend on an accurate supply voltage in order to properly meter the fuel.

3   Check the air filter element - a dirty or partially blocked filter will severely impede performance and economy (see Chapter 1).

4   If a blown fuse is found, replace it and see if it blows again. If it

does, search for a grounded wire in the harness related to the system.

5   Check the air intake duct from the air cleaner housing to the intake manifold for leaks, which will result in an excessively lean mixture. Also check the condition of the vacuum hoses connected to the intake manifold.

6   Remove the air intake duct from the throttle body and check for carbon and residue build-up. If it's dirty, clean it with aerosol carburetor cleaner (make sure the can says it's safe for use with oxygen sensors and catalytic converters) and a shop towel **(see illustration)**.

7   With the engine running, place a stethoscope against each injector, one at a time, and listen for a clicking sound, indicating operation **(see illustration)**. If you don't have an automotive stethoscope you can use a long screwdriver; just place the tip of the screwdriver against the injector body and press your ear against the handle.

8   If there is a problem with an injector, purchase a special injector test light ("noid" light) and install it into the injector electrical connector **(see illustration)**. Start the engine and make sure that each injector connector flashes the noid light. This will test for the proper voltage signal to the injector.

9   With the engine OFF and the fuel injector electrical connectors disconnected, measure the resistance of each injector **(see illustration)**. Each injector should measure about 13.4 to 14.2 ohms. If not, the injector is probably faulty.

10   The remainder of the system checks should be left to a dealer service department or other qualified repair shop, as there is a chance that the control unit may be damaged if not performed properly.

11   Electronic Fuel Injection (EFI) system - component check and replacement

Warning: *Gasoline is extremely flammable, so take extra precautions when you work on any part of the fuel system. Don't smoke or allow open flames or bare light bulbs near the work area, and don't work in a garage where a natural gas-type appliance (such as a water heater or a clothes dryer) with a pilot light is present. Since gasoline is carcinogenic, wear latex gloves when there's a possibility of being exposed to fuel, and, if you spill any fuel on your skin, rinse it off immediately with soap and water. Mop up any spills immediately and do not store fuel-soaked rags where they could ignite. The fuel system is under constant pressure, so, if any fuel lines are to be disconnected, the fuel pressure in the system must be relieved first. When you perform any kind of work on the fuel system, wear safety glasses and have a Class B type fire extinguisher on hand.*

Caution: *If the stereo in your vehicle is equipped with an anti-theft system, make sure you have the correct activation code before disconnecting the battery.*

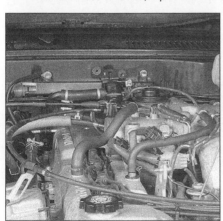

10.7 Use a stethoscope or a screwdriver to determine if the injectors are working properly - they should make a steady clicking sound that rises and falls with engine speed changes

10.8 Install the "noid" light into the fuel injector electrical connector and check to see that it blinks with the engine running

10.9 Using an ohmmeter, measure the resistance across the terminals of the injector

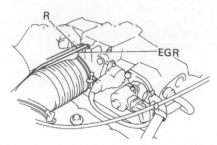

Port No.	At idling	Other than idling
EGR	No vacuum	Vacuum
R	No vacuum	Vacuum

11.2a Throttle body vacuum port guide and table for 1988 through 1992 models

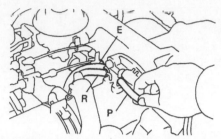

Port name	At idle	At 3,500 rpm
P	No vacuum	Vacuum
E	No vacuum	Vacuum
R	No vacuum	Vacuum

11.2b Throttle body vacuum port guide and table for 1993 and later models

Throttle body

Refer to illustrations 11.2a, 11.2b, 11.11a and 11.11b

Check

1   Verify that the throttle linkage operates smoothly.
2   Start the engine, detach each vacuum hose and, using a vacuum gauge, check the vacuum at each port on the throttle body with the engine at idle and above idle, then compare your observations with the accompanying tables **(see illustrations)**.

Replacement

Warning: *Wait until the engine is completely cool before beginning this procedure.*
3   Detach the cable from the negative terminal of the battery (see the **Caution** at the beginning of this Section).
4   Loosen the hose clamps and remove the air intake duct.
5   Detach the accelerator cable from the throttle lever (see Section 8).
6   Detach the throttle cable bracket and set it aside (it's not necessary to detach the throttle cable from the bracket).
7   If your vehicle is equipped with an automatic transmission, detach the throttle valve (TV) cable from the throttle linkage (see Chapter 7B), detach the TV cable brackets from the engine and set the cable and brackets aside.
8   Clearly label, then detach, all vacuum and coolant hoses from the throttle body. Plug the coolant hoses to prevent coolant loss.
9   Disconnect the electrical connector from the throttle position sensor (TPS).
10   Remove the throttle body mounting bolts.

11   Detach the throttle body and gasket **(see illustrations)** from the intake manifold.
12   Using a soft brush and aerosol carburetor cleaner, thoroughly clean the throttle body casting, then blow out all passages with compressed air. **Caution:** *Do not clean the throttle position sensor with anything. Just wipe it off carefully with a clean, soft cloth.*
13   Installation of the throttle body is the reverse of removal.
14   Be sure to tighten the throttle body mounting bolts to the torque listed in this Chapter's Specifications.

Throttle position sensor (TPS)

Refer to illustrations 11.16, 11.17a, 11.17b and 11.20

Check

15   Disconnect the electrical connector from the throttle position sensor (TPS). On California models, apply vacuum to the throttle positioner.
16   Insert a feeler gauge of the specified thickness between the throttle stop screw and the throttle lever **(see illustration)**.
17   Using an ohmmeter, measure the resistance between the indicated terminal pairs **(see illustrations)**.
18   If the resistance is not as specified, insert a 0.93 mm (0.0366 in.) feeler gauge on 1988 through 1992 models or a 0.62 mm (0.024 in.) feeler gauge on 1993 and later models, between the throttle stop screw and lever, and connect the ohmmeter to terminals IDL and E2. Loosen the TPS mounting screws and slowly rotate the sensor clockwise until the ohmmeter needle (or readout) just deflects, then stop.
19   Tighten the mounting screws, and using the proper feeler gauge, recheck the continuity between the specified terminals.

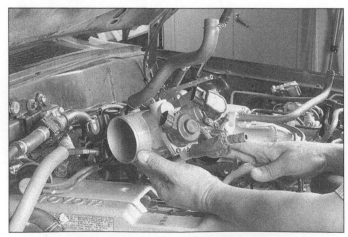

11.11a Remove the throttle body assembly

11.11b Carefully remove the throttle body sealing gasket and replace it

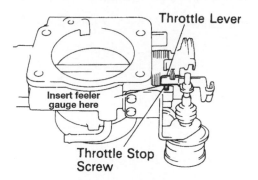

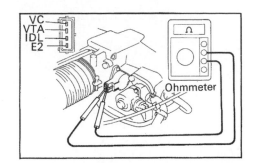

11.16 To check the throttle position sensor (TPS), insert a feeler gauge of the specified thickness between the throttle stop screw and the throttle lever, then measure the resistance between the proper terminals

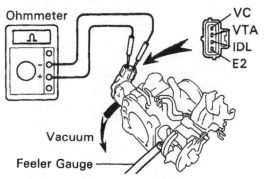

Clearance between lever and stop screw	Between terminals	Resistance
0 mm (0 in.)	VTA – E2	0.3 – 6.3 kΩ
0.77 mm (0.0303 in.)	IDL – E2	Less than 2.3 kΩ
1.09 mm (0.0429 in.)	IDL – E2	Infinity
Throttle valve fully opened position	VTA – E2	3.5 – 10.3 kΩ
—	VC – E2	4.25 – 8.25 kΩ

11.17a TPS terminal guide and continuity table for 1988 through 1992 models

Clearance between lever and stop screw	Between terminals	Resistance
0 mm (0 in.)	VTA – E2	0.2 – 5.7 kΩ
0.50 mm (0.020 in.)	IDL – E2	2.3 kΩ or less
0.75 mm (0.030 in.)	IDL – E2	Infinity
Throttle valve fully open	VTA – E2	2.0 – 10.2 kΩ
—	VC – E2	2.5 – 5.9 kΩ

11.17b TPS terminal guide and continuity table for 1993 and later models

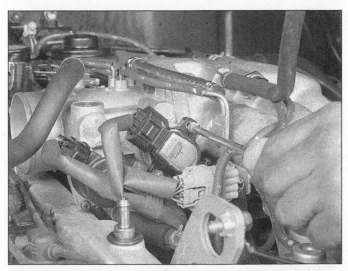

11.20 Remove the two mounting screws to remove the TPS from the throttle body

Replacement

20 If adjustment doesn't bring the sensor within specifications, disconnect it, remove the screws and replace it with a new one **(see illustration)**, then adjust it as described in Step 18.

Fuel pressure regulator

Refer to illustrations 11.26a and 11.26b

Check

21 Refer to the fuel pump/fuel pressure check procedure (see Section 3).

Replacement

22 Relieve the fuel pressure (see Section 2) and detach the cable from the negative terminal of the battery (see the **Caution** at the beginning of this Section).
23 Detach the vacuum sensing hose from the regulator.
24 Place a metal container or shop towel under the fuel return hose.
25 On 1988 through 1992 models, remove the banjo fitting and detach the fuel return line and sealing washers from the fuel pressure regulator. On 1993 and later models, squeeze the hose clamp and detach the hose from the regulator.

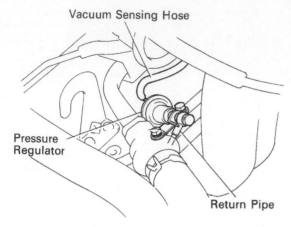

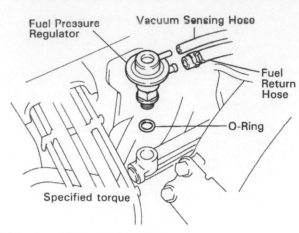

11.26a To remove the fuel pressure regulator from the fuel rail, detach the fuel return line, remove the two regulator bolts and separate the regulator from the fuel rail (1988 through 1992 models)

11.26b On 1993 and later models, use a large open-end wrench to unscrew the fuel pressure regulator

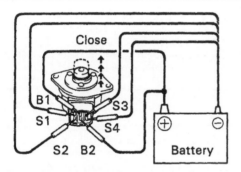

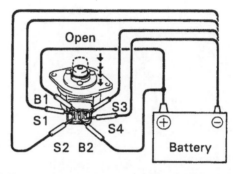

11.31 With battery voltage applied, ground terminals S1, S2, S3 and S4 alternately to close the pintle

11.32 To open the pintle, ground terminals S4, S3, S2 and S1 alternately

26   On 1988 through 1992 models, remove the pressure regulator mounting bolts and detach the pressure regulator from the fuel rail. On 1993 and later models, unscrew the regulator and retrieve the O-ring **(see illustrations)**.

27   Use a new O-ring and make sure that the pressure regulator is installed properly on the fuel rail.

28   The remainder of installation is the reverse of removal.

Idle Speed Control (ISC) valve or Idle Air Control (IAC) valve

Refer to illustrations 11.31, 11.32, 11.34 and 11.37

Note 1: *The minimum idle speed is pre-set at the factory and should not require adjustment under normal operating conditions; however if the throttle body has been replaced or you suspect the minimum idle speed has been tampered with (for example, if the idle speed screw was removed from the throttle body) have the vehicle checked by a dealer service department or other qualified automotive repair shop.*

Note 2: *On 1988 through 1992 models this valve is called the Idle Speed Control (ISC) valve. On 1993 and later models it's called the Idle Air Control (IAC) valve.*

Check

29   Check for a distinct sound from the ISC valve. Have an assistant turn the ignition key off while listening carefully to the ISC valve. It should click noticeably directly after shut-down.

30   Disconnect the ISC valve harness connector.

31   Using fused jumper wires, connect positive battery voltage to terminals B1 and B2 at the same time. Ground terminal S1, then S2, then S3, then S4; the ISC valve pintle should extend (closed position) **(see illustration)**.

32   Using fused jumper wires, connect positive battery voltage to terminals B1 and B2 at the same time. Now, ground terminal S4, then S3, then S2, then S1; the valve pintle should retract (open position) **(see illustration)**.

33   If the valve does not move as specified, replace the valve with a new part.

34   Measure the resistance between B1 to S1 and B1 to S3; then B2 to S2 and B2 to S4 **(see illustration)**. Compare your results to the ISC/IAC valve resistance in this Chapter's Specifications.

a) *If the resistance is as specified, the valve is okay (but there may be a problem with the wiring or the ECM).*

b) *If the resistance is not as specified, replace the valve.*

35   Connect the electrical connector to the valve.

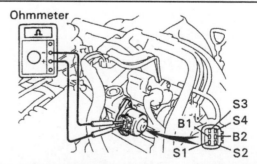

11.34 Using an ohmmeter, check the resistance between B1 to S1 and B1 to S3; then B2 to S2 and B2 to S4. It should be between 10 and 30 ohms

11.37 Remove the mounting screws (arrows) that retain the ISC/IAC valve to the throttle body (1993 and later model shown)

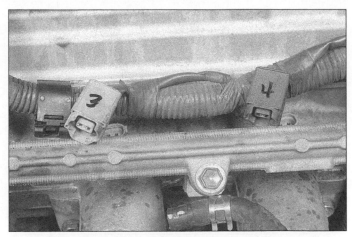

11.45 Number each injector electrical connector before disconnecting them from the fuel rail

Replacement

36   If you find it difficult to access the ISC/IAC valve screws, remove the throttle body (see Steps 3 through 11).

37   Remove the mounting screws and detach the valve and gasket **(see illustration)**.

38   If you're replacing the valve, be sure to install the Temperature Vacuum Valve (TVV) from the original assembly into the new unit (if equipped).

39   Installation of the valve is the reverse of removal. Be sure to use a new gasket.

Fuel rail and fuel injectors

Refer to illustrations 11.45, 11.48, 11.51a, 11.51b and 11.51c

Check

40   Refer to the fuel injection system checking procedure (see Section 10).

Replacement

41   Relieve the fuel pressure (see Section 2).

42   Detach the cable from the negative terminal of the battery (see the **Caution** at the beginning of this Section).

43   Remove the PCV hose from the cylinder head and intake manifold.

44   Remove the air intake plenum (see Steps 70 through 79).

45   Carefully mark each injector connector with a felt pen or paint **(see illustration)**. Disconnect the fuel injector electrical connectors and set the injector wire harness aside.

46   Detach the vacuum sensing hose from the fuel pressure regulator.

47   Disconnect the fuel lines from the fuel pressure regulator and the fuel rail.

11.48 Remove the bolts (arrows) that retain the fuel rail to the intake manifold

48   Remove the fuel rail mounting bolts **(see illustration)**.

49   Remove the fuel rail with the fuel injectors attached.

50   Remove the fuel injector(s) from the fuel rail and set them aside in a clearly labeled storage container.

51   If you intend to re-use the same injectors, replace the grommets and O-rings **(see illustrations)**.

52   Installation of the fuel injectors is the reverse of removal.

53   Tighten the fuel rail mounting bolts to the torque listed in this Chapter's Specifications.

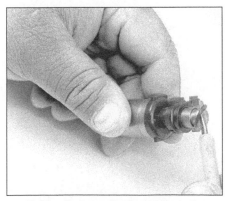

13.51a Remove the O-ring from the injector

11.51b Remove the grommet from the top of the injector

11.51c Remove the O-rings from the bores in the intake manifold

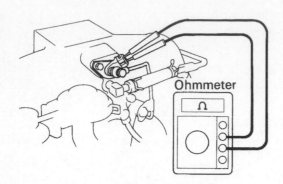

11.55 Disconnect the cold start injector and check the resistance

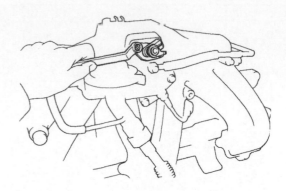

11.60 Remove the cold start injector mounting bolts

Cold start injector (1988 through 1992 models)

Refer to illustrations 11.55 and 11.60

Check

54   Disconnect the electrical connector from the cold start injector.

55   Using an ohmmeter, measure the resistance between the injector terminals **(see illustration)**. Compare your measurement to the resistance listed in this Chapter's Specifications.

 a) *If the indicated resistance is within the specified range, the cold start injector is okay. Check the start injector time switch for correct operation* (see Steps 38 through 41).

 b) *If the indicated resistance isn't within the specified range, replace the cold start injector.*

56   Connect the cold start injector electrical connector.

Removal

57   Relieve the fuel pressure (see Section 2). Disconnect the cable from the negative terminal of the battery (see **Caution** at the beginning of this Section).

58   Disconnect the cold start injector electrical connector.

59   Place a metal container or shop towel under the banjo fitting and remove the banjo bolt and sealing washers. Discard the washers.

60   Remove the cold start injector mounting nuts, the injector and the gasket **(see illustration)**.

Bench test

61   The cold start injector can be bench tested (for spray pattern) but the test requires special equipment. If you are in any doubt as to the status of the cold start injector, take it to a dealer service department or other repair shop and have it tested.

Installation

62   Installation of the cold start injector is the reverse of removal. Be sure to use new sealing washers on each side of the banjo fitting.

Cold start injector time switch (1988 through 1992 models)

Refer to illustration 11.64

Check

63   Disconnect the electrical connector from the start injector time switch and using an ohmmeter, check the resistance between each terminal. **Note:** *The cold start injector time switch is located next to the thermostat housing.*

64   First, check the resistance of the switch with the engine cold (below 59-degrees F). It should have 30 to 50 ohms resistance **(see illustration)**.

65   Next, warm up the engine (above 86-degrees F) and check the resistance of the switch. It should read 70 to 90 ohms resistance.

66   Check the resistance of terminal STA to ground. It should read 30 to 90 ohms. **Note:** *Terminal STA is the terminal which corresponds with the black/red wire on the wiring harness.*

Replacement

Warning: *Wait until the engine is completely cool before beginning this procedure.*

67   Drain the coolant from the radiator (see Chapter 1).

68   Use a deep socket and remove the switch from the engine.

69   Installation is the reverse of removal.

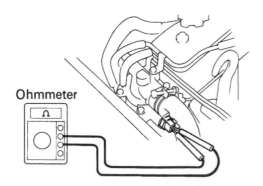

11.64 First check the resistance of the cold start injector time switch below 59-degrees F, then warm-up the engine and check the resistance above 86-degrees F

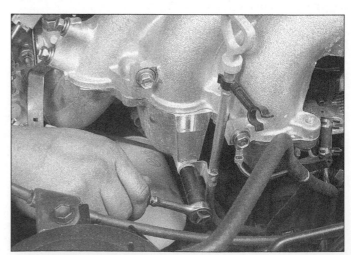

11.73 Remove the oil dipstick mounting bolt from the intake manifold (1993 and later models)

11.74 Carefully label each vacuum line

11.78 Remove the bolts that retain the air intake plenum to the intake manifold

Air intake plenum

Refer to illustrations 11.73, 11.74, 11.78 and 11.80

Removal

Warning: *Wait until the engine is completely cool before beginning this procedure.*

70   Detach the cable from the negative terminal of the battery (see the **Caution** at the beginning of this Section).

71   Disconnect the electrical connectors at the ISC/IAC valve, throttle position sensor (TPS) and EGR/EVAP canister control solenoid.

72   Detach the accelerator cable (see Section 8) and transmission linkage (see Chapter 7) from the throttle body assembly.

73   Remove the oil dipstick tube bracket bolt from the intake manifold **(see illustration)**.

74   Clearly label, then detach, the vacuum lines from the air intake plenum, the EGR valve and the fuel pressure regulator **(see illustration)**.

75   Detach the PCV system by disconnecting the hose from the fitting on the air intake plenum.

76   Remove the plenum bracket and bolt (if equipped).

77   Detach the coolant hoses from the throttle body and plug them.

78   Remove the air intake plenum retaining bolts and nuts **(see illustration)** and the lower (side) retaining bolts.

79   Remove the air intake plenum and throttle body as an assembly from the lower intake manifold.

Installation

80   Be sure to clean and inspect the mounting surface of the lower intake manifold and the air intake plenum before positioning the new

11.80 Install the two intake manifold-to-plenum gaskets on the intake manifold

gasket onto the lower intake mounting face **(see illustration)**. Install the air intake plenum and throttle body assembly onto the intake manifold. Ensure the gasket remains in place. Install the retaining bolts and nuts and tighten the bolts to the torque listed in this Chapter's Specifications. Installation is otherwise the reverse of removal.

Notes

Chapter 5
Engine electrical systems

Contents

	Section			Section
Alternator - removal and installation	13		Ignition coil - check and replacement...	7
Alternator components (1993 and later models) - check			Ignition system - check..	6
and replacement...	14		Ignition system - general information and precautions..................	5
Battery cables - check and replacement	4		Ignition timing - check and adjustment........................... See Chapter 1	
Battery check, maintenance and charging..................... See Chapter 1			Pick-up coil - check and replacement	10
Battery - emergency jump starting	2		Spark plug replacement .... See Chapter 1	
Battery - removal and installation	3		Spark plug wire, distributor cap and rotor check	
Charging system - check..	12		and replacement .... See Chapter 1	
Charging system - general information and precautions..............	11		Starter motor - removal and installation.......................................	17
CHECK ENGINE light .... See Chapter 6			Starter motor - testing in vehicle...	16
Distributor - removal and installation	8		Starter solenoid (1980 through 1987 models) - removal	
Drivebelt check, adjustment and replacement............... See Chapter 1			and installation..	18
General information..	1		Starting system - general information and precautions.................	15
Igniter - replacement..	9			

Specifications

Ignition system

Ignition timing	See Chapter 1
Ignition coil resistance	
1980 through 1987 models	
Primary resistance	0.50 to 0.70 ohms
Secondary resistance	11.5 to 15.5 K-ohms
1988 through 1990 models	
Primary resistance	0.56 to 0.64 ohms
Secondary resistance	11.5 to 15.5 K-ohms
1991 models	
Primary resistance	0.41 to 0.50 ohms
Secondary resistance	10.2 to 13.8 K-ohms
1992 models	
Primary resistance	0.30 to 0.60 ohms
Secondary resistance	9.0 to 15.0 K-ohms
1993 and later models	
Cold (coil temperature below 122-degrees F)	
Primary resistance..	0.36 to 0.55 ohms
Secondary resistance...	9.0 to 15.4 K-ohms
Warm (coil temperature above 122-degrees F)	
Primary resistance..	0.45 to 0.65 ohms
Secondary resistance...	11.4 to 18.1 K-ohms
Distributor	
Air gap	0.008 to 0.016 inch
Pick-up coil resistance	
1980 through 1987 models...	Not available
1988 through 1991 models	
G to G- terminals..	140 to 180 ohms
NE to G- terminals..	140 to 180 ohms
1992 models	
G to G- terminals..	185 to 265 ohms
NE to G- terminals..	185 to 265 ohms
1993 and later models	
Cold (below 103-degrees F)	
G1 to G- terminals...	185 to 275 ohms
G2 to G- terminals...	185 to 275 ohms
NE to G- terminals...	185 to 275 ohms
Hot (above 104-degrees F)	
G1 to G- terminals...	240 to 325 ohms
G2 to G- terminals...	240 to 325 ohms
NE to G- terminals...	240 to 325 ohms

Charging system

Charging voltage .... 13.9 to 15.1 volts
Standard amperage
    All lights and accessories turned off .... Less than 10 amps
    Headlights (hi-beam) and heater blower motor turned on .... 30 amps or more
Alternator brush length (exposed)
    Standard 13/32-inch
    Minimum 1/16-inch

1   General information

The engine electrical systems include all ignition, charging and starting components. Because of their engine related functions, these components are discussed separately from chassis electrical devices such as the lights, the instruments, etc. (which are included in Chapter 12).

Always observe the following precautions when working on the electrical systems:

a) *Be extremely careful when servicing engine electrical components. They are easily damaged if checked, connected or handled improperly.*

b) Never leave the ignition switch on for long periods of time (10 minutes maximum) with the engine off.

c) *Don't disconnect the battery cables while the engine is running.*

d) *Maintain correct polarity when connecting a battery cable from another vehicle during jump starting.*

e) *Always disconnect the negative cable first and hook it up last or the battery may be shorted by the tool being used to loosen the cable clamps.*

It's also a good idea to review the safety-related information regarding the engine electrical systems located in the *Safety first* section near the front of this manual before beginning any operation included in this Chapter.

2   Battery - emergency jump starting

Refer to the *Booster battery (jump) starting* procedure at the front of this manual.

3   Battery - removal and installation

Refer to illustration 3.1

1   Starting with the negative battery cable **(see illustration)**, disconnect both cables from the battery terminals. **Caution:** *If the stereo in your vehicle is equipped with an anti-theft system, make sure you have the correct activation code before disconnecting the battery.*

2   Remove the battery hold-down clamp.

3   Lift out the battery. Be careful, it's heavy.

4   While the battery is out, inspect the carrier (tray) for corrosion.

5   If you are replacing the battery, make sure that you get one that's identical, with the same dimensions, amperage rating, cold cranking rating, etc. as the original.

6   Installation is the reverse of removal.

4   Battery cables - check and replacement

Caution: *If the stereo in your vehicle is equipped with an anti-theft system, make sure you have the correct activation code before disconnecting the battery.*

1   Periodically inspect the entire length of each battery cable for damage, cracked or burned insulation and corrosion. Poor battery cable connections can cause starting problems and decreased engine performance.

2   Check the cable-to-terminal connections at the ends of the cables for cracks, loose wire strands and corrosion. The presence of white, fluffy deposits under the insulation at the cable terminal connection is a sign that the cable is corroded and should be replaced. Check the terminals for distortion, missing mounting bolts and corrosion.

3   When removing the cables, always disconnect the negative cable first and hook it up last or the battery may be shorted by the tool used to loosen the cable clamps. Even if only the positive cable is being replaced, be sure to disconnect the negative cable from the battery first (see Chapter 1 for further information regarding battery cable removal).

4   Disconnect the old cables from the battery, then trace each of them to their opposite ends and detach them from the starter solenoid and ground terminals. Note the routing of each cable to ensure correct installation.

5   If you are replacing either or both of the old cables, take them with you when buying new cables. It is vitally important that you replace the cables with identical parts. Cables have characteristics that make them easy to identify: positive cables are usually red, larger in cross-section and have a larger diameter battery post clamp; ground cables are usually black, smaller in cross-section and have a slightly smaller diameter clamp for the negative post.

6   Clean the threads of the solenoid or ground connection with a wire brush to remove rust and corrosion. Apply a light coat of battery terminal corrosion inhibitor, or petroleum jelly, to the threads to prevent future corrosion.

7   Attach the cable to the solenoid or ground connection and tighten the mounting nut/bolt securely.

8   Before connecting a new cable to the battery, make sure that it reaches the battery post without having to be stretched.

9   Connect the positive cable first, followed by the negative cable.

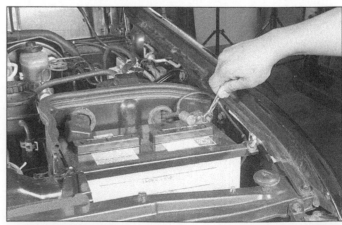

3.1 To remove the battery, detach the negative battery cable first, then the positive cable, remove the hold-down strap nut and bolt and remove the hold-down strap

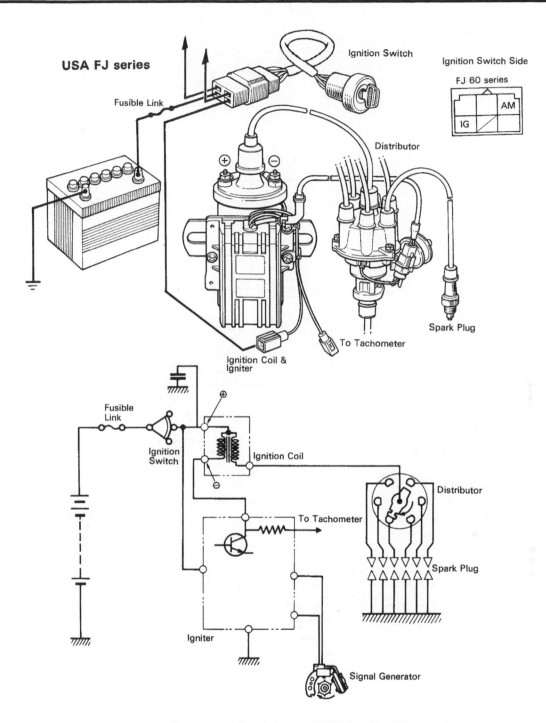

5.3 Schematic of the ignition system on 1980 through 1987 models

5   Ignition system - general information and precautions

Refer to illustrations 5.3, 5.4 and 5.5

1   The electronic ignition system includes the ignition switch, the battery, the igniter, a hall-effect type signal generator (1980 through 1987 models) or a pick-up coil (1988 and later models), the ignition coil, the primary (low voltage) and secondary (high voltage) wiring circuits, the distributor and the spark plugs. The ignition system on 1988 and later models is controlled by the Electronic Control Unit (ECU) or the Electronic Control Module (ECM). Using data provided by informa-

tion sensors which monitor various engine functions (such as rpm, intake air volume, engine temperature, etc.), the system ensures a perfectly timed spark under all conditions. This system is known as Electronic Spark Advance (ESA).

2   The ignition coil is remotely mounted.

3   Early models (1980 through 1987) are equipped with centrifugal and vacuum advance mechanisms. The igniter is mounted externally on the ignition coil. A signal generator is used instead of a pick-up coil and it is mounted in the distributor on the breaker plate along with the vacuum advance arm **(see illustration)**.

4   1988 through 1992 models are also equipped with Electronic

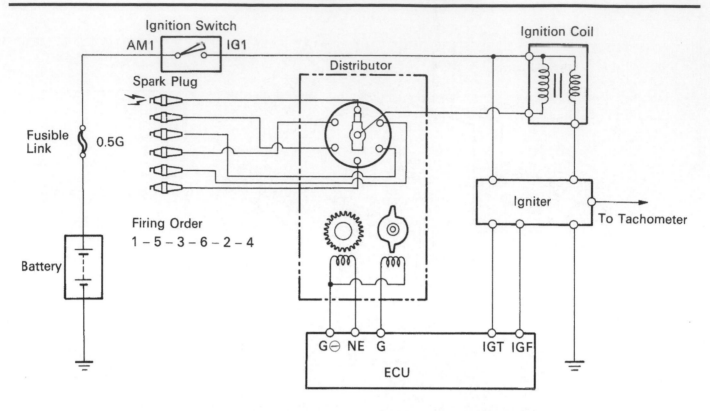

5.4 Schematic of the ignition system on 1988 through 1992 models

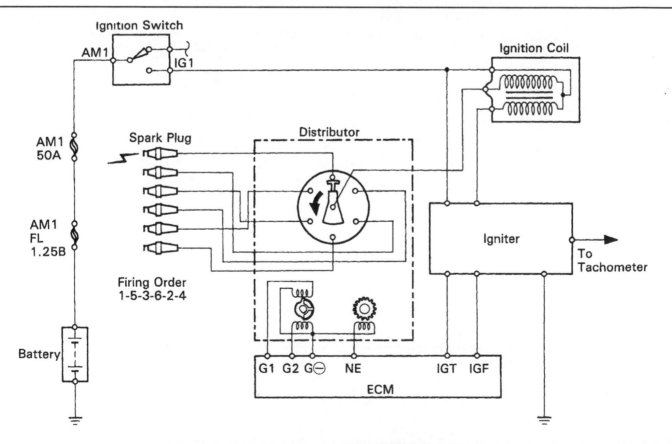

5.5 Schematic of the ignition system on 1993 and later models

6.1 To use a calibrated ignition tester (available at most auto parts stores), remove an ignition wire from a cylinder, connect the spark plug boot to the tester and clip the tester to a good ground - if there is enough voltage to fire the plug, sparks will be clearly visible between the electrode tip and the tester body as the engine is turned over

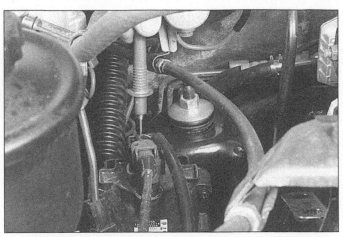

6.7 Check for battery voltage to the positive side (+) of the ignition coil

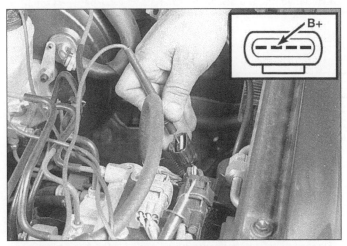

6.9 Disconnect the igniter electrical connector and with the ignition key ON (engine not running), check for battery voltage to the igniter (1993 and later models shown)

Spark Advance (ESA). Unlike earlier models, the distributor uses a pick-up coil arrangement. The igniter is still mounted on the coil **(see illustration)**.

5   1993 and later models are similar to 1988 through 1992 models but the pick-up coil uses a dual G1 and G2 circuit **(see illustration)**.

6   When working on the ignition system, take the following precautions:

a) *Because of the high voltage generated by the ignition system, extreme care should be taken whenever an operation is performed involving ignition components. This not only includes the igniter, coil, distributor and spark plug wires, but related components such as plug connectors, tachometer and other test equipment also.*

b) *Do not keep the ignition switch on for more than 10 seconds if the engine will not start.*

c) *Always connect a tachometer in accordance with the manufacturer's instructions. Some tachometers may be incompatible with this ignition system. Consult a dealer service department before buying a tachometer for use with this vehicle.*

d) *Never allow the ignition coil terminals to touch ground. Grounding the coil could result in damage to the igniter and/or the ignition coil.*

e) *Do not disconnect the battery when the engine is running.*

f) *Make sure the igniter is properly grounded.*

---

6   Ignition system - check

Refer to illustrations 6.1, 6.7 and 6.9

Warning: *Because of the high voltage generated by the ignition system, extreme care should be taken whenever an operation is performed involving ignition components. This not only includes the igniter, coil, distributor and spark plug wires, but related components such as plug connectors, tachometer and other test equipment also.*

1   If the engine turns over but won't start, disconnect the spark plug wire from any spark plug and attach it to a calibrated tester (available at most auto parts stores) **(see illustration)**. Connect the clip on the tester to a bolt or metal bracket on the engine. If you're unable to obtain a calibrated ignition tester, remove the wire from one of the spark plugs and using an insulated tool, pull back the boot and hold the end of the wire about 1/4-inch from a good ground.

2   Crank the engine and watch the end of the tester or spark plug wire to see if a bright blue, well-defined sparks occur.

3   If sparks occur, sufficient voltage is reaching the plug to fire it

(repeat the check at the remaining plug wires to verify that the distributor cap and rotor are OK). However, the plugs themselves may be fouled, so remove and check them as described in Chapter 1.

4   If no sparks or intermittent sparks occur, remove the distributor cap and check the cap and rotor as described in Chapter 1. If moisture is present, dry out the cap and rotor, then reinstall the cap and repeat the spark test.

5   If there's still no spark, detach the coil secondary wire from the distributor cap and hook it up to the tester (reattach the plug wire to the spark plug), then repeat the spark check. Again, if you don't have a tester, hold the end of the wire about 1/4-inch from a good ground.

6   If sparks now occur, the distributor cap, rotor or plug wire(s) may be defective.

7   If no sparks occur, check the primary wire connections at the coil to make sure they're clean and tight. Check for voltage to the coil on the primary circuit from the ignition switch **(see illustration)**. Check the ignition coil (see Section 7) and the distributor pick-up coil (see Section 10). Make any necessary repairs, then repeat the check again.

8   If there's still no spark, the coil-to-cap wire may be bad (check the resistance with an ohmmeter and compare it to the spark plug wire resistance Specifications found in Chapter 1). If a known good wire doesn't make any difference in the test results, the igniter may be defective.

9   Check for battery voltage to the igniter **(see illustration)** with the ignition key ON, engine not running. If voltage is available and there is still no spark, replace the igniter (see Section 9).

7.4a To check the primary resistance of the coil, measure the resistance between the positive and the negative terminals (1993 and later model shown)

7.4b To check the secondary resistance of the coil, measure the resistance between the positive terminal and the high tension terminal (1993 and later model shown)

7.9 Remove the retaining screws (arrows) and lift the coil from the engine compartment (1993 and later model shown)

7   Ignition coil - check and replacement

Check

Refer to illustrations 7.4a and 7.4b

1   Detach the cable from the negative terminal of the battery **Caution:** *If the stereo in your vehicle is equipped with an anti-theft system, make sure you have the correct activation code before disconnecting the battery..*
2   Remove the plastic shield from the coil, if equipped.
3   Remove the mounting bolts from the coil electrical connectors and install the ohmmeter probes.
4   Using an ohmmeter:
  a) *Measure the resistance between the positive and negative terminals of the coil* **(see illustration)**. *Compare your reading with the specified coil primary resistance listed in this Chapter's Specifications.*
  b) *Measure the resistance between the positive terminal and the high tension terminal* **(see illustration)**. *Compare your reading with the specified coil secondary resistance listed in this Chapter's Specifications.*
5   If either of the above tests yield resistance values outside the specified amount, replace the coil.

Replacement

Refer to illustration 7.9

6   Detach the cable from the negative terminal of the battery (see the **Caution** above).
7   Remove the heat shield from the coil.
8   Label and disconnect the wires from the coil terminals.
9   Remove the coil mounting screws **(see illustration)**.
10 Installation is the reverse of removal.

8   Distributor - removal and installation

Removal

Refer to illustrations 8.5 and 8.7

1   Detach the cable from the negative battery terminal. **Caution:** *If the stereo in your vehicle is equipped with an anti-theft system, make sure you have the correct activation code before disconnecting the battery.*
2   Disconnect the electrical connectors from the distributor.
3   Look for a raised "1" on the distributor cap. This marks the location for the number one cylinder spark plug wire terminal. If the cap does not have a mark for the number one terminal, locate the number

8.5 Marking the relationship of the distributor to the cylinder head

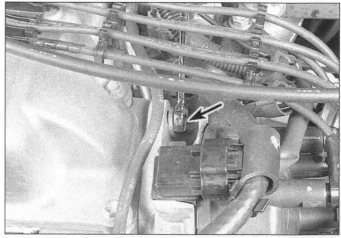

8.7 Remove the hold-down bolt from the distributor body and pull the distributor straight out

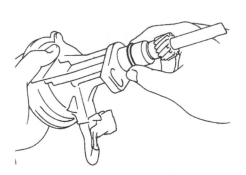

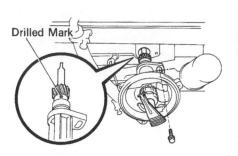

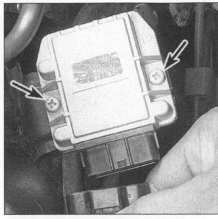

8.8 Install a new O-ring in the groove in the distributor housing

8.9 If you have positioned the engine at TDC compression for number one cylinder, turn the drilled mark on the distributor gear straight up

9.3 Remove the igniter mounting screws (arrows) and separate the igniter from the engine compartment firewall (1993 and later model shown)

one spark plug and trace the wire back to the terminal on the cap.

4   Remove the distributor cap (see Chapter 1) and turn the engine over until the rotor is pointing toward the number one spark plug terminal (see locating TDC procedure in Chapter 2A).

5   Make a mark on the edge of the distributor base directly below the rotor tip and in line with it. Also, mark the distributor base and the engine block to ensure that the distributor is installed correctly **(see illustration)**.

6   If equipped with collar bolts, loosen but do not remove the two bolts in the distributor collar. This will give the distributor shaft clearance.

7   Remove the distributor hold-down bolt **(see illustration)**, then pull the distributor straight out to remove it. **Caution:** *DO NOT turn the crankshaft while the distributor is out of the engine, or the alignment marks will be useless.*

Installation

Refer to illustrations 8.8 and 8.9

Note: *If the crankshaft has been moved while the distributor is out, locate Top Dead Center (TDC) for the number one piston (see Chapter 2A) and position the distributor and the rotor accordingly.*

8   Install a new O-ring onto the distributor housing **(see illustration)**.

9   Align the drilled mark facing up before installing the distributor **(see illustration)**. **Note:** *1993 and later models are equipped with a groove in the distributor housing and a protrusion on the drive gear that must be aligned before installing.*

10   Insert the distributor into the engine in exactly the same relationship to the block that it was in when removed.

11   If the distributor does not seat completely, recheck the alignment marks between the distributor base and the block to verify that the distributor is in the same position it was in before removal. Also check the rotor to see if it's aligned with the mark you made on the edge of the distributor base.

12   Install the distributor hold-down bolt, but don't fully tighten it yet.

13   The remainder of the installation is the reverse of removal.

14   Check the ignition timing (see Chapter 1) and tighten the distributor hold-down bolt securely.

9   Igniter - replacement

Refer to illustration 9.3

Note: *On 1980 through 1987 models, the igniter is attached directly to the ignition coil. On 1988 through 1992 models, the igniter is located on the engine compartment fenderwell. On 1993 and later models, the igniter is attached to the firewall near the brake master cylinder.*

1   Detach the cable from the negative terminal of the battery.

Caution: *If the stereo in your vehicle is equipped with an anti-theft system, make sure you have the correct activation code before disconnecting the battery.*

2   Disconnect the electrical connector from the igniter.

3   Remove the screws and pull the igniter/bracket assembly out of the engine compartment **(see illustration)**.

4   Installation is the reverse of removal.

10   Pick-up coil - check and replacement

Note: *1980 through 1987 models are equipped with a hall-effect type signal generator instead of a pick-up coil. Specifications are not available for testing the signal generator. If you suspect the signal generator is defective, have it diagnosed by a dealer service department or other qualified automotive repair facility.*

Resistance check

Refer to illustration 10.1

1   Disconnect the electrical connector at the distributor and using an ohmmeter, measure the resistance between the pick-up coil terminals **(see illustration)**.

2   Compare the measurements to those listed in this Chapter's Specifications. If the resistance is not as specified, replace the pick-up coil.

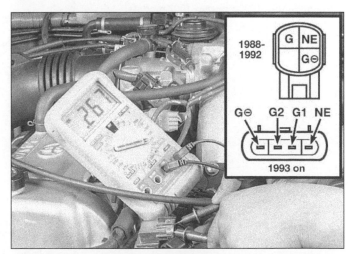

10.1 Check the resistance between the terminals on the pick-up coil indicated in this Chapter's Specifications

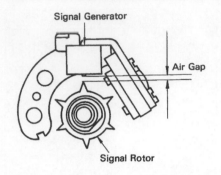

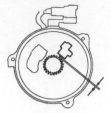

10.5a Checking the signal generator air gap on 1980 through 1987 models

NE Pickup   G Pickup

10.5b Pick-up coil arrangement and air gap check on 1988 through 1992 models

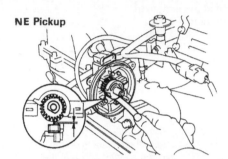
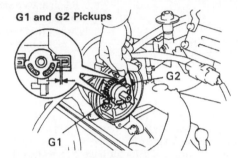

10.5c Pick-up coil arrangement and air gap check on 1993 and later models

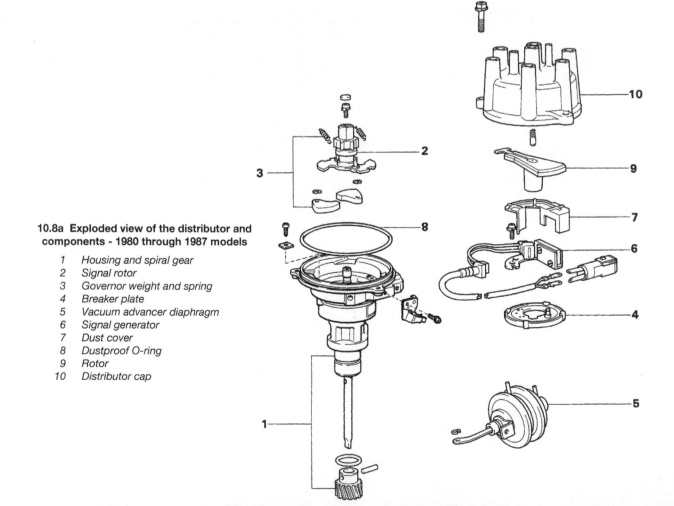

10.8a Exploded view of the distributor and components - 1980 through 1987 models

1   Housing and spiral gear
2   Signal rotor
3   Governor weight and spring
4   Breaker plate
5   Vacuum advancer diaphragm
6   Signal generator
7   Dust cover
8   Dustproof O-ring
9   Rotor
10 Distributor cap

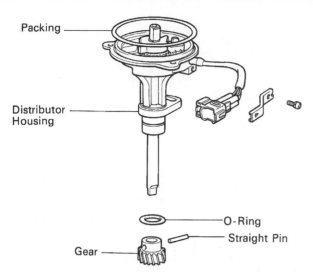

10.8b Exploded view of the distributor and components - 1988 through 1992 models

Air gap check

Refer to illustrations 10.5a, 10.5b and 10.5c

3   Detach the cable from the negative terminal of the battery. **Caution:** *If the stereo in your vehicle is equipped with an anti-theft system, make sure you have the correct activation code before disconnecting the battery.*

4   Remove the distributor from the engine (see Section 8).

5   Using a brass feeler gauge, measure the gap between the signal rotor and the pick-up coil projection **(see illustrations)**. Compare your measurement to the air gap listed in this Chapter's Specifications. If the air gap is not as specified, loosen the signal generator screws and reposition it to provide the proper clearance (1980 through 1987 models). On 1988 and later models replace the distributor, as the air gap is not adjustable. Excessive air gap is usually a sign of wear in the distributor shaft bushing.

Replacement

Refer to illustrations 10.8a, 10.8b and 10.9

6   Detach the cable from the negative terminal of the battery.

10.9 Remove the pick-up coil mounting screws (arrows) (1993 and later model shown)

Caution: *If the stereo in your vehicle is equipped with an anti-theft system, make sure you have the correct activation code before disconnecting the battery.*

7   Remove the distributor from the engine (see Section 8).

8   Remove the set screws that retain the pick-up coil to the distributor **(see illustrations)**.

9   Remove the pick-up coil from the distributor **(see illustration)**.

10   Installation is the reverse of removal. On 1980 through 1997 models, adjust the air gap.

11   Charging system - general information and precautions

Refer to illustrations 11.1a through 11.1e

The charging system includes the alternator, an internal voltage regulator, a charge indicator light, the battery, a fusible link and the wiring between all the components **(see illustrations)**. The charging system supplies electrical power for the ignition system, the lights, the radio, etc. The alternator is driven by a drivebelt at the front of the

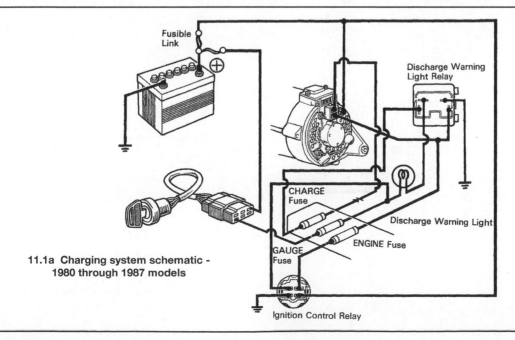

11.1a Charging system schematic - 1980 through 1987 models

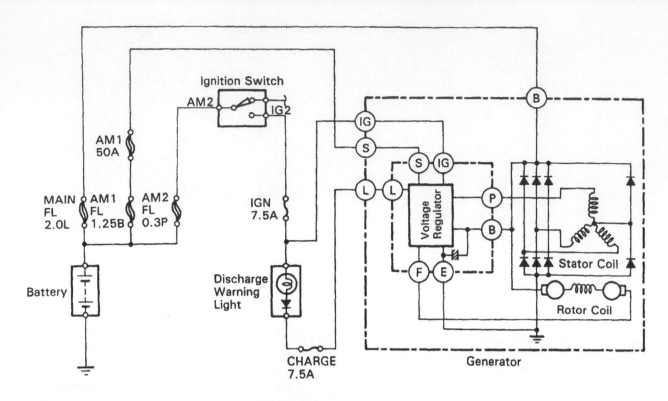

11.1b Charging system schematic - 1993 and later models

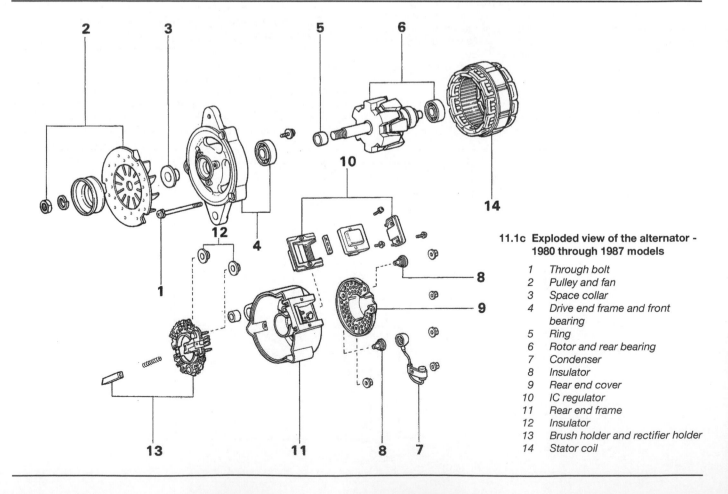

**11.1c Exploded view of the alternator -
1980 through 1987 models**

1   Through bolt
2   Pulley and fan
3   Space collar
4   Drive end frame and front
    bearing
5   Ring
6   Rotor and rear bearing
7   Condenser
8   Insulator
9   Rear end cover
10 IC regulator
11 Rear end frame
12 Insulator
13 Brush holder and rectifier holder
14 Stator coil

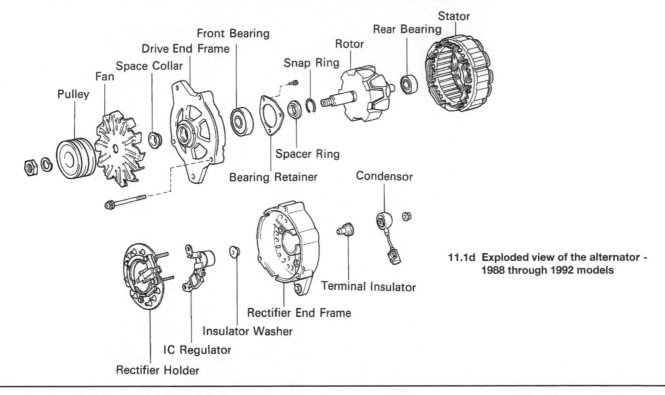

11.1d Exploded view of the alternator - 1988 through 1992 models

engine. **Note:** *1980 through 1987 models are equipped with an integral voltage regulator. The voltage regulator is mounted on the side of the alternator body and it is accessible with the alternator in the vehicle* **(see illustration)**. *1988 through 1992 models are equipped with a true internal regulator that is attached to the rectifier assembly within the alternator body* **(see illustration)**. *1993 and later models are equipped with an integral voltage regulator and brush assembly that is accessible by removing the rear cover from the back of the alternator* **(see illustration)**. *The alternator must be removed from the vehicle first on this type.*

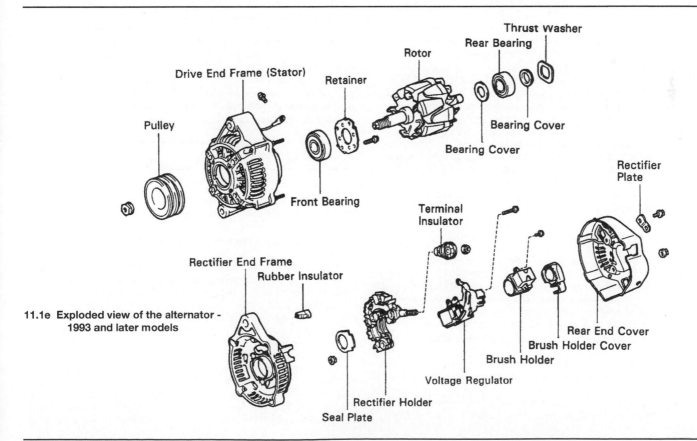

11.1e Exploded view of the alternator - 1993 and later models

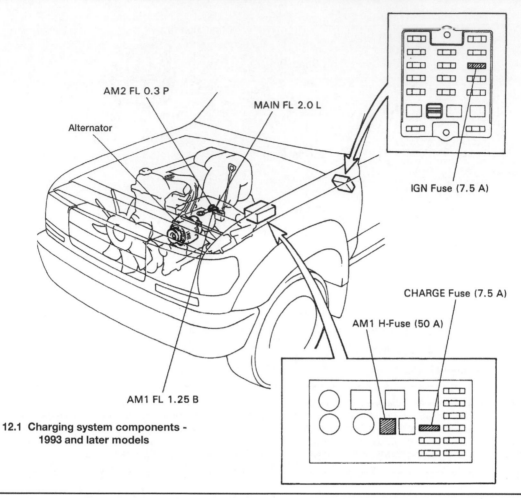

AM2 FL 0.3 P

Alternator

MAIN FL 2.0 L

IGN Fuse (7.5 A)

CHARGE Fuse (7.5 A)

AM1 H-Fuse (50 A)

AM1 FL 1.25 B

**12.1 Charging system components -
1993 and later models**

The purpose of the voltage regulator is to limit the alternator's voltage to a preset value. This prevents power surges, circuit over-loads, etc., during peak voltage output.

1980 through 1987 models are equipped with an ignition control relay. When activated, this relay allows power to the ignition and charging system fuses. 1980 through 1992 models are equipped with a discharge warning light relay that provides power to the warning light if the system does not charge properly. 1993 and later are equipped with various fuses and fusible links to govern circuit integrity.

The fusible link is a short length of insulated wire integral with the engine compartment wiring harness (see Chapter 12). The fusible link is several wire gauges smaller in diameter than the circuit it protects. See Chapter 12 for additional information regarding fusible links.

The charging system doesn't ordinarily require periodic mainte-nance. However, the drivebelt, battery and wires and connections should be inspected at the intervals outlined in Chapter 1.

The dashboard warning light should come on when the ignition key is turned to Start, then should go off immediately. If it remains on, there is a malfunction in the charging system. Some vehicles are also equipped with a voltage gauge. If the voltage gauge indicates abnor-mally high or low voltage, check the charging system (see Section 12).

Be very careful when making electrical circuit connections to a vehicle equipped with an alternator and note the following:

a) *When reconnecting wires to the alternator from the battery, be sure to note the polarity.*
b) *Before using arc welding equipment to repair any part of the vehi-cle, disconnect the wires from the alternator and the battery termi-nals.*
c) *Never start the engine with a battery charger connected.*
d) *Always disconnect both battery leads before using a battery charger.*

e) *The alternator is driven by an engine drivebelt which could cause serious injury if your hand, hair or clothes become entangled in it with the engine running.*
f) *Because the alternator is connected directly to the battery, it could arc or cause a fire if overloaded or shorted out.*
g) *Wrap a plastic bag over the alternator and secure it with rubber bands before steam cleaning the engine.*

12 Charging system - check

Refer to illustrations 12.1, 12.3, 12.7 and 12.8

1   If a malfunction occurs in the charging circuit, don't automatically assume that the alternator is causing the problem. First check the fol-lowing items:

a) *Check the drivebelt tension and its condition. Replace it if worn or deteriorated.*
b) Make sure the alternator mounting and adjustment bolts are tight.
c) *Inspect the alternator wiring harness and the electrical connectors at the alternator and voltage regulator. They must be in good con-dition and tight.*
d) *Check the fusible link located at the positive battery cable or the large main fuses in the engine compartment. If it's burned, deter-mine the cause, repair the circuit and replace the link or fuse (the vehicle won't start and/or the accessories won't work if the fusible link or fuse blows).*
e) *Check all the fuses that are in series with the charging system circuit* **(see illustration)**. *The location of these fuses and fusible links may vary from year and model so refer to Chapter 12 and the wiring schematics at the end of Chapter 12 for additional information.*

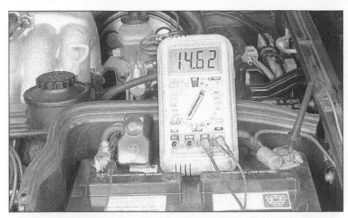

12.3 Voltage should increase to approximately 14.5 volts after the engine has been started

f) *Start the engine and check the alternator for abnormal noises (a shrieking or squealing sound indicates a bad bushing).*

g) *Check the specific gravity of the battery electrolyte. If it's low, charge the battery (doesn't apply to maintenance free batteries).*

h) *Make sure that the battery is fully charged (one bad cell in a battery can cause overcharging by the alternator).*

i) *Disconnect the battery cables (negative first, then positive).* **Caution:** *If the stereo in your vehicle is equipped with an anti-theft system, make sure you have the correct activation code before disconnecting the battery. Inspect the battery posts and the cable clamps for corrosion. Clean them thoroughly if necessary (see Section 4 and Chapter 1). Reconnect the positive cable, then the negative cable.*

j) *Check the operation of the charging system warning light relay. 1980 through 1992 models are equipped with a charging system warning light relay. This relay is located in different locations according to the year and model. Refer to Chapter 12 for additional information on the location and check of the charging system warning light relay.*

2   Using a voltmeter, check the battery voltage with the engine off. It should be approximately 12 volts.

3   Start the engine and check the battery voltage again. It should now be approximately 13.5 to 15.1 volts **(see illustration)**.

4   Turn on the headlights. The voltage should drop and then come back up, if the charging system is working properly.

5   If the voltage reading is greater than the specified charging voltage, replace the voltage regulator (see Section 14).

6   If the voltmeter reading is less than standard voltage, check the regulator and alternator as follows.

7   Ground terminal F at the rear of the alternator **(see illustration)**, start the engine, check the voltage at the battery and compare your reading to the standard voltage. **Note:** *On some models, you may have to remove a cover to access terminal F.*

a) *If the voltmeter reading is greater than standard voltage, replace the regulator.*

b) *If the voltmeter reading is less than standard voltage, check the alternator (or have it checked by a dealer service department if you do not have an ammeter).*

8   If you have an ammeter, hook it up to the charging system as shown **(see illustration)**. If you don't have an ammeter, you can also use an inductive-type current indicator. This device is inexpensive, readily available at auto parts stores and accurate enough to perform simple amperage checks like the following test.

9   With the engine running at 2,000 rpm, check the reading on the ammeter with all accessories and lights off, then again with the high-beam headlights on and the heater blower switch turned to the HI position. Compare your readings to the standard amperage listed in this Chapter's Specifications.

10   If the ammeter reading is less than standard amperage, repair or replace the alternator.

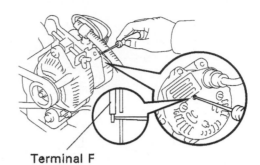

Terminal F

12.7 If the alternator is putting out less than standard voltage, ground terminal F, start the engine and check the voltage at the battery - if the reading is greater than standard voltage, replace the regulator; if the reading is less than standard, check the alternator or have it checked by a dealer service department or other qualified repair shop

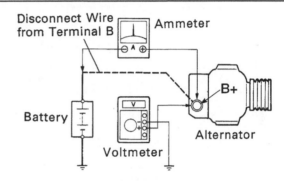

12.8 Hook up an ammeter as shown to check alternator output

13   Alternator - removal and installation

Refer to illustrations 13.2 and 13.3

1   Detach the cable from the negative terminal of the battery. **Caution:** *If the stereo in your vehicle is equipped with an anti-theft system, make sure you have the correct activation code before disconnecting the battery.*

2   Detach the electrical connectors from the alternator **(see illustration)**.

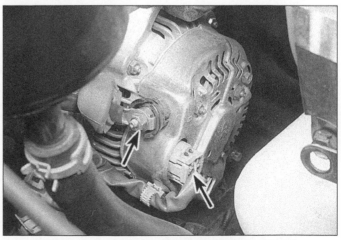

13.2 Disconnect the electrical connections from the alternator (arrows)

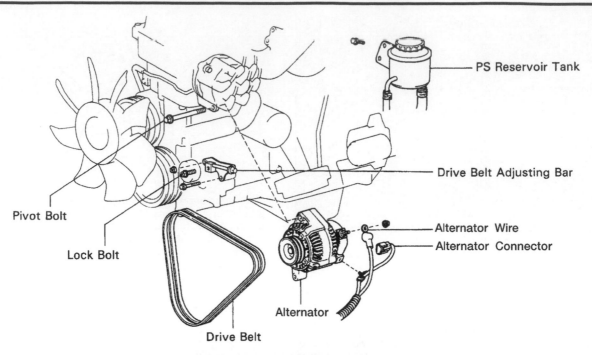

PS Reservoir Tank

Drive Belt Adjusting Bar

Alternator Wire
Alternator Connector

Pivot Bolt

Lock Bolt

Alternator

Drive Belt

13.3 Alternator installation details

3   Loosen the alternator adjustment, pivot and lock bolts **(see illus-tration)** and detach the drivebelt.
4   Remove the adjustment and lock bolts from the alternator adjustment bracket.
5   Separate the alternator and bracket from the engine.
6   If you are replacing the alternator, take the old alternator with you when purchasing a replacement unit. Make sure that the new/rebuilt unit is identical to the old alternator. Look at the terminals - they should be the same in number, size and locations as the terminals on the old alternator. Finally, look at the identification markings - they will be stamped in the housing or printed on a tag or plaque affixed to the housing. Make sure that these numbers are the same on both alternators.
7   Many new/rebuilt alternators do not have a pulley installed, so you may have to switch the pulley from the old unit to the new/rebuilt one. When buying an alternator, find out the shop's policy regarding installation of pulleys - some shops will perform this service free of charge.
8   Installation is the reverse of removal.
9   After the alternator is installed, adjust the drivebelt tension

(see Chapter 1).
10   Check the charging voltage to verify proper operation of the alternator (see Section 12).

14   Alternator components (1993 and later models) - check and replacement

Note: *1980 through 1987 models are equipped with an integral voltage regulator. The voltage regulator is mounted on the side of the alternator body and it is accessible with the alternator in the vehicle. 1988 through 1992 models are equipped with a true internal regulator that is attached to the rectifier assembly within the alternator body. 1993 and later models are equipped with an integral voltage regulator and brush assembly that is accessible by removing the rear cover from the back of the alternator. The alternator must be removed from the vehicle first. This procedure covers only the 1993 and later models. Although the alternators are basically the same, they differ in the voltage regulator designs.*

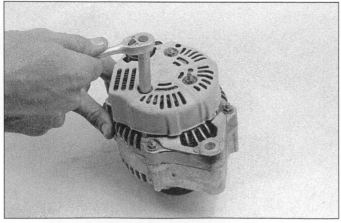

14.2a Remove the three nuts from the rear cover

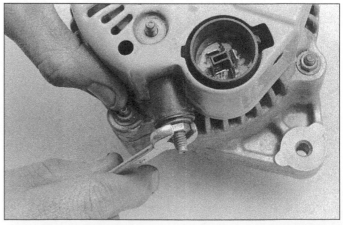

14.2b Remove the nut, washer and insulator from terminal B and remove the alternator end cover

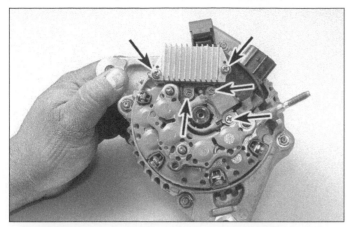

14.3 After removing the rear cover, remove the five screws (arrows) that retain the voltage regulator and the brush holder

14.4a Remove the brush holder

14.4b Remove the voltage regulator

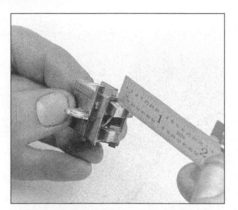

14.5 Measure the exposed length of the brushes and compare your measurements to the specified minimum length to determine if replacement is necessary

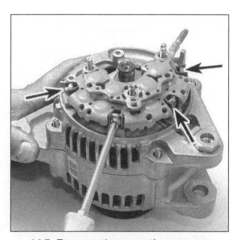

14.7 Remove the mounting screws (arrows) retaining the rectifier assembly

Disassembly

Refer to illustrations 14.2a, 14.2b, 14.3, 14.4a, 14.4b, 14.5 and 14.7

1   Remove the alternator (see Section 13) and place it on a clean workbench.

2   Remove the rear cover nuts, the nut and terminal insulator and the rear cover **(see illustrations)**.

3   Remove the five voltage regulator and brush holder mounting screws **(see illustration)**.

4   Remove the brush holder and the regulator from the rear end frame **(see illustrations)**. If you are only replacing the regulator, proceed to Step 8, install the new unit, reassemble the alternator and install it on the engine (see Section 13). If you are going to replace the brushes, proceed with the next Step.

5   Measure the exposed length of each brush **(see illustration)** and compare it to the minimum length listed in this Chapter's Specifications. If the length of either brush is less than the specified minimum, replace the brushes and brush holder assembly. **Note:** *On some models, it may be necessary to solder the new brushes in place.*

6   Make sure that each brush moves smoothly in the brush holder.

7   Remove the rectifier assembly **(see illustration)**. Remove the four rubber insulators and the seal plate.

8   Scribe or paint marks on the front and rear end frame housings of the alternator to facilitate reassembly.

9   Remove the nut retaining the pulley to the rotor shaft and remove the pulley.

10   Remove the four nuts retaining the front and rear end frames together, then separate the rear end frame assembly from the front end frame.

11   Remove the thrust washer and remove the rotor from the front end frame.

Component checks

Refer to illustrations 14.12a, 14.12b, 14.13, 14.14a, 14.14b, 14.14c and 14.14d

12   Check for an open between the two slip rings **(see illustration)**. There should be 2 to 4 ohms resistance between the slip rings. Check

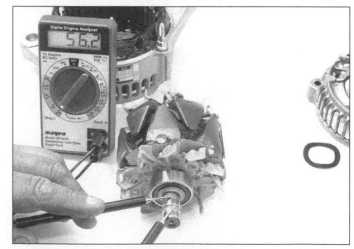

14.12a Continuity should exist between the rotor slip rings

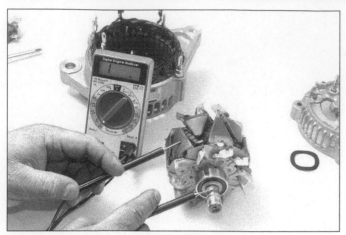

14.12b Check for a short between the rotor and the slip rings.
There should be NO continuity

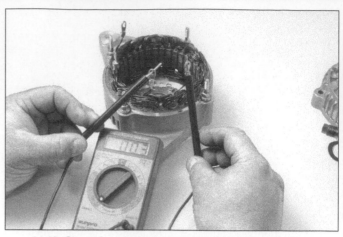

14.13 Check for continuity between the stator windings

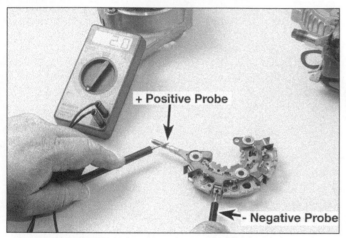

14.14a Position the positive probe of the ohmmeter onto the
diode assembly positive post and the negative probe to the
ground terminal. Continuity should exist

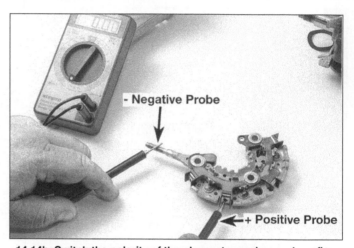

14.14b Switch the polarity of the ohmmeter probes and confirm
that now there is NO continuity within the diodes. Check each
diode (four total) individually

for grounds between each slip ring and the rotor (see illustration).
There should be no continuity (infinite resistance) between the rotor
and either slip ring. If the rotor fails either test, or if the slip rings are
excessively worn, the rotor is defective.

13   Check for opens between each end terminal of the stator windings (see illustration). If either reading is high (infinite resistance), the
stator is defective. Check for a grounded stator winding between each
stator terminal and the frame. If there's continuity between any stator
winding and the frame the stator is defective.

14   Check the positive and negative rectifiers.

a) First start the checks on the positive diode assembly by touching
    the positive probe of the ohmmeter onto the positive terminal and
    the negative probe onto one of the rectifier diode terminals (see
    illustration). Then reverse the probes and check again (see illus-
    tration). The diode should have continuity with the ohmmeter one
    way and no continuity when the probes are reversed (see illustra-
    tion). Check each of the terminals in this manner. If any of the
    diodes fail the test, the diode assembly is defective.
b) Now, check the negative diode assembly by touching the negative
    probe of the ohmmeter onto the NEGATIVE TERMINALS (see

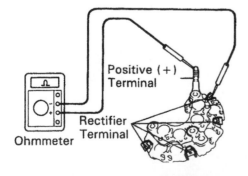

14.14c Check the positive rectifier diodes

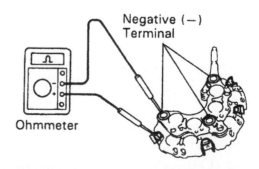

14.14d Check each diode on the negative rectifier diode assembly

14.16 To facilitate installation of the brush holder, depress each brush with a small screwdriver to clear the shaft

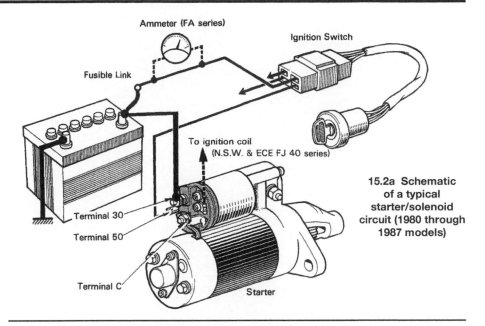

15.2a Schematic of a typical starter/solenoid circuit (1980 through 1987 models)

illustration) and the other probe onto each rectifier terminal. Reverse the polarity (reverse probes) and check to make sure there is no continuity in one position and continuity in the other position. Check each of the terminals in this manner. If any of the diodes fail the test, the diode assembly is defective.

Reassembly

Refer to illustration 14.16

15 Install the components in the reverse order of removal, noting the following:

16 Install the brush holder by depressing each brush with a small screwdriver to clear the shaft **(see illustration).**

17 Install the voltage regulator and brush holder screws into the rear frame.

18 Install the rear cover and tighten the three nuts securely.
19 Install the terminal insulator and tighten it with the nut.
20 Install the alternator (see Section 13).

15 Starting system - general information and precautions

Refer to illustrations 15.2a and 15.2b

The sole function of the starting system is to turn over the engine quickly enough to allow it to start.

The starting system consists of the battery, the starter motor, the starter solenoid and the electrical circuit connecting the components. The solenoid is mounted directly on the starter motor **(see illustrations). Note:** *1980 through 1987 models are equipped with a lever-*

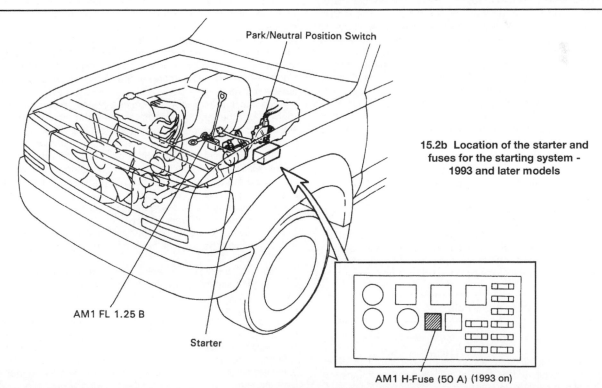

15.2b Location of the starter and fuses for the starting system - 1993 and later models

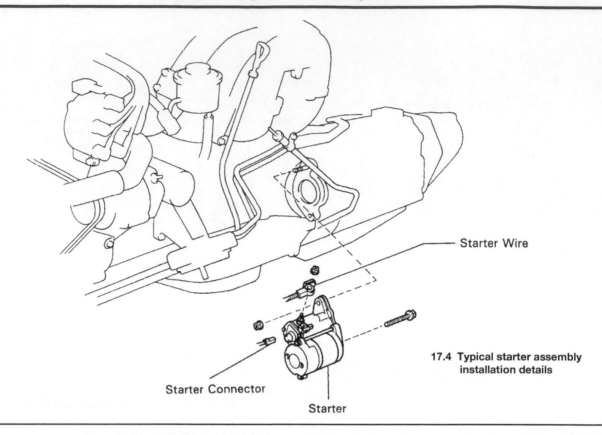

Starter Wire

**17.4 Typical starter assembly
installation details**

Starter Connector

Starter

and-pinion gear style solenoid mounted on the starter body. 1988 and
later models are equipped with two different types of gear reduction
starters. If the gear reduction starter is defective, it is recommended
that the entire starter/solenoid assembly be replaced as a unit.

The solenoid/starter motor assembly is installed on the rear of the
engine, next to the transmission bellhousing.

When the ignition key is turned to the START position, the starter
solenoid is actuated through the starter control circuit. The starter
solenoid then connects the battery to the starter. The battery supplies
the electrical energy to the starter motor, which does the actual work
of cranking the engine.

The starter motor on a vehicle equipped with a manual transmission can be operated only when the clutch pedal is depressed;
the starter on a vehicle equipped with an automatic transmission
can be operated only when the transmission selector lever is in Park
or Neutral.

Always observe the following precautions when working on the
starting system:

a) *Excessive cranking of the starter motor can overheat it and cause
serious damage. Never operate the starter motor for more than 15
seconds at a time without pausing to allow it to cool for at least
two minutes.*

b) *The starter is connected directly to the battery and could arc or
cause a fire if mishandled, overloaded or short circuited.*

c) *Always detach the cable from the negative terminal of the battery
before working on the starting system.* **Caution:** *If the stereo in
your vehicle is equipped with an anti-theft system, make sure you
have the correct activation code before disconnecting the battery.*

16 Starter motor - testing in vehicle

Note: *Before diagnosing starter problems, make sure the battery is fully
charged.*

1   If the starter motor does not turn at all when the switch is operated,
make sure that the shift lever is in Neutral or Park (automatic transmission) or that the clutch pedal is depressed (manual transmission).

2   Make sure that the battery is charged and that all cables, both at
the battery and starter solenoid terminals, are clean and secure.

3   If the starter motor spins but the engine is not cranking, the over-
running clutch in the starter motor is slipping and the starter motor
must be replaced.

4   If, when the switch is actuated, the starter motor does not operate
at all but the solenoid clicks, then the problem lies with either the battery, the main solenoid contacts or the starter motor itself (or the
engine is seized).

5   If the solenoid cannot be heard when the switch is actuated,
the battery is bad, the fusible link is burned (the circuit is open), the
starter relay is defective or the starter solenoid itself is defective.
Note: *Follow the relay testing procedure in Chapter 12 to diagnose
a defective relay.*

6   To check the solenoid, connect a jumper lead between the battery (+) and the ignition switch terminal (the small terminal) on the
solenoid. If the starter motor now operates, the solenoid is OK and the
problem is in the ignition switch, Neutral start switch or in the wiring.

7   If the starter motor still does not operate, remove the
starter/solenoid assembly for disassembly, testing and repair.

8   If the starter motor cranks the engine at an abnormally slow
speed, first make sure that the battery is charged and that all terminal
connections are tight. If the engine is partially seized, or has the wrong
viscosity oil in it, it will crank slowly.

9   Run the engine until normal operating temperature is reached,
then disconnect the coil wire from the distributor cap and ground it on
the engine.

10   Connect a voltmeter positive lead to the battery positive post and
connect the negative lead to the negative post.

11   Crank the engine and take the voltmeter readings as soon as a
steady figure is indicated. Do not allow the starter motor to turn for
more than 15 seconds at a time. A reading of nine volts or more, with
the starter motor turning at normal cranking speed, is normal. If the
reading is nine volts or more but the cranking speed is slow, the motor
is faulty. If the reading is less than nine volts and the cranking speed is
slow, the solenoid contacts are probably burned, the starter motor is
bad, the battery is discharged or there is a bad connection.

17   Starter motor - removal and installation

Refer to illustrations 17.4 and 17.5

1   Detach the cable from the negative terminal of the battery. **Caution:** *If the stereo in your vehicle is equipped with an anti-theft system, make sure you have the correct activation code before disconnecting the battery.*
2   Remove the battery from the engine compartment.
3   Disconnect and remove the cruise control assembly from the engine compartment (see Chapter 12).
4   Detach the electrical connectors from the starter/solenoid assembly **(see illustration)**.
5   Remove the starter motor mounting bolts **(see illustration)**.
6   Remove the bracket from the upper section of the starter/solenoid assembly. **Note:** *It is necessary to loosen one or two of the bracket bolts to allow the starter/solenoid assembly to partially drop down to gain access to the remaining bracket bolts.*
7   Installation is the reverse of removal.

18   Starter solenoid (1980 through 1987 models) - removal and installation

Refer to illustration 18.3
Note: *1980 through 1987 models are equipped with a lever and pinion gear style solenoid mounted on the starter body. 1988 and later models are equipped with two different types of gear reduction starters. If the*

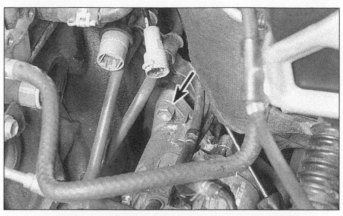

17.5 Remove the top bolt from the starter assembly (1993 and later model shown)

gear reduction starter is defective, it is recommended that the entire starter/solenoid assembly be replaced as a unit.

1   Detach the cable from the negative terminal of the battery. **Caution:** *If the stereo in your vehicle is equipped with an anti-theft system, make sure you have the correct activation code before disconnecting the battery.*
2   Remove the starter assembly from the engine (see Section 17).
3   Remove the solenoid mounting bolts **(see illustration)**.
4   Separate the solenoid from the starter body.
5   Installation is the reverse of removal.

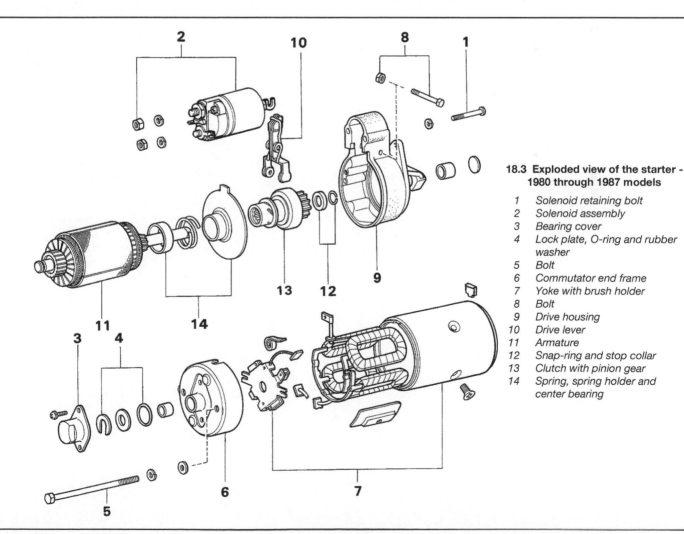

18.3 Exploded view of the starter - 1980 through 1987 models

1   Solenoid retaining bolt
2   Solenoid assembly
3   Bearing cover
4   Lock plate, O-ring and rubber washer
5   Bolt
6   Commutator end frame
7   Yoke with brush holder
8   Bolt
9   Drive housing
10   Drive lever
11   Armature
12   Snap-ring and stop collar
13   Clutch with pinion gear
14   Spring, spring holder and center bearing

Notes

Chapter 6
Emissions and engine control systems

Contents

	Section
Air Injection Reactor (AIR) system	3
Catalytic converter	18
CHECK ENGINE light	See Section 3
Choke breaker system	10
Choke opener system	6
Deceleration fuel cut system	7
Electronic Fuel Injection (EFI) system - general information and ECM removal and installation	11
Evaporative Emission Control (EVAP) system	14
Exhaust Gas Recirculation (EGR) system	15
Fuel tank cap gasket replacement	See Chapter 1
General information	1

	Section
Heat control valve system	8
High Altitude Compensation (HAC) system	4
Hot Air Intake (HAI) system	5
Hot Idle Compensation (HIC) system	9
Information sensors	13
On Board Diagnosis (OBD) system - description trouble code access	12
Pulse Air Injection (PAIR) system	17
PCV valve check and replacement	See Chapter 1
Positive Crankcase Ventilation (PCV) system	16
Spark Control (SC) system	2
Thermostatic air cleaner check	See Chapter 1

Specifications

EGR gas temperature sensor resistance
112 degrees F .... 69 to 89 K-ohms
212 degrees F .... 11 to 15 K-ohms
302 degrees F .... 2 to 4 K-ohms

1   General information

Refer to illustration 1.1

To prevent pollution of the atmosphere from burned and evaporating gases, a number of major and auxiliary emission control systems are incorporated on the vehicles covered by this manual. The combination of systems used depends on the year in which the vehicle was manufactured, the locality to which it was originally delivered and the engine type. Check the Vehicle Emissions Control Information (VECI) label **(see illustration)** in your engine compartment to determine which systems are used on your vehicle. The major systems incorporated include:

Air Injection Reactor system
Pulse Air Injection system
Catalytic Converter system
Evaporative Emission Control system
Exhaust Gas Recirculation system
High Altitude Compensation system
Hot Air Intake system
Positive Crankcase Ventilation system
Spark Control system

The auxiliary systems incorporated include:

Choke Breaker system
Choke Opener System

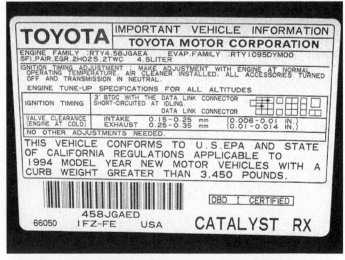

1.1 The Vehicle Emission Control Information (VECI) label is a handy reference guide for tune-up information and for the types of emission devices used on your vehicle - another label (not shown) near this one provides a vacuum schematic showing the location of all emissions hoses and devices

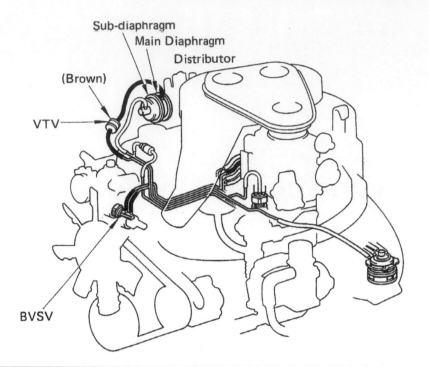

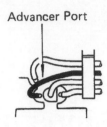

2.1 Schematic of the Spark Control system

Deceleration Fuel Cut system
Heat Control Valve system
Hot Idle Compensation system
Oxidation Catalyst system

Note: *To verify which emission control system is installed on the year and model Land Cruiser in question, check the VECI label for a list of systems or refer to a dealer service department for additional information.*

The Sections in this Chapter include general descriptions, checking procedures (where possible) and component replacement procedures (where applicable) for each of the systems listed above.

Before assuming that an emission control system is malfunctioning, check the fuel and ignition systems carefully. In some cases, special tools and equipment, as well as specialized training, are required to accurately diagnose the causes of a rough running or difficult to start engine. If checking and servicing become too difficult or if a procedure is beyond the scope of the home mechanic, consult a dealer service department. This does not necessarily mean however, that the emission control systems are particularly difficult to maintain and repair. You can quickly and easily perform many checks and do most (if not all) of the regular maintenance at home with common tune-up and hand tools. **Note:** *The most frequent cause of emission system problems is simply a loose or broken vacuum hose or wiring connection.* Therefore, always check hose and wiring connections first.

Pay close attention to any special precautions outlined in this Chapter (particularly those concerning the catalytic converter). It should be noted that the illustrations of the various systems may not exactly match the systems installed in your particular vehicle, due to changes made by the manufacturer during production or from year to year.

---

2   Spark Control (SC) system

General description

Refer to illustrations 2.1 and 2.3

1   The Spark Control system **(see illustration)** is designed to reduce HC and NOX emissions by delaying the ignition timing only when the engine is cold. By delaying the vacuum advance, combustion tempera-

tures are not increased. These systems are equipped on certain models from 1980 through 1987 depending upon state and emission standards.

2   The SC system includes a distributor mounted vacuum unit, a Bi-metal Vacuum Switching Valve (BVSV) and a Vacuum Transmitting Valve (VTV) and various hoses and fittings, although not all components are incorporated on all systems.

3   Depending on the engine coolant temperature, altitude and the position of the throttle, vacuum is applied to either one or both sides of the diaphragms in the distributor vacuum unit and the ignition timing is changed to reduce emissions and improve cold engine driveability **(see illustration)**.

Check

Refer to illustrations 2.7 and 2.8

4   Install a T-fitting between the vacuum gauge and the main diaphragm and the BVSV and observe that with the engine cold (coolant temperature 86 degrees F or less) there is no vacuum to the distributor.

5   Warm the engine to normal operating temperature and observe that vacuum is allowed past the BVSV and to the vacuum advance. If no vacuum is available, remove the BVSV and check it.

6   With the BVSV cold (below 86 degrees F), apply air pressure through the top port and observe that no air is allowed out of the bottom port.

7   Place the BVSV in a pan of hot water (above 111 degrees F) and observe that air is allowed through the bottom port when pressure is applied through the top port **(see illustration)**.

8   Also check the operation of the VTV. Check that air flows easily from B to A but it does not flow from A to B **(see illustration)**.

9   If vacuum is correct, check the operation of the vacuum advance (see Chapter 5).

Component replacement

Vacuum advance unit

10   To replace the distributor components, see Chapter 5.

Bi-metal Vacuum Switching Valve (BVSV)

11   Drain the coolant from the engine block (see Chapter 1), remove the vacuum hoses from the BVSV, then remove the BVSV and replace

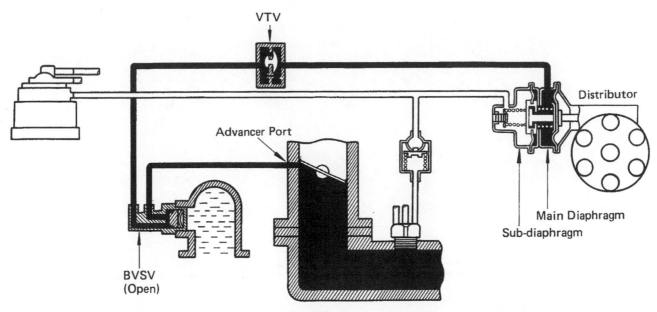

Coolant Temp.	BVSV	Throttle Valve Opening	Vacuum Ignition Timing
Cold Below 30°C (86°F)	CLOSED	—	NOT ADVANCED
Hot Above 44°C (111°F)	OPEN	Positioned below advancer port	NOT ADVANCED
		Positioned above advancer port	DELAYED by VTV

2.3 Operating parameters of the Spark Control system

it with a new one. Installation is the reverse of the removal procedures. **Note:** *Be sure to apply liquid sealer to the threads of the new BVSV before installing it.*

Vacuum Transmitting Valve (VTV)

12   Disconnect the hoses from both ends of the faulty unit and replace it with a new one, then reconnect the hoses. Make sure that you replace the faulty unit with one of the same color coding.

3   Air Injection Reactor (AIR) system

Refer to illustrations 3.2a and 3.2b

General description

1   This system supplies air under pressure to the exhaust ports to promote combustion of unburned hydrocarbons and carbon monoxide

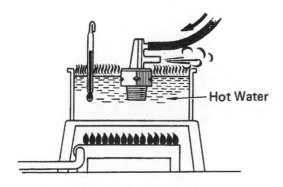

2.7 Air is allowed through the bottom port on the BVSV above 111 degrees F

2.8 Check that air flows through the VTV from black to white but is restricted when going from white to black

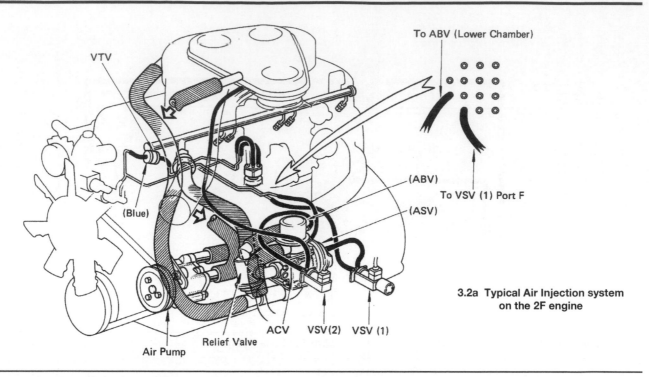

3.2a Typical Air Injection system
on the 2F engine

before they are allowed to exit the exhaust system.
2   This system **(see illustrations)** is composed of various configurations of any of the following devices: an air pump, check valve(s), thermo switch(es), thermo sensor(s), an oxygen sensor, restrictor jets, plus Air By-pass Valve (ABV), Air Control Valve (ACV), Air Switching Valve (ASV), Vacuum Switching Valve (VSV) and Vacuum Transmitting Valve (VTV) components. Note that not all components are included on every system. To determine which components are employed on the AIR system on your vehicle, refer to the VECI label (see Section 1).
3   The check valve mounted on the manifold prevents the reverse flow of exhaust gases into the system. The other components control

the injection of the air into the ports based on catalytic converter temperatures and engine load.

Check

4   Visually check the hoses, tubes and connections for cracks, loose fittings and separated parts.
5   Check the drivebelt condition and tension (refer to Chapter 1 for this procedure).

Air pump, ABV, ACV and ASV

6   These components require special tools and/or checking proce-

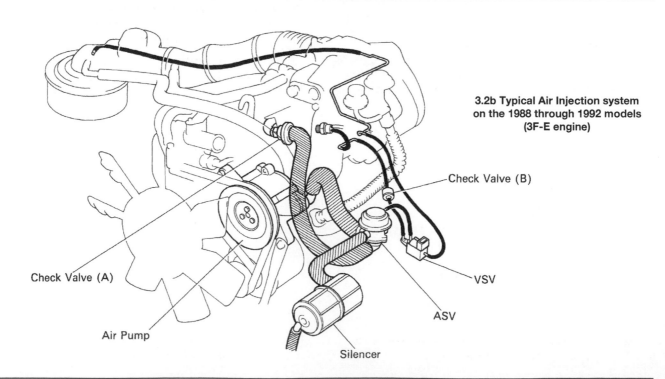

3.2b Typical Air Injection system
on the 1988 through 1992 models
(3F-E engine)

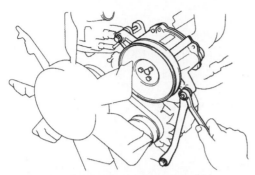

3.13 Remove the air pump mounting brackets (1988 through 1992 models)

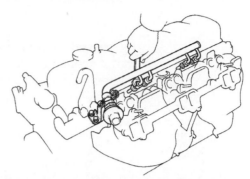

3.16 Remove the air injection manifold fittings from the cylinder head (1988 through 1992 models)

dures for checking and servicing. It is therefore recommended that you have these components inspected by your Toyota dealer.

Check valves

7   Inspect the check valve that is located between the ACV and the manifold, by disconnecting the hose from the inlet side and removing the ACV or ASV from the manifold. Next, blow into the check valve from the manifold side and verify that air does not flow through the check valve, then reverse the check valve and verify that air does flow through the valve when blowing into the inlet side.

8   If a problem is found, replace the check valve with a new one.

9   Inspect the check valve that is found between the ASV and the manifold by disconnecting the hoses from both ends. Next blow air into the white pipe of the valve and verify that air flows through the valve. Reverse the valve and verify that air does not flow through the valve when blowing air into it from the black side.

10   If a problem is found, replace the check valve with a new one.

Component replacement

Manifold check valve

11   Replace the valve by disconnecting the hose and unscrewing the check valve, then reverse the procedure, installing the new part.

In-line check valve

12   Replace these parts by disconnecting the hoses and/or wires from the faulty part, then reverse the procedure, installing the new part.

Air pump

Refer to illustration 3.13

13   Remove the mounting brackets the bolts **(see illustration)** from

the air pump and remove the assembly from the engine.

14   Installation is the reverse of removal.

Air injection manifold

Refer to illustration 3.16

15   Remove the intake and exhaust manifolds from the engine (see Chapter 2A).

16   Using a special line wrench, loosen the injection line bolts from the cylinder head **(see illustration)**.

17   Lift the injection manifold off the engine.

18   Installation is the reverse of removal.

---

4   High Altitude Compensation (HAC) system

Note: *This section applies only to carbureted engines on 1980 through 1987 models. 1988 through 1996 models are equipped to compensate for high altitudes using the ECM and the output actuators and sensors of the fuel injection system.*

General description

Refer to illustrations 4.1 and 4.2

1   The High Altitude Compensation (HAC) system ensures that the proper air/fuel mixture is supplied by the carburetor at altitudes of 3930 feet and above. The HAC system also advances ignition timing to improve driving performance at high altitudes **(see illustration)**.

2   The main components of this system include the carburetor altitude compensator ports (primary high and low speed ports), a sec-

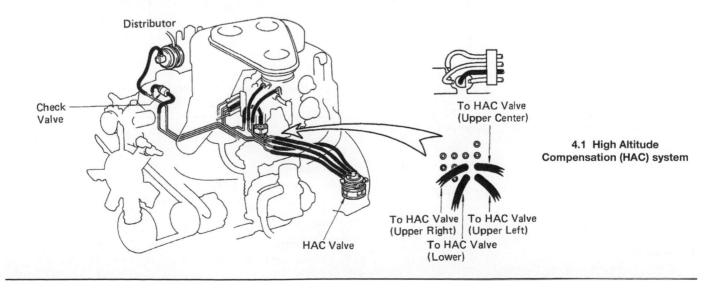

4.1 High Altitude Compensation (HAC) system

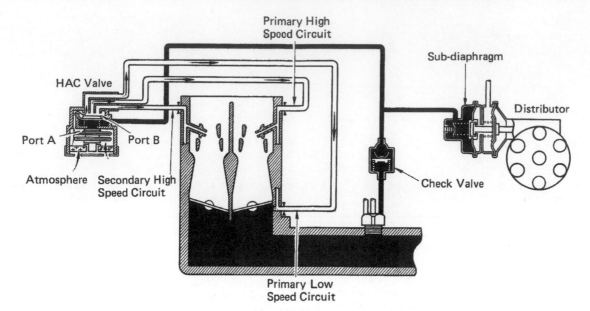

Altitude	Bellows in HAC Valve	Port A in HAC Valve	Distributor Sub-diaphragm	Port B in HAC Valve	Air from HAC Valve	Sub-vacuum Advancer
HIGH Above 1,198 m (3,930 ft)	EXPANDED	CLOSED	PULLED	OPEN	Led into primary low and high speed circuits.	ADVANCED (+6°)
LOW Below 783 m (2,570 ft)	CONTRACTED	OPEN	NOT PULLED	CLOSED	STOPPED	NOT/ADVANCED

4.2 Working parameters of the High Altitude Compensation (HAC) system

ondary high speed port and a fuel cut port, a HAC valve and a check valve, plus connecting hoses **(see illustration)**.

3   The high altitude compensation is accomplished by two methods: Additional air is supplied to the primary high and low and secondary high speed circuits of the carburetor and the ignition timing is advanced for improved driveabilty above 3930 feet. At lower altitudes, the additional air is cut off and the initial advance occurs only at idle.

Check

HAC valve

Refer to illustration 4.4

4   First check the position of the HAC valve by blowing into any one of the top three hoses with the engine idling. If the HAC is in high altitude mode, air will pass through out the bottom **(see illustration)**. The system will be difficult to check if the vehicle is at low altitude.

5   Above 3,930 feet, check that air flows into either of the two ports on top of the HAC valve with the engine idling. Below 2,570 feet, check that air does not flow into either of the two ports on top of the HAC valve with the engine idling.

6   Depending on the atmospheric pressure between 2,570 feet and 3,930 feet, the HAC valve may be either opened or closed. Therefore, attempt to check the valve at either altitude listed in paragraphs 4 and 5.

7   Remove the cover from the bottom of the HAC valve and clean and visually check the filter.

Carburetor

8   Disconnect the hoses from the pipes on top of the HAC valve.

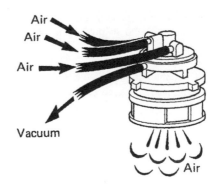

4.4 Blow air into any of the three top ports and observe air come out the bottom

9   Blow air into each hose and make sure that air flows into the carburetor.

Check valve

10   Disconnect the hoses from the check valve.
11   Check the air flows through the valve from the inlet side.
12   Check that air does not flow through the valve from the outlet side.

Distributor

13   Disconnect the hose from the distributor sub-diaphragm and plug the hose end.

14   Check the ignition timing at idle (see Chapter 1).
15   Reconnect the hose to the sub-diaphragm.
16   With the engine still idling at 950 rpm, verify that the ignition advances slightly (approximately 7 degrees).
17   Disconnect the vacuum hose between the check valve and the vacuum pipe at the vacuum pipe side and plug the pipe end.
18   Check that the ignition timing remains stationary for more than one minute.
19   Stop the engine and reconnect the hose to the vacuum pipe, then go to Step 24.
20   If the HAC valve is in the low altitude position, check the ignition timing and advance as described above in Steps 15 through 17.
21   Disconnect the vacuum hose from the lower port of the HAC valve, plug the hose end and perform the checks below.
22   Disconnect the vacuum hose between the check valve and the vacuum pipe at the vacuum pipe side and plug the pipe end.
23   Check that the ignition timing remains stationary for more than one minute.
24   Stop the engine and reconnect the hoses to their proper locations.
25   Disconnect the three hoses from the pipes on top of the HAC valve.
26   Blow air into each hose and make sure that air flows into the carburetor.
27   Reconnect the hoses to their proper locations.
28   Remove the distributor cap and rotor (see Chapter 5, if necessary).
29   Apply vacuum to the diaphragm and make sure that the vacuum advancer moves in accordance with the vacuum.
30   Reinstall the rotor and distributor cap.
31   If no problems have been encountered in these tests, the HAC system is okay.

Component replacement

32   The HAC valve and check valve can be replaced by removing the hoses from the faulty component and replacing it with a new one.
33   If problems are encountered with the distributor, see Chapter 5.

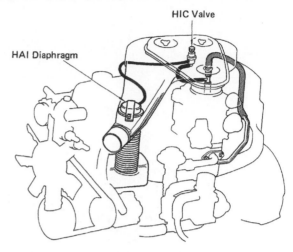

5.1 Hot Air Intake (HAI) system

5   Hot Air Intake (HAI) system

Refer to illustrations 5.1 and 5.3

General description

1   This system **(see illustration)** is designed to improve driveability and prevent carburetor icing in extremely cold weather by directing hot air from around the exhaust manifold to the air cleaner intake.
2   The system is composed of a diaphragm-activated air control valve located in the air cleaner intake, a thermo valve, an exhaust manifold shroud and interconnecting hoses and ducts.
3   When the underhood temperature is below 81 degrees F, the thermo valve allows manifold vacuum to act on the air cleaner diaphragm, which closes the control valve, or baffle, and allows hot air

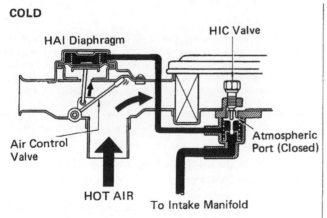

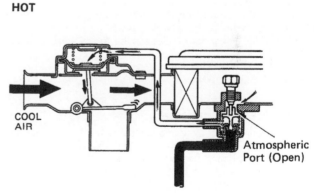

Temperature in Air Cleaner	HIC Valve	Air Control Valve	Intake Air
Cool Below 27°C (81°F)	Atmospheric port is CLOSED	Hot air passage OPEN	HOT
Hot Above 33°C (91°F)	Atmospheric port is OPEN	Cool air passage OPEN	COOL

5.3 Operating parameters of the Hot Air Intake (HAI) system

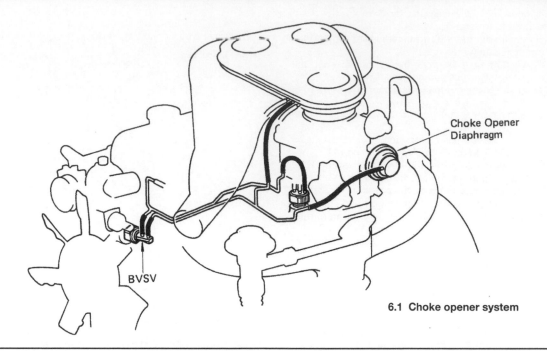

Choke Opener
Diaphragm

BVSV

6.1 Choke opener system

to enter the air cleaner. As temperature rises above 91 degrees F, the thermo valve cuts off the vacuum to the diaphragm and spring pressure opens the air control valve. Intake air is then drawn from under the hood, rather than from around the exhaust manifold **(see illustration)**.

Check

Air control valve

4   Remove the air cleaner cover.
5   Make sure the underhood temperature is below 81 degrees F.
6   Verify that the air control valve closes the cool air passage with the engine running at idle.
7   Reinstall the air cleaner cover and warm up the engine until the underhood temperature is above 91 degrees F.
8   Verify that the air control valve opens the cool air passage at idle.

Hoses and connections

9   Visually check the hoses and connections for cracks, leaks or damage.

Component replacement

10   Replacement of all of the component(s) in this system is a simple matter of removing the faulty component and replacing it with a new one. Make sure the hoses are correctly installed and tighten all hose clamps and mounting bolts securely. If a new air control valve is installed, make sure it moves freely before checking the system operation.

---

6   Choke opener system

General description

Refer to illustrations 6.1 and 6.2.
1   The choke opener system **(see illustration)** holds the choke valve open after the engine has warmed-up to prevent an over rich fuel mixture. The choke opening system includes: a choke opener diaphragm, a Bi-metal Vacuum Switching Valve (BVSV) and connecting hoses and linkage.
2   The choke opener diaphragm opens the choke valve slightly when the engine starts, preventing an overly rich mixture and the resulting

increase in emissions discharge to the atmosphere **(see illustration)**.
3   After the engine reaches a pre-determined temperature the choke opener holds the choke valve open and releases the fast idle cam to the 4th step which lowers the engine speed and prevents an overly rich mixture condition.

Check

4   Refer to Chapter 1, for the general carburetor choke check procedures. If the choke is determined to be malfunctioning, check the hoses for cracks, kinks and broken sections and make sure all wiring connections are tight before checking the other system components.

Choke opener

5   With the engine cold (below 41 degrees F), start the engine.
6   Disconnect the hose from the choke opener diaphragm and reconnect it again. The choke opener linkage should not move.
7   Warm the engine up to operating temperature and perform the same test as in Step 6. Observe that vacuum pulls the choke opener. If there is no action, check the vacuum source.
8   If the choke opener does not perform as described, apply vacuum to the diaphragm and verify that the linkage moves. If it doesn't, replace the diaphragm.

Bi-metal Vacuum Switching Valve (BVSV)

9   Remove the BVSV and check it in a pan of cold water:

a) *The water temperature must be below 40 degrees F.*
b) *Apply air pressure to the top port and observe the lower port is closed and no air passes through.*
c) *Heat the pan of water above 66 degrees F and observe that the BVSV bottom port allows air to pass through.*
d) *Installation is the reverse of removal.*

Component replacement

10   Because replacement of the choke opener and/or fast idle cam breaker requires partial disassembly of the carburetor, refer to the instructions included with the carburetor rebuild kit for your carburetor. Be sure to make any adjustments required after the components are replaced.
11   To replace the BVSV, remove the hoses from the pipes, then use a wrench to remove the BVSV from the intake manifold. Installation is the reverse of the removal procedure.

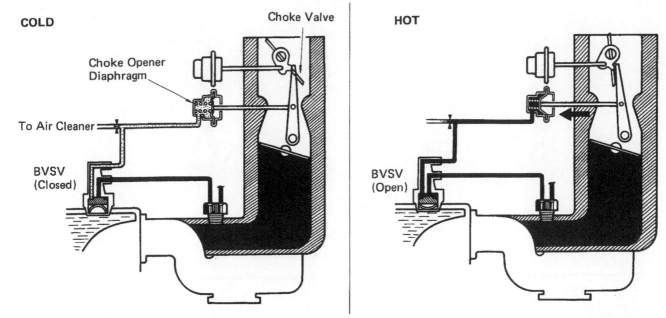

Coolant Temp.	BVSV	Diaphragm	Choke Valve
Below 5°C (41°F)	CLOSED	Released by spring tension.	CLOSED
Above 19°C (66°F)	OPEN	Pulled by intake manifold vacuum.	OPEN

6.2 Operating parameters of the Choke Opener system

7 Deceleration fuel cut system

General description

Refer to illustrations 7.1 and 7.3

1   This system **(see illustration)** serves to prevent overheating and after-burning in the exhaust system.

2   The system is made up of a fuel cut solenoid valve, a vacuum switch and attaching vacuum hose.

3   The system cuts off part of the fuel in the slow speed circuit of the carburetor at low rpm under high and low vacuum conditions and at high rpm under low vacuum conditions. At high rpm under high vacuum conditions, the system is off and the slow speed circuit in the carburetor is closed **(see illustration)**.

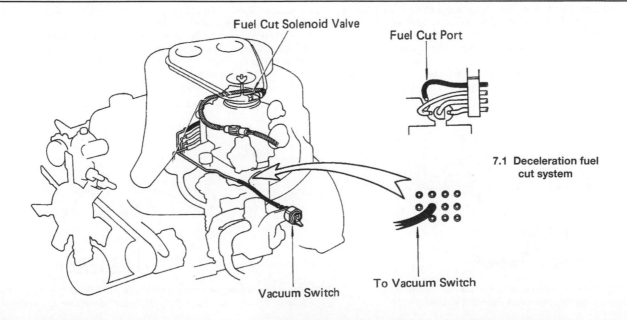

7.1 Deceleration fuel cut system

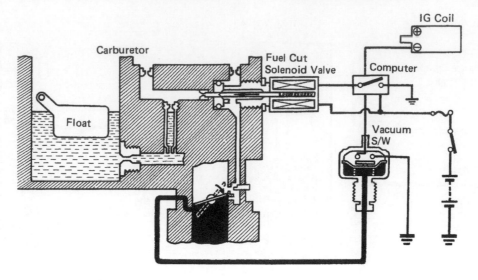

Engine RPM	Vacuum in the Vacuum S/W	Vacuum S/W	Computer	Fuel Cut Solenoid Valve	Slow Circuit in Carburetor
Below 1,330 rpm	Low vacuum below 290 mmHg (11.42 in.Hg)	ON	ON	ON	OPEN
	High vacuum above 355 mmHg (13.97 in.Hg)	OFF	ON	ON	OPEN
Above 1,800 rpm	Low vacuum below 290 mmHg (11.42 in.Hg)	ON	ON	ON	OPEN
	High vacuum above 355 mmHg (13.97 in.Hg)	OFF	OFF	OFF	CLOSED

7.3 Operating parameters of the deceleration fuel cut system

Check

General

Refer to illustrations 7.7 and 7.16

4   For vehicles with HAC systems, disconnect the vacuum hoses from the lower port of the HAC valve and plug the hose end (refer to Section 4).

5   Connect a tachometer to the engine according to the instructions supplied by the tachometer manufacturer.

6   Start the engine and observe that it runs normally.

7   Pinch off the hose to the vacuum switch **(see illustration)**.

8   Gradually increase the engine speed and observe that the engine misfires slightly between 1800 and 3000 rpm. **Caution:** *Perform this check quickly to avoid overheating the catalytic converter.*

9   Release the pinched hose.

10   Gradually increase the rpm to 3000 again and observe that the engine operation returns to normal.

11   With the engine idling, unplug the wiring connector to the solenoid valve and observe that the engine idles roughly or dies. **Note:** *Perform this check quickly to avoid overheating the catalytic converter.*

12   *Stop the engine and reconnect the wiring.*

13   Remove the tachometer.

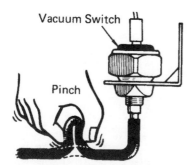

7.7 Pinch off the vacuum hose to the vacuum switch to check the Deceleration Fuel Cut system

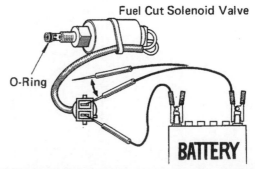

7.16 Using fused jumper wires, connect a 12 volt battery to the fuel cut solenoid and listen for a distinct "click" as power is supplied

COLD ENGINE

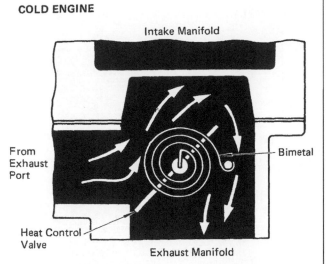

Intake Manifold

From Exhaust Port

Bimetal

Heat Control Valve

Exhaust Manifold

HOT ENGINE

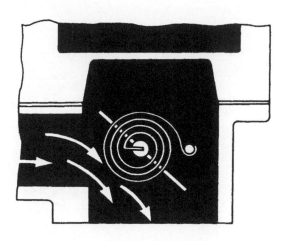

8.2 Heat control valve system

14   If no problem was encountered in the checks above, the system is okay, otherwise inspect the components as follows.

Fuel cut solenoid valve

15   Disconnect the connector and remove the fuel cut solenoid valve.
16   Connect the two terminals inside the connector to a 12 volt battery **(see illustration)**.
17   Observe that you can feel a click from the solenoid when the battery is connected and disconnected.
18   Check the O-ring for damage.
19   Reinstall the valve and hook up the wiring connector.
20   If a problems was encountered, replace the solenoid valve or O-ring with a new one.

Vacuum switch

21   Using an ohmmeter, check for continuity between the switch terminal and body.
22   Start the engine.
23   Using an ohmmeter, check that there is no continuity between the switch terminal and the body.
24   If a problem is encountered, replace the vacuum switch with a new one.

Component replacement

25   The fuel cut solenoid valve and vacuum switch can be replaced by disconnecting the connector and hoses, removing the faulty component and installing a new one, reversing the removal procedures.

8   Heat control valve system

General description

Refer to illustration 8.2

1   To reduce cold engine emissions and improve driveability, the intake manifold is heated by the heat control valve during cold engine operation to accelerate vaporization of the fuel.
2   The heat control valve system **(see illustration)** is composed of a bi-metal heat control valve mounted inside the exhaust duct directly under the intake manifold.
3   With the engine running and the coolant below 109 degrees F, the valve remains closed, trapping warm exhaust gasses and allows them to swirl in a chamber below the intake manifold. The manifold warms and consequently warms the air directed into the engine.

Check

4   Start the engine.
5   Verify that the coolant temperature is below 109 degrees F.
6   Make sure the bi-metal heat valve remains shut.
7   Warm up the engine to above 131 degrees F.
8   Check that the heat valve is open.

Component replacement

9   To replace the bi-metal heat control valve, replace the exhaust manifold (see Chapter 2A).
10   Installation is the reverse of removal.

9   Hot Idle Compensation (HIC) system

General information

Refer to illustrations 9.1 and 9.2

1   The HIC system allows controlled air to enter the intake manifold to maintain proper air/fuel mixture during high temperatures at idle conditions **(see illustration)**.

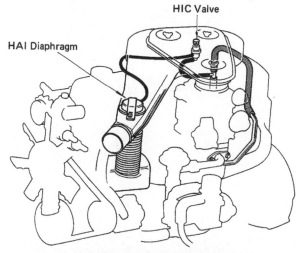

HIC Valve

HAI Diaphragm

9.1 Hot Idle Compensation (HIC) system

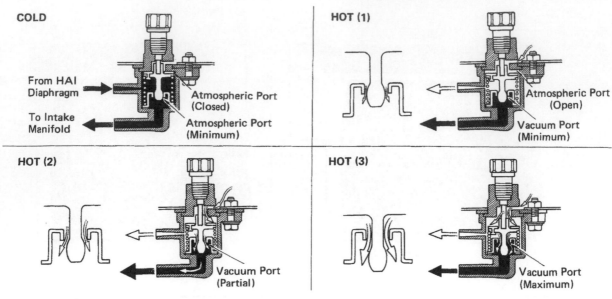

Temperature in Air Cleaner	HIC Valve Atmospheric Port	HIC Valve Vacuum Port Opening	HIC System
HOT (1) Between 27°C (81°F) and 50°C (122°F)	OPEN	MINIMUM	OFF
HOT (2) Between 50°C (122°F) and 85°C (185°F)	OPEN	PARTIAL	ON Air volume is controlled by HIC valve
HOT (3) Above 85°C (185°F)	OPEN	MAXIMUM	ON

9.2 Operating parameters of the Hot Idle Compensation (HIC) system

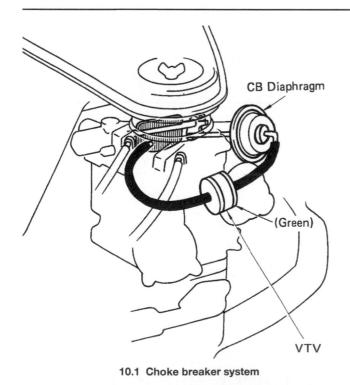

10.1 Choke breaker system

2   The HIC valve will vary as the temperature of the air warms from cold (81 to 122 degrees F) to hot (above 185 degrees F). The HIC valve is closed (cold), then gradually opens (hot) as the heated air is passed over the valve (see illustration).

Check

3   Using the tip of the finger, close the HIC valve pipe to the intake manifold and check that air does not flow from the diaphragm to the atmospheric port when pressure is applied through the top port. At temperatures below 81 degrees F, the HIC valve should be closed.
4   Remove the HIC valve, turn it upside down and place the top section in a pan of heated water. With the valve temperature above 185 degrees F, air should pass out the atmospheric port when air is applied through the lower port as described in Step 3.
5   If the test results are incorrect, replace the HIC valve.

Replacement

6   Remove the mounting clip and lift the HIC valve from the air cleaner.

10 Choke breaker system

General information

Refer to illustrations 10.1 and 10.2

1   The choke breaker system opens the choke valve slightly to prevent a rich mixture at cold starting conditions (see illustration).

(1)

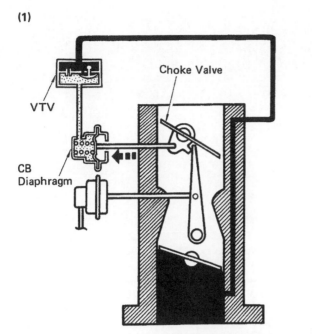

(2)

10.2 Operating characteristics of the choke breaker system

2   The choke breaker system consists of a choke breaker diaphragm, a Vacuum Transmitting Valve (VTV), a check valve and the vacuum hose routing **(see illustration)**. The VTV delays the choke valve opening.

Check

3   Start the engine and disconnect the vacuum hose between the carburetor and the VTV on the side of the carburetor.
4   Check that the choke breaker returns back to normal position by spring tension.
5   Reconnect the vacuum hose and observe that the choke breaker linkage is pulled back in within 5 to 15 seconds.
6   If the test results are incorrect, check the vacuum source and the choke breaker linkage.

Replacement

7   Remove the mounting brackets for the choke breaker diaphragm and remove the assembly from the carburetor.
8   Installation is the reverse of removal.

---

11   Electronic Fuel Injection (EFI) system - general information and ECM removal and installation

General information

1   The fuel injection system is controlled by means of a microcomputer known as the Electronic Control Module (ECM).
2   The ECM receives signals from various sensors which monitor changing engine operating conditions such as intake air volume, intake air temperature, coolant temperature, engine rpm, acceleration/deceleration, exhaust oxygen content, etc. These signals are utilized by the ECM to determine the correct fuel injection duration.
3   The system is analogous to the central nervous system in the human body: The sensors (nerve endings) constantly relay signals to the ECM (brain), which processes the data and, if necessary, sends out a command to change the operating parameters of the engine (body).
4   Here's a specific example of how one portion of this system operates: An oxygen sensor, located in the exhaust manifold, constantly monitors the oxygen content of the exhaust gas. If the percentage of oxygen in the exhaust gas is incorrect, an electrical signal is sent to the ECM. The ECM takes this information, processes it and then sends a command to the fuel injection system demanding a change in the air/fuel mixture. This happens in a fraction of a second and it goes on continuously when the engine is running. The end result is an air/fuel mixture ratio which is constantly maintained at a predetermined ratio, regardless of driving conditions.
5   In the event of a sensor malfunction, a backup circuit will take over to provide driveability until the problem is identified and fixed.

Precautions

6   Follow these steps:
a) *Always disconnect the power by either turning off the ignition switch or disconnecting the battery terminals before unplugging any electrical connectors.* **Warning:** *These models are equipped with airbags. The airbag is armed and can deploy (inflate) anytime the battery is connected. To prevent accidental deployment (and possible injury), turn the ignition key to LOCK and disconnect the negative battery cable whenever working near airbag components. After the battery is disconnected, wait at least two minutes before beginning work. This system has a back-up capacitor that must fully discharge. For more information, see Chapter 12.* **Caution:** *If the stereo in your vehicle is equipped with an anti-theft system, make sure you have the correct activation code before disconnecting the battery.*
b) *When installing a battery, be particularly careful to avoid reversing the positive and negative battery cables.*
c) *Do not subject EFI components, emissions-related components or the ECM to severe impact during removal or installation.*
d) *Do not be careless during troubleshooting. Even slight terminal contact can invalidate a testing procedure and damage one of the numerous transistor circuits.*
e) *Never attempt to work on the ECM or open the ECM cover. The ECM is protected by a government-mandated extended warranty that will be nullified if you tamper with or damage the ECM.*
f) *If you are inspecting electronic control system components during rainy weather, make sure that water does not enter any part. When washing the engine compartment, do not spray these parts or their electrical connectors with water.*

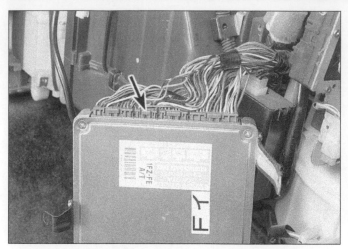

11.10 The ECM is located under the dash area - CAREFULLY disconnect the electrical connectors (arrow)

11.11 Remove the ECM bracket mounting bolts (arrows)

ECM removal and installation

Refer to illustrations 11.10 and 11.11

7   Disconnect the negative cable from the battery (see Chapter 5). **Caution:** *If the stereo in your vehicle is equipped with an anti-theft system, make sure you have the correct activation code before disconnecting the battery.*

8   Remove the lower finish panel on the passenger side under the glove compartment (see Chapter 11).

9   Remove the center console from the passenger compartment (see Chapter 11).

10   Disconnect the electrical connectors from the ECM. Each connector has a locking tab which must be disengaged before the connector is unplugged. **Note:** *If necessary, remove the ECM and then disconnect the harness connector(s) to ease removal* **(see illustration).**

11   Remove the bolts from the ECM brackets **(see illustration).**

12   Lift the ECM from the vehicle.

13   Installation is the reverse of removal.

12 On Board Diagnosis (OBD) system - description and trouble code access

Note: *This procedure does not include the diagnostic codes or the code extracting procedure for 1995 and 1996 models, which are equipped with the OBD II system. These models require a special SCAN tool to read out the various levels of coded information. Have the 1995 and 1996 models diagnosed by a dealer service department or other qualified repair shop in the event of emissions systems or computer failure.*

General information

1   The ECM contains a built-in self-diagnosis system which detects and identifies malfunctions occurring in the network. When the ECM detects a problem, three things happen: the CHECK ENGINE light comes on, the trouble is identified and a diagnostic code is recorded and stored. The ECM stores the failure code assigned to the specific problem area until the diagnosis system is canceled by removing the EFI fuse with the ignition switch off. **Note:** *There are several codes that will not set the CHECK ENGINE light when the code is stored in the ECM It is a good idea to check for any trouble codes when the engine exhibits driveability problems.*

2   The CHECK ENGINE warning light, which is located on the instrument panel, comes on when the ignition switch is turned to ON and the engine is not running. When the engine is started, the warning light should go out. If the light remains on, the self-diagnosis system has detected a malfunction.

Retrieving a diagnostic code

Refer to illustration 12.5

3   To retrieve a diagnostic code, verify first that the battery voltage is above 11 volts, the throttle is fully closed, the transmission is in Neutral, the accessory switches are off and the engine is at normal operating temperature.

4   Turn the ignition switch to ON (engine not running). Do not start the engine.

5   Use a jumper wire to bridge terminals TE1 and E1 of the test connector **(see illustration). Note:** *The self-diagnosis system can be accessed by using either test terminal number 1 (engine compartment) or test terminal number 2 (under driver's dash).*

6   Read the diagnosis code as indicated by the number of flashes of the "CHECK ENGINE" light on the dash. Normal system operation is indicated by Code No. 1 (no malfunctions) for all models. The "CHECK ENGINE" light displays a Code No. 1 by blinking once every 0.25 seconds. Each code will be displayed by first blinking the first digit of the code, then pause, and blink the second digit of the code. For example; Code 24 (IAT sensor) will flash two times, pause, and then flash four times. Each flash will be the exact same length but the distinction will be the pause that separates the digits of the code. Only code 1 (normal operation) will flash continuously without a pause.

7   If there are any malfunctions in the system, their corresponding trouble codes are stored in computer memory and the light will blink the requisite number of times for the indicated trouble codes. If there's

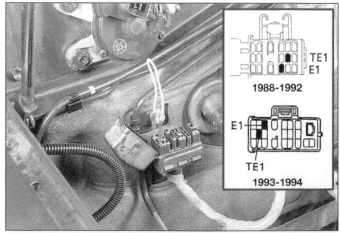

12.5 To access the self diagnosis system, locate the test connector in the engine compartment and using a jumper wire, bridge terminals TE1 and E1

more than one trouble code in the memory, they'll be displayed in numerical order (from lowest to highest) with a pause between each one. After the code with the largest number of flashes has been displayed, there will be another pause and then the sequence will begin all over again. **Note:** *The diagnostic trouble codes 25, 26, 27 and 71 use a special diagnostic capability called "two-trip detection logic". With this system, when a malfunction is first detected, it is temporarily stored into the ECM on the first trip. The engine must be turned off and the vehicle taken on another trip to allow the malfunction to be stored permanently in the ECM. This will distinguish a true problem from a false alarm on vehicles with these particular codes entered into the ECM. Normally the self-diagnosis system will detect the malfunctions, but in the event you want to double-check the diagnosis by canceling the codes and rechecking, then it will be necessary to go on two test drives to confirm any malfunctions with these particular codes.*

8   To ensure correct interpretation of the flashing "CHECK ENGINE" light, watch carefully for the interval between the end of one code and the beginning of the next; otherwise, you will become confused by the apparent number of flashes and misinterpret the display (the length of this interval varies with the model year).

Canceling a diagnostic code

Refer to illustration 12.9

9   After the malfunctioning component has been repaired/replaced, the trouble codes stored in computer memory must be canceled. To accomplish this, simply remove the 15A EFI fuse **(see illustration)** for at least 10 seconds with the ignition switch off.

10   A stored code can also be canceled by removing the cable from the negative battery terminal, but other memory systems (such as the clock and radio presets) will also be canceled. **Caution:** *If the stereo in your vehicle is equipped with an anti-theft system, make sure you have the correct activation code before disconnecting the battery.*

11   If the diagnostic code is not canceled, it will be stored by the ECM and appear with any new codes in the event of future trouble.

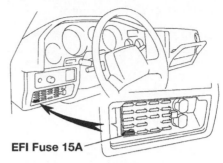

EFI Fuse 15A

1988-1990

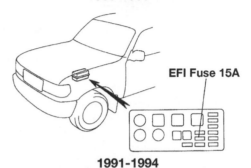

EFI Fuse 15A

1991-1994

12.9 Location of the 15A EFI fuse that cancels the diagnostic codes

12   Should it become necessary to work on engine components requiring removal of the battery terminal, always check to see if a diagnostic code has been recorded before disconnecting the battery.

Diagnostic Trouble Codes

Code	Circuit or system	Diagnosis	Trouble area
Code 1	Normal	The CHECK ENGINE light flashes on and off rapidly when no codes are identified	
Code 11	ECM voltage	Power supply to ECM is momentarily interrupted	Battery connections Main relay; ECM or circuit
Code 12	RPM signal	No rpm signal to the ECM within several seconds after the engine is cranked	Distributor or circuit Crankshaft position sensor or circuit ECM or circuit
Code 13	RPM signal	No rpm signal to the ECM with engine speed above 1,500 rpm	Distributor or circuit; Crankshaft position sensor or circuit ECM or circuit
Code 14	Ignition signal	No ignition signal to the ECM	Igniter or circuit; Ignition switch or circuit Ignition coil; ECM or circuit
Code 21	Main oxygen sensor	Problem in the main oxygen sensor circuit	Main oxygen sensor or circuit ECM or circuit
Code 22	Coolant temperature	Open or short in the coolant temperature sensor circuit	Coolant temperature sensor or circuit ECM or circuit
Code 24	Intake air temperature	Open or short in the intake air sensor temperature sensor circuit	Intake air temperature sensor or circuit ECM or circuit
Code 25	Oxygen sensor or circuit	An excessively lean air/fuel ratio has been indicated by the oxygen sensor circuit	Injector or circuit; Fuel pressure regulator Oxygen sensor or circuit; ECM or circuit Coolant temperature sensor or circuit Intake air temperature sensor or circuit Vacuum or exhaust leak; Contaminated fuel Ignition system
Code 26	Oxygen sensor or circuit	An overly rich air/fuel ratio has been indicated by the oxygen sensor circuit	Injector or injector circuit; Fuel pressure regulator Coolant temperature sensor or circuit Oxygen sensor or circuit; MAP sensor or circuit Intake air temperature sensor or circuit; Air intake system EVAP system; EGR system; ECM or circuit

Diagnostic Trouble Codes (continued)

Code	Circuit or system	Diagnosis	Trouble area
Code 28	Post catalytic converter	Open or shorted circuit in the post catalytic converter oxygen sensor circuit	Sub-oxygen sensor or circuit ECM or circuit
Code 31	Airflow sensor	Open or short in Airflow sensor circuit	Airflow sensor or circuit ECM or circuit
Code 35	HAC sensor signal	Open or short in the altitude compensation circuit	ECM or circuit
Code 41	Throttle position sensor	Open or short in the throttle position sensor circuit	Throttle position sensor or circuit ECM or circuit
Code 42	Vehicle speed sensor	No speed signal for 8 seconds when the engine speed is between 3,000 and 5,000 rpm and the transmission is in gear	Vehicle speed sensor or circuit ECM or circuit Speedometer Instrument panel printed circuit
Code 43	Starter signal	No starter signal to the ECM until engine speed reaches 800 rpm with the vehicle not moving	Starter signal circuit Ignition switch ECM or circuit
Code 51	Switch condition signal	No throttle position signal, gear selector signal or air conditioning signal to the ECM	Air conditioning switch or circuit Air conditioning amplifier Neutral Start switch; Throttle Position sensor ECM or circuit
Code 52 (1993 and 1994)	Knock sensor no. 1 signal	Open or short circuit in knock sensor circuit	Knock sensor no. 1 or circuit ECM or circuit
Code 53	Knock sensor signal	Open or short circuit in knock sensor circuit	Knock sensor or circuit ECM or circuit
Code 55 (1993 and 1994)	Knock sensor no. 2 signal	Open or short circuit in knock sensor circuit	Knock sensor no. 2 or circuit ECM or circuit
Code 71	EGR system	EGR temperature signal is too low	EGR system (EGR valve, hoses, etc.) EGR temperature sensor or circuit EGR vacuum switching valve; ECM or circuit
Code 81; Code 83 Code 84; Code 85 (1993 and 1994)	ECM Communication	Open or short in ECM circuit	ECM or circuit

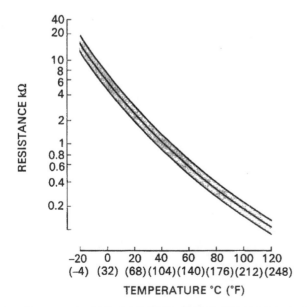

13.1 Compare the indicated resistance values specified on this graph - note that as the temperature increases (as the engine warms up) the resistance decreases

13   Information sensors

Note 1: *Most of the components described in this Section are protected by a Federally mandated extended warranty. See your dealer for the details regarding your vehicle.*
Note 2: *Refer to Chapters 4B and 5 for additional information on the location and the diagnostic procedures for the sensors that are not directly covered in this Section.*

Coolant temperature sensor

General description
Refer to illustration 13.1

1   The coolant temperature sensor is a thermistor (a resistor which varies the value of its voltage output in accordance with temperature changes). As the sensor temperature DECREASES, the resistance values will INCREASE. As the sensor temperature INCREASES, the resistance values will DECREASE **(see illustration)**. A failure in this sensor circuit should set a Code 22. This code indicates a failure in the coolant temperature sensor circuit, so in most cases the appropriate solution to the problem will be either repair of a connector or wire, or replacement of the sensor.

Check
Refer to illustrations 13.2 and 13.3

2   To check the sensor, disconnect the electrical connector and measure the resistance across the sensor terminals **(see illustration)**. With the engine completely cold (68-degrees F) the resistance should

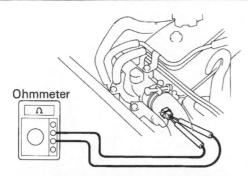

13.2 To check the coolant temperature sensor, use an ohmmeter to measure the resistance between the two sensor terminals (1988 through 1992 model shown)

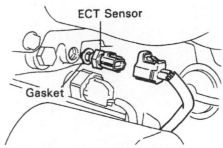

13.3 Use a voltmeter and probe the coolant temperature sensor connector for reference voltage with the ignition key ON (engine not running). It should be approximately 5.0 volts (1993 through 1996 model shown)

be 2,000 to 3,000 ohms. Next, start the engine and warm it up until it reaches operating temperature (180-degrees F) - the resistance should be 200 to 400 ohms. **Note:** *If necessary, remove the sensor and perform the tests in a pan of heated water to simulate the conditions. Compare the resistance values with the accompanying graph.*

3   If the resistance values of the coolant temperature sensor are correct, check the circuit for the proper reference voltage. Turn the ignition key ON (engine not running) and check for reference voltage **(see illustration)**. It should be approximately 5 volts.

Replacement

4   To remove the sensor, depress the locking tabs, unplug the electrical connector, then carefully unscrew the sensor. **Caution:** *Handle the coolant sensor with care. Damage to this sensor will affect the operation of the entire fuel injection system.*

5   Before installing the new sensor, wrap the threads with Teflon sealing tape to prevent leakage and thread corrosion.

6   Installation is the reverse of removal.

Oxygen sensor

General description

7   These models are equipped with either a single oxygen sensor system or a dual-stage oxygen sensor system. On dual-stage systems, the main oxygen sensor is mounted ahead of the front catalytic converter and monitors the exhaust gases exiting the engine. The post catalytic converter oxygen sensor monitors the exhaust gases after they have passed through the front catalytic converter. Each oxygen sensor monitors the oxygen content of the exhaust gas stream. The oxygen content in the exhaust reacts with the oxygen sensor to produce a voltage output which varies from 0.1-volt (high oxygen, lean mixture) to 0.9-volts (low oxygen, rich mixture). The ECM constantly monitors this variable voltage output to determine the ratio of oxygen to fuel in the mixture. The ECM alters the air/fuel mixture ratio by controlling the pulse width (open time) of the fuel injectors. A mixture ratio of 14.7 parts air to 1 part fuel is the ideal mixture ratio for minimizing exhaust emissions, thus allowing the catalytic converter to operate at maximum efficiency. It is this ratio of 14.7 to 1 which the ECM and the oxygen sensor attempt to maintain at all times.

8   The oxygen sensor produces no voltage when it is below its normal operating temperature of about 600-degrees F. During this initial period before warm-up, the ECM operates in open loop mode.

9   If the engine reaches normal operating temperature and/or has been running for two or more minutes, and if the main oxygen sensor is producing a steady signal voltage below 0.70-volts at 1,500 or more rpm, the ECM will set a Code 21. Code 28 will indicate a problem with the post catalytic converter oxygen sensor.

10   When there is a problem with the oxygen sensor or its circuit, the ECM operates in the open loop mode - that is, it controls fuel delivery in accordance with a programmed default value instead of feedback information from the oxygen sensor.

11   The proper operation of the oxygen sensor depends on four conditions:

a)   **Electrical** - The low voltages generated by the sensor depend upon good, clean connections which should be checked whenever a malfunction of the sensor is suspected or indicated.

b)   **Outside air supply** - The sensor is designed to allow air circulation to the internal portion of the sensor. Whenever the sensor is removed and installed or replaced, make sure the air passages are not restricted.

c)   **Proper operating temperature** - The ECM will not react to the sensor signal until the sensor reaches approximately 600-degrees F. This factor must be taken into consideration when evaluating the performance of the sensor.

d)   **Unleaded fuel** - The use of unleaded fuel is essential for proper operation of the sensor. Make sure the fuel you are using is of this type.

12   In addition to observing the above conditions, special care must be taken whenever the sensor is serviced.

a)   *The oxygen sensor has a permanently attached pigtail and electrical connector which should not be removed from the sensor. Damage to or removal of the pigtail or electrical connector can adversely affect operation of the sensor.*

b)   *Grease, dirt and other contaminants should be kept away from the electrical connector and the louvered end of the sensor.*

c)   *Do not use cleaning solvents of any kind on the oxygen sensor.*

d)   *Do not drop or roughly handle the sensor.*

e)   *The silicone boot must be installed in the correct position to prevent the boot from being melted and to allow the sensor to operate properly.*

Check

Refer to illustrations 13.13, 13.16, 13.17 and 13.18

13   To check the oxygen sensor use a digital voltmeter to monitor the millivolt signal from the oxygen sensor during actual operating condi-

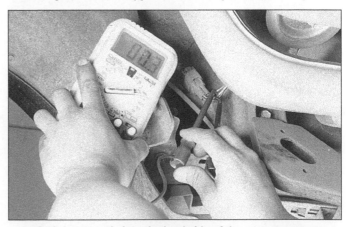

13.13 Insert a pin into the backside of the oxygen sensor connector on the correct terminal and check for a millivolt output signal generated by the sensor

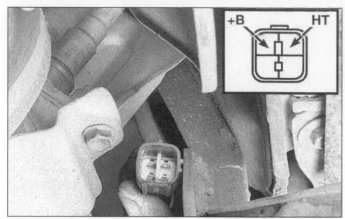

13.16 To test the oxygen sensor heater, disconnect the harness connector and check the resistance across terminals HT and +B of the oxygen sensor connector

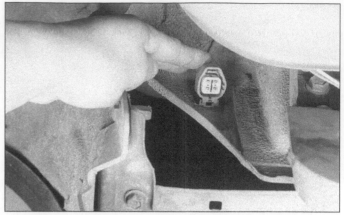

13.17 Check for the voltage to the heater . It should be approximately 12 volts (battery voltage)

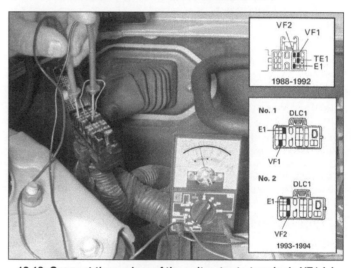

13.18 Connect the probes of the voltmeter to terminals VF1 (+) and E1 (-), raise the engine speed to 2,500 rpm and jump terminals TE1 and E1 with a jumper wire or paper clip

tions. Locate the oxygen sensor electrical connector and backprobe the SIGNAL (+) wire on the harness side of the oxygen sensor connector **(see illustration)**. To properly backprobe the connector insert a long straight pin (a T-pin is preferred) alongside the wire until the pin contacts the metal wire terminal inside the connector. Connect the positive (+) probe of a voltmeter onto the pin and the negative (-) probe to ground. Refer to the wiring schematics at the end of Chapter 12 for the correct wire color and designation for your year and model vehicle.

14   Start the engine and monitor the voltage signal (millivolts) of the main oxygen sensor as the engine warms-up. The oxygen sensor will produce a steady voltage signal at first (open loop) of approximately 0.1 to 0.2 volts with the engine cold. After a period of approximately two minutes, the engine will reach operating temperature and the oxygen sensor voltage will fluctuate between 0.1 to 0.9 volts (closed loop). If the oxygen sensor fails to operate as described, replace it.

15   If equipped, check the post catalytic converter-oxygen sensor. Locate the electrical connector and check it in the same manner as the main oxygen sensor. Post catalytic converter oxygen sensors will exhibit the same millivolt range but in a much slower and deliberate reaction.

16   Also check the post catalytic converter-oxygen sensor heater (if equipped) as follows: Disconnect the oxygen sensor electrical connector and connect an ohmmeter between the +B and HT terminals on the oxygen sensor side of the connector **(see illustration)**. It should measure approximately 5.0 to 6.5 ohms. **Note:** *Not all models are equipped with a heated oxygen sensor. Models with heated oxygen sensors will be equipped with a four-wire electrical connector.*

17   Check for proper supply voltage to the oxygen sensor heater. With the ignition key ON (engine not running), check for battery voltage on the correct terminals on the harness side of the connector **(see illustration)**. Refer to the wiring schematics at the end of Chapter 12 for the correct wire color and designation for your year and model vehicle.

18   It's also possible to check the oxygen sensor in another manner. With the engine completely warmed up and the oxygen sensor connected, connect a voltmeter to the test connector VF1 (positive probe) and E1 (negative probe) **(see illustration)**. **Note:** *Use only an analog type voltmeter because it will be necessary to watch the needle fluctuations.*

19   Run the engine at 2,500 rpm for approximately two minutes and then jump terminals TE1 and E1 of the test connector. Check the number of times the needle fluctuates in 10 seconds. It should fluctuate eight times or more. If it does not, warm the engine up again and repeat the test.

20   If the voltmeter still does not fluctuate eight times or more, remove the jumper wire from terminals TE1 and E1 of the test connector. Maintain engine speed at 2,500 rpm and measure the voltage between terminals VF1 or VF2 and E1. If the voltage reading is more than 0 volts, then replace the oxygen sensor with a new part. If the

voltage reading is 0 volts, access the self diagnostic codes (see Section 12) and check for any malfunctions.

21   If codes 21, 25 or 26 are obtained, then remove the PCV hose from the valve cover (see Section 8) and measure the voltage between VF1 and E1. If the voltage is 0 volts, replace the oxygen sensor. If the voltage reading is more than 0 volts, repair the over-rich running condition.

22   If codes other than 21, 25 or 26 are obtained, repair the particular sensor or circuit.

Replacement

Refer to illustration 13.26

Note: *Because it is installed in the exhaust manifold or pipe, which contracts when cool, the oxygen sensor may be very difficult to loosen when the engine is cold. Rather than risk damage to the sensor (assuming you are planning to reuse it in another manifold or pipe), start and run the engine for a minute or two, then shut it off. Be careful not to burn yourself during the following procedure.*

23   Disconnect the cable from the negative terminal of the battery. **Caution:** *If the stereo in your vehicle is equipped with an anti-theft system, make sure you have the correct activation code before disconnecting the battery.*

24   Raise the vehicle and place it securely on jackstands.

25   Carefully disconnect the electrical connector from the sensor pigtail lead.

26   Remove the oxygen sensor from the exhaust system **(see illus-**

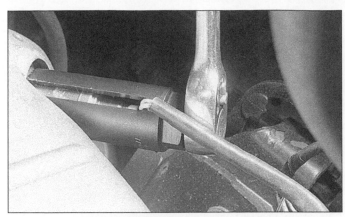

13.26 Slotted sockets are available for easing oxygen sensor removal

tration). **Caution:** *Excessive force may damage the threads.* **Note:** *Some oxygen sensors are threaded directly into the exhaust manifold while others are mounted in the exhaust manifold or pipe with two bolts.*

27 Anti-seize compound must be used on the threads of the sensor to facilitate future removal. The threads of new sensors will already be coated with this compound, but if an old sensor is removed and rein-stalled, recoat the threads.

28 Install the sensor and tighten it securely.

29 Reconnect the electrical connector of the pigtail lead to the main engine wiring harness.

30 Lower the vehicle and reconnect the cable to the negative termi-nal of the battery.

Throttle Position Sensor (TPS)

General description

31 The Throttle Position Sensor (TPS) is located on the end of the throttle shaft on the throttle body (see Chapter 4B). By monitoring the output voltage from the TPS, the ECM can alter fuel delivery based on throttle valve angle (driver demand). A broken or loose TPS can cause intermittent bursts of fuel from the injector and an unstable idle because the ECM thinks the throttle is moving. All the checks and replacement procedures are covered in Chapter 4B.

Airflow sensor

General Information

32 The Airflow sensor measures the volume of air that enters the

13.34b Check the designated terminals of the airflow sensor for the correct resistance (1993 and 1994 models) (use the chart shown in illustration 13.34a for the correct resistance values)

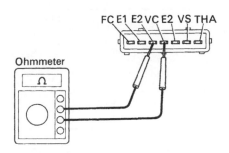

Between terminals	Resistance	Temperature
E2 – VS	200 – 600 Ω	–
E2 – VC	200 – 400 Ω	–
E2 – THA	10 – 20 kΩ	–20°C (–4°F)
	4 – 7 kΩ	0°C (32°F)
	2 – 3 kΩ	20°C (68°F)
	0.9 – 1.3 kΩ	40°C (104°F)
	0.4 – 0.7 kΩ	60°C (140°F)
E1 – FC	Infinity	–

13.34a Check the designated terminals of the airflow sensor for the correct resistance (1988 through 1992 models)

intake system. 1998 through 1994 models use a vane-type air flow sensor. As air enters the air by-pass passage, the measuring plate (vane) swings open and allows an electrical device to gather informa-tion using the position of the measuring plate. 1995 and 1996 models use a vortex-type airflow sensor, using a hot-wire to measure air mass. This information is relayed to the computer to inject the correct amount of the fuel into the combustion chamber for the volume of air (load) that is demanded.

Check

Refer to illustrations 13.34a, 13.34b, 13.34c, 13.35a and 13.35b

33 Disconnect the airflow sensor harness connector.

34 Measure the resistance of the airflow sensor at closed or idle position. Follow the terminal designations **(see illustrations)**.

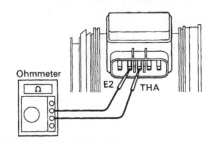

Between terminals	Resistance	Temperature
THA – E2	10 – 20 kΩ	–20°C (–4°F)
THA – E2	4 – 7 kΩ	0°C (32°F)
THA – E2	2 – 3 kΩ	20°C (68°F)
THA – E2	0.9 – 1.3 kΩ	40°C (104°F)
THA – E2	0.4 – 0.7 kΩ	60°C (140°F)
THA – E2	0.2 – 0.4 kΩ	80°C (176°F)

13.34c Check the designated terminals of the airflow sensor for the correct resistance (1995 and 1996 models)

Between Terminals	Resistance (Ω)	Measuring plate opening
E1 – FC	Infinity	Fully closed
	Zero	Other than closed
E2 – VS	200 – 600	Fully closed
	20 – 1,200	Fully open

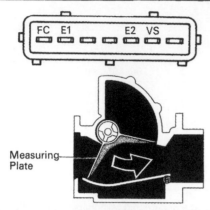

13.35a Measure the designated terminals with the measuring plate open and closed (1988 through 1992 models)

13.35b Measure the resistance between terminals VS and E2. It should read 200 to 600 ohms with the measuring plate fully closed and 20 to 1,200 ohms with the measuring plate fully open (1993 and 1994 models)

13.44 Disconnect the EGR gas temperature sensor connector and check the resistance of the sensor

35   On 1998 through 1994 models, remove the air intake system and check the resistance of the airflow sensor as you manually move the measuring plate from closed to open position. Follow the terminal designations **(see illustrations)**.

36   If the resistance readings are correct, check the wiring harness for open circuits or a damaged harness (see Chapter 12).

Replacement

37   Disconnect the electrical connector from the airflow sensor.

38   Remove the air cleaner assembly (see Chapter 4B).

39   Remove the four bolts and lift the airflow sensor from the engine compartment.

40   Installation is the reverse of removal.

Intake Air Temperature (IAT) sensor

General description

41   The intake air temperature sensor is located inside the airflow sensor. This sensor is a resistor which changes value according to the temperature of the air entering the engine. Low temperatures produces a high resistance value (for example, at 68-degrees F the resistance is 2,000 to 3,000 ohms) while high temperatures produce low resistance values (at 176-degrees F the resistance is 200 to 400 ohms). The ECM supplies approximately 5-volts (reference voltage) to the air temperature sensor. The IAT sensor alters the voltage according to the temperature of the incoming air. The signal voltage sent back to the ECM will be high when the air temperature is cold and low when the air temperature is warm. Any problems with the air temperature sensor will usually set a code 24.

Check

42   To check the air temperature sensor, follow the resistance checks for the airflow sensor in this section.

EGR temperature sensor (California models)

General Description

43   The EGR temperature sensor is mounted near the EGR valve. This sensor detects the temperature of the exhaust as it moves through the EGR valve. The information is sent to the ECM and in turn the EGR

on/off time is regulated precisely and more efficiently. Any malfunction with the EGR temperature sensor will set a code 71.

Check

Refer to illustration 13.44

44   Disconnect the harness connector for the EGR temperature sensor **(see illustration)** and measure the resistance of the sensor at various temperatures. Refer to the Specifications listed in this Chapter for a list of the temperatures and the resistance values.

Replacement

45   Disconnect the harness connector for the EGR temperature sensor and using an open-end wrench, remove the sensor from the intake manifold.

46   Installation is the reverse of removal.

Vehicle speed sensor

General description

Refer to illustration 13.47

47   The Vehicle Speed Sensor (VSS) is located on the output section of the transmission (see Chapter 7B) **(see illustration)**. The sensor is electronically controlled and sends a pulsing voltage signal to the ECM, which the ECM converts to miles per hour.

48   Any problems with the VSS and its circuit will set a code 42. Have the vehicle speed sensor, circuit and the ECM diagnosed by a dealership service department or other qualified repair shop.

13.47 Location of the VSS on 1993 through 1996 models

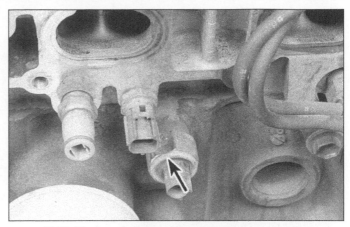

13.50 The knock sensor (arrow) is located under the intake manifold

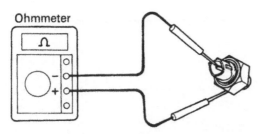

13.51 Check that NO continuity exists between the sensor terminal and the body of the sensor

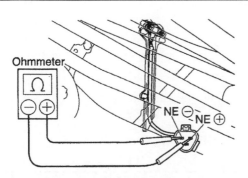

13.54 Check the resistance of the crankshaft sensor on the electrical connector

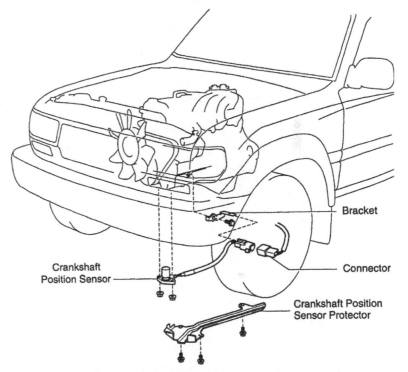

13.53 Location of the crankshaft sensor under the timing chain cover

Knock sensor

General Description

49   Irregular octane levels in modern gasoline can cause detonation in an engine. Detonation is sometimes referred to as "spark knock". The knock sensor sends a voltage signal to the ECM when no spark knock is occurring and the ECM provides normal advance. When the knock sensor detects abnormal vibration (spark knock), it turns off the circuit to the ECM, and distributor timing is retarded until the knock is eliminated. Any problems with the knock sensor or sensor circuit will set a code 52.

Check and replacement

Refer to illustrations 13.50 and 13.51

50   The knock sensor is located on the firewall side of the engine block, under the intake manifold **(see illustration)**. It will be necessary to remove the air intake plenum (see Chapter 4) and the intake manifold (see Chapter 2A) to gain access to the knock sensor.

51   Using an ohmmeter, check that there is no continuity between the terminal on the knock sensor and the body **(see illustration)**.
52   If continuity exists, replace the sensor.

Crankshaft Position Sensor (1995 and 1996 models)

Refer to illustrations 13.53 and 13.54

General Description

53   The crankshaft position sensor is located under the timing chain cover near the crankshaft pulley **(see illustration)**. The crankshaft position sensor relays a signal to the ECM to indicate the exact position (angle) of the crankshaft.

Check and replacement

54   Using an ohmmeter, measure the resistance of the crankshaft position sensor **(see illustration)**. It should be between 1,630 to 3,225 ohms

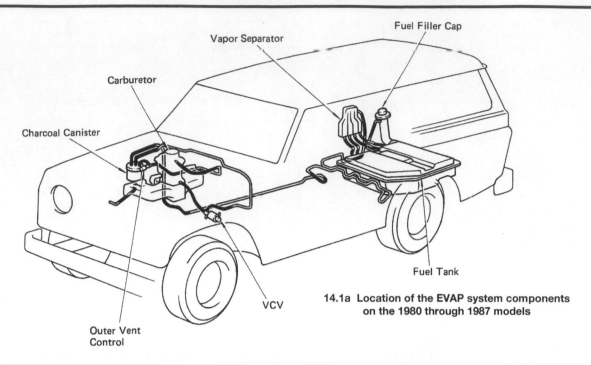

14.1a Location of the **EVAP** system components
on the 1980 through 1987 models

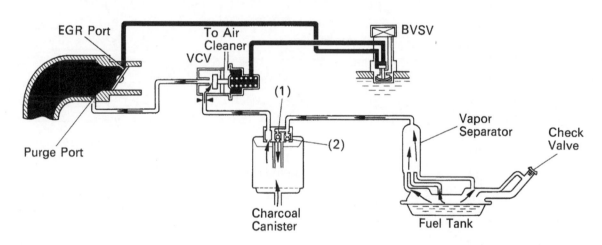

To reduce HC emission, evaporated fuel from the fuel tank is routed through the charcoal canister to the intake manifold for combustion in the cylinders.							
Coolant Temp.	BVSV	Vacuum at EGR Port	VCV	Check Valve		Check Valve in Cap	Evaporated Fuel (HC)
				(1)	(2)		
Below 45°C (113°F)	CLOSED	—	CLOSED	—	—	—	HC from tank is absorbed into the canister.
Above 64°C (147°F)	OPEN	Blow 50 mmHg. (1.97 in.Hg)	CLOSED	—	—	—	
		Above 70 mmHg. (2.76 in.Hg)	OPEN	—	—	—	HC from canister is led into air intake chamber.
Hight pressure in tank	—	—	—	OPEN	CLOSED	CLOSED	HC from tank is absorbed into the canister.
High vacuum in tank	—	—	—	CLOSED	OPEN	OPEN	Air is led into the fuel tank.

14.1b Typical EVAP system and operation chart for 1988 through 1992 models

depending on the temperature; the warmer the temperature of the sensor, the higher the resistance value. If the resistance is not within the specified range, replace the sensor with a new part.

55   To replace the sensor, remove the crankshaft protector plate, disconnect the electrical connector and remove the bolts from the crankshaft position sensor.

56   Installation is the reverse of removal.

14 Evaporative Emission Control (EVAP) system

General description

Refer to illustrations 14.1a through 14.1e

1   This system is designed to trap and store fuel that evaporates from the fuel tank, throttle body and intake manifold that would nor-

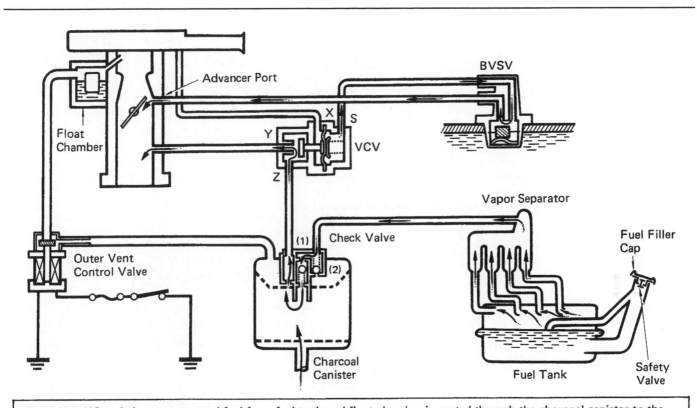

IG S/W	Outer Vent Control Valve	Coolant Temp.	BVSV	Vacuum at Advancer Port	VCV	Check Valve		Safety Valve in Cap	Evaporated Fuel (HC)
						(1)	(2)		
OFF	OPEN	—	—	—	CLOSED	—	—	—	HC from tank and float chamber is absorbed into the canister.
ON	CLOSED	Below 30°C (86°F)	CLOSED	—	CLOSED	—	—	—	HC from tank is absorbed into the canister.
		Above 44° (111°F)	OPEN	Below 50 mmHg (1.97 in.Hg)	CLOSED	—	—	—	
				Above 70 mmHg (2.76 in.Hg)	OPEN	—	—	—	HC from canister is led into intake manifold.
High pressure in tank	—	—	—	—	—	OPEN	CLOSED	CLOSED	HC from tank is absorbed into the canister.
High vacuum in tank	—	—	—	—	—	CLOSED	OPEN	OPEN	(Air is led into the tank.)

To reduce HC emissions, evaporated fuel from fuel tank and float chamber is routed through the charcoal canister to the carburetor for combustion in the cylinders.

14.1c Typical EVAP system and operation chart for 1980 through 1987 models

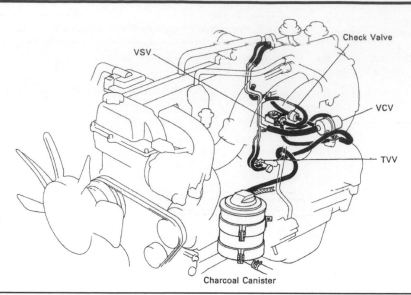

14.1d Typical EVAP system schematic for 1993 through 1996 models

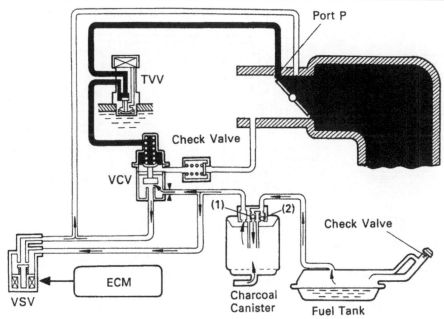

ECT	TVV	Throttle Position	VCV	VSV	Check Valve		Check Valve in Cap	Evaporated Fuel (HC)
					(1)	(2)		
Below 45°C (113°F)	CLOSED	—	CLOSED	—	—	—	—	HC From tank is absorbed into the canister.
Above 64°C (147°F)	OPEN	Below port P	CLOSED	—	—	—	—	
		Above port P	OPEN	—	—	—	—	HC from canister is led into air intake chamber.
Above 35°C (95°F)	—	Idling (A/C idle-up)	—	*VALIABLE OPEN	—	—	—	HC from canister is led into air intake chamber.
		Others	—	CLOSE	—	—	—	HC from tank is absorbed into the canister.
High pressure in tank	—	—	—	—	OPEN	CLOSED	CLOSED	HC from tank is absorbed into the canister.
High vacuum in tank	—	—	—	—	CLOSED	OPEN	OPEN	Air is led into the fuel tank.

14.1e Typical EVAP system and operation chart for 1993 through 1996 models

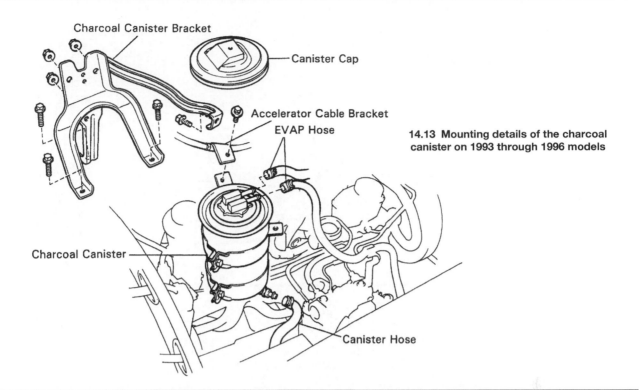

14.13 Mounting details of the charcoal canister on 1993 through 1996 models

mally enter the atmosphere in the form of hydrocarbon (HC) emissions **(see illustrations)**. The systems varied from year to year with fuel system changes and stricter emissions laws.

2   The Evaporative Emission Control (EVAP) system consists of a charcoal-filled canister, the lines connecting the canister to the fuel tank, the Temperature Vacuum Valve (TVV) Vacuum Control Valve (VCV), Vacuum Switching Valve (VSV) and a check valve.

3   Fuel vapors are transferred from the fuel tank and throttle body to a canister where they're stored when the engine isn't running. When the engine is running, the fuel vapors are purged from the canister by intake airflow and consumed in the normal combustion process.

4   The charcoal canister is equipped with a check valve that incorporates three check balls. Depending upon the running conditions and the pressure in the fuel tank, the check balls open and close the passageways to the TVV (consequently the throttle body) and fuel tank.

Check

5   Poor idle, stalling and poor driveability can be caused by an inoperative check valve, a damaged canister, split or cracked hoses or hoses connected to the wrong fittings. Check the fuel filler cap for a damaged or deformed gasket (see Chapter 1).

6   Evidence of fuel loss or fuel odor can be caused by liquid fuel leaking from fuel lines, a cracked or damaged canister, an inoperative check valve, disconnected, misrouted, kinked, deteriorated or damaged vapor or control hoses.

7   Inspect each hose attached to the canister for kinks, leaks and cracks along its entire length. Repair or replace as necessary.

8   Look for fuel leaking from the bottom of the canister. If fuel is leaking, replace the canister and check the hoses and hose routing.

9   Inspect the canister. If it's cracked or damaged, replace it.

10   Check for a clogged filter or a stuck check valve. Using low pressure compressed air, blow into the canister tank pipe. Air should flow freely from the other pipes. If a problem is found, replace the canister.

11   Check the operation of the TVV. With the engine completely cold, use a hand-held pump and direct air into the TVV from the manifold side. Air should not pass through the TVV. Now warm the engine to operating temperature (above 129-degrees F) and observe that air passes through the TVV. Replace the valve if the test results are incorrect.

Charcoal canister replacement

Refer to illustration 14.13

12   Clearly label, then detach the vacuum hoses from the canister.

13   Remove the mounting clamp bolts **(see illustration)**, lower the canister with the bracket, disconnect the hoses from the check valve and remove it from the vehicle.

14   Installation is the reverse of removal.

15   Exhaust Gas Recirculation (EGR) system

General description

Refer to illustrations 15.1a, 15.1b and 15.1c

1   To reduce NOX emissions, part of the exhaust gases are recirculated through the EGR valve into the intake manifold to lower the maximum combustion temperature **(see illustrations on following pages)**.

2   The main component of the system is the EGR valve. It operates in conjunction with a wide variety of devices, such as the EGR vacuum modulator, the Bi-metal Vacuum Switching Valve (BVSV), the Vacuum Switching Valve (VSV) and the Vacuum Transmitting Valve (VTV), although not all components are incorporated on all models.

3   At low engine temperatures, the VSV or BVSV and EGR valves are shut and the exhaust gas is not being recirculated. At higher engine temperatures, the VSV or BVSV opens. When the throttle valve is pivoted open enough to expose the EGR port, and the pressure in the EGR valve is low, the pressure increases, closing the modulator and causing the EGR valve to open. The pressure then drops, reopening the modulator and closing the EGR valve, cutting off exhaust gas recirculation. The VSV(s) and VTV, where incorporated, serve the EGR system in various capacities, depending on coolant temperature, exhaust gas pressure, fuel flow pressure and ignition switch position.

Check

4   If the engine runs roughly at idle, hesitates under acceleration, accelerates poorly or gets poor mileage, the EGR system is probably not shutting off.

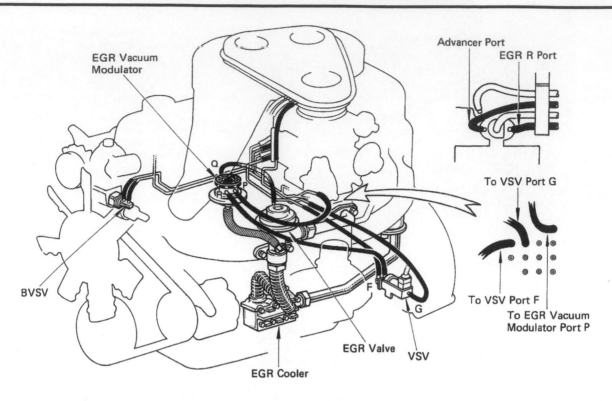

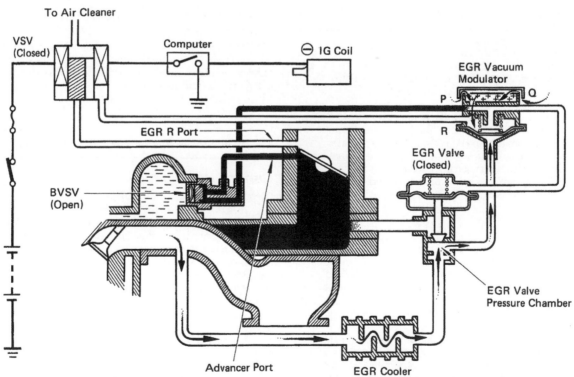

15.1a Typical EGR system and cross-sectional diagram for 1980 through 1987 models

5   Before checking the EGR valve, always inspect the condition of all vacuum hoses in the system. Make sure they're all properly attached and are in good condition. If any of the hoses are cracked or otherwise damaged, replace them

6   Also inspect the EGR vacuum modulator filter for contamination or damage. If it's dirty, clean it with compressed air.

7   Disconnect the vacuum hose from the EGR valve and connect a vacuum pump to it.

8   Apply vacuum directly to the EGR valve. The engine should run roughly or stall. If it doesn't, replace the EGR valve.

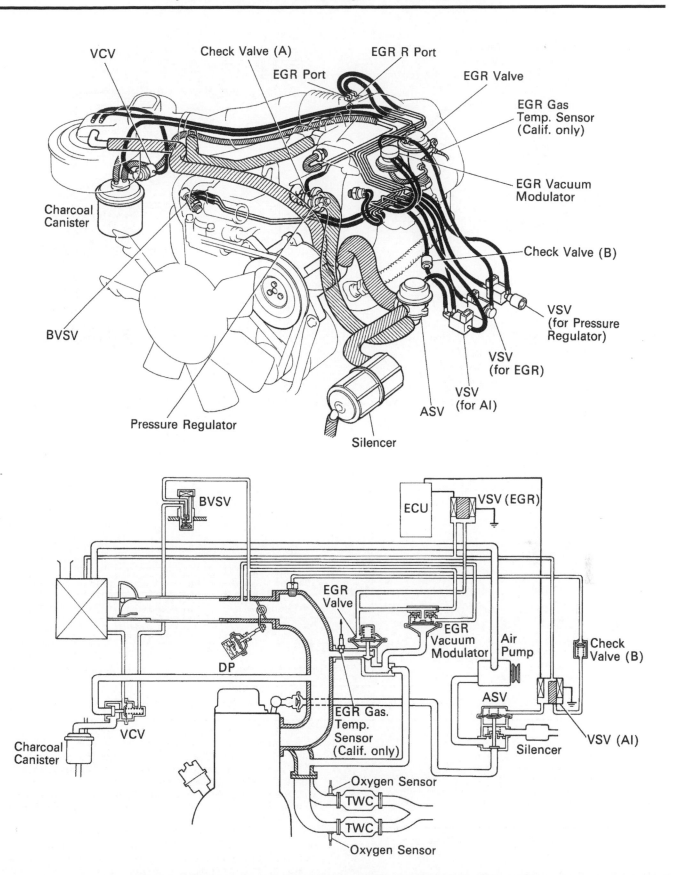

15.1b Typical EGR system and cross-sectional diagram for 1988 through 1992 models

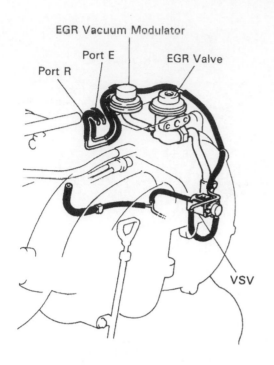

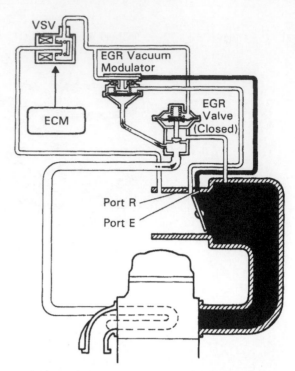

ECT	RPM	Driving Condition	Intake Air Volume	VSV	Throttle Position	Pressure in the EGR Valve Pressure Chamber		EGR Vacuum Modulator	EGR Valve	Exhaust Gas
Below 47°C (117°F)	–	–	–	OFF	–	–		–	CLOSED	Not recirculated
Above 53°C (127°F)	Above 3,500 rpm	–	–	OFF	–	–		–	CLOSED	Not recirculated
	Below 3,500 rpm	Deceleration	–	OFF	–	–		–	CLOSED	Not recirculated
		Ex. deceleration	LOW	OFF	–	–		–	CLOSED	Not recirculated
			HIGH	OFF	–	–		–	CLOSED	Not recirculated
				ON	Below port E	–		–	CLOSED	Not recirculated
				ON	Between port E and port R	(1) LOW	* Pressure constantly alternating between low and high	OPENS passage to atmosphere	CLOSED	Not recirculated
						(2) HIGH		CLOSES passage to atmosphere	OPEN	Recirculated
					Above port R	(3) HIGH	**	CLOSES passage to atmosphere	OPEN	Recirculated (increase)

* Pressure increase → Modulator closes → EGR valve opens → Pressure drops → EGR valve closes ← Modulator opens ←

** When the throttle valve is positioned above the R port, the EGR vacuum modulator will close the atmosphere passage and open the EGR valve to increase the EGR gas, even if the exhaust pressure is insufficiently low.

15.1c Typical EGR system and cross-sectional diagram for 1993 through 1996 models

Replacement

Refer to illustration 15.9

9   Disconnect the vacuum hose, remove the retaining bolts and the valve **(see illustration)**, replace the faulty valve with a new one, install the bolts and reconnect the vacuum hoses.

16   Positive Crankcase Ventilation (PCV) system

General information

Refer to illustration 16.1

1   The Positive Crankcase Ventilation (PCV) system reduces hydro-carbon emissions by scavenging crankcase vapors. It does this by circulating fresh air from the air cleaner through the crankcase, where it mixes with blow-by gases and is then rerouted through a PCV valve to the intake manifold **(see illustration)**.

2   The main components of the PCV system are the PCV valve, a fresh air intake and the vacuum hoses connecting these components to the engine.

3   To maintain idle quality, the PCV valve restricts the flow when the intake manifold vacuum is high. If abnormal operating conditions (such as piston ring problems) arise, the system is designed to allow excessive amounts of blow-by gases to flow back through the crankcase vent tube into the air cleaner to be consumed by normal combustion.

4   This system directs the blow-by into the throttle body which, over time, can cause an oily residue build up in the area near the throttle

15.9 Remove the three bolts that retain the EGR valve assembly to the engine

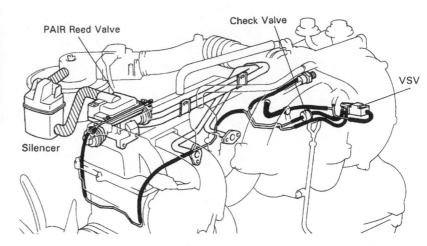

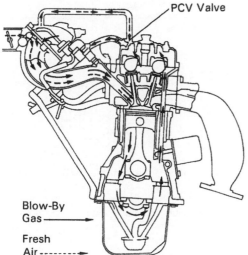

16.1 Diagram of the PCV system

17.1 Schematic of the PAIR system

---

plate. Consequently, it's a good idea to periodically clean this residue from the throttle body. Refer to Chapter 4 for this cleaning procedure.

Check

5   To check the valve, first pull it out of the grommet in the valve cover and shake the valve. It should rattle, indicating that it's not clogged with deposits. If the valve does not rattle, replace it with a new one.

6   Start the engine and allow it to idle, then place your finger over the valve opening. If vacuum is felt, the PCV valve is working properly. If no vacuum is felt, the PCV valve may be bad or the hose may be plugged. Also, check for vacuum leaks at the valve, filler cap and all the hoses.

Replacement

7   Pull straight up on the valve to remove it. Check the rubber grommet for cracks and distortion. If it's damaged, replace it.

8   If the valve is clogged, the hose is also probably plugged. Remove the hose and clean it with solvent.

9   After cleaning the hose, inspect it for damage, wear and deterioration. Make sure it fits snugly on the fittings.

10   If necessary, install a new PCV valve.

11   Install the clean PCV hose. Make sure that the PCV valve and hose are secure.

17   Pulse Air Injection (PAIR) system

Note: *Because of a federally mandated extended warranty which covers emissions-related components such as the catalytic converter, check with a dealer service department before replacing any of the components of the PAIR system at your own expense.*

General description

Refer to illustration 17.1

1   To reduce hydrocarbon and carbon monoxide emissions, the PAIR system draws fresh air into the exhaust ports to increase oxidation and reduction. The fresh air is drawn into the exhaust manifold by vacuum generated by the exhaust pulsation from the combustion process **(see illustration)**.

2   With the engine coolant temperature below 95 degrees F and under normal driving situations, the VSV will switch the PAIR system ON when the engine rpm is below 3,150 and then OFF above 3,150 rpm. Under deceleration and with the engine warmed to operating temperature, the VSV will switch the PAIR system OFF below 1,200 rpm or ON if the rpm is above 1,400.

Check

Refer to illustrations 17. 4 and 17.6

3   Periodically inspect the hoses for cracks, damage or loose fit-

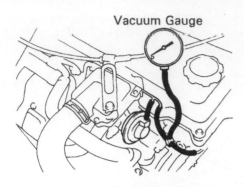

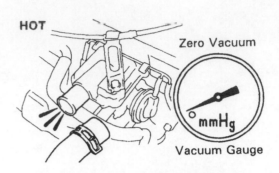

17.4 Install a vacuum gauge between the reed valve and the number 2 water bypass pipe

17.6 With the engine at operating temperature, there should be no vacuum and the reed valve should be quiet

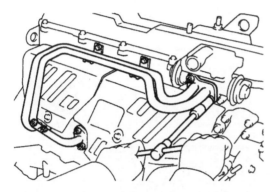

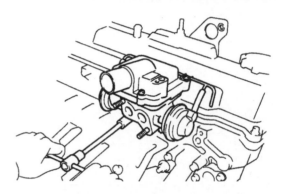

17.9 Remove the bolts and nuts from the air pipe

17.10 Remove the mounting bolts from the PAIR reed valve

tings. Make sure that there are no loose bolts and no leaks.
4   Install a vacuum gauge between the reed valve and the number 2 water bypass pipe **(see illustration)**.
5   With the engine coolant temperature below 95 degrees F (35 degrees C), disconnect the number 2 air hose from the PAIR reed valve, start the engine and check that the vacuum gauge indicates vacuum at idle. There should be a bubbling noise coming from the reed valve.
6   Warm the engine to operating temperature and perform the same test. The vacuum gauge should indicate NO vacuum and the PAIR reed valve should be quiet **(see illustration)**.
7   Raise the engine rpm and quickly close the throttle. There should

be a momentary lapse in vacuum and there should be no noise from the reed valve.
8   If the test results are incorrect, replace the PAIR reed valve.

Replacement

Refer to illustrations 17. 9 and 17.10
9   Remove the bolts and nuts that retain the air pipe to the PAIR reed valve **(see illustration)**.
10   Remove the bolts from the PAIR reed valve and separate the assembly from the engine **(see illustration)**.
11   Installation is the reverse of removal.

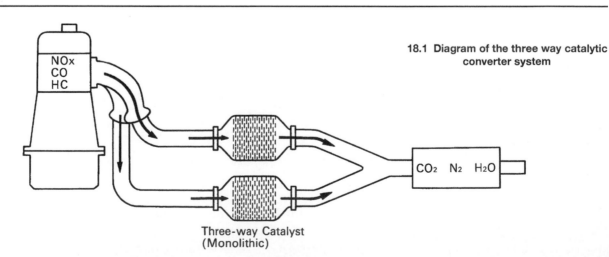

18.1 Diagram of the three way catalytic converter system

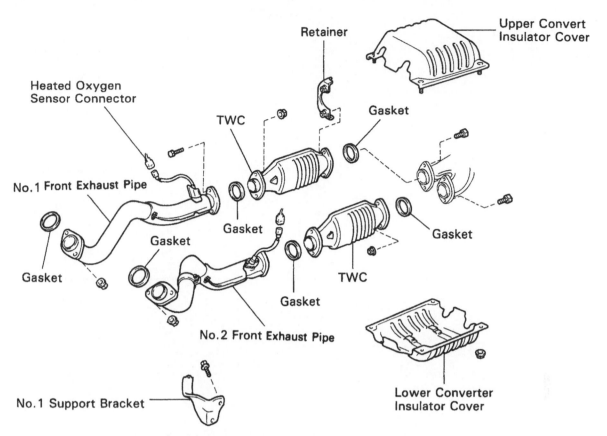

Retainer

Upper Convert
Insulator Cover

Heated Oxygen
Sensor Connector

Gasket

TWC

No.1 Front Exhaust Pipe

Gasket

Gasket

Gasket

Gasket

TWC

Gasket

No.2 Front Exhaust Pipe

No.1 Support Bracket

Lower Converter
Insulator Cover

18.4 Exploded view of a late model catalytic converter system

18 Catalytic converter

Refer to illustration 18.1
Note: *Because of a federally mandated extended warranty which covers emissions-related components such as the catalytic converter, check with a dealer service department before replacing the converter at your own expense.*
Note: *Some early carbureted models are equipped with the oxidation catalyst system. This simplified two catalyst converter works similar to later style three-way catalytic converters. If the catalytic converter is overheated above 1,445 degrees F (785 degrees C), a thermo sensor in the cat will signal the Air Injection system OFF. Have the thermo sensor and catalytic converter checked by a exhaust system repair facility.*

General description

1   To reduce hydrocarbon, carbon monoxide and oxides of nitrogen emissions, all vehicles are equipped with a three-way catalyst system which oxidizes and reduces these chemicals, converting them into harmless nitrogen, carbon dioxide and water **(see illustration)**.
2   The catalytic converter is mounted in the exhaust system much like a muffler.

Check

Refer to illustration 18.4
3   Periodically inspect the catalytic converter-to-exhaust pipe mating flanges and bolts. Make sure that there are no loose bolts and no leaks between the flanges.
4   Look for dents in or damage to the catalytic converter protector **(see illustration)**. If any part of the protector is damaged or dented enough to touch the converter, repair or replace it.
5   Inspect the heat insulator for damage. Make sure that there is adequate clearance between the heat insulator and the catalytic converter.

Replacement

6   To replace the catalytic converter, refer to Chapter 4A.

Notes

Chapter 7 Part A
Manual transmission

Contents

	Section
General information	1
Manual transmission lubricant level change	See Chapter 1
Manual transmission lubricant level check	See Chapter 1

	Section
Transmission mount - check and replacement	See Chapter 7B
Transmission overhaul - general information	3
Transmission - removal and installation	2

Specifications

Torque specifications
Ft-lbs

Clutch housing to engine bolts	47
Transmission-to-clutch housing bolts	47
Transmission-mount to rear crossmember bolts	43
Rear crossmember to frame rail bolts	43

1   General information

All vehicles covered in this manual are equipped with either a 4 or 5-speed manual transmission or an automatic transmission. All information on the manual transmission is included in this Part of Chapter 7. Information on the automatic transmission can be found in Part B. Information on the transfer case can be found in Part C.

Due to the complexity, unavailability of replacement parts and the special tools necessary, internal repair procedures for the manual transmission is not recommended for the home mechanic. The information contained within this Section will be limited to general information and removal and installation procedures.

Depending on the expense involved in having a faulty transmission overhauled, it may be an advantage to consider replacing the unit with either a new or rebuilt one. Your local dealer or transmission shop should be able to supply you with information concerning cost, availability and exchange policy. Regardless of how you decide to remedy a transmission problem, you can still save considerable expense by removing and installing the unit yourself.

2   Transmission - removal and installation

Refer to illustrations 2.3, 2.15 and 2.16
Note: *This procedure is for removal and installation of the transmission and transfer case as one unit.*
1   Disconnect the negative cable at the battery. **Caution:** *If the radio*

in your vehicle is equipped with an anti-theft system, make sure you have the correct activation code before disconnecting the battery.
2   Unscrew the shift lever knob, then detach the screws securing the shift lever boot and remove the boot from the vehicle.
3   Working at the base of the shift lever, press downward on the shift lever retaining ring while rotating counterclockwise then remove the shift lever from the vehicle **(see illustration)**.
4   Raise the vehicle and support it securely on jackstands.
5   Remove the transmission rock shield (if equipped) and drain the lubricant from the transmission and transfer case.

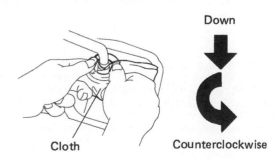

2.3 Press downward on the shift lever retaining ring while rotating it counterclockwise to remove the shift lever

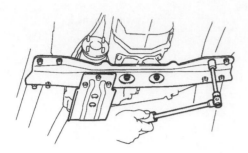

2.15 Remove the rear crossmember mounting nuts and bolts

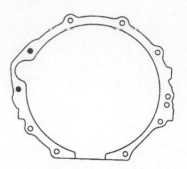

2.16 Remove the transmission-to-engine mounting bolts, then pull the transmission and transfer case assembly towards the rear of the vehicle while slowly lowering the transmission jack

6   Remove the driveshafts (see Chapter 8).
7   Detach all electrical connectors and vacuum hoses from the transmission and transfer case.
8   Detach the speedometer cable, then remove the starter motor (see Chapter 5).
9   Unbolt the clutch release cylinder and fasten it out of the way (see Chapter 8). **Caution:** *Do not depress the clutch pedal while the release cylinder is detached.*
10   Detach the shift lever from the transfer case.
11   Remove the exhaust system components as necessary for clearance (see Chapter 4).
12   Support the engine. This can be done from above by using an engine hoist, or by placing a jack (with a block of wood as an insulator) under the engine oil pan. The engine must remain supported at all times while the transmission is out of the vehicle.
13   Support the transmission with a jack - preferably a special jack made for this purpose. Safety chains will help steady the transmission on the jack.
14   Remove the bolts securing the transmission mount to the rear crossmember.
15   Raise the transmission slightly, then detach the nuts and bolts securing the rear crossmember to the frame rails. Remove the crossmember **(see illustration)**.
16   Remove the bolts securing the transmission to the engine **(see illustration)**.
17   Make a final check that all wires and hoses have been disconnected from the transmission and transfer case and then move the transmission and jack toward the rear of the vehicle until the clutch housing is clear of the engine dowel pins. Keep the transmission level as this is done.
18   Lower the transmission and transfer case assembly and remove it from under the vehicle.
19   The clutch components now can be inspected (Chapter 8). In most cases, new clutch components should be installed as a matter of course if the transmission is removed.

Installation

20   If removed, install the clutch components (Chapter 8).
21   With the transmission secured to the jack as on removal, raise it into position behind the engine and then carefully slide it forward, engaging the clutch housing over the dowel pins. Do not use excessive force to install the transmission. If it does not slide into place, readjust the angle of the transmission so it is level.
22   Install the transmission/clutch housing-to-engine bolts. Tighten the bolts to the specified torque.

23   Install the crossmember and transmission support. Tighten all nuts and bolts securely.
24   Remove the jacks supporting the transmission and the engine.
25   Install the various items removed previously, referring to Chapter 8 for the installation of the driveshaft and Chapter 4 for information regarding the exhaust system components.
26   Make a final check that all wires, hoses and the speedometer cable have been connected and that the transmission and transfer case have been filled with lubricant to the proper level (Chapter 1). Lower the vehicle.
27   From inside the vehicle reinstall the shift lever.
28   Connect the negative battery cable. Road test the vehicle for proper operation and check for leakage.

3   Transmission overhaul - general information

Overhauling a manual transmission is a difficult job for a do-it-yourselfer. It involves the disassembly and reassembly of many small parts. Numerous clearances must be precisely measured and, if necessary, changed with select fit spacers and snap-rings. As a result, if transmission problems arise, it can be removed and installed by a competent do-it-yourselfer, but overhaul should be left to a transmission repair shop. Rebuilt transmissions may be available - check with your dealer parts department and auto parts stores. At any rate, the time and money involved in an overhaul is almost sure to exceed the cost of a rebuilt unit.

Nevertheless, it's not impossible for an inexperienced mechanic to rebuild a transmission if the special tools are available and the job is done in a deliberate step-by-step manner so nothing is overlooked.

The tools necessary for an overhaul include internal and external snap-ring pliers, a bearing puller, a slide hammer, a set of pin punches, a dial indicator and possibly a hydraulic press. In addition, a large, sturdy workbench and a vise or transmission stand will be required.

During disassembly of the transmission, make careful notes of how each piece comes off, where it fits in relation to other pieces and what holds it in place. This will make it much easier to get the transmission back together.

Before taking the transmission apart for repair, it will help if you have some idea what area of the transmission is malfunctioning. Certain problems can be closely tied to specific areas in the transmission, which can make component examination and replacement easier. Refer to the *Troubleshooting* Section at the front of this manual for information regarding possible sources of trouble.

Chapter 7 Part B
Automatic transmission

Contents

	Section			Section
Automatic transmission - removal and installation	8		Neutral start switch - check and adjustment	6
Automatic transmission fluid change	See Chapter 1		Shift linkage - check and adjustment	3
Automatic transmission fluid level check	See Chapter 1		Shift lock system - description and check	4
Diagnosis - general	2		Throttle Valve (TV) cable - check and adjustment	5
General information	1		Transmission mount - check and replacement	7

Specifications

Torque specifications

	Ft-lbs
Transmission-to-engine bolts	
17 mm	53
14 mm	27
12 mm	13
Torque converter bolts	40
Transmission mount-to-rear crossmember bolts	43
Rear crossmember-to-frame rail bolts	43

1 General information

All models which are equipped with automatic transmissions use a 4-speed automatic overdrive transmission with a lock-up torque converter. 1993 and later transmissions are equipped with a self-diagnostic system which is activated with the Overdrive button.

Because of the complexity of the clutch mechanisms and the hydraulic control systems, and because of the special tools and expertise needed to rebuild an automatic transmission, this work is usually beyond the scope of the home mechanic. The procedures in this Chapter are therefore limited to general diagnosis, routine maintenance, adjustments and transmission removal and installation.

Should the transmission require major repair work, take it to a dealer service department or a transmission repair shop. You can, however, save the expense of removing and installing the transmission by doing it yourself.

2 Diagnosis - general

Note: *Automatic transmission malfunctions may be caused by five general conditions: poor engine performance, improper adjustments, hydraulic malfunctions, mechanical malfunctions or malfunctions in the computer or its signal network. Diagnosis of these problems should always begin with a check of the easily repaired items: fluid level and condition (see Chapter 1), shift linkage adjustment and throttle linkage adjustment. Next, perform a road test to determine if the problem has been corrected or if more diagnosis is necessary. If the problem persists after the preliminary tests and corrections are completed, additional diagnosis should be done by a dealer service department or transmission repair shop. Refer to the Troubleshooting section at the front of this manual for information on symptoms of transmission problems.*

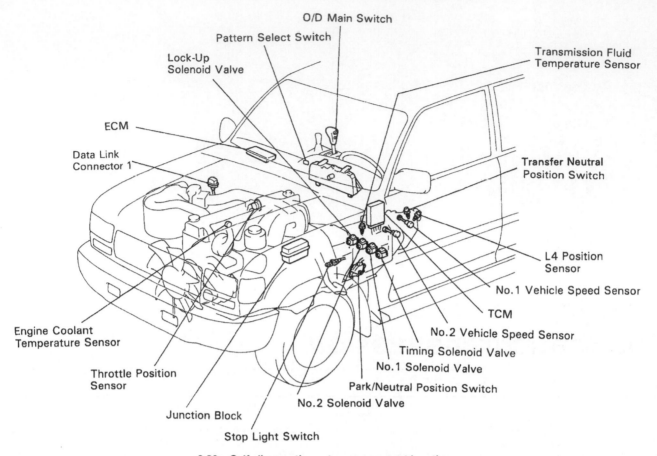

2.30a Self-diagnostic system component location

Preliminary checks

1   Drive the vehicle to warm the transmission to normal operating temperature.

2   Check the fluid level as described in Chapter 1:

a) *If the fluid level is unusually low, add enough fluid to bring the level within the designated area of the dipstick, then check for external leaks (see below).*

b) *If the fluid level is abnormally high, drain off the excess, then check the drained fluid for contamination by coolant. The presence of engine coolant in the automatic transmission fluid indicates that a failure has occurred in the internal radiator walls that separate the coolant from the transmission fluid (see Chapter 3).*

c) *If the fluid is foaming, drain it and refill the transmission, then check for coolant in the fluid or a high fluid level.*

3   Check the engine idle speed. **Note:** *If the engine is malfunctioning, do not proceed with the preliminary checks until it has been repaired and runs normally.*

4   Inspect the shift linkage (see Section 3). Make sure that it's properly adjusted and that the linkage operates smoothly.

Fluid leak diagnosis

5   Most fluid leaks are easy to locate visually. Repair usually consists of replacing a seal or gasket. If a leak is difficult to find, the following procedure may help.

6   Identify the fluid. Make sure it's transmission fluid and not engine oil or brake fluid (automatic transmission fluid is a deep red color).

7   Try to pinpoint the source of the leak. Drive the vehicle several miles, then park it over a large sheet of cardboard. After a minute or two, you should be able to locate the leak by determining the source of the fluid dripping onto the cardboard.

8   Make a careful visual inspection of the suspected component and the area immediately around it. Pay particular attention to gasket mating surfaces. A mirror is often helpful for finding leaks in areas that are hard to see.

9   If the leak still cannot be found, clean the suspected area thoroughly with a degreaser or solvent, then dry it.

10   Drive the vehicle for several miles at normal operating temperature and varying speeds. After driving the vehicle, visually inspect the suspected component again.

11   Once the leak has been located, the cause must be determined before it can be properly repaired. If a gasket is replaced but the sealing flange is bent, the new gasket will not stop the leak. The bent flange must be straightened.

12   Before attempting to repair a leak, check to make sure that the following conditions are corrected or they may cause another leak. **Note:** *Some of the following conditions cannot be fixed without highly specialized tools and expertise. Such problems must be referred to a transmission repair shop or a dealer service department.*

Gasket leaks

13   Check the pan periodically. Make sure the bolts are tight, no bolts are missing, the gasket is in good condition and the pan is flat (dents in the pan may indicate damage to the valve body inside).

14   If the pan gasket is leaking, the fluid level or the fluid pressure may be too high, the vent may be plugged, the pan bolts may be too tight, the pan sealing flange may be warped, the sealing surface of the transmission housing may be damaged, the gasket may be damaged or the transmission casting may be cracked or porous. If sealant instead of gasket material has been used to form a seal between the pan and the transmission housing, it may be the wrong sealant.

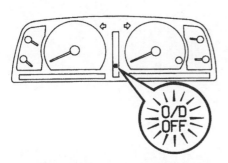

2.30b The diagnostic codes are indicated by the number of flashes from O/D OFF light on the dashboard

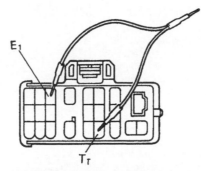

2.32 On 1993 and 1994 models the service connector is located in the engine compartment - use a jumper wire to connect the TT and E1 terminals

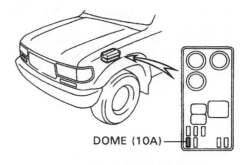

2.35 Remove the 10 amp DOME fuse for 10 seconds or more to clear the codes from the ECU memory

Seal leaks

15   If a transmission seal is leaking, the fluid level or pressure may be too high, the vent may be plugged, the seal bore may be damaged, the seal itself may be damaged or improperly installed, the surface of the shaft protruding through the seal may be damaged or a loose bearing may be causing excessive shaft movement.

16   Make sure the dipstick tube seal is in good condition and the tube is properly seated. Periodically check the area around the speedometer gear for leakage. If transmission fluid is evident, check the O-ring for damage.

Case leaks

17   If the case itself appears to be leaking, the casting is porous and will have to be repaired or replaced.

18   Make sure the oil cooler hose fittings are tight and in good condition.

Fluid comes out of the vent pipe or fill tube

19   If this condition occurs, the transmission is overfilled, there is coolant in the fluid, the case is porous, the dipstick is incorrect, the vent is plugged or the drain back holes are plugged.

Road test

DRIVE range

20   While stopped, position the lever in the DRIVE range and accelerate. Check for a 1-2 shift, 2-3 shift and 3-OD shift. Also, the converter should lock up in second, third or OD gear, depending on calibration. Check for part throttle downshift by depressing the accelerator 3/4 of the way - the transmission should downshift. Check for full detent downshift by depressing the throttle all the way - the transmission should immediately downshift.

Range 2

21   At speed, shift the transmission to Range 2. The transmission should downshift immediately. It should not shift back to Overdrive. Also check for part throttle and full throttle downshifts in this range.

Low range

22   Position the lever in Low position and check the operation.

Overrun braking

23   This can be checked by manually shifting to a lower range. Engine rpm should increase and a braking effect should be noticed.

Reverse

24   Position the shifter in Reverse and check operation.

General checks

25   Verify that the engine is not at fault. If the engine has not received a tune-up recently, refer to Chapter 1 and make sure all engine components are functioning properly.

26   Check the adjustment of the throttle valve cable (see Section 5).

27   Check the condition of all vacuum and electrical lines and fittings at the transmission, or leading to it.

28   Check for proper adjustment of the shift linkage (see Section 3).

29   If at this point a problem remains, there is one final check before the transmission is removed for overhaul. The vehicle should be taken to a transmission specialist who will connect a special oil pressure gauge and check the line pressure in the transmission.

Self-diagnosis system (1993 and 1994 models)

Refer to illustrations 2.30a, 2.30b, 2.32 and 2.35

30   On 1993 and 1994 models a self-diagnostic system is built into the electronic control system for the automatic transmission. A malfunction is indicated by the O/D OFF warning light on the dashboard **(see illustrations)**.

31   To access the self-diagnostic codes for the automatic transmission, turn the ignition switch to ON (do not start the engine) and then press the O/D switch to ON. The O/D switch must be on or else the O/D OFF light will illuminate constantly.

32   Use a jumper wire and connect the TT and E1 terminals of the diagnostic connector (located in the engine compartment) together **(see illustration)**.

33   Read the diagnostic codes directly from the O/D OFF light on the dashboard. If the system is operating correctly, the light will flash two times per second constantly. If there is a malfunction, the first digit of the code will flash, followed by a pause of 1.5 seconds, then the second digit of the code will flash. Count the number of flashes for each code digit. For example, four flashes, a pause, then two flashes would be a code 42.

34   There are nine codes with designated numbers and symptoms for diagnosing driveability problems with the automatic transmission.

Code 41	Defective throttle position sensor or circuit.
Code 42	Defective number 1 speed sensor in the combination meter.
Code 61	Defective number 2 speed sensor in the automatic transmission.
Code 62	Defective number 1 solenoid or circuit.
Code 63	Defective number 2 solenoid or circuit.
Code 64	Defective lock-up solenoid or circuit.
Code 65	Defective timing solenoid or circuit.
Code 86	Defective engine speed sensor or circuit.
Code 88	Defective ECM and TCM or circuit.

35   After the repair has been made, it is necessary to cancel the code(s) from the memory of the ECU **(see illustration)**. Remove the 10A DOME fuse from the passenger compartment fuse panel for 10 seconds or more (see Chapter 12). **Note:** *It is possible to erase the codes from the ECU by simply detaching the negative terminal from the battery, but other memory systems will be erased as well.*

Self-diagnosis system (1995 and 1996 models)

36   On 1995 and 1996 models the transmission is governed by an OBD II self-diagnostic system which requires a special scan tool to access trouble codes and diagnose transmission faults. OBD II scan tools are not yet readily available to the home mechanic but since they may be available in the near future we've elected to furnish you with the following trouble codes.

37   When a malfunction has occurred the check engine light will illuminate and the O/D off light on the dashboard will blink.

38   To access the self-diagnostic codes for the automatic transmission, connect an OBD II scan tool to the data link connector in the passenger compartment fuse panel. Turn the ignition switch to ON (do not start the engine) and then press the OBD II scan tool to the ON position.

39   Consult the OBD II scan tool instruction booklet and record the stored trouble codes.

40   There are ten codes with designated numbers and symptoms for diagnosing driveability problems with the automatic transmission.

Code P0710   Defective fluid temperature sensor or circuit.

Code P0720   Defective number 1 speed sensor in the combination meter.

Code P0750   Defective number 1 shift solenoid. Possible stuck open or closed condition.

Code P0753   Electrical malfunction number 1 shift solenoid or circuit.

Code P0755   Defective number 2 shift solenoid or circuit.

Code P0758   Electrical malfunction number 2 shift solenoid or circuit.

Code P0770   Defective lock-up solenoid. Possible stuck open or closed condition.

Code P0773   Electrical malfunction lock-up solenoid or circuit.

Code P1700   Defective number 2 vehicle speed sensor or circuit.

Code P1780   Defective park neutral switch or circuit.

---

3   Shift linkage - check and adjustment

Refer to illustration 3.4

1   This adjustment should not be considered routine and does not have to be done unless wear in the linkage or misalignment of the shift

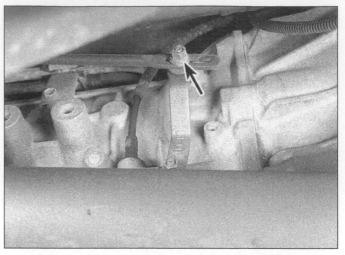

3.4 Loosen the cross shaft nut to adjust the shift linkage

position indicator occurs.

Check

2   Move the shift lever through the gear ranges and make sure the transmission responds accordingly. Position the lever in Neutral and see if the transmission shifts to Neutral. Check to see if the shift position indicator registers correctly with the lever in each detent position.

3   If adjustment is required, the vehicle must be raised and supported securely on jackstands.

Adjustment

4   Loosen the cross shaft nut **(see illustration)**. Push the cross shaft all the way towards the front of the vehicle then back two notches to the Neutral position.

5   With the shift selector in N, but held lightly towards R (reverse), adjust the cross shaft nut, then tighten it securely.

6   Lower the vehicle and drive it to check the operation of the linkage in each gear.

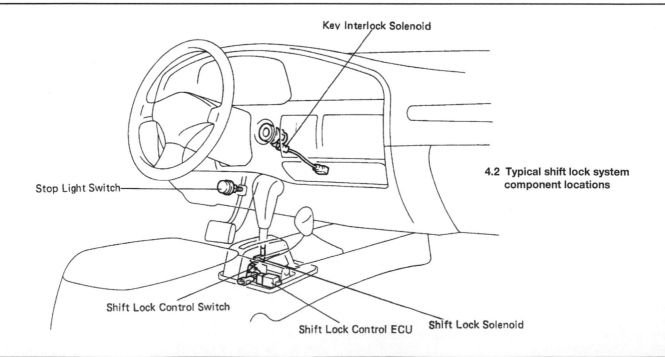

Key Interlock Solenoid

Stop Light Switch

4.2 Typical shift lock system component locations

Shift Lock Control Switch

Shift Lock Control ECU   Shift Lock Solenoid

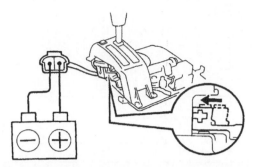

4.3 Apply battery voltage to the shift lock solenoid terminals and verify that there's an audible "click"

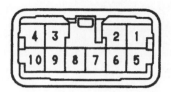

4.4a On 1991 through 1995 models measure the resistance between terminals 1 and 5 to test the shift lock solenoid

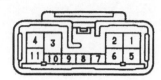

4.4b On 1996 models measure the resistance between terminals 7 and 8 to test the shift lock solenoid

4   Shift lock system - description and check

Refer to illustrations 4.2, 4.3, 4.4a, 4.4b and 4.5

Description

1   The shift lock system prevents the shift lever from being shifted out of Park or Neutral until the brake pedal is applied. Other than the following simple component checks, diagnosis of the shift lock system should be left to a dealer service department.

Check

2   Most of the major components used in the shift lock system are located under the center floor console **(see illustration)**. To access these components the center floor console must first be removed (see Chapter 11).
3   Unplug the electrical connector from the shift lock solenoid. Using jumper wires, apply battery voltage to the solenoid terminals **(see illustration)** and verify that there's an audible "click". If the shift lock solenoid doesn't click when energized by battery voltage, replace it.
4   Unplug the electrical connector from the key interlock solenoid. Using an ohmmeter measure the resistance between the indicated solenoid terminals **(see illustrations)**. Resistance should be approximately 12 to 17 ohms.
5   Unplug the electrical connector from the shift lock control switch. Using an ohmmeter, check for continuity between the indicated terminals **(see illustration)**.

a) *With the shift lever in the Park, there should be continuity between P1 and P.*
b) *With the shift lever in the Park and the release button depressed, there should be continuity between P1 and P and between P2 and P.*
c) *With the shift lever in the remaining positions R, N, D, 2 and L there should be continuity between P2 and P.*

6   If any of the above components fail the indicated tests, replace the component(s) and retest the system.

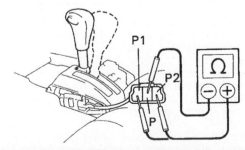

4.5 Check for continuity between the shift lock control switch terminals in each of the indicated positions

5   Throttle Valve (TV) cable - check and adjustment

Refer to illustration 5.3
1   Make sure the throttle cable bracket is not bent or loose before attempting to make this adjustment. Also, the rubber boot must be seated properly on the adjuster.
2   Press the accelerator pedal all the way to the floor. **Note:** *The throttle valve in the throttle body must be completely open. Check and adjust the accelerator cable if necessary.*
3   Check the distance the stopper (or painted mark) on the cable extends from the boot **(see illustration)**.

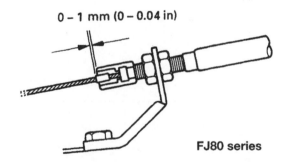

FJ80 series

FJ60 and F62 series

5.3 Check the distance between the end of the rubber boot and the stopper - adjust if necessary to attain the proper length

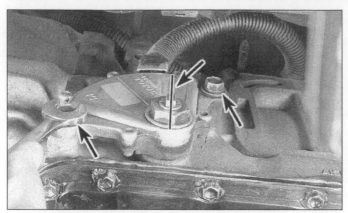

6.5 Loosen the neutral start switch bolt(s) and adjust the switch until the groove in the switch lines up with the neutral basic line on the housing

Shift position \ Terminal	C	RL	N	B	PL
P range	O				O
			O—O		
R range	O—O				
N range			O—O		

6.8a   1988 through 1990 neutral start switch terminal guide and continuity chart

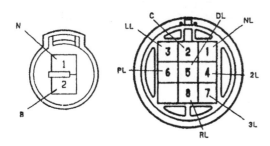

Terminal \ Switch position	B	N	C	PL	RL	NL	DL	3L	2L	LL
P range	O—O		O—O							
R range			O——O							
N range	O—O		O———O							
D range			O———————O							
3 range			O—————————O							
2 range			O———————————O							
L range			O—————————————O							

6.8b   1991 and 1992 neutral start switch terminal guide and continuity chart

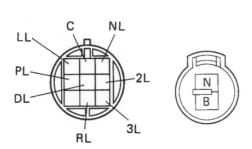

O—O : Continuity

Terminal \ Shift position	B	N	C	PL	RL	NL	DL	3L	2L
P	O—O		O—O						
R			O——O						
N	O—O		O———O						
D			O—————————O						
2			O———————————O						
L			O—————————————O						

6.8c   1993 and 1994 neutral start switch terminal guide and continuity chart

4   If the boot-to-stopper clearance is not as specified, loosen and back off the outer adjusting nut. Turn the inner adjusting nut to move the cable housing as required to produce the specified clearance. Be sure to tighten the outer nut to lock the housing in position.

6   Neutral start switch - check and adjustment

Refer to illustration 6.5, 6.8a, 6.8b, 6.8c and 6.8d

1   The neutral start switch is located on the right side of the transmission housing. Its purpose is to allow the starter to operate only when the selector lever is in Neutral or Park. It also operates the back-up lights when the lever is in Reverse.

2   If the engine can be started with the shift lever in any position other than Park or Neutral, the neutral start switch should be checked and adjusted and, if necessary, replaced.

3   Place the shift lever in Neutral.

4   Raise the vehicle and place it securely on jackstands.

5   Loosen the neutral start switch bolts **(see illustration)**.

6   Align the groove in the end of the control shaft with the vertical index line on the switch by pivoting the switch.

7   Tighten the switch bolts.

8   To verify that you've adjusted the switch properly, check for continuity between the indicated terminals of the switch electrical connector **(see illustrations)**.

9   If the switch fails the continuity check, replace it.

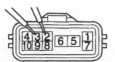

Terminal Position	6	5	4	7	8	10	9	2	3
P	O—O		O—O						
R			O—	—O					
N	O—O		O—			—O			
D			O—				—O		
2			O—					—O	
L			O—						—O

6.8d 1995 and 1996 neutral start switch terminal guide and continuity chart

7 Transmission mount - check and replacement

Refer to illustration 7.3

1   Insert a large screwdriver or prybar into the space between the transmission mount and the bracket and try to pry the transmission up slightly.
2   The transmission should not move away from the bracket much at all.
3   If it is necessary to replace the mount, support the transmission with a jack and remove the mount-to-crossmember bolts and nuts, then separate the mount and lower it from the vehicle **(see illustration)**.
4   Installation is the reverse of removal.

8 Automatic transmission - removal and installation

Refer to illustrations 8.12, 8.14, 8.15, 8.20a and 8.20b

Removal

1   Disconnect the cable from the negative battery terminal. On airbag-equipped models wait two minutes before proceeding any further. **Caution:** *If the stereo in your vehicle is equipped with an anti-theft system, make sure you have the correct activation code before disconnecting the battery.*
2   On later models loosen the fan shroud to avoid damage to the fan.

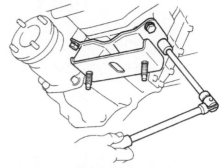

7.3 With the transmission supported with a jack, remove the nuts and bolts and detach the transmission mount

3   Disconnect the throttle valve linkage cable at the throttle-body end and pull it down, toward the transmission, so the cable does not become entangled when the transmission is lowered. **Note:** *Disconnecting the cable at the transmission end requires removing the fluid pan and valve body.*
4   Remove the transfer case shift lever knob and boot. **Note:** *The center floor console, if equipped, must be removed first* (see Chapter 11).
5   Raise the vehicle and support it on jackstands (if not already done).
6   Remove the transmission rock shield (if equipped) and drain the fluid from the transmission and transfer case (see Chapter 1).
7   Disconnect the front and rear driveshafts (see Chapter 8).
8   Unbolt the front exhaust pipe and remove it from the vehicle.
9   Disconnect the oil filler tube and oil cooler lines from the transmission case and plug them.
10   Disconnect the speedometer cable on 1992 and earlier models.
11   Unplug all electrical connectors and vacuum hoses from the transmission and transfer case.
12   Disconnect the shift linkage from transmission and the transfer case **(see illustration)**.
13   Remove the starter motor (see Chapter 5).
14   Remove the torque converter bolt access cover **(see illustration)**. Mark the relationship of the torque converter to the driveplate so they can be assembled in the same relative position on installation.
15   Working through the opening in the torque converter housing, remove the bolts which join the driveplate and converter together **(see illustration)**. They can be removed one at a time by rotating the driveplate. To do this, turn the crankshaft with a wrench attached to the pulley bolt at the front of the engine.

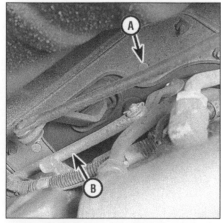

8.12 Remove the shift linkage from the transmission (A) and the transfer case (B)

8.14 Pry out the cover to access the torque converter bolts

8.15 Remove the torque converter to driveplate bolts - rotate the crankshaft to bring each bolt into the access cover window

8.20a Unscrew the torque converter housing-to-engine bolts on the passenger side of the vehicle (arrows) . . .

8.20b . . . then remove the remaining bolts (arrows) on the driver side of the vehicle - don't forget to remove the two bolts securing then top of the torque converter housing to engine (not shown)

16   Support the engine. This can be done from above by using an engine hoist, or by placing a jack (with a block of wood as an insulator) under the engine oil pan. The engine must remain supported at all times while the transmission is out of the vehicle.

17   Support the transmission with a jack - preferably a special jack made for this purpose. Safety chains will help steady the transmission on the jack.

18   Remove the bolts securing the transmission mount to the rear crossmember **(see illustration 7.3)**.

19   Raise the transmission slightly, then detach the nuts and bolts securing the rear crossmember to the frame rails. and remove the crossmember.

20   Remove the remaining bolts securing the transmission to the engine **(see illustrations)**.

21   Make a final check that all wires and hoses have been disconnected from the transmission and transfer case and then move the transmission and jack toward the rear of the vehicle until the torque converter housing is clear of the engine dowel pins. Make sure the torque converter is detached from the driveplate. Secure the torque converter to the transmission so that it will not fall out during removal. Lower the transmission and remove it from under the vehicle.

Installation

22   Make sure the torque converter is fully engaged in the transmission prior to installation. To do this, rotate the converter while pushing it toward the transmission. If it wasn't already fully in place, you'll feel it "clunk" into position as it engages with the input shaft and front pump. It may even "clunk" more than once.

23   The remainder of the installation the reverse of removal. When joining the torque converter and driveplate, be sure to align the previously applied matchmarks. Refill the cooling system and the transmission with the required fluids (see Chapter 1). Adjust the shift and throttle linkage as described in this Chapter before road testing the vehicle.

Chapter 7 Part C
Transfer case

Contents

	Section		Section
General information	1	Shift motor actuator (later models) - check and replacement	4
Oil change	See Chapter 1	Transfer case oil seals (early models) - replacement	2
Oil level check	See Chapter 1	Transfer case - removal and installation	5
Shift diaphragm (early models) - check and replacement	3		

Specifications

Torque specifications

	Ft-lbs
Shift diaphragm	13
Shift motor actuator bolts	14
Transfer case companion flange retaining nut	94
Transfer case-to-transmission bolts	
10 mm bolt	27
12 mm bolt	47

---

1 General information

All models are equipped with a transfer case mounted on the rear of the transmission. Drive is passed from the engine through the transmission and the transfer case to the front and rear wheels by the driveshafts.

Because of the special tools and techniques required, disassembly and overhaul of the transfer case should be left to a dealer or properly equipped shop. You can, however, remove and install the transfer case yourself and save the expense, even if the repair work is done by a specialist.

2 Transfer case oil seals (early models) - replacement

Refer to illustrations 2.2 and 2.5

1   Raise the vehicle and support it on jackstands. Remove the transfer case rock guard and drain the oil.

2   Mark the position of the appropriate driveshaft flange in relation to the transfer case flange **(see illustration)**. Remove the bolts and disconnect the driveshaft.

3   Remove the locknut from the transfer case companion flange. To keep the flange from turning, reinstall two of the driveshaft mounting bolts and engage a pry bar between them.

4   Using a puller, remove the companion flange.

5   Carefully pry the oil seal out with a screwdriver, being sure not to scratch or damage the seal bore in the case **(see illustration)**. A block of wood placed between the screwdriver and case can provide additional leverage.

6   Lubricate the lip of the seal with gear oil and tap it into position with a socket or block of wood.

7   Place the companion flange into position. Install the locknut and tighten it to the specified torque.

8   Reinstall the driveshaft, being sure the flange marks made during removal are aligned.

9   Refill the transfer case with oil, referring to Chapter 1 if necessary.

10   Reinstall the rock guard and lower the vehicle to the ground.

3   Shift diaphragm (early models) - check and replacement

Check

Refer to illustrations 3.2 and 3.3

1   Remove the shift diaphragm from the diaphragm cylinder body (see Step 6).

2.2 Mark the relationship of the driveshaft to the transfer case companion flange before removing the bolts (this will ensure correct alignment of the shaft during installation)

2   Remove the shift link lever and pin, then connect a hand operated vacuum pump to the shift diaphragm upper port **(see illustration)**.

3   Apply a vacuum of 16 in-Hg to the shift diaphragm. Make sure the diaphragm rod moves 7/8-inch (22 mm) **(see illustration)**. If not, replace the diaphragm.

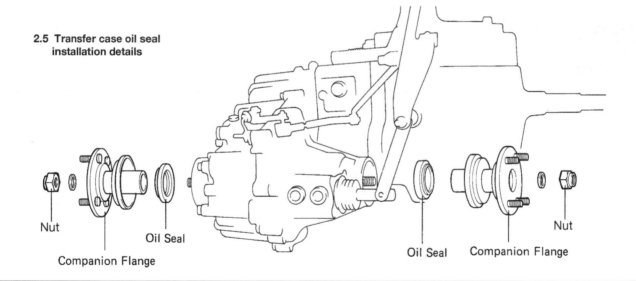

2.5 Transfer case oil seal installation details

Nut   Oil Seal   Oil Seal   Companion Flange   Nut

Companion Flange

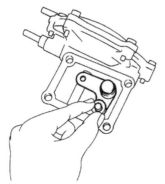

3.2 Pull out the shift link lever and pin from the shift diaphragm body

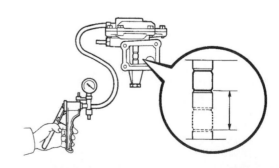

3.3 Using a vacuum pump, apply vacuum to the shift diaphragm and measure the distance the diaphragm rod moves

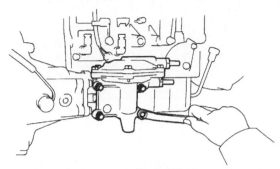

3.6 Disconnect the vacuum hoses and detach the retaining bolts to remove the shift diaphragm

4   The shift diaphragm should also maintain the vacuum applied to the upper port for at least thirty seconds. If it doesn't, replace the diaphragm.

Replacement

Refer to illustration 3.6

5   Disconnect the vacuum hoses leading to the shift diaphragm.
6   Detach the shift diaphragm retaining bolts **(see illustration)** and remove the diaphragm from the transfer case.
7   Installation is the reverse of removal.

4   Shift motor actuator (later models) - check and replacement

Check

Refer to illustration 4.2

1   Trace the wire from the shift motor actuator to the main wiring harness connector and disconnect the connectors.
2   Using an ohmmeter, check the resistance between terminals 2

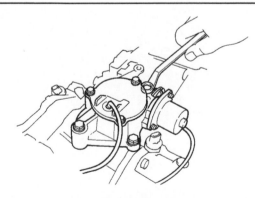

4.5 Disconnect the breather hose and the electrical connector, then detach the retaining bolts to remove the shift motor actuator

Motor Actuator Side

4.2 Shift motor actuator terminal guide

and 3 of the shift motor actuator connector **(see illustration)**. Resistance should be 0.3 to 100 ohms. If the resistance isn't as specified, replace the shift motor actuator.
3   Check the resistance between terminal 2 or 3 and ground. It should be greater than 0.5 M-ohms. If the resistance isn't as specified, replace the shift motor actuator.

Replacement

Refer to illustration 4.5

4   Disconnect the breather hose and the electrical connector leading to the shift motor actuator assembly.
5   Detach the shift motor retaining bolts **(see illustration)** and remove the motor from the transfer case.
6   Installation is the reverse of removal.

5   Transfer case - removal and installation

FJ60 and FJ62 series

Note: *Removing the transfer case from the transmission on FJ60 and FJ62 series vehicles requires major disassembly of the transfer case unit. Because of the complexity of the transfer case, and because special tools are required, this procedure is beyond the scope of this manual.*

FJ80 series

1   Remove the transmission and transfer case from the vehicle as a unit, referring to the removal procedures in either Part A or Part B of this Chapter.
2   Carefully clean the assembly, removing all mounts, dampers and dust covers from the transfer case. If equipped, engage the center differential lock.
3   Stand the transfer case/transmission in a vertical position, with the transfer case pointed up. Remove the transfer case-to-transmission bolts.
4   Pull the transfer case straight up and remove it from the transmission, taking care not to damage the rear oil seal.
5   Before installation, inspect the transmission rear oil seal and replace it if necessary.
6   To install the transfer case, turn the transfer case rear drive flange until the splines on the transmission output shaft align with the splines on the transfer case input shaft.
7   The remainder of the installation is the reverse of removal.

Notes

Chapter 8
Clutch and driveline

Contents

	Section		Section
Axle assembly - removal and installation	16	Clutch release bearing - replacement	4
Axleshaft and oil seal (front) - removal, overhaul and installation	15	Clutch release cylinder - removal, overhaul and installation	7
Axleshaft and oil seal (rear) (full-floating axle) - removal and installation	19	Driveline inspection	11
		Driveshafts - removal and installation	12
Axleshaft, bearing and seal (rear) (semi-floating axle) - removal and installation	18	Driveshafts, differentials and axles - general information	10
		Freewheel hub - removal and installation	14
Clutch - description and check	2	General information	1
Clutch components - removal, inspection and installation	3	Hub and wheel bearing assembly(rear) (full-floating axle) - removal, installation and adjustment	20
Clutch hydraulic system - bleeding	8	Pilot bearing - inspection and replacement	5
Clutch master cylinder - removal, overhaul and installation	6	Pinion oil seal - replacement	17
Clutch pedal assembly - removal and installation	9	Universal joint - replacement	13

Specifications

Clutch

Fluid type	DOT 3 brake fluid
Clutch disc maximum runout	0.031 inch (0.8 mm)
Clutch disc to rivet head depth minimum	0.012 inch (0.3 mm)
Diaphragm spring maximum tip out-of-alignment	0.020 inch (0.5 mm)
Flywheel maximum runout	0.004 inch (0.1 mm)

Rear axle shaft

Maximum shaft runout	0.031 inch (0.8 mm)

Differential

Drive pinion preload (within backlash)	7.8 to 11.3 in-lbs

Rear hub (full-floating axle)

Hub bearing preload	5.7 to 12.8 lbs

Torque specifications

Ft-lbs (unless otherwise indicated)

Note: *One foot-pound (ft-lb) of torque is equivalent to 12 inch-pounds (in-lbs) of torque. Torque values below approximately 15 ft-lbs are expressed in inch-pounds, because most foot-pound torque wrenches are not accurate at these smaller values.*

Clutch pressure plate-to-flywheel	14
Clutch master cylinder reservoir mounting bolt	18
Clutch master cylinder mounting bolt	108 in-lbs
Clutch release cylinder mounting bolt	108 in-lbs
Driveshaft-to-transfer case companion flanges	54
Driveshaft-to-companion flange on differential	54
Freewheel hub body-to-axle hub	23
Manual locking hub cover-to-hub body	84 in-lbs
Spindle mounting bolts	34
Rear axleshaft-(full floating) to hub body	25
Differential carrier cover bolts	108 in-lbs
Drive pinion nut	181 to 325

1   General information

The information in this Chapter deals with the components from the rear of the engine to the drive wheels, except for the transmission and transfer case, which are dealt with in the previous Chapter. For the purposes of this Chapter, these components are grouped into three categories; clutch, driveshaft and axles. Separate Sections within this Chapter offer general descriptions and checking procedures for components in each of the three groups.

Since nearly all the procedures covered in this Chapter involve working under the vehicle, make sure it's securely supported on sturdy jackstands or on a hoist where the vehicle can be easily raised and lowered.

2   Clutch - description and check

Refer to illustration 2.1

1   All models equipped with a manual transmission feature a single dry plate, diaphragm spring-type clutch **(see illustration)**. The actuation is through a hydraulic system.

2   When the clutch pedal is depressed, hydraulic fluid (under pressure from the clutch master cylinder) flows into the release cylinder. Because the release cylinder is connected to the clutch fork, the fork moves the release bearing into contact with the pressure plate release fingers, disengaging the clutch plate.

3   The hydraulic system locates the clutch pedal and provides clutch adjustment automatically, so no adjustment of the linkage is required.

4   Terminology can be a problem regarding the clutch components because common names have in some cases changed from that used by the manufacturer. For example, the driven plate is also called the clutch plate or disc, the clutch release bearing is sometimes called a throw-out bearing, and the release cylinder is sometimes called the operating or slave cylinder.

5   Due to the slow wearing qualities of the clutch, it is not easy to decide when to go to the trouble of removing the transmission in order to check the wear on the friction lining. The only positive indication that something should be done is when it starts to slip or when squealing noises during engagement indicate that the friction lining has worn down to the rivets. In such instances it can only be hoped that the friction surfaces on the flywheel and pressure plate have not been badly worn or scored.

6   A clutch will wear according to the way in which it is used. Much intentional slipping of the clutch while driving - rather than the correct selection of gears - will accelerate wear. It is best to assume, however, that the disc will need replacement at about 40,000 miles (64,000 km).

7   Because of the clutch's location between the engine and transmission, it cannot be worked on without removing either the engine or transmission. If repairs which would require removal of the engine are not needed, the quickest way to gain access to the clutch is by removing the transmission, as described in Chapter 7A.

8   Other than to replace components with obvious damage, some preliminary checks should be performed to diagnose a clutch system failure .

a) *The first check should be of the fluid level in the clutch master cylinder. If the fluid level is low, add fluid as necessary and re-test. If the master cylinder runs dry, or if any of the hydraulic components are serviced, bleed the hydraulic system as described in Section 8.*

b) *To check "clutch spin down time", run the engine at normal idle speed with the transmission in Neutral (clutch pedal up-engaged). Disengage the clutch (pedal down), wait nine seconds and shift the transmission into Reverse. No grinding noise should be heard. A grinding noise would indicate component failure in the pressure plate assembly or the clutch disc.*

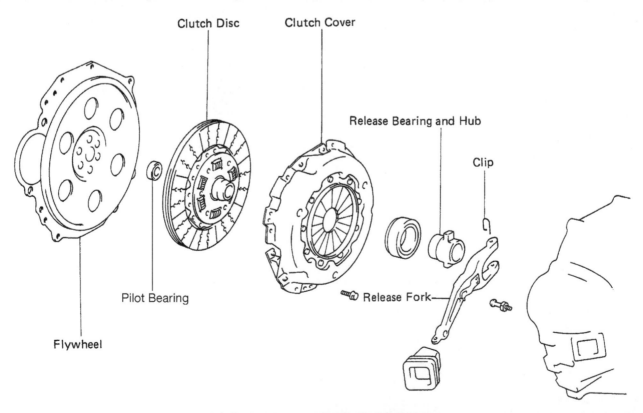

2.1 Exploded view of the clutch components

c) To check for complete clutch release, run the engine (with the brake on to prevent movement) and hold the clutch pedal approximately 1/2-inch from the floor mat. Shift the transmission between 1st gear and Reverse several times. If the shift is not smooth, component failure is indicated. Measure the release cylinder pushrod travel. With the clutch pedal completely depressed the release cylinder pushrod should extend substantially. If the pushrod will not extend very far or not at all, check the fluid level in the clutch master cylinder.

d) Visually inspect the clutch pedal bushing at the top of the clutch pedal to make sure there is no sticking or excessive wear.

e) Under the vehicle, check that the clutch fork is solidly mounted on the ballstud.

Note: *Because access to the clutch components is an involved process, any time either the engine or transmission Is removed, the clutch disc, pressure plate assembly and release bearing should be carefully inspected and, if necessary, replaced with new parts. Since the clutch disc is normally the item of highest wear, it should be replaced as a matter of course if there is any question about its condition.*

3 Clutch components - removal, inspection and installation

Warning: *Dust produced by clutch wear and deposited on clutch components may contain asbestos, which is hazardous to your health. DO NOT blow it out with compressed air and DO NOT inhale it. DO NOT use gasoline or petroleum-based solvents to remove the dust. Brake system cleaner should be used to flush the dust into a drain pan. After the clutch components are wiped clean with a rag, dispose of the contaminated rags and cleaner in a covered container.*

Removal

Refer to illustrations 3.4 and 3.5

1   Access to the clutch components is normally accomplished by removing the transmission, leaving the engine in the vehicle. If, of course, the engine is being removed for major overhaul, then the opportunity should always be taken to check the clutch for wear and replace worn components as necessary. The following procedures assume that the engine will stay in place.

2   Remove the clutch release cylinder (see Section 7).

3   Referring to Chapter 7A, remove the transmission from the vehicle. Support the engine while the transmission is out. Preferably, an engine hoist should be used to support it from above. However, if a jack is used underneath the engine, make sure a piece of wood is used between the jack and oil pan to spread the load. **Caution:** *The pickup*

3.4 A clutch alignment tool can be used to prevent the disc from dropping out as the pressure plate is removed

for the oil pump is very close to the bottom of the oil pan. If the pan is bent or distorted in any way, engine oil starvation could occur.

4   To support the clutch disc during removal, install a clutch alignment tool through the clutch disc hub **(see illustration)**.

5   Carefully inspect the flywheel and pressure plate for indexing marks. The marks are usually an X, an O or a white letter. If they cannot be found, apply marks yourself so the pressure plate and the flywheel will be in the same alignment during installation **(see illustration)**.

6   Turning each bolt only 1/2-turn at a time, slowly loosen the pressure plate-to-flywheel bolts. Work in a diagonal pattern and loosen each bolt a little at a time until all spring pressure is relieved. Then hold the pressure plate securely and completely remove the bolts, followed by the pressure plate and clutch disc.

Inspection

Refer to illustrations 3.8, 3.10, 3.12a and 3.12b

7   Ordinarily, when a problem occurs in the clutch, it can be attributed to wear of the clutch driven disc assembly. However, all components should be inspected at this time.

8   Inspect the flywheel for cracks, heat checking, grooves or other signs of obvious defects **(see illustration)**. If the imperfections are slight, a machine shop can machine the surface flat and smooth, which is highly recommended regardless of the surface appearance. Refer to Chapter 2 for the flywheel removal and installation procedure.

9   Inspect the pilot bearing (see Section 5).

3.5 Be sure to mark the pressure plate and flywheel in order to ensure proper alignment during installation

3.8 Check the flywheel for cracks, hot spots (as seen in this photo) and other obvious defects. Slight imperfections can be removed by a machine shop

3.10 Once the clutch disc is removed, the rivet depth can be measured and compared to the Specifications

3.12a Examine the pressure plate friction surface for score marks, cracks and evidence of overheating

10   Inspect the lining on the clutch disc. There should be at least 1/32-inch of lining above the rivet heads. Check for loose rivets, warpage, cracks, distorted springs or damper bushings and other obvious damage **(see illustration)**. As mentioned above, ordinarily the clutch disc is replaced as a matter of course, so if in doubt about the condition, replace it with a new one.

11   Ordinarily, the clutch release bearing is also replaced along with the clutch disc (see Section 4).

12   Check the machined surfaces and the diaphragm spring fingers

of the pressure plate **(see illustrations)**. If the surface is grooved or otherwise damaged, replace the pressure plate. Also check for obvious damage, distortion, cracking, etc. Light glazing can be removed with sandpaper or emery cloth. If a new pressure plate is indicated, new or factory-rebuilt units are available.

Installation

Refer to illustration 3.14

13   Before installation, carefully wipe the flywheel and pressure plate machined surfaces clean with a rubbing-alcohol dampened rag. It's important that no oil or grease is on these surfaces or the lining of the clutch disc. Handle these parts only with clean hands.

14   Position the clutch disc and pressure plate with the clutch held in place with an alignment tool **(see illustration)**. Make sure it's installed properly (most replacement clutch discs will be marked "flywheel side" or something similar - if not marked, install the clutch with the damper springs or bushings toward the transmission).

15   Tighten the pressure plate-to-flywheel bolts only finger tight, working around the pressure plate.

16   Center the clutch disc by ensuring the alignment tool is through the splined hub and into the pilot bearing in the crankshaft. Wiggle the tool up, down or side-to-side as needed to bottom the tool in the pilot bushing. Tighten the pressure plate-to-flywheel bolts a little at a time, working in a criss-cross pattern to prevent distorting the cover. After all of the bolts are snug, tighten them to the specified torque. Remove the alignment tool.

NORMAL FINGER WEAR

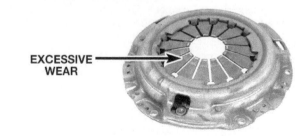

EXCESSIVE WEAR →

EXCESSIVE FINGER WEAR

BROKEN OR BENT FINGERS

3.12b Replace the pressure plate if excessive wear is noted

3.14 Insert a clutch alignment tool through the middle of the clutch, install the pressure plate, then move the disc until it is centered

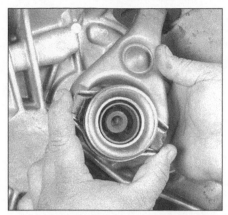

4.3 Disengage the clutch release lever from the ballstud, then remove the release bearing and release lever

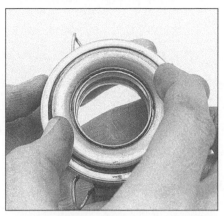

4.4 To check the bearing, hold it by the outer race and rotate the inner race while applying pressure; if the bearing doesn't turn smoothly or if it's noisy, replace the bearing.

4.5 Apply a light coat of high-temperature grease to the bearing surface of the retainer (before installing the transmission, apply the same grease to the input shaft splines to help the shaft slide through the clutch hub)

17   Using high temperature grease, lubricate the inner groove of the clutch release bearing (refer to Section 4). Also place grease on the fork fingers.

18   Install the clutch release bearing as described in Section 4.

19   Install the transmission, clutch release cylinder and all components removed previously, tightening all fasteners to the proper torque specifications.

4   Clutch release bearing - replacement

Warning: *Dust produced by clutch wear and deposited on clutch components may contain asbestos, which is hazardous to your health. DO NOT blow it out with compressed air and DO NOT inhale it. DO NOT use gasoline or petroleum-based solvents to remove the dust. Brake system cleaner should be used to flush it into a drain pan. After the clutch components are wiped clean with a rag, dispose of the contaminated rags and cleaner in a labeled, covered container.*

Removal

Refer to illustration 4.3

1   Disconnect the negative cable from the battery. **Caution:** *If the stereo in your vehicle is equipped with an anti-theft system, make sure you have the correct activation code before disconnecting the battery.*

2   Remove the transmission (see Chapter 7A).

3   Disengage the clutch release lever from the ballstud, then remove the bearing and lever **(see illustration)**.

Inspection

Refer to illustration 4.4

4   Hold the bearing by the outer race and rotate the inner race while applying pressure **(see illustration)**. If the bearing doesn't turn smoothly or if it's noisy, replace the bearing/hub assembly with a new one. Wipe the bearing with a clean rag and inspect it for damage, wear and cracks. Don't immerse the bearing in solvent; it's sealed for life, to do so would ruin it. Also check the release lever for cracks and bends.

Installation

Refer to illustrations 4.5, 4.6a and 4.6b

5   Lubricate the bearing surface of the input shaft retainer with high-temperature grease **(see illustration)**.

6   Lubricate the release lever ball socket, lever ends and release cylinder pushrod socket with high-temperature grease **(see illustrations)**.

7   Attach the retaining clips and the release bearing to the release lever **(see illustration 2.1)**.

8   Slide the release bearing onto the transmission input shaft bearing retainer while passing the end of the release lever through the opening in the clutch housing. Push the clutch release lever onto the ballstud until it's firmly seated.

9   Apply a light coat of high-temperature grease to the face of the release bearing where it contacts the pressure plate diaphragm fingers.

10   The remainder of installation is the reverse of removal.

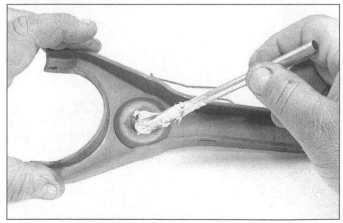

4.6a Use high-temperature grease to lubricate the ballstud socket in the back of the release lever . . .

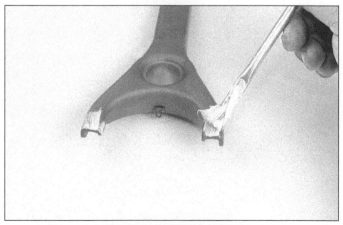

4.6b . . . the lever ends and the depression for the release cylinder pushrod

5   Pilot bearing - inspection and replacement

Refer to illustrations 5.6, 5.9, 5.10 and 5.11

Inspection

1   The clutch pilot bearing is a ball-type bearing which is pressed into the rear of the crankshaft. Its primary purpose is to support the front of the transmission input shaft. The pilot bearing should be inspected whenever the clutch components are removed from the engine. Due to its inaccessibility, if you are in doubt as to its condition, replace it with a new one. **Note:** *If the engine has been removed from the vehicle, disregard the following steps which do not apply.*
2   Remove the transmission (see Chapter 7A)
3   Remove the clutch components (see Section 3).
4   Using a clean rag, wipe the bearing clean and inspect for any excessive wear, scoring or obvious damage. A flashlight will be helpful to direct light into the recess.
5   Check to make sure the pilot bearing turns smoothly and quietly by hand, while pressing in on it, if it is rough or noisy or if the transmission input shaft contact surface is worn or damaged a new one should be installed. **Note:** *Ordinarily, the pilot bearing is also replaced along with the clutch disc. A bearing is fairly inexpensive and provides insurance against having to do the job over again.*

5.6 The pilot bearing can be removed with a slide hammer and puller attachment

Replacement

6   Removal can be accomplished with a special puller **(see illustration)** but an alternative method also works very well.
7   Find a solid steel bar which is slightly smaller in diameter than the bearing. Alternatives to a solid bar would be a wood dowel or a socket with a bolt fixed in place to make it solid.
8   Check the bar for fit - it should just slip into the bearing with very little clearance.
9   Pack the bearing and the area behind it (in the crankshaft recess) with heavy grease **(see illustration)**. Pack it tightly to eliminate as much air as possible.
10   Insert the bar into the bearing bore and hammer on the bar, which will force the grease to the backside of the bearing and push it out **(see illustration)**. Remove the bearing and clean all grease from the crankshaft recess.
11   To install the new bearing, lubricate the outside surface with oil, then drive it into the recess with a hammer and a socket with an outside diameter that matches the bearing outer race **(see illustration)**.
12   Install the clutch components, transmission and all other components removed to gain access to the pilot bearing.

6   Clutch master cylinder - removal, overhaul and installation

Refer to illustrations 6.2, 6.3a and 6.3b
Caution: *Do not allow brake fluid to contact any painted surfaces of the vehicle, as damage to the finish may result.*
Note: *Before beginning this procedure, contact local parts stores and dealer service departments concerning the purchase of a rebuild kit or a new release cylinder. Availability and cost of the necessary parts may dictate whether the cylinder is rebuilt or replaced with a new one. If you decide to rebuild the cylinder, inspect the bore as described in Step 7 before purchasing parts.*

Removal

1   Disconnect the cable from negative terminal of the battery. **Caution:** *If the stereo in your vehicle is equipped with an anti-theft system, make sure you have the correct activation code before disconnecting the battery.*
2   Disconnect the master cylinder pushrod from the clutch pedal **(see illustration)**. Disconnect the hydraulic line from the master cylinder and drain the fluid into a suitable container. If available, use a flare-nut wrench on the fitting, which will prevent the fitting from being rounded off.
3   Remove the master cylinder flange mounting nuts and withdraw the unit from the engine compartment **(see illustrations)**.

5.9 To remove the pilot bearing without a puller, fill the opening behind the bearing with grease . . .

5.10 . . . then force the bearing out hydraulically with a steel rod that just fits through the bore in the bearing - when the hammer strikes the rod, the grease will transmit force to the backside of the bearing and push it out

5.11 Using a hammer and socket to carefully drive the new bearing into place

6.2   To release the clutch pushrod from the clutch pedal, remove the clip and clevis pin from the clutch pedal

Overhaul

4   Detach the hold-down bolt located inside reservoir tank and remove the reservoir. Newer models use a slotted spring pin to attach the reservoir. To remove, drive out the pin with a small punch **(see illustrations 6.3a and 6.3b)**.

5   Pull back the boot and remove the snap-ring.

6   Pull out the pushrod, washer and piston.

7   Examine the inner surface of the cylinder bore. If it is scored or exhibits bright wear areas, the entire master cylinder should be replaced.

8   If the cylinder bore is in good condition, obtain a clutch master cylinder rebuild kit.

9   Replace all parts included in the rebuild kit, following any instructions in the kit. Clean all reused parts with new brake fluid, brake system cleaner or denatured alcohol. **Warning:** *Do not use any petroleum-based solvents. During reassembly, lubricate all parts liberally with clean brake fluid.*

10   Installation of the parts in the cylinder is the reverse of removal.

Installation

11   Installation is the reverse of removal, but be sure to bleed the hydraulic system (**see** Section 8) and check the pedal height and freeplay as described in Chapter 1.

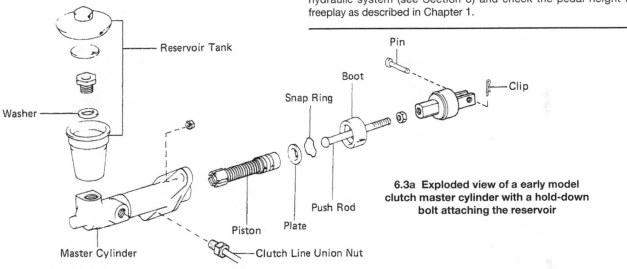

6.3a   Exploded view of a early model clutch master cylinder with a hold-down bolt attaching the reservoir

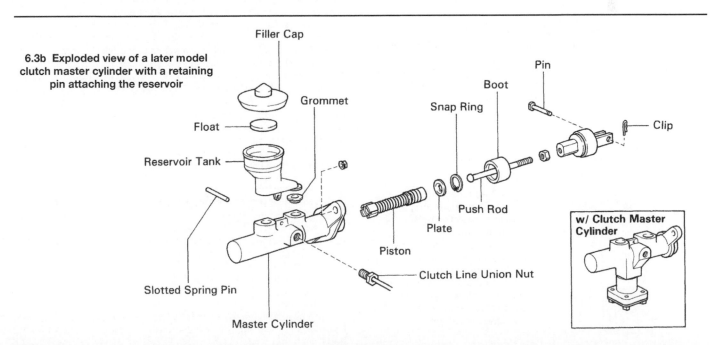

6.3b   Exploded view of a later model clutch master cylinder with a retaining pin attaching the reservoir

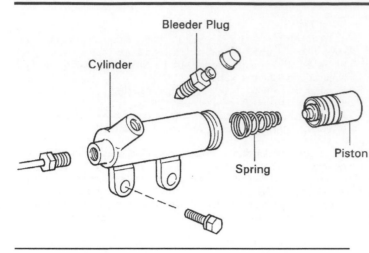

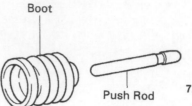

7.3 Exploded view of a the clutch release cylinder components

7 Clutch release cylinder - removal, overhaul and installation

Note: *Before beginning this procedure, contact local parts stores and dealer service departments concerning the purchase of a rebuild kit or a new release cylinder. Availability and cost of the necessary parts may dictate whether the cylinder is rebuilt or replaced with a new one. If you decide to rebuild the cylinder, inspect the bore as described in Step 8 before purchasing parts.*

Removal

Refer to illustration 7.3

1   Disconnect the cable from negative terminal of the battery. **Caution:** *If the stereo in your vehicle is equipped with an anti-theft system, make sure you have the correct activation code before disconnecting the battery.*
2   Raise the vehicle and support it securely on jackstands.
3   Disconnect the hydraulic line at the release cylinder **(see illustration)**. If available, use a flare-nut wrench on the fitting, which will prevent the fitting from being rounded off. Have a small can and rags handy, as some fluid will be spilled as the line is removed.
4   Remove the release cylinder mounting bolts.
5   Remove the release cylinder.

Overhaul

6   Remove the pushrod and the boot **(see illustration 7.3)**.
7   Tap the cylinder on a block of wood to eject the piston and seal. Remove the spring from inside the cylinder.
8   Inspect the bore of the cylinder. Check for deep scratches, score marks and ridges. The bore must be smooth to the touch. If any imperfections are found, the release cylinder must be replaced with a new one.
9   Using the new parts in the rebuild kit, assemble the components using plenty of fresh brake fluid for lubrication. Note the installed direction of the spring and the seal.

8.4 When bleeding the clutch hydraulic system, a hose is connected to the bleeder valve at the release cylinder and then submerged in brake fluid. When the pedal is depressed and the valve is opened, air will be seen as bubbles in the hose and container

Installation

10   Install the release cylinder on the clutch housing. Make sure the pushrod is seated in the release fork pocket.
11   Connect the hydraulic line to the release cylinder. Tighten the connection.
12   Fill the clutch master cylinder with brake fluid (conforming to DOT 3 specifications).
13   Bleed the system (see Section 8).
14   Lower the vehicle and connect the cable to the negative terminal of the battery.

8 Clutch hydraulic system - bleeding

Refer to illustration 8.4

1   The hydraulic system should be bled of all air whenever any part of the system has been removed or if the fluid level has been allowed to fall so low that air has been drawn into the master cylinder. The procedure is very similar to bleeding a brake system.
2   Fill the master cylinder with new brake fluid conforming to DOT 3 specifications. **Caution:** *Do not re-use any of the fluid coming from the system during the bleeding operation or use fluid which has been inside an open container for an extended period of time.*
3   Raise the vehicle and place it securely on jackstands to gain access to the release cylinder, which is located on the left side of the clutch housing.
4   Remove the dust cap which fits over the bleeder valve and push a length of plastic hose over the valve. Place the other end of the hose into a clear container with about two inches of brake fluid in it. The hose end must be submerged in the fluid **(see illustration)**.
5   Have an assistant depress the clutch pedal and hold it. Open the bleeder valve on the release cylinder, allowing fluid to flow through the hose. Close the bleeder valve when fluid stops flowing from the hose. Once closed, have your assistant release the pedal.
6   Continue this process until all air is evacuated from the system, indicated by a full, solid stream of fluid being ejected from the bleeder valve each time and no air bubbles in the hose or container. Keep a close watch on the fluid level inside the clutch master cylinder reservoir; if the level drops too low, air will be sucked back into the system and the process will have to be started all over again.
7   Install the dust cap and lower the vehicle. Check carefully for proper operation before placing the vehicle in normal service.

9 Clutch pedal assembly - removal and installation

1   Remove the pedal return spring(s).
2   Disconnect the master cylinder pushrod from the pedal by removing the spring clip and pulling out the pushrod pin.
3   Remove the pedal shaft.
4   Remove the clutch pedal with its bushings and collar.
5   Clean the parts in solvent and replace any that are damaged or excessively worn.
6   Installation is the reverse of the removal procedure. During installation, apply multi-purpose grease to the pedal boss, return spring, pedal shaft and pushrod pin.

10.5 Full floating axle designs are easily identifiable by the ring of bolts on the center hub

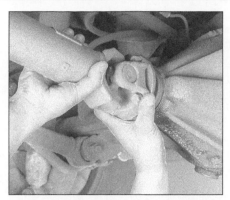

11.5 Use a twisting motion to check for freeplay in the U-joint - any movement is evidence of wear

12.2a Make sure you mark the U-joint, companion flanges . . .

10 Driveshafts, differentials and axles - general information

Refer to illustration 10.5

All 4WD vehicles use two driveshafts; the primary shaft runs between the transfer case and the front differential and the rear driveshaft runs between the transfer case and the rear differential.

The driveshaft universal joints are of the cross type and can be replaced separate from the driveshaft.

The driveshafts are finely balanced during production and whenever they are removed or disassembled, they must be reassembled and reinstalled in the exact manner and positions they were originally in, to avoid excessive vibration.

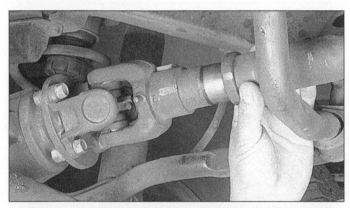

12.2b . . . and the slip joint to ensure the driveshaft retains its balance when you reinstall it

12.3 Jam a large screwdriver into the U-joint to immobilize the driveshaft when you break the flange nuts loose

Both front and rear axles have a 'banjo' style differential which transfers the turning force of the driveshaft to the axleshafts. The axleshafts are splined at their inner ends to fit into the splines in the differential gears and the outer ends are either of the semi-floating or full floating design depending on the year and model of the vehicle. All front axles are of the full floating design.

Full floating axles are easily identifiable from a semi-floating axle by the ring of bolts in the center of the hub **(see illustration)**.

Because of the complexity and critical nature of the differential adjustments, as well as the special equipment needed to perform the operations, we recommend any disassembly of the differential be done by a Toyota dealer service department or other qualified repair facility.

11 Driveline inspection

Refer to illustration 11.5

1   Raise the vehicle and support it securely on jackstands.
2   Slide under the vehicle and visually inspect the condition of the driveshafts. Look for any dents or cracks in the tubing. If any are found, the driveshaft must be replaced.
3   Check for any oil leakage at the front and rear of the driveshaft. Leakage at the transfer case indicates a defective output shaft seal (see Chapter 7C). Leakage from the differential indicates a defective pinion seal (see Section 18).
4   While you're under the vehicle, have an assistant turn a rear wheel so the driveshaft will rotate. As it rotates, make sure the U-joints are functioning properly; they should not be binding, noisy or loose.
5   The U-joints can also be checked with the driveshaft motionless. First, make sure the U-joint bolts at both ends are tight, then grip either side of each joint and try to twist it **(see illustration)**. Any movement in the U-joint is evidence of considerable wear. Lifting up on the shaft will also indicate movement in the universal joints.
6   Inspect for leakage around both sliding yokes. Grease in this area indicates failure of the yoke seal.
7   While you're under the vehicle, inspect both ends of both axle tubes for leaks. Gear oil leakage onto the brake backing plate or drum indicates a leaking axleshaft seal.
8   Remove the jackstands and lower the vehicle.

12 Driveshafts - removal and installation

Refer to illustrations 12.2a, 12.2b and 12.3

1   Raise the vehicle and support it securely on jackstands. Place the transmission and transfer case in Neutral with the parking brake off.
2   Use a scribe, white paint or a hammer and punch to place alignment marks on all driveshaft flanges and slip joint **(see illustrations)** to ensure that the driveshaft is reinstalled in the same position so its dynamic balance is maintained.
3   Detach the bolts from the differential and transfer case flanges and remove the drive shaft(s) from the vehicle **(see illustration)**.

4   While the driveshaft is removed, check the U-joints: While pushing and pulling on them, try to move them back and forth and side to side simultaneously. If there's any play in the U-joints, replace them (see Section 13).

5   Installation is the reverse of removal.

a) *Make sure you match up the marks you made on all flanges.*

b) *Tighten all fasteners to the torque listed in this Chapter's Specifications.*

13 Universal joint - replacement

Refer to illustrations 13.2, 13.3a and 13.3b

Note: *A press or large vise will be required for this procedure. It may be a good idea to take the driveshaft to a repair or machine shop where the U-joints can be replaced for you, usually at a reasonable charge.*

1   Remove the driveshaft (see Section 12).

2   Place the driveshaft in a sturdy vise, mark the shaft and yoke for proper reassembly and, using a hammer and punch, drive the snap-rings off the bearing cups **(see illustration)**.

3   Place a piece of pipe or a large socket with the same inside diameter over one of the bearing cups. Position a socket which is of slightly smaller diameter than the cup on the opposite bearing cup **(see illustration)** and use the vise to force the cup out (inside the pipe or large socket), stopping just before it comes completely out of the yoke. Use the vise or large pliers to work the cup the rest of the way out **(see illustration)**.

4   Transfer the sockets to the other side and press the opposite bearing cup out in the same manner.

5   After the bearing cups have been removed, lift the U-joint from the yoke and thoroughly clean all dirt and debris from the yokes on both ends of the driveshaft. Be sure to remove any metal burrs from the yoke bores.

6   Pack the new U-joint bearing cups with grease, this will allow the needle bearings to be held in place while your installing the bearing cups. Ordinarily, specific instructions for lubrication will be included with the U-joint servicing kit and should be followed carefully.

7   Position the U-joint body in the yoke and partially install one bearing cup in the yoke. If the U-joint is equipped with a grease fitting, be sure it points in the same direction as the grease fitting on the opposite end of the driveshaft.

8   Start the U-joint body into the bearing cup and then partially install the other cup. Align the U-joint body between the bearing cups and press the bearing cups into position, being careful not to damage the dust seals.

9   Install the snap-rings. If difficulty is encountered in seating the snap-rings, strike the driveshaft yoke sharply with a hammer. This will spring the yoke ears slightly and allow the snap-rings to seat in the

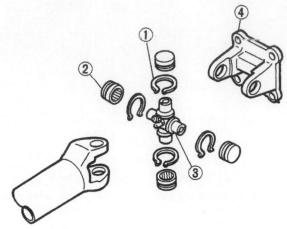

13.2 An exploded view of a U-joint assembly

1	Snap-ring	3	U-joint body
2	Bearing cup	4	U-joint yoke

groove. This should also be done to center the U-joint after assembly.

Note: *If you still have difficulty seating the snap-rings, one of the small needle bearings may have become stuck between the bearing cap and the end of the spider. Disassemble and inspect the joint.*

10   Install the driveshaft (see Section 12). If the U-joint is equipped with a grease fitting, lubricate it as described in Chapter 1.

14 Freewheel hub - removal and installation

Manual locking hubs

Refer to illustrations 14.2 and 14.7

1   Set the hub cover to the Free position.

2   Remove the hub cover mounting bolts and pull off the cover with the clutch **(see illustration)**.

3   Using snap-ring pliers, remove the snap-ring from the end of the axle shaft **(see illustration 14.2)**.

4   Remove the mounting nuts from the freewheel hub body **(see illustration 14.2)**.

5   Using a brass hammer, tap on the hub body to loosen the cone washers and remove them from the hub.

6   Pull the freewheel hub body from the axle hub.

7   The freewheel hub clutch assemblies can become clogged with dirt and water which can make them inoperable. If further disassembly and inspection of the clutch assembly is required pay very close atten-

13.3a Press out the U-joint bearing cups with a vise and sockets

13.3b Extract the bearing cup with locking pliers

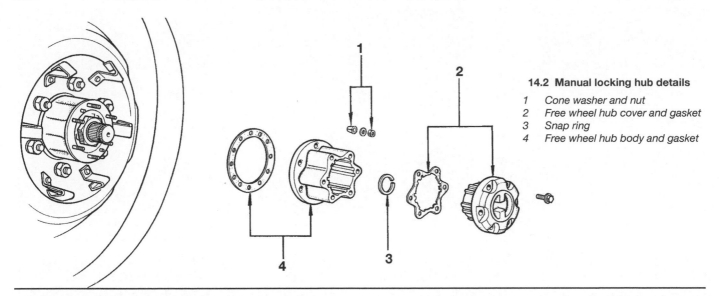

14.2 Manual locking hub details

1 Cone washer and nut
2 Free wheel hub cover and gasket
3 Snap ring
4 Free wheel hub body and gasket

tion to the way the parts fit together when disassembling. Lay all the parts out in sequence in which they were removed. Clean the parts one at a time and lay them back out in the same sequence. Lubricate the parts with a light coat of multi-purpose grease, then reassemble the clutch assembly in the reverse order **(see illustration).**

8 Installation of the hub assembly on to the front axle is the reverse of the removal. Use new gaskets and apply multi-purpose grease to the inner hub splines. The control handle should be set to the Free position and the cover should be attached to the body with the following pawl tabs aligned with the non-toothed portions of the body.

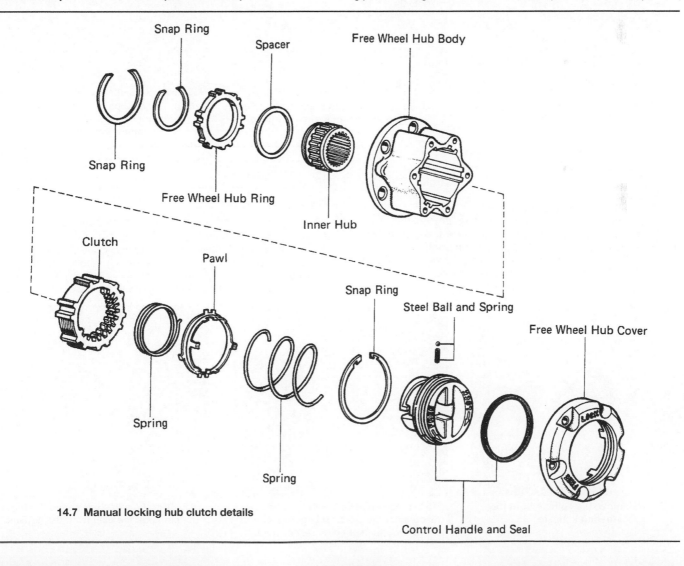

14.7 Manual locking hub clutch details

14.9 Remove the dust cap on automatic
locking freewheel hubs

14.10 Remove the hub cover
mounting bolts

14.11 Use a brass hammer to dislodge
the cone shaped washers

14.12 Remove the snap-ring from the end
of the axleshaft

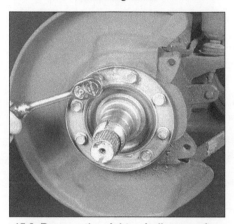

15.3 Remove the eight spindle mounting
bolts then remove the dust seal and cover

15.4a Tap the spindle with a soft-faced
hammer to separate it from the
steering knuckle

Automatic locking hubs

Refer to illustrations 14.9, 14.10, 14.11 and 14.12

Removal

9   Remove the dust cap **(see illustration)**.
10   Remove the hub cover mounting bolts **(see illustration)**.
11   Using a brass hammer, tap on the hub cover to loosen the cone
washers and remove them from the hub **(see illustration)**.
12   Using snap-ring pliers, detach the snap-ring from the end of the
axle shaft **(see illustration)** and remove the hub disc.
13   Installation is the reverse of removal.

15   Axleshaft and oil seal (front) - removal, overhaul and installation

Removal

Refer to illustration 15.3, 15.4a, 15.4b, 15.5, 15.6 and 15.7
1   Remove the freewheel hub (see Section 14).
2   Remove the front wheel bearing assembly (see Chapter 1).
3   Remove bolts attaching the brake backing plate and spindle to the
steering knuckle and remove the backing plate **(see illustration)**. If
equipped with front drum brakes, remove the complete brake assembly

15.4b Remove the spindle from the
steering knuckle

15.5 If equipped with anti-lock brakes,
disconnect the electrical connector and
remove the speed sensor (arrow)

15.6 Turn the axleshaft so the flat faces
up, then pull the axle from the housing

15.7 Front axleshaft oil seal location (arrow)

15.9 Dislodge the outer joint and shaft with a brass drift and hammer (be careful not to let the joint fall)

15.10 Tilt the inner race and cage and remove the balls from the joint one at a time

and tie it aside without disconnecting the hydraulic line (don't let it hang by the line).

4   Remove the spindle **(see illustrations)**. Inspect the spindle bushing for wear and replace if necessary.

5   Remove the anti-lock brake sensor if equipped **(see illustration)**.

6   Turn the axleshaft until the flat spot on the CV housing lines up with the top of the steering knuckle, then pull it straight out of the axle housing to remove it **(see illustration)**. Be extremely careful not to damage the axleshaft seal, which is located in the tube.

7   If you are replacing the axleshaft oil seal, use a suitable puller to remove the oil seal from the housing **(see illustration)**. A slide hammer with an internal puller attachment works well.

Overhaul

Refer to illustrations 15.9, 15.10, 15.11, 15.12, 15.16 and 15.19

8   Place the inner axleshaft in a vise with soft jaws or blocks of wood, so as not to mar the shaft.

9   Place a brass drift against the inner race of the joint and drive off the joint and outer shaft with a hammer **(see illustration)**.

10   Tilt the inner race and cage and remove the ball bearings one at a time **(see illustration)**.

11   Tilt the two large openings in the cage around the lands of the outer shaft and pull out the cage and inner race **(see illustration)**.

12   Remove the inner race from the cage by positioning it so that two of its lands line up with the large openings in the cage, turning the race 90-degrees and pulling it out **(see illustration)**.

13   Clean and inspect the inner parts of the joint for wear or damage. If necessary, replace any worn or damaged parts with new ones.

14   To reassemble the joint, coat the inner parts of the joint and the inside of the outer shaft with molybdenum disulfide grease.

15   Insert the inner race into the cage by reversing Step 12.

15.11 Tilt the inner race and cage 90-degrees, then align the windows of the cage with the lands of the housing and pull out to remove them from the housing

16   Position the protruding end of the inner race toward the wide side of the cage **(see illustration)**.

17   Assemble the cage and inner race to the outer shaft by reversing Step 11. **Note:** *Make sure to position the wide side of the cage and the protruding end of the inner race facing out.*

18   Install new snap-rings on the inner axleshaft.

19   Place the CV joint in the vise (still lined with soft jaws or wood) and install the axleshaft while compressing the inner snap-ring with a screwdriver **(see illustration)**.

20   Verify that the axleshaft can't be pulled out of the joint.

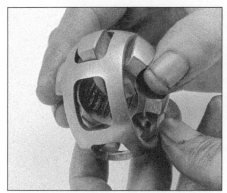

15.12 Align the inner race lands with the cage windows and rotate the inner race out of the cage

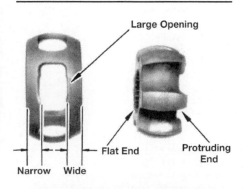

Large Opening

Flat End

Protruding End

Narrow   Wide

15.16 The protruding end of the inner race and the wide side of the cage must be on the same side when assembled

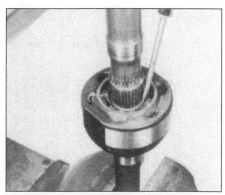

15.19 Compress the inner snap-ring with a screwdriver and slide the axleshaft into the inner race of the joint

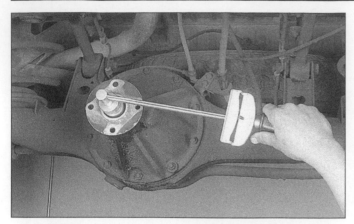

17.5 Using an inch-pound torque wrench, measure and record the torque required to turn the pinion nut (within the pinion's backlash) before replacing the pinion oil seal

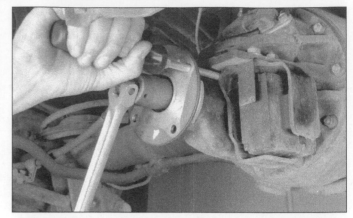

17.6a A large screwdriver, positioned as shown, can be used to hold the companion flange stationary while the nut is loosened

Installation

21   Install a new oil seal if necessary and lubricate the splines on the inner end of the axleshaft with grease. Keep the axleshaft straight as you insert it through the steering knuckle and into the axle tube. Make sure the splined inner end of the axleshaft is properly engaged with the differential.
22   The remainder of the installation is the reverse of removal.

16   Axle assembly - removal and installation

Note: *This procedure applies to both front and rear axles.*

Removal

1   Loosen the wheel lug nuts, raise the vehicle and support it securely on jackstands positioned under the frame rails. Remove the wheels.
2   Remove the brake components (see Chapter 9).
3   On front axles, disconnect the tie-rod ends and center link from the steering knuckles (see Chapter 10).
4   Disconnect the driveshaft from the pinion flange (see Section 12).
5   Remove the stabilizer bar if equipped (see Chapter 10).
6   If the axle is equipped with a track bar, disconnect it from the axle (see Chapter 10).
7   Disconnect the parking brake cables from the spring clips on the rear axle assembly and detach the cable brackets from the lower suspension arms (see Chapter 9).
8   If equipped, unplug the electrical connector from the anti-lock brake system speed sensors.
9   Detach the flexible brake hose(s) from its bracket on top of the axle housing where it attaches to the metal brake line (see Chapter 9).
10   If equipped, disconnect the load leveling sensor from the rear axle.
11   Support the axle with two floor jacks and raise the axle assembly slightly.
12   Disconnect the lower ends of the shock absorbers from the axle assembly (see Chapter 10).
13   Remove the coil springs, if so equipped (see Chapter 10).
14   If equipped with leaf springs, unbolt the U-bolts and tie plates from the axle (see Chapter 10). Remove the bolts attaching the front spring shackles to the springs and lower the springs to the ground. **Caution:** *Once the leaf springs are disconnected from the axle assembly, it's no longer stable. Have an assistant help balance the axle on the floor jack pad, if only one jack is being used.*
15   On coil spring-equipped vehicles, Disconnect the upper and or lower suspension arms from the axle assembly (see Chapter 10). **Caution:** *Once the suspension arms are disconnected from the axle assembly, it's no longer stable. Have an assistant help balance the axle on the floor jack pad.*
16   Slowly lower the axle assembly and wheel it out from under the vehicle.

Installation

17   Installation is the reverse of the removal procedure. When installing the axle, make sure it's positioned properly. For fasteners with specified torque values, be sure they are tightened to the torque listed in this Chapter's Specifications, as well as the Chapter 9 and 10 Specifications.

17   Pinion oil seal - replacement

Refer to illustrations 17.5, 17.6a, 17.6b, 17.7, 17.8 and 17.9
1   A pinion shaft oil seal failure results in the leakage of differential gear lubricant past the seal and onto the driveshaft yoke or flange. The seal can be replaced without removing or disassembling the differential.
2   Raise the vehicle and support it securely on jackstands.
3   Drain the differential lubricant (see Chapter 1). After draining is complete, install the drain plug and tighten it securely.
4   Disconnect the driveshaft from the pinion shaft yoke (see Section 12). Support the driveshaft out of the way with a piece of wire.
5   Using an inch-pound torque wrench, measure the torque required to move the pinion shaft within its backlash (break-loose torque). Record this figure, as it will be used to set the pinion shaft preload during installation **(see illustration)**.
6   Using a large screwdriver or prybar, hold the companion flange stationary, then remove the companion flange nut **(see illustration)**. Mark the relationship of the pinion shaft to the companion flange **(see illustration)**.
7   Remove the companion flange from the pinion shaft with a two-jaw puller **(see illustration)**.
8   Carefully pry the seal out of the differential with a screwdriver or prybar. It may be necessary to knock the seal out using a hammer and

17.6b Mark the relationship of the flange to the pinion shaft

17.7 Remove the companion flange from the pinion shaft with a two-jaw puller

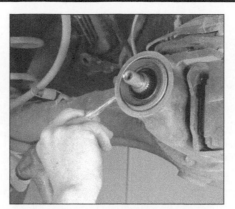

17.8 It'll probably be necessary to drive the seal out with a hammer and punch - be careful not to damage the splines or threads on the pinion shaft

17.9 A hammer and a socket of the proper diameter can be used to drive the seal into place

a punch **(see illustration)**. Be careful not to damage the splines on the pinion shaft.

9   Lubricate the new seal lip with multi-purpose grease or differential lubricant and carefully install it in position in the differential. Using a hammer and a seal driver, large socket or a short section of pipe of the proper diameter, carefully drive the seal into place **(see illustration)**.

10   Clean the sealing lip contact surface of the companion flange. Apply a thin coat of multi-purpose grease to the seal contact surface and the shaft spines. Slide the companion flange onto the shaft, making sure the match marks line up.

11   Coat the threads of the pinion shaft with multi-purpose grease. Install a new nut and, holding the flange stationary, tighten the nut to the initial (lower) torque listed in this Chapter's Specifications.

12   Turn the companion flange several times to seat the bearing.

13   Using an inch-pound torque wrench, see how much torque is required to turn the pinion shaft within its backlash (break-loose torque). The desired preload is the previously recorded torque value plus five inch-pounds. If the preload is less than desired, retighten the nut in small increments until the desired preload is reached. If the maximum torque on the pinion shaft nut (see this Chapter's Specifications) is reached before the desired preload is obtained, the bearing spacer must be replaced by a repair shop. **Note:** *Do not back-off the pinion*

nut to reduce the preload. After the preload is properly adjusted, proceed to the next Step.

14   Connect the driveshaft to the companion flange (see Section 12).

15   Fill the differential with the recommended lubricant (see Chapter 1).

16   Lower the vehicle to the ground and test drive the vehicle. Check around the companion flange for evidence of leakage.

18   Axleshaft, bearing and seal (rear) (semi-floating axle) - removal and installation

Refer to illustrations 18.1, 18.6a, 18.6b, 18.7 and 18.8

1   The axleshafts can be removed without disturbing the differential assembly. They must be removed in order to replace the bearings and oil seals and when removing the differential carrier from the rear axle housing **(see illustration)**. **Note:** *Read the entire Section before starting work.*

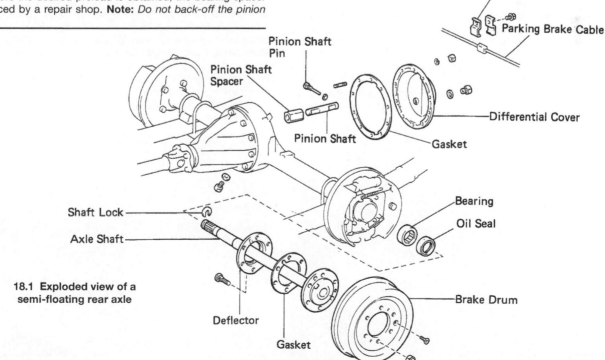

18.1 Exploded view of a semi-floating rear axle

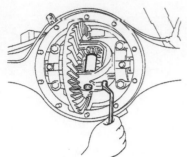

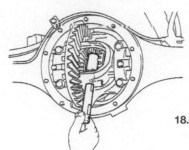

18.6a Rotate the differential until the pinion shaft pin can be accessed, then remove it

18.6b Pull out the pinion shaft and spacer

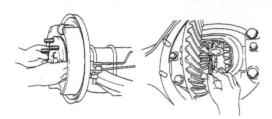

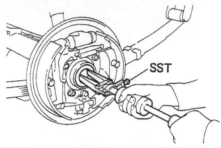

18.7 Push inward on the axleshaft to allow removal of the axleshaft locking clip

18.8 A slide hammer with an internal puller jaw attachment can be used to remove the axle bearing and oil seal

2   Raise the rear of the vehicle and support it securely on jackstands. Block the front wheels to keep the vehicle from rolling.

3   Remove the rear wheels and release the parking brake, then remove the brake drums (see Chapter 9 for details).

4   Remove the drain plug and drain the differential oil into a suitable container. When the draining is complete, finger-tighten the drain plug in place.

5   Remove the rear differential cover.

6   Remove the pinion shaft pin, pinion shaft and the pinion shaft spacer **(see illustrations)**.

7   Have an assistant push in on the outer flanged end of the axleshaft while you remove the axle shaft lock clip from the groove in the inner end of the axleshaft **(see illustration)**. Pull straight out on the axleshaft to remove it.

8   Inspect the axleshaft bearing and seal for wear. If replacement is necessary the bearing and oil seal can be removed using the slide hammer with an internal puller jaw attachment **(see illustration)**. **Note:** *Always lubricate the inner lip of the new oil seal with a small amount of multi-purpose grease before reinstalling the axleshaft.*

9   Refer to the beginning of this Chapter for the torque specifications, then reinstall the components in the reverse order of removal. Install the differential cover and fill the rear axle with the lubricant specified in Chapter 1.

19   Axleshaft and oil seal (rear) (full-floating axle) - removal and installation

Refer to illustrations 19.1, 19.2, 19.3 and 19.4

1   Remove the axleshaft flange nuts and lock washers **(see illustration)**.

2   Using a brass punch or hammer, tap on center of the flange studs to loosen the cone washers and remove them from the hub **(see illustration)**.

3   Tighten two bolts into the axle flange service holes until the axleshaft separates from the rear hub assembly, then remove the axleshaft **(see illustration)**.

4   Inspect the axleshaft seal for wear. If replacement is necessary the oil seal can be removed using a slide hammer with an internal puller jaw attachment, or it can be carefully pried out with a seal removal tool **(see illustration)**. **Note:** *Always lubricate the inner lip of the new oil seal with a small amount of multi purpose grease before reinstalling the axleshaft.*

5   When installing the axleshaft, clean the gasket sealing area and place a new gasket on the flange.

6   Slide the axleshaft into the housing, engaging the axleshaft splines into the differential.

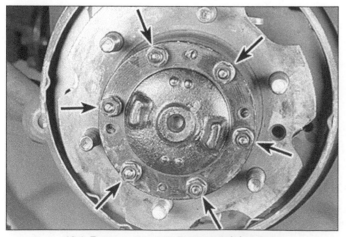

19.1 Remove the axle flange retaining nuts

19.2 Use a brass punch or hammer to dislodge the cone washers

19.3 Tighten two bolts in the flange service holes until the axleshaft separates from the rear hub

19.4 Use a hooked shaped tool to pry out the oil seal

20.3a Remove the locknut retaining screws

20.3b Unscrew the locknut . . .

7   Install the cone washers, lock washers and nuts. Tighten the nuts to the torque listed in this Chapter's Specifications.

20   Hub and wheel bearing assembly(rear) (full-floating axle) - removal, installation and adjustment

Removal

Refer to illustrations 20.3a, 20.3b, 20.3c, 20.4 and 20.6

1   Raise the vehicle, support it securely on jackstands and remove the rear wheel.

2   Remove the axleshaft (see Section 19).

3   Remove the screws securing the locknut, then remove the locknut and lockring **(see illustrations)**.

4   Slide the hub off the axle slightly and remove the outer wheel bearing **(see illustration)**.

5   Pull the hub assembly straight off the axle tube and place it on a clean flat surface with the inner bearing facing up.

6   Use a large screwdriver, prybar or seal removal tool to pry out the seal **(see illustration)**.

7   Remove the inner wheel bearing from the hub.

20.3c . . . then remove the lockring

20.4 Pull the hub assembly out slightly, then push it back in to remove the outer wheel bearing

8   Use solvent to wash the bearings, hub and axle tube. A small brush may prove useful; make sure no bristles from the brush embed themselves in the bearing rollers. Allow the parts to air dry.

9   Carefully inspect the bearings for cracks, wear and damage. Check the axle tube flange, studs, and hub splines for damage and corrosion. Check the bearing cups (races) for pitting or scoring. Worn or damaged components must be replaced with new ones. If necessary, tap the bearing cups from the hub with a hammer and brass drift and install new ones with the appropriate size bearing driver.

Installation

10   Lubricate the bearings and the axle tube contact areas with wheel bearing grease. Work the grease completely into the bearings, forcing it between the rollers, cone and cage. Place the inner bearing into the hub and install the seal (with the seal lip facing into the hub). Using a seal driver or block of wood, drive the seal in until it's flush with the hub. Lubricate the seal with gear oil or grease.

11   Place the hub assembly on the axle tube, taking care not to damage the oil seal. Place the outer bearing into the hub.

12   Install the lockring and locknut and adjust the bearings as described below.

Adjustment

Refer to illustration 20.15

13   Install the brake disc and hold it in place with two lug nuts. Tighten the locknut to 43 ft-lbs while rotating the hub assembly to seat the bearings.

14   Back off the locknut just until it can be turned by hand.

15   Attach a spring tension gauge to one of the wheel studs and measure the amount of effort to turn the hub **(see illustration)**. Tighten the hub locknut in small increments until the hub preload (breakaway torque) is within the range listed in this Chapter's Specifications.

16   Install the locknut retaining screws 180-degrees apart from each other.

17   Check that the hub assembly spins freely with no noticeable freeplay. If freeplay exists, repeat Steps 13, 14 and 15 until proper adjustment is obtained.

18   Install the axleshaft (see Section 19). Install the rear wheel and lower the vehicle.

20.6 Use a screwdriver or a seal removal tool to pry out the seal

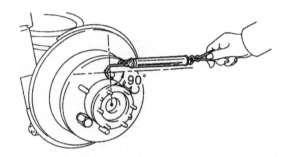

20.15 Use a spring scale to measure the hub bearing preload

Chapter 9   Brakes

Contents

	Section
Anti-lock Brake System (ABS) - general information	2
Brake check	See Chapter 1
Brake disc - inspection, removal and installation	5
Brake fluid level check	See Chapter 1
Brake hoses and lines - inspection and replacement	9
Brake hydraulic system - bleeding	10
Brake light switch - removal, installation and adjustment	15
Brake pedal adjustment	See Chapter 1
Disc brake caliper - removal, overhaul and installation	4
Disc brake pads - replacement	3
Drum brake shoes - replacement	6

	Section
General information	1
Load sensing proportioning and by-pass valve - general information	16
Master cylinder - removal, overhaul and installation	8
Parking brake - adjustment	13
Parking brake cables - replacement	14
Parking brake shoes (rear disc brakes only) - inspection and replacement	12
Power brake booster - check, removal and installation	11
Wheel cylinder - removal, overhaul and installation	7

Specifications

General

Brake fluid type	See Chapter 1
Power brake booster pushrod-to-master cylinder piston clearance	0.0 inch (0.0 mm)

Disc brakes

Brake pad thickness	
Standard	0.394 inch (10 mm)
Minimum	0.157 inch (4.0 mm)
Parking brake shoe thickness (all models with rear disc brakes)	
Standard	0.158 inch (4.0 mm)
Minimum	0.039 inch (1.0 mm)
Disc runout	0.0059 inch (0.15 mm)
Disc minimum thickness	Refer to the dimension cast into the disc
Parking brake disc inside diameter (all models with rear disc brakes)*	
Standard	9.06 inches (230 mm)
Maximum	9.09 inches (231 mm)

Drum brakes

Shoe lining thickness	
Standard	0.265 inch (6.5 mm)
Minimum	0.059 inch (1.5 mm)
Drum inside diameter (maximum)	Refer to the dimension cast into the drum

Torque specifications

Ft-lbs (unless otherwise indicated)

Note: One foot-pound (ft-lb) of torque is equivalent to 12 inch-pounds (in-lbs) of torque. Torque values below approximately 15 ft-lbs are expressed in inch-pounds, because most foot-pound torque wrenches are not accurate at these smaller values.

Disc brake caliper mounting bolts	
Front caliper	90
Rear caliper	65
Front caliper half retaining bolts	43
Rear caliper torque plate bolts	76
Disc-to-front hub bolts	
1990 and earlier	34
1991 and later	54
Brake hose-to-caliper banjo fitting bolt	22
Wheel cylinder mounting bolts	84 in-lbs
Master cylinder-to-brake booster nuts	108 in-lbs
Power brake booster mounting nuts	108 in-lbs
Load Sensing Proportioning Valve (LSPV) mounting bolts	108 in-lbs
Wheel lug nuts	See Chapter 1

1   General information

The vehicles covered by this manual are equipped with hydraulically operated front and rear brake systems. The front and rear brakes are either drum or disc type. Both the front and rear brakes are self adjusting. The disc brakes automatically compensate for pad wear, while the rear drum brakes incorporate an adjustment mechanism which is activated as the parking brake is applied.

Hydraulic system

The hydraulic system consists of two separate circuits. The master cylinder is designed for the split system and incorporates a primary piston for one circuit and a secondary piston for the other, and, in the event of a leak or failure in one hydraulic circuit, the other circuit will remain operative. A dual proportioning valve on the firewall provides brake balance between the front and rear brakes. A load sensing proportioning and bypass valve adjusts brake fluid pressure to the rear brakes according to how much weight is in the rear of the vehicle.

Power brake booster

The power brake booster, utilizing engine manifold vacuum and atmospheric pressure to provide assistance to the hydraulically operated brakes, is mounted on the firewall in the engine compartment.

Parking brake

The parking brake operates the rear brakes only, through cable actuation. It's activated by the parking brake lever which is mounted next to transmission shift lever.

Service

After completing any operation involving disassembly of any part of the brake system, always test drive the vehicle to check for proper braking performance before resuming normal driving. When testing the brakes, perform the tests on a clean, dry, flat surface. Conditions other than these can lead to inaccurate test results.

Test the brakes at various speeds with both light and heavy pedal pressure. The vehicle should stop evenly without pulling to one side or the other. Avoid locking the brakes, because this slides the tires and diminishes braking efficiency and control of the vehicle.

Tires, vehicle load and wheel alignment are factors which also affect braking performance.

2   Anti-lock Brake System (ABS) - general information

1   The Anti-lock Brake System (ABS), which is available on 1993 and later 80 series vehicles, is designed to maintain vehicle steerabilty, directional stability and optimum deceleration under severe braking conditions and on most road surfaces. It does so by monitoring the rotational speed of each wheel and controlling the brake line pressure to each wheel during braking. This prevents the wheel from locking up.

Components

Refer to illustration 2.2

Actuator assembly

2   The actuator assembly consists of the master cylinder, an electric hydraulic pump and four solenoid valves **(see illustration)**.

a) *The electric pump provides hydraulic pressure to charge the reservoirs in the actuator, which supplies pressure to the braking system. The pump and reservoirs are housed in the actuator assembly.*

b) *The solenoid valves modulate brake line pressure during ABS operation. The body contains four valves - one for each wheel.*

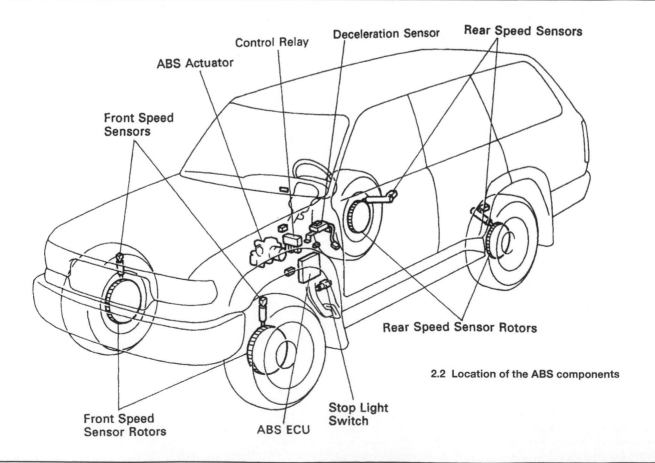

2.2 Location of the ABS components

Speed sensors

3   These sensors are located at each wheel and generate small electrical pulsations when the toothed sensor rings are turning, sending a signal to the electronic controller indicating wheel rotational speed.

4   The front speed sensors **(see illustration 2.2)** are mounted to the front steering knuckle in close relationship to the toothed sensor rings, which are integral with the front axle shaft assemblies.

5   The rear wheel sensors are bolted to the axle carriers **(see illustration 2.2)**. The sensor rings are integral with the rear hub assemblies.

ABS computer (ECU)

6   The ABS computer is mounted under the dashboard and is the "brain" for the ABS system. The function of the computer is to accept and process information received from the wheel speed sensors to control the hydraulic line pressure, avoiding wheel lock up. The computer also constantly monitors the system, even under normal driving conditions, to find faults within the system.

7   If a problem develops within the system, an "ABS" light will glow on the dashboard. A diagnostic code will also be stored in the computer and, when retrieved, will indicate the problem area or component.

Diagnostic codes

Refer to illustration 2.10

8   The ABS system control unit (computer) has a built-in self-diagnosis system which detects malfunctions in the system sensors and alerts the driver by illuminating an ABS warning light in the instrument panel. The computer stores the failure code until the diagnostic system is cleared or malfunction is repaired.

9   The ABS warning light should come on when the ignition switch is placed in the On position. When the engine is started, the warning light should go out. If the light remains on, the diagnostic system has detected a malfunction or abnormality in the system. **Note:** *When the transfer case is placed in the LOW (center differential lock) position, the ABS system becomes disabled and the ABS warning light will stay on.*

10   The codes for the ABS can be accessed by turning the ignition key to the ON position (engine not running). Install a jumper wire or paper clip onto terminals E1 and Tc and remove the small pin from the WA and WB terminals of the data link connector located in the engine compartment **(see illustration)**. Make sure that the transfer case is placed in the 2WD position.

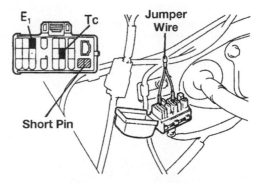

2.10 The data link connector is located in the engine compartment - use a jumper wire or paper clip and connect terminals E1 and Tc, then remove the pin from the WA and WB terminals

11   The diagnostic code is the number of flashes indicated on the ABS light. If any malfunction has been detected, the light will blink the first digit(s) of the code, pause 1.5 seconds, then blink the second digit of the code. For example, a code 34 (left rear wheel sensor) will first blink three flashes, pause 1.5 seconds and blink four flashes. If there is more than one code stored in the ECM, the ECM will pause 2.5 seconds before flashing the next code. If the system is operating normally (no malfunctions), the warning light will blink once every 0.5 seconds.

12   The accompanying tables explain the code that will be flashed for each of the malfunctions. The accompanying charts indicate the diagnostic code - in blinks - along with the system, diagnosis and specific areas.

13   After the diagnosis check, clear the trouble codes by depressing the brake pedal eight or more times within three seconds. Remove the jumper wire and reinstall the cap on the data link connector. Check the indicated system or component or take the vehicle to a dealer service department to have the malfunction repaired.

Code chart for the ABS system

Code number	Light pattern	Trouble area	Action to take
Code 11	1 flash pause 1 flash	Open circuit in solenoid relay circuit	*Check the solenoid relay and the wire
Code 12	1 flash pause 2 flashes	Short circuit in solenoid relay circuit	harness of the solenoid relay circuit
Code 13	1 flash pause 3 flashes	Open circuit in pump motor relay circuit	*Check the pump motor relay and the wire
Code 14	1 flash pause 4 flashes	Short circuit in pump motor relay circuit	harness of the pump motor relay circuit
Code 21	2 flashes pause 1 flash	Problem in circuit of solenoid right front wheel	*Check the actuator solenoid and the
Code 22	2 flashes pause 2 flashes	Problem in circuit of solenoid left front wheel	connector of the actuator solenoid circuit
Code 23	2 flashes pause 3 flashes	Problem in circuit of solenoid rear wheel	
Code 31	3 flashes pause 1 flash	Sensor signal problem right front wheel	*Check the indicated speed sensors,
Code 32	3 flashes pause 2 flashes	Sensor signal problem front left wheel	sensor rotors, wire harness and
Code 33	3 flashes pause 3 flashes	Sensor signal problem right rear wheel	connector of the speed sensors
Code 34	3 flashes pause 4 flashes	Sensor signal problem left rear wheel	
Code 35	3 flashes pause 5 flashes	Open circuit in left front or right rear sensor	
Code 36	3 flashes pause 6 flashes	Open circuit in right front or left rear sensor	
Code 41	4 flashes pause 1 flash	Abnormally high or low battery positive pressure	*Check the charging system (alternator, battery and voltage regulator) for any problems (see Chapter 5)
Code 43	4 flashes pause 3 flashes	Defective deceleration sensor	*Check the sensor, sensor installation,
Code 44	4 flashes pause 4 flashes	Problem in circuit of the deceleration sensor	wire harness and connector of the deceleration sensor

Code chart for the ABS system (continued)

Code number	Light pattern	Trouble area	Action to take
Code 48	4 flashes pause 8 flashes	Problem in circuit of center differential	*Check the center differential lock, lock switch, indicator light, and wire harness
Code 51	5 flashes pause 1 flashes	Pump motor of actuator locked or open circuit in pump motor circuit in actuator	*Check the pump motor, relay and battery for shorts or abnormalities
Always ON	Continuous ON	ECU malfunction	*ECU problem

3.5a Detach the pin retaining clip

3.5b Remove the lower pad retaining pin

3.5c Lift out the anti-rattle spring
(1992 and later models)

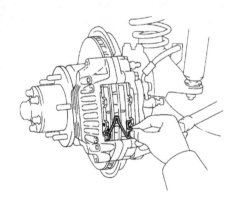

3.5d On 1991 and earlier models, remove
the W-shaped anti-rattle spring

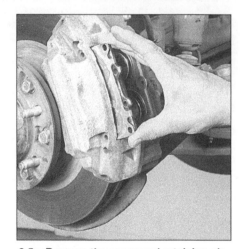

3.5e Remove the upper pad retaining pin,
then remove the outer brake pad

3   Disc brake pads - replacement

Refer to illustrations 3.5a through 3.5q

Warning: *Disc brake pads must be replaced on both front or rear wheels at the same time - never replace the pads on only one wheel. Also, the dust created by the brake system may contain asbestos, which is harmful to your health. Never blow it out with compressed air and don't inhale any of it. An approved filtering mask should be worn when working on the brakes. Do not, under any circumstances, use petroleum-based solvents to clean brake parts. Use brake cleaner or denatured alcohol only!*

Note: *This procedure applies to both the front and rear disc brakes.*

1   Remove the cap from the brake fluid reservoir.
2   Loosen the wheel lug nuts, raise the front or rear of the vehicle and support it securely on jackstands. Block the wheels at the opposite end.
3   Remove the wheels. Work on one brake assembly at a time, using the assembled brake for reference if necessary.
4   Inspect the brake disc carefully as outlined in Section 5. If machining is necessary, follow the information in that Section to remove the disc, at which time the pads can be removed as well.
5   Follow the accompanying photos for the actual pad replacement procedure. Use **illustrations 3.5a through 3.5i** to replace the front brake pads and **3.5j through 3.5q** to replace the rear brake pads. Be sure to stay in order and read the caption under each illustration.

3.5f Push the pistons back into their bores to provide room for the new brake pad (replace one pad at a time)

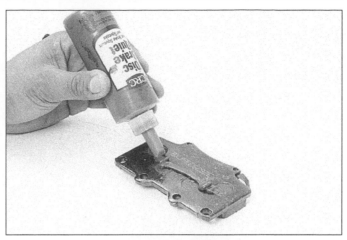

3.5g Apply anti-squeal compound to the back of the brake pads . . .

3.5h . . . then between the number 1 and number 2 anti-squeal shims . . .

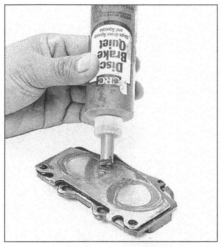

3.5i . . . and to the back of the number 2 shim - reinstall the pads and the remaining hardware in the reverse order of removal

3.5j On rear disc brakes, a large C-clamp can be used to push the piston back into the caliper bore - note that one end of the clamp is on the flat area near the brake hose fitting and the other end (screw end) is pressing against the outer pad

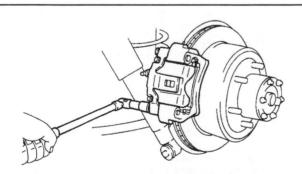

3.5k Remove the rear caliper mounting bolts . . .

3.5l . . . then lift off the rear caliper - use a piece of wire to hang the caliper out of the way - DO NOT let the caliper hang by the brake hose

Note1: *As the caliper pistons are depressed, the fluid in the master cylinder will rise. Make sure that it doesn't overflow. If necessary, siphon off some of the fluid.* **Note 2:** *When working on rear disc brakes, remove the pad support plates from the torque plate - they should be replaced with new ones if distorted in any way. Also, be sure to transfer the wear indicators from the old brake pads onto the new pads. If they are worn or bent, replace them with new parts.*

3.5m Remove the pads from the torque plate

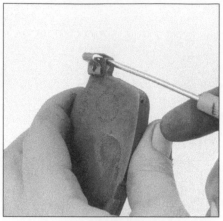

3.5n Pry the wear indicators and the anti-squeal shim (outer pad) off the brake pads and transfer them to the new pads - if they are worn or bent, replace them

3.5o Remove the pad support plates from the torque plate for inspection - they should be replaced with new ones if distorted in any way - if **OK**, reinstall them in the torque plate

3.5p Install the new brake pads equipped with wear indicators (outer pad) and anti-squeal shim into the torque plate

3.5q Apply anti-squeal compound to the outside of the anti-squeal shim (outer pad) and a small amount to the inner pad where it will contact the piston - reinstall the caliper in the reverse order of removal

6   When reinstalling the caliper, be sure to tighten the mounting bolts to the torque listed in this Chapter's Specifications. After the job has been completed, firmly depress the brake pedal a few times to bring the pads into contact with the disc. Check the level of the brake fluid, adding some if necessary. Check the operation of the brakes carefully before placing the vehicle into normal service.

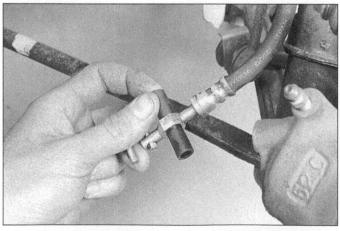

4.2 Using a piece of rubber hose of the appropriate size, plug the brake line

4   Disc brake caliper - removal, overhaul and installation

Refer to illustrations 4.2, 4.3, 4.4, 4.5, 4.6a, 4.6b, 4.8, 4.9 and 4.10

Warning: *Dust created by the brake system may contain asbestos, which is harmful to your health. Never blow it out with compressed air and don't inhale any of it. An approved filtering mask should be worn when working on the brakes. Do not, under any circumstances, use petroleum-based solvents to clean brake parts. Use brake cleaner or denatured alcohol only!*

Note1: *If an overhaul is indicated (usually because of fluid leakage), explore all options before beginning the job. New and factory rebuilt calipers are available on an exchange basis, which makes this job quite easy. If it's decided to rebuild the calipers, make sure a rebuild kit is available before proceeding. Always rebuild the calipers in pairs - never rebuild just one of them.*

Note 2: *This procedure is illustrated with a four piston type caliper. The procedures are the same for the single piston type except where noted. This procedure applies to both the front and rear disc brakes.*

Removal

1   Remove the banjo fitting bolt and disconnect the brake hose from the caliper. Discard the sealing washers - new ones should be used on installation.

2   Plug the brake hose to keep contaminants out of the brake sys-

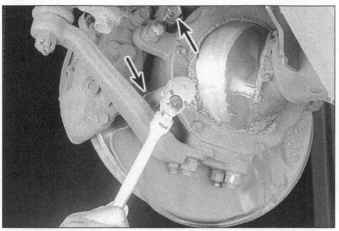

4.3 Remove the front caliper mounting bolts (arrows)

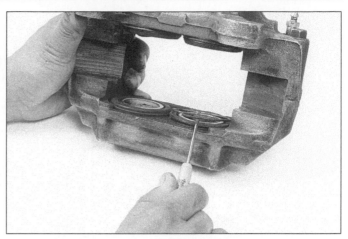

4.4 Using a screwdriver, remove the cylinder boot set ring

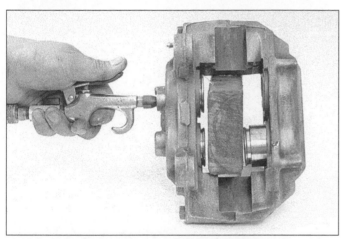

4.5 With the caliper padded to catch the piston, use compressed air to force the piston out of its bore - make sure your fingers are not between the piston and caliper

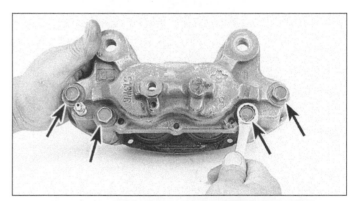

4.6a Remove the bolts (arrows) to separate the halves (four-piston type calipers)

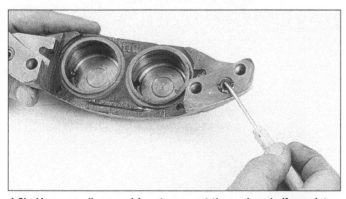

4.6b Use a small screwdriver to pry out the end seals (four-piston type calipers) - be careful not to damage the metal surfaces

4.8 The piston seal(s) should be removed with a plastic or wooden tool to avoid damage to the bore and seal groove - a pencil will do the job

tem and to prevent losing any more brake fluid than necessary **(see illustration)**.

3   Slightly depress the piston, remove the upper and lower mounting bolts and remove the caliper **(see accompanying illustration and illustration 3.5k)**.

Overhaul

4   To overhaul the caliper, remove the boot set ring and the boot from each piston **(see illustration)**.

5   Place a wood block between the piston and caliper to prevent

damage as it is removed. To remove the piston from the caliper, apply compressed air to the brake fluid hose connection on the caliper body **(see illustration)**. Use only enough pressure to ease the piston out of its bore. **Warning:** *Be careful not to place your fingers between the piston and the caliper as the piston may come out with some force.*

6   On four-piston calipers, remove the bolts securing the caliper halves together, then pry out the end seals **(see illustrations)**. **Note:** *This procedure is used on four-piston type calipers only.*

7   Inspect the mating surfaces of the piston and caliper bore wall. If there is any scoring, rust, pitting or bright areas, replace the complete caliper unit with a new one.

8   If these components are in good condition, remove the piston

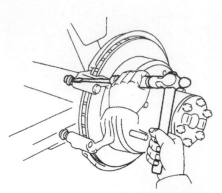

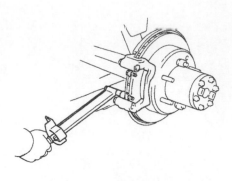

4.9 Tap out the pin boot and sliding bushing from the torque plate (single-piston type calipers)

4.10 Wash all parts thoroughly with brake cleaner and allow them to dry

5.2 The rear torque plate (caliper mounting bracket) is secured to the axle with two bolts (arrows)

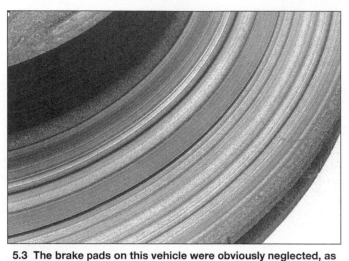

5.3 The brake pads on this vehicle were obviously neglected, as they wore down to the rivets and cut deep grooves into the disc - wear this severe will require replacement of the disc

5.4a Use a dial indicator to check disc runout - if the reading exceeds the maximum allowable runout limit, the disc will have to be machined or replaced

seal from the caliper bore using a wooden or plastic tool **(see illustration)**. Metal tools may damage the cylinder bore.

9   On single-piston type calipers, push the sliding bushings out of the torque plate and remove the dust boots from both ends **(see illustration)**.

10   Wash all the components in clean brake fluid or brake cleaner **(see illustration)**.

11   Submerge the new piston seal(s) and the piston in brake fluid and install them into the caliper bore. Do not force the piston into the bore, but make sure that it is squarely in place, then apply firm (but not excessive) pressure to install it.

12   Install the new piston dust boot and set ring.

13   On four-piston calipers, submerge the new end seals in brake fluid and install them into there bore. Install the bolts and tighten the caliper halves together to the specified torque listed at the beginning of this Chapter. **Note:** *Make sure the end seals stay seated in their bores when assembling the caliper halves.*

14   On single-piston calipers, apply silicone-based grease (supplied with the rebuild kit) to the sliding pin, sliding bushing and the torque plate bore. Install the dust boots.

Installation

15   Install the caliper by reversing the removal procedure. Remember to replace the copper sealing washers (gaskets) at the brake hose-to-caliper connection (new washers normally come with the rebuild kit).

16   Bleed the brake circuit according to the procedure in Section 10.

Make sure there are no leaks from the hose connections. Test the brakes carefully before returning the vehicle to normal service.

---

5   Brake disc - inspection, removal and installation

Note: *This procedure applies to both front and rear brake discs.*

Inspection

Refer to illustrations 5.2, 5.3, 5.4a, 5.4b, 5.5a and 5.5b

1   Loosen the wheel lug nuts, raise the vehicle and support it securely on jackstands. Remove the wheel and, if you're working on a rear disc brake, install the lug nuts to hold the disc in place (you may have to install washers under the lug nuts). If the rear brake disc is being worked on, release the parking brake.

2   Slightly depress the caliper piston, remove the caliper mounting bolts and remove the caliper **(see illustrations 3.5k and 4.3)**. Remove the brake pads as outlined in Section 3. It isn't necessary to disconnect the brake hose. After removing the caliper bolts, suspend the caliper out of the way with a piece of wire. If the rear brake disc is being worked on, remove the two torque plate bolts **(see illustration)** and detach the torque plate.

3   Visually inspect the disc surface for score marks and other damage. Light scratches and shallow grooves are normal after use and may not always be detrimental to brake operation, but deep scoring - over 0.039-inch (1.0 mm) - requires disc removal and refinishing by an

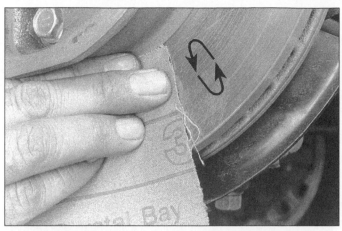

5.4b Using a swirling motion, remove the glaze from the disc surface with sandpaper or emery cloth

5.5b Use a micrometer to measure disc thickness

5.5a The minimum wear dimension is cast into the hub area of the disc (this is a rear disc - note the dimension for the parking brake drum maximum diameter on the right)

7 To remove a rear disc, remove the lug nuts which were put on to hold the rear disc in place and remove the disc from the hub. If the rear disc is stuck to the hub and won't come off, loosen the parking brake adjuster screw, then tap on the disc with a rubber mallet until the disc is free.

Installation

8 On front disc brakes, refer to Chapter 1 for the front wheel bearing repack and adjustment procedures.
9 Place the rear disc in position over the wheel studs. Install the torque plate and tighten the bolts to the torque listed in this Chapter's Specifications.
10 Install the brake pads, shims and springs.
11 Install the caliper and tighten the mounting bolts to the torque listed in this Chapter's Specifications.
12 Install the wheel and lug nuts, then lower the vehicle to the ground. Tighten the lug nuts to the torque listed in the Chapter 1 Specifications. Depress the brake pedal a few times to bring the brake pads into contact with the disc. Bleeding won't be necessary unless the brake hose was disconnected from the caliper. Check the operation of the brakes carefully before driving the vehicle.

6 Drum brake shoes - replacement

Refer to illustrations 6.3a, 6.3b, 6.3c, 6.4a through 6.4p and 6.5
Warning: *Drum brake shoes must be replaced on both wheels at the same time - never replace the shoes on only one wheel. Also, the dust created by the brake system may contain asbestos, which is harmful to your health. Never blow it out with compressed air and don't inhale any of it. An approved filtering mask should be worn when working on the brakes. Do not, under any circumstances, use petroleum-based solvents to clean brake parts. Use brake cleaner or denatured alcohol only!*
Caution: *Whenever the brake shoes are replaced, the return and hold-down springs should also be replaced. Due to the continuous heating/cooling cycle the springs are subjected to, they lose tension over a period of time and may allow the shoes to drag on the drum and wear at a much faster rate than normal.*
1 Loosen the wheel lug nuts, raise the front or rear of the vehicle and support it securely on jackstands. Block the wheels at the opposite end to keep the vehicle from rolling.
2 Release the parking brake.
3 Remove the wheel and brake drum. If the brake drum cannot be easily pulled off the axle and shoe assembly, make sure that the parking brake is completely released, then apply some penetrating oil at the hub-to-drum joint. Allow the oil to soak in and try to pull the drum off. If the drum still cannot be pulled off, the brake shoes will have to be retracted.

automotive machine shop. Be sure to check both sides of the disc **(see illustration)**. If pulsating has been noticed during application of the brakes, suspect disc runout.
4 To check disc runout, place a dial indicator at a point about 1/2-inch from the outer edge of the disc **(see illustration)**. Set the indicator to zero and turn the disc. The indicator reading should not exceed the specified allowable runout limit. If it does, the disc should be refinished by an automotive machine shop. **Note:** *The discs should be resurfaced regardless of the dial indicator reading, as this will impart a smooth finish and ensure a perfectly flat surface, eliminating any brake pedal pulsation or other undesirable symptoms related to questionable discs. At the very least, if you elect not to have the discs resurfaced, remove the glaze from the surface with sandpaper or emery cloth using a swirling motion* **(see illustration)**.
5 It's absolutely critical that the disc not be machined to a thickness under the minimum allowable refinish thickness. The minimum wear (or discard) thickness is cast into the hub area of the disc **(see illustration)**. The disc thickness can be checked with a micrometer **(see illustration)**.

Removal

6 To remove the front disc refer to Chapter 1 for the front wheel bearing repack and adjustment procedures. Once the disc and hub assembly have been removed, the disc can be unbolted from the hub. **Note:** *On 1990 and earlier models it is also necessary to remove the wheel studs from the hub with a hydraulic press. Take the assembly to an automotive machine shop if you don't have the necessary equipment.* When installing the disc to the hub, be sure to tighten the bolts to the torque listed in this Chapter's Specifications.

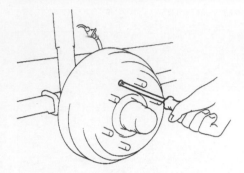

6.3a Mark the relationship of the drum to the hub, so the balance will be retained, then remove the retaining screw (if equipped)

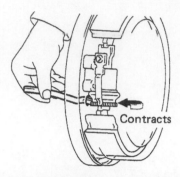

6.3b If the brake shoes are hanging up on the drum (because of excessive wear), insert a screwdriver or an adjusting tool through the hole in the backing plate to turn the star wheel to retract the brake shoes (dual leading type brake)

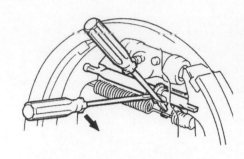

6.3c On leading/trailing type brakes, it'll be necessary to push the adjuster lever off the star wheel with a screwdriver before turning the star wheel with another screwdriver (or brake adjusting tool)

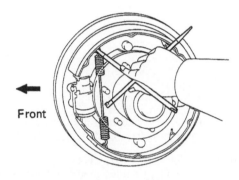

6.4a A pair of locking pliers or a brake spring removal tool can be used to unhook the front return spring from the brake assembly (dual leading type brake)

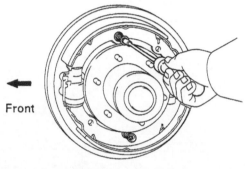

6.4b Remove the hold-down springs - depress the retainer and turn it 90-degrees to release it - a pair of pliers will work, but this special hold-down spring removal tool makes this much easier (they are available at most auto parts stores and aren't very expensive)

This is accomplished by first removing the plug from the backing plate. With the plug removed, turn the adjusting screw star wheel with a screwdriver, moving the shoes away from the drum **(see illustrations)**. On leading/trailing type brakes, it will be necessary to push the adjusting lever off the star wheel with one narrow screwdriver while turning the star wheel with another narrow screwdriver. The drum should now come off **(see illustration)**. **Warning:** *Before removing anything, clean the brake assembly with brake cleaner and allow it to dry - position a drain pan under the brake to catch the fluid and residue - DO NOT USE COMPRESSED AIR TO BLOW THE DUST FROM THE PARTS!*

4 Follow **illustrations 6.4a through 6.4e** for the inspection and replacement of dual leading type brake shoes. Follow **illustrations 6.4f through 6.4p** for the inspection and replacement of leading/trailing type brake shoes. Be sure to stay in order and read the caption under each illustration. Pay close attention to the way the parts are installed during the removal procedures as this will help aid the assembly process.

5 Before reinstalling the drum, it should be checked for cracks, score marks, deep scratches and hard spots, which will appear as small discolored areas. If the hard spots cannot be removed with

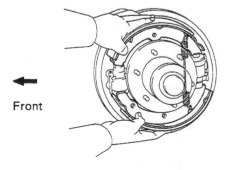

6.4c Remove both brake shoes and the rear return spring at the same time. Remove the return spring from the inner side of the shoes and reassemble install it on the new shoes

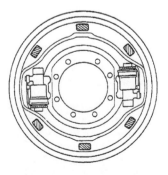

6.4d Lubricate the brake shoe contact area on the backing plate with high-temperature grease - reinstall the brake shoes in the reverse order of removal

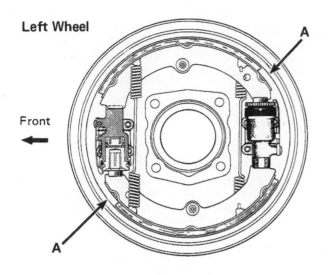

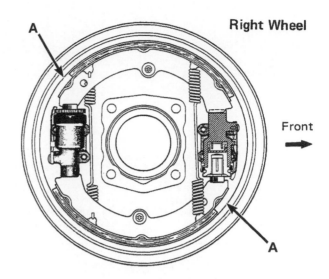

6.4e Dual leading type shoe details - be sure to assemble the short side (A) of the shoe in the correct position

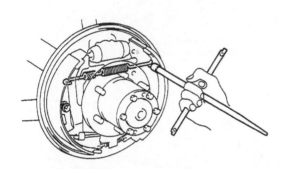

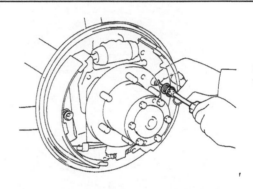

6.4f A pair of locking pliers or a brake spring removal tool can be used to unhook the upper tension spring from the brake assembly (leading/trailing type brake)

6.4g Detach the rear hold-down spring and remove the rear brake shoe - depress the retainer and turn it 90-degrees to release it

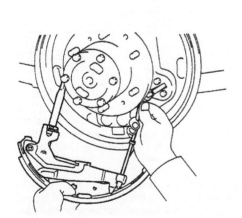

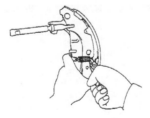

6.4i Remove the adjusting lever spring . . .

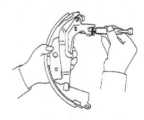

6.4j . . . and the adjuster from the front shoe

6.4h Detach the front hold-down spring and remove the front shoe and adjuster assembly from the backing plate - hold the end of the parking brake cable with a pair of pliers and pull it out of the parking brake lever

6.4k Pry off the E-clip and remove the adjusting lever

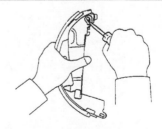

6.4l Pry the C-clip apart and remove the parking brake lever

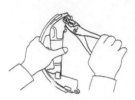

6.4m Assemble the parking brake lever to the new front shoe and crimp the C-clip closed with a pair of pliers, then install the adjuster lever and E-clip (always use new retaining clips)

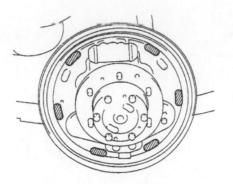

6.4n Lubricate the brake shoe contact area on the backing plate . . .

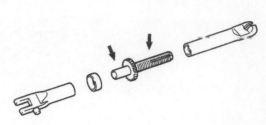

6.4o . . . and the moving parts of the adjuster screw with a light coat of high-temperature grease - the screw portion of the adjuster will need to be threaded in further than before to allow the drum to fit over the new shoes. Reinstall the brake shoes in the reverse order of removal

Left Wheel   **Right Wheel**

Front   Front

6.4p Leading/trailing type shoe assembly details

6.5 The maximum drum diameter is cast into the inside of the brake drums

emery cloth or if any of the other conditions listed above exist, the drum must be taken to an automotive machine shop to have it resurfaced. **Note:** *Professionals recommend resurfacing the drums each time a brake job is done. Resurfacing will eliminate the possibility of out-of-round drums. If the drums are worn so much that they can't be resurfaced without exceeding the maximum allowable diameter (stamped into the drum), then new ones will be required* (**see illustration**). *At the very least, if you elect not to have the drums resurfaced, remove the glaze from the surface with sandpaper or emery cloth using a swirling motion.*

6   Install the brake drum on the axle flange and check the preliminary brake adjustment. If the brake drum is fits loosely over the shoes, remove the drum again and rotate the adjuster until the brake drum just barely slides over the brake shoes. Now, turn the adjuster until the shoes can't be heard dragging on the drum as the drum is rotated.

7   Mount the wheel, install the lug nuts, then lower the vehicle.

8   Make a number of forward and reverse stops and operate the parking brake to adjust the brakes until satisfactory pedal action is obtained.

9   Check the operation of the brakes carefully before driving the vehicle.

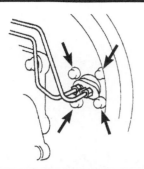

7.4a Disconnect the brake line fittings, then remove the wheel cylinder mounting bolts (arrows) - dual leading type brake

7 Wheel cylinder - removal, overhaul and installation

Note: *If an overhaul is indicated (usually because of fluid leaks or sticky operation), explore all options before beginning the job. New wheel cylinders are available, which makes this job quite easy. If it's decided to rebuild the wheel cylinder, make sure a rebuild kit is available before proceeding. Never overhaul or replace only one wheel cylinder - always rebuild or replace both of them at the same time.*

Removal

Refer to illustration 7.4a and 7.4b

1   Raise the front or rear of the vehicle and support it securely on jackstands. Block the wheels at the opposite end to keep the vehicle from rolling.
2   Remove the brake shoe assembly (see Section 6).
3   Remove all dirt and foreign material from around the wheel cylinder.
4   Disconnect the brake line(s) with a flare-nut wrench, if available **(see illustrations)**. Don't pull the brake line away from the wheel cylinder.
5   Remove the wheel cylinder mounting bolts.
6   Detach the wheel cylinder from the brake backing plate and place it on a clean workbench. Immediately plug the brake line(s) to prevent fluid loss and contamination.

Overhaul

Refer to illustrations 7.7a and 7.7b

7   Remove the bleeder screw, boots, pistons, cups and spring from the wheel cylinder body **(see illustrations)**.
8   Clean the wheel cylinder with brake fluid, denatured alcohol or brake system cleaner. **Warning:** *Do not, under any circumstances, use*

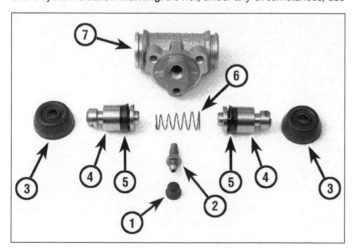

7.7b An exploded view of the wheel cylinder used with the leading/trailing type brake

1	Bleeder screw cap
2	Bleeder screw
3	Boot
4	Piston

5	Cup
6	Spring
7	Cylinder body

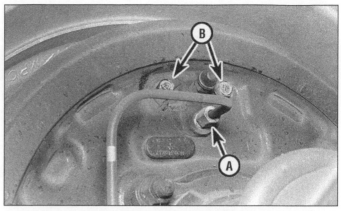

7.4b Disconnect the brake line fitting (A), then remove the wheel cylinder mounting bolts (B) - leading/trailing type brake

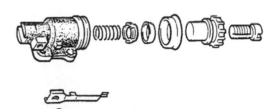

7.7a An exploded view of the wheel cylinders used with the dual leading type brake

petroleum-based solvents to clean brake parts!
9   Use compressed air to dry the wheel cylinder and blow out the passages.
10   Check the bore for corrosion and score marks. Crocus cloth can be used to remove light corrosion and stains, but the cylinder must be replaced with a new one if the defects cannot be removed easily, or if the bore is scored.
11   Lubricate the new cups with brake fluid.
12   Assemble the brake cylinder components **(see illustrations 7.7a and 7.7b)**. Make sure the cup lips face in.

Installation

Refer to illustration 7.13

13   Place the wheel cylinder(s) in position and install the bolts finger tight **(see illustration)**. Connect the brake line(s) to the cylinder, being careful not to cross-thread the fitting(s). Tighten the wheel cylinder bolts to the torque listed in this Chapter's Specifications.

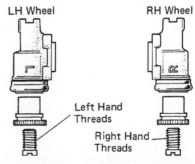

7.13 On vehicles equipped with the dual leading type brake, the wheel cylinders must be installed on the proper side of the vehicle, as indicated

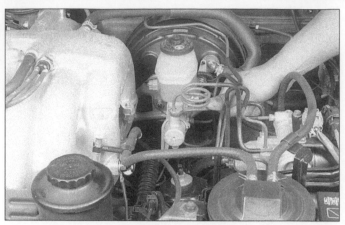

8.4 Unplug the electrical connector (if equipped) and completely loosen the brake line fittings at the master cylinder

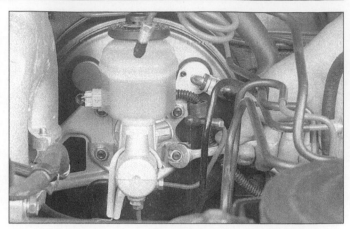

8.6 Remove the mounting nuts from the master cylinder

8.8a On master cylinders equipped with a small reservoir, remove the set bolt inside the reservoir . . .

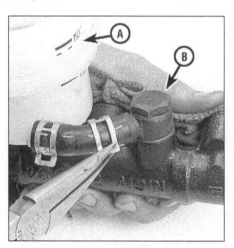

8.8b . . . then use pliers to release the hose clamp - separate the reservoir (A) and the reservoir union fitting (B) from the master cylinder body

8.8c On master cylinders equipped with a large reservoir, remove the set screw (arrow) and pull straight up to remove the reservoir from the master cylinder body

14   Tighten the brake line(s) securely and install the brake shoe assembly (see Section 6).

15   Bleed the brakes (see Section 10).

16   Check the operation of the brakes carefully before driving the vehicle.

8   Master cylinder - removal, overhaul and installation

Note: *Before deciding to overhaul the master cylinder, check on the availability and cost of a new or factory rebuilt unit and also the availability of a rebuild kit.*

Removal

Refer to illustrations 8.4 and 8.6

1   The master cylinder is located in the engine compartment, mounted on the power brake booster.

2   Remove as much fluid as possible from the reservoir with a syringe.

3   Place rags under the fittings and prepare caps or plastic bags to cover the ends of the lines once they're disconnected. **Caution:** *Brake fluid will damage paint. Cover all body parts and be careful not to spill fluid during this procedure.*

4   Loosen the fittings at the ends of the brake lines where they enter the master cylinder **(see illustration)**. To prevent rounding off the flats, use a flare-nut wrench, which wraps around the fitting hex.

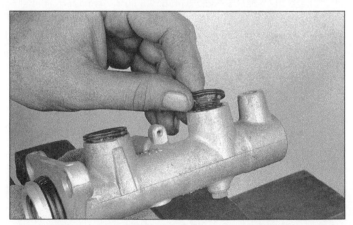

8.8d After the reservoir has been removed, pull the grommets from the master cylinder body

5   Pull the brake lines away from the master cylinder and plug the ends to prevent contamination.

6   Disconnect the electrical connector (if equipped) at the master cylinder, then remove the nuts attaching the master cylinder to the power booster **(see illustration)**. Pull the master cylinder off the studs to remove it. Again, be careful not to spill the fluid as this is done.

8.9 Using a Phillips screwdriver, depress the pistons and remove the stopper bolt. Be sure to replace the copper washer (gasket) on the stopper bolt when reassembling

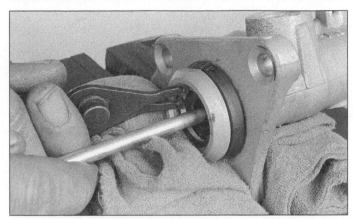

8.10 Depress the pistons again and remove the snap-ring with a pair of snap-ring pliers

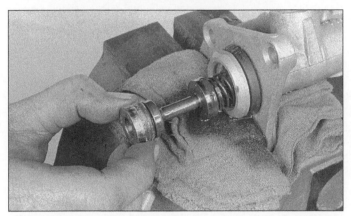

8.11a After the snap-ring has been removed, the primary (no. 1) piston assembly can be removed

8.11b Remove the cylinder from the vise and tap it against a block of wood until the secondary (no. 2) piston is exposed. Pull the piston assembly STRAIGHT OUT - if it becomes even slightly cocked, the bore may be damaged

Overhaul

Refer to illustrations 8.8a, 8.8b, 8.8c, 8.8d, 8.9, 8.10, 8.11a, 8.11b, 8.11c and 8.11d

7   Two different types of master cylinders where used during the production years covered by this manual. Earlier models used a small reservoir and later models were equipped with large reservoir. **Note**: *This procedure is illustrated with a large reservoir master cylinder. The procedures are the same for the small reservoir type except where noted. Before attempting the overhaul of the master cylinder, obtain the proper rebuild kit, which will contain the necessary replacement*

parts and also any instructions which may be specific to your model.
8   Remove the reservoir and inspect the reservoir grommet for indications of leakage near the base of the reservoir **(see illustrations)**.
9   Place the cylinder in a vise and use a punch or Phillips screwdriver to fully depress the pistons until they bottom against the other end of the master cylinder **(see illustration)**. Hold the pistons in this position and remove the stop bolt on the side of the master cylinder.
10   Depress the pistons again and remove the snap-ring at the end of the master cylinder **(see illustration)**.
11   The internal components can now be removed from the bore **(see illustrations)**. Make a note of the proper order of the components so

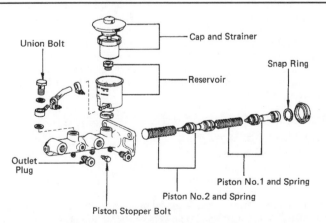

8.11c An exploded view of an early model (small reservoir type) master cylinder

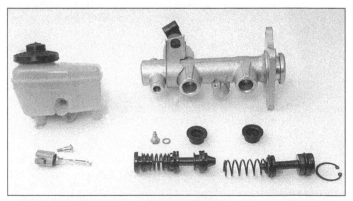

8.11d An exploded view of a late model (large reservoir type) master cylinder

8.27 Have an assistant depress the brake pedal and hold it down, then loosen the fitting nut, allowing the air and fluid to escape. Repeat this procedure on both fittings until the fluid is clear of air bubbles

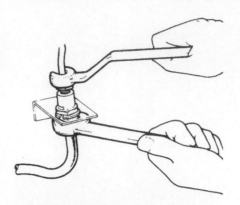

9.3 Hold the hose fitting with a wrench to prevent twisting the line, then loosen the tube nut with a flare-nut wrench to prevent rounding off the corners of the nut

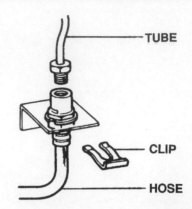

9.4 Once the tube nut has been completely loosened, remove the clip with a pair of pliers - some models may be equipped with an E-ring, as shown here, which can be removed with a screwdriver

they can be returned to their original locations. **Note:** *The two springs are different, so pay particular attention to their installed order.*

12   Carefully inspect the bore of the master cylinder. Any score marks or other damage will mean a new master cylinder is required. DO NOT attempt to hone the bore.

13   Replace all parts included in the rebuild kit, following any instructions in the kit. Clean all reused parts with new brake fluid, brake system cleaner or denatured alcohol. **Warning:** *Do not use any petroleum-based solvents. During reassembly, lubricate all parts liberally with clean brake fluid.*

14   Push the assembled components into the bore, bottoming them against the end of the master cylinder, then install the stop bolt.

15   Install the new snap-ring, making sure it's seated properly in the groove.

16   Install new reservoir grommets and the reservoir in the reverse order of removal.

17   Before installing the master cylinder, it should be bench bled. Since you'll have to apply pressure to the master cylinder piston and, at the same time, control flow from the brake line outlets, the master cylinder should be mounted in a vise, with the jaws of the vise clamping on the mounting flange.

18   Insert threaded plugs into the brake line outlet holes and snug them down so no air will leak past them, but not so tight that they can't be easily loosened.

19   Fill the reservoir with brake fluid of the recommended type (see Chapter 1).

20   Remove one plug and push the piston assembly into the bore to expel the air from the master cylinder. A large Phillips screwdriver can be used to push on the piston assembly.

21   To prevent air from being drawn back into the master cylinder, the plug must be replaced and snugged down before releasing the pressure on the piston.

22   Repeat the procedure until only brake fluid is expelled from the brake line outlet hole. When only brake fluid is expelled, repeat the procedure at the other outlet hole and plug. Be sure to keep the master cylinder reservoir filled with brake fluid to prevent the introduction of air into the system.

23   Since high pressure isn't involved in the bench bleeding procedure, an alternative to the removal and replacement of the plugs with each stroke of the piston assembly is available. Before pushing in on the piston assembly, remove the plug as described in Step 20. Before releasing the piston, however, instead of replacing the plug, simply put your finger tightly over the hole to keep air from being drawn back into the master cylinder. Wait several seconds for brake fluid to be drawn from the reservoir into the bore, then depress the piston again, removing your finger as brake fluid is expelled. Be sure to put your finger back over the hole each time before releasing the piston, and when the

bleeding procedure is complete for that outlet, replace the plug and tighten it before going on to the other port.

Installation

Refer to illustration 8.27

24   Install the master cylinder over the studs on the power brake booster and tighten the nuts only finger-tight at this time.

25   Thread the brake line fittings into the master cylinder. Since the master cylinder is still a bit loose, it can be moved slightly so the fittings thread in easily. Don't strip the threads as the fittings are tightened.

26   Tighten the mounting nuts and the brake line fittings.

27   Fill the master cylinder reservoir with fluid, then bleed the master cylinder and the brake system (see Section 10). To bleed the master cylinder on the vehicle, have an assistant depress the brake pedal and hold it down. Loosen the fitting to allow air and fluid to escape. Tighten the fitting, then allow your assistant to return the pedal to its rest position. Repeat this procedure on both fittings until the fluid is free of air bubbles **(see illustration)**. Check the operation of the brake system carefully before driving the vehicle.

---

9   Brake hoses and lines - inspection and replacement

Inspection

1   About every six months, with the vehicle raised and supported securely on jackstands, the rubber hoses which connect the steel brake lines with the front and rear brake assemblies should be inspected for cracks, chafing of the outer cover, leaks, blisters and other damage. These are important and vulnerable parts of the brake system and inspection should be complete. A light and mirror will be helpful for a thorough check. If a hose exhibits any of the above conditions, replace it with a new one.

Replacement

Front brake hose

Refer to illustrations 9.3 and 9.4

2   Loosen the wheel lug nuts, raise the vehicle and support it securely on jackstands. Remove the wheel.

3   At the frame bracket, hold the hose fitting with an open-end wrench and unscrew the brake line fitting from the hose **(see illustration)**. Use a flare-nut wrench to prevent rounding off the corners.

4   Remove the U-clip and E-ring, if equipped, from the female fitting at the bracket with a pair of pliers, then pass the hose through the bracket **(see illustration)**.

10.8 When bleeding the brakes, a hose is connected to the bleeder valve at the caliper or wheel cylinder and then submerged in brake fluid. Air will be seen as bubbles in the tube and container. All air must be expelled before moving to the next wheel

5   At the caliper end of the hose, remove the banjo fitting bolt, then separate the hose from the caliper. Note that there are two copper sealing washers on either side of the fitting - they should be replaced with new ones during installation.
6   To install the hose, connect the fitting to the caliper with the banjo bolt and copper washers. Make sure the locating lug on the fitting is engaged with the hole in the caliper, then tighten the bolt to the torque listed in this Chapter's Specifications.
7   Route the hose into the frame bracket, making sure it isn't twisted, then connect the brake line fitting, starting the threads by hand. Install the clip (and E-ring, if equipped), then tighten the fitting securely.
8   Bleed the caliper (see Section 10).
9   Install the wheel and lug nuts, lower the vehicle and tighten the lug nuts to the torque specified in Chapter 1.

Rear brake hose

10   Perform Steps 2, 3 and 4 above, then repeat Steps 3 and 4 at the other end of the hose. Be sure to bleed the wheel cylinder (or caliper) (see Section 10).

Metal brake lines

11   When replacing brake lines, be sure to use the correct parts. Don't use copper tubing for any brake system components. Purchase steel brake lines from a dealer or auto parts store.
12   Prefabricated brake line, with the tube ends already flared and fittings installed, is available at auto parts stores and dealer parts departments. These lines will probably have to be bent to the proper shape. If so, use the old line as a pattern and make the bends with a tubing bender.
13   When installing the new line, make sure it's securely supported in the brackets and has plenty of clearance between moving or hot components.
14   After installation, check the master cylinder fluid level and add fluid as necessary. Bleed the brake system (see Section 10) and test the brakes carefully before driving the vehicle in traffic.

10   Brake hydraulic system - bleeding

Refer to illustrations 10.8 and 10.14
Warning: *Wear eye protection when bleeding the brake system. If the fluid comes in contact with your eyes, immediately rinse them with water and seek medical attention.*
Note: *Bleeding the hydraulic system is necessary to remove any air*

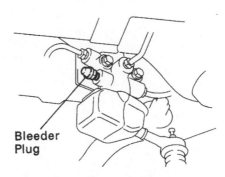

10.14 The load sensing proportioning valve is located above the rear axle assembly

that manages to find its way into the system when it's been opened during removal and installation of a hose, line, caliper or master cylinder.
1   You'll probably have to bleed the system at all four brakes if air has entered it due to low fluid level, or if the brake lines have been disconnected at the master cylinder.
2   If a brake line was disconnected only at a wheel, then only that caliper or wheel cylinder must be bled.
3   If a brake line is disconnected at a fitting located between the master cylinder and any of the brakes, that part of the system served by the disconnected line must be bled.
4   Remove any residual vacuum from the brake power booster by applying the brake several times with the engine off.
5   Remove the master cylinder reservoir cover and fill the reservoir with brake fluid. Reinstall the cover. **Note:** *Check the fluid level often during the bleeding operation and add fluid as necessary to prevent the fluid level from falling low enough to allow air bubbles into the master cylinder.*
6   Have an assistant on hand, as well as a supply of new brake fluid, a clear plastic container partially filled with clean brake fluid, a length of plastic, rubber or vinyl tubing to fit over the bleeder valve and a wrench to open and close the bleeder valve.
7   Beginning at the right rear wheel, loosen the bleeder valve slightly, then tighten it to a point where it's snug but can still be loosened quickly and easily.
8   Place one end of the tubing over the bleeder valve and submerge the other end in brake fluid in the container **(see illustration)**.
9   Have the assistant pump the brakes slowly a few times to get pressure in the system, then hold the pedal down firmly.
10   While the pedal is held down, open the bleeder valve just enough to allow a flow of fluid to leave the valve. Watch for air bubbles to exit the submerged end of the tube. When the fluid flow slows after a couple of seconds, close the valve and have your assistant release the pedal.
11   Repeat Steps 9 and 10 until no more air is seen leaving the tube, then tighten the bleeder valve and proceed to the left rear wheel, the right front wheel and the left front wheel, in that order, and perform the same procedure. Be sure to check the fluid in the master cylinder reservoir frequently.
12   Never use old brake fluid. It contains moisture which can boil, rendering the brakes useless.
13   Refill the master cylinder with fluid at the end of the operation.
14   Bleed the load sensing proportioning and by-pass valve. Using the same procedure as described above, insert the tube and clear plastic container onto the bleeder valve **(see illustration)** of the load sensing proportioning valve and bleed the system until there are no signs of air bubbles. For additional information on the load sensing proportioning and by-pass valve, refer to Section 16.
15   Check the operation of the brakes. The pedal should feel solid when depressed, with no sponginess. If necessary, repeat the entire process. **Warning:** *Do not operate the vehicle if you're in doubt about the effectiveness of the brake system.*

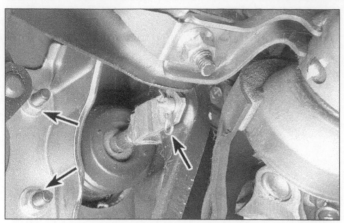

11.7 Remove this retaining clip (arrow), pull out the clevis pin and detach the pushrod from the brake pedal; the two left power brake booster mounting nuts (arrows) are visible in this photo (the two right nuts, not visible in this photo, are to the right of the booster mounting bracket)

11   Power brake booster - check, removal and installation

Operating check

1   Depress the brake pedal several times with the engine off and make sure there's no change in the pedal reserve distance.
2   Depress the pedal and start the engine. If the pedal goes down slightly, operation is normal.

Airtightness check

3   Start the engine and turn it off after one or two minutes. Depress the brake pedal slowly several times. If the pedal depresses less each time, the booster is airtight.
4   Depress the brake pedal while the engine is running, then stop the engine with the pedal depressed. If there's no change in the pedal reserve travel after holding the pedal for 30 seconds, the booster is airtight.

Removal

Refer to illustration 11.7

5   Power brake booster units shouldn't be disassembled. They require special tools not normally found in most automotive repair stations or shops. They're fairly complex and, because of their critical relationship to brake performance, should be replaced with a new or rebuilt one.
6   To remove the booster, first remove the brake master cylinder (see Section 8).
7   Remove the driver side knee bolster (see Chapter 11). Locate the

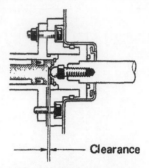

11.14a There should be no clearance between the booster pushrod and the master cylinder pushrod, but no interference either - if there is interference between the two, the brakes may drag; if there is clearance, there will be excessive brake pedal travel

11.14b Measure the distance that the pushod protrudes from the master cylinder mounting surface (including the gasket, if equipped) of the power brake booster

pushrod clevis connecting the booster to the brake pedal **(see illustration)**. It's accessible from inside the vehicle, under the dashboard.
8   Remove the clevis pin retaining clip with pliers.
9   Holding the clevis with pliers, loosen the locknut with a wrench. **Note:** *It isn't necessary to remove the clevis, unless the new booster unit isn't equipped with one.* Remove the clevis pin.
10   Disconnect the hose from the brake booster. Be careful not to damage the hose when removing it from the booster fitting.
11   Remove the four nuts and washers holding the brake booster to the firewall (you may need a light to see them).
12   Slide the booster straight out from the firewall until the studs clear the holes.

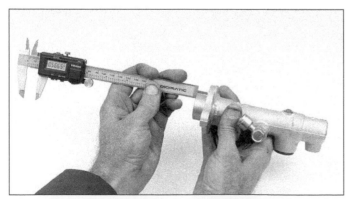

11.14c Measure the distance from the mounting flange to the end of the master cylinder

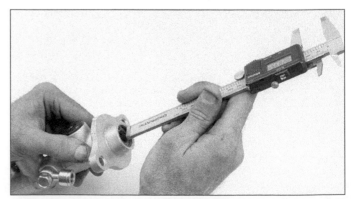

11.14d Measure the distance from the piston pocket to the end of the master cylinder

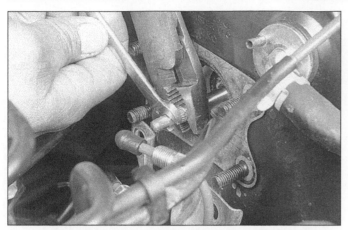

11.15 To adjust the length of the booster pushrod, hold the serrated portion of the rod with a pair of pliers and turn the adjusting screw in or out, as necessary, to achieve the desired setting

12.3 Unhook the parking brake shoe return springs - a special brake spring tool makes this much easier and is available at most auto parts stores

Installation

Refer to illustrations 11.14a, 11.14b, 11.14c, 11.14d and 11.15

13 Installation procedures are basically the reverse of removal. Tighten the clevis locknut securely and the booster mounting nuts to the torque listed in this Chapter's Specifications.

14 If a new power brake booster unit is being installed, check the pushrod clearance **(see illustration)** as follows:

  a) *Measure the distance that the pushrod protrudes from the master cylinder mounting surface on the front of the power brake booster. Jot down this measurement **(see illustration)**. This will be called "dimension A."*

  b) *Measure the distance from the mounting flange to the end of the master cylinder **(see illustration)**. Jot down this measurement. This will be called "dimension B."*

  c) *Measure the distance from the end of the master cylinder to the bottom of the pocket in the piston **(see illustration)**. Jot down this measurement. This will be called "dimension C."*

  d) *Subtract measurement B from measurement C, then subtract measurement A from the difference between B and C. This is the actual pushrod clearance.*

  e) *Compare your calculation to the pushrod clearance listed in this Chapter's Specifications. If necessary, adjust the pushrod length to achieve the correct clearance.*

15 To adjust the pushrod length, hold the knurled part of the pushrod with a pair of pliers and turn the pushrod end with a wrench **(see illustration)**. Recheck the clearance. Repeat this step as often as neces-

sary until the clearance is correct.

16 After the final installation of the master cylinder and brake hoses and lines, the brake pedal height and freeplay must be adjusted and the system must be bled (see Chapter 1).

12 Parking brake shoes (rear disc brakes only) - inspection and replacement

Refer to illustrations 12.3, 12.4, 12.5, 12.7, 12.8a, 12.8b, 12.10, 12.13 and 12.15

Warning: *Dust created by the brake system may contain asbestos, which is hazardous to your health. Never blow it out with compressed air and don't inhale any of it. An approved filtering mask should be worn when working on the brakes. Do not, under any circumstances, use petroleum-based solvents to clean brake parts. Use brake cleaner only!*

1 Remove the brake disc (see Section 5). Clean the parking brake assembly with brake system cleaner.

2 Inspect the thickness of the lining material on the shoes. If the lining has worn down to 0.039 inch (1.0 mm) or less, the shoes must be replaced.

3 Remove the parking brake shoe return springs from the anchor pin **(see illustration)**.

4 Remove the shoe strut from between the shoes **(see illustration)**.

5 Remove the rear shoe hold-down spring, then remove the shoe and adjuster **(see illustration)**.

12.4 Remove the shoe strut and spring assembly

12.5 Detach the rear hold-down spring, then remove the rear shoe

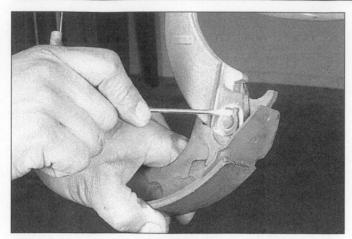

12.7 Use a screwdriver to pry the C-clip apart - always use a new C-clip when reinstalling the parking brake lever on the new shoes

12.8a Apply a light coat of high-temperature grease to the parking brake shoe contact areas on the backing plate

12.8b Clean the adjuster screw and apply high-temperature grease to the indicated areas (arrows)

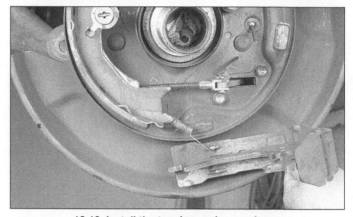

12.10 Install the tension spring as shown

6   Remove the front shoe hold-down spring, disconnect the parking brake cable from the lever and remove the shoe.

7   Spread the C-clip on the parking brake lever pivot pin with a screwdriver, then remove the lever, shim and pin. Transfer the parts to the new front shoe and crimp the C-clip to the pin using a pair of pliers **(see illustration)**.

8   Apply a thin coat of high-temperature grease to the shoe contact surfaces of the backing plate and to the threads and sliding portion of the adjuster **(see illustrations)**.

9   Connect the parking brake cable to the lever and mount the front shoe to the backing plate. Install the hold-down spring.

10   Connect the tension spring to the lower ends of both shoes and install the adjuster **(see illustration)**.

11   Position the rear shoe on the plate and install the hold-down spring. Install the parking brake cable return spring to the rear shoe.

12   Install the parking brake strut, with the spring facing rearward,

between the two shoes.

13   Install the shoe return springs **(see illustration)**.

14   Install the brake disc. Temporarily thread three of the wheel lug nuts onto the studs to hold the disc in place.

15   Remove the hole plug from the brake disc. Adjust the parking brake shoe clearance by turning the adjuster star wheel with a brake adjusting tool or screwdriver until the shoes contact the disc and the disc can't be turned **(see illustration)**. Back-off the adjuster eight notches, then install the hole plug.

16   Install the torque plate, brake pads and brake caliper. Be sure to

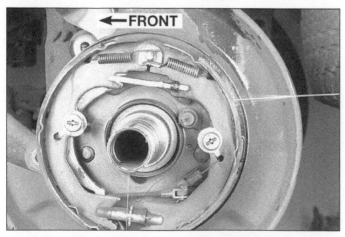

12.13 Rear disc parking brake assembly details (left side shown)

12.15 Use a brake adjusting tool or a screwdriver to turn the adjuster star wheel until the disc will not turn, then back off the adjuster eight notches and install the plug (disc removed for clarity)

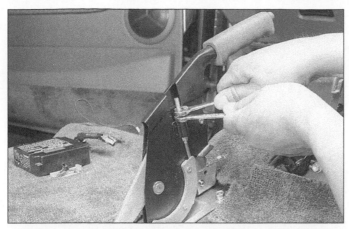

13.3 Loosen the locknut, then turn the adjusting nut until the desired handle travel is obtained

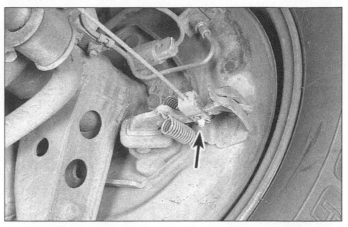

14.5 Remove the clevis pin and clip from the bellcrank

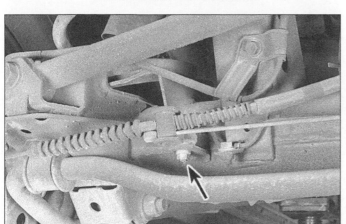

14.6 Detach the cable bracket (arrow) from the equalizer

15.1 The brake light switch is located on the chassis above the brake pedal

tighten the bolts to the torque listed in this Chapter's Specifications.
17   Install the wheel and tighten the lug nuts to the torque specified in Chapter 1.
18   Pull up on the parking brake handle and count the number of clicks that it travels. It should be between four and five clicks - if it's not, adjust the parking brake as described in the next Section.
19   To bed the shoes to the drum, drive the vehicle at approximately 30 mph on a dry, level road. Push in on the parking brake release button and pull up slightly on the lever with about 20 pounds of force. Drive the vehicle for 1/4-mile with the parking brake applied like this.
20   Repeat this procedure two or three times, allowing the brakes to cool between applications.

13 Parking brake - adjustment

Refer to illustration 13.3

1   The parking brake lever, when properly adjusted, should travel four to five clicks when a moderate pulling force is applied. If it travels less than four clicks, there's a chance the parking brake might not be releasing completely and might be dragging on the drum or disc. If the lever can be pulled up more than five clicks, the parking brake may not hold adequately on an incline, allowing the car to roll.
2   To gain access to the parking brake cable adjuster, remove the center console.
3   Loosen the locknut (the upper nut) while holding the adjusting nut (lower nut) with a wrench **(see illustration)**. Tighten the adjusting nut until the desired travel is attained. Tighten the locknut.
4   Recheck the parking brake travel.

14 Parking brake cables - replacement

Refer to illustrations 14.5 and 14.6

1   Remove the center console to gain access to the parking brake cable adjuster.
2   With the lever in the down (off) position, remove the locknut and the adjusting nut and detach the cable from the lever.
3   Working under the vehicle, pull the cable to the rear. Pull the cable through the hole in the floorpan.
4   Unbolt the cable brackets from the frame.
5   Remove the clevis pin and clip from the bellcrank **(see illustration)**.
6   Detach the bolt from the clamp located on the equalizer **(see illustration)** and remove the cable from the vehicle.
7   Installation is the reverse of removal. Apply a light coat of grease to the portion of the cable end that contacts the equalizer. Adjust the parking brake lever as outlined previously (see Section 13).

15 Brake light switch - removal, installation and adjustment

Refer to illustration 15.1

Removal and installation

1   The brake light switch is located on a bracket at the top of the brake pedal **(see illustration)**. The switch activates the brake lights at

the rear of the vehicle when the pedal is depressed.
2   Disconnect the negative battery cable from the battery.
3   Disconnect the electrical connector from the brake light switch.
4   Loosen the locknut and detach the switch from the pedal bracket.
5   Installation is the reverse of removal.

Adjustment

6   Loosen the locknut, adjust the switch until the clearance is as specified, then tighten the locknut.
7   Connect the wires at the switch and the battery. Make sure the brake lights are functioning properly.

16   Load sensing proportioning and by-pass valve - general information

Refer to illustration 16.1
1   The load sensing and proportioning by-pass valve is a device incorporated into the brake system that senses the amount of weight (passengers, equipment, etc.) the vehicle is holding and adjusts the brake fluid pressure to compensate for this additional load. This system is located above the rear axle assembly and it consists of a valve, spring shaft and adjusting bolt **(see illustration).**
2   When bleeding the brake system, it is necessary to bleed the load sensing proportioning and by-pass valve last, after all the wheel cylinders and calipers are bled (see Section 10).

16.1 Load sensing proportioning and by-pass valve

3   In the event of poor braking performance, have the system adjusted by a dealer service department.
4   If it becomes necessary to remove the assembly for replacement of another component, be sure to mark the position of the adjusting nut. It must be installed in exactly the same position as before removal.

Chapter 10
Suspension and steering systems

Contents

	Section		Section
Coil spring (FJ80 series) - removal and installation	5	Steering gear - removal and installation	11
Front end alignment - general information	15	Steering knuckle - removal and installation	8
General information	1	Steering linkage - inspection, removal and installation	10
Leaf springs and bushings (FJ60 and FJ62 series) -		Steering wheel - removal and installation	9
removal and installation	4	Suspension and steering check	See Chapter 1
Power steering pump - removal and installation	12	Suspension arms (FJ80 series) - removal and installation	7
Power steering system - bleeding	13	Tire and tire pressure checks	See Chapter 1
Shock absorbers (front and rear) - removal and installation	3	Tire rotation	See Chapter 1
Stabilizer bar and bushings (front and rear) - removal		Track bar (FJ80 series) - removal and installation	6
and installation	2	Wheels and tires - general information	14

Specifications

Torque specifications

Ft-lbs (unless otherwise indicated)

Note: One foot-pound (ft-lb) of torque is equivalent to 12 inch-pounds (in-lbs) of torque. Torque values below approximately 15 ft-lbs are expressed in inch-pounds, because most foot-pound torque wrenches are not accurate at these smaller values.

Suspension

Leaf spring	
Hanger bolt	67
Shackle bolt	67
U-bolt-to-tie plate nuts	90
Shock absorber	
Front (FJ60 and FJ62 series)	
Upper nut	19
Lower bolt	47
Rear (FJ60 and FJ62 series)	
Upper bolt	47
Lower bolt	27
Front (FJ80 series)	
Upper nut	51
Lower nut	51
Rear (FJ80 series)	
Upper nut	51
Upper bracket bolts	37
Lower bolt	47
Stabilizer bar	
Stabilizer-to-end link bolt	19
Stabilizer link-to-axle housing	19
Stabilizer bushing bracket bolts	13
Steering knuckle	
Upper bolts	71
Lower nuts	71
Bearing preload	72 to 120 in-lbs
Track bar	
Track bar-to-frame bracket nut/bolt	130
Track bar-to-axle bracket nut/bolt	181

Torque specifications (continued)

Steering   Ft-lbs

Pitman arm-to-steering shaft nut	130
Steering damper nuts	54
Steering gear mounting bolts	90
Steering wheel-to-steering shaft hub nut	25
Tie-rod adjustment sleeve clamp bolt	33
Tie-rod end ballstud nut	33
Wheel lug nuts	80 to 110

---

1   General information

Front suspension

Refer to illustrations 1.1a and 1.1b

All of the vehicles covered by this manual utilize a solid front axle. The front axle consists of a one-piece steel housing and axle tube assembly (for more information on the innards of the front axle, see Chapter 8). FJ60 and FJ62 models use a pair of shock absorbers and a pair leaf springs to control vertical and lateral movement of the front axle. FJ80 models use two control arms, a pair of shock absorbers and coil springs to control vertical movement of the front axle while lateral movement is prevented by a track bar. A stabilizer bar controls body roll on all models. Each steering knuckle is positioned by a pair of tapered roller bearings to upper and lower ends of a yoke welded to the end of the axle **(see illustrations)**.

Rear suspension

Refer to illustrations 1.2a and 1.2b

The rear axle also consists of a one piece steel housing and axle tube assembly (for more information on the innards of the rear axle, see Chapter 8). FJ60 and FJ62 models use a pair of shock absorbers and a pair leaf springs to control vertical and lateral movement of the rear axle. FJ80 models use four control arms, a pair of shock absorbers and coil springs to control vertical movement of the rear

axle while lateral movement is prevented by a track bar. A stabilizer bar controls body roll on all models. **(see illustrations)**.

Steering

The power steering system uses a recirculating-ball type steering gearbox which transmits turning force through the steering linkage (Pitman arm, center link and tie-rods) to the steering knuckle arms. A steering damper between the axle and the center link reduces unwanted "bump steer" (the slight turning or steering of a wheel away from its normal direction of travel as it moves through its suspension travel). A small U-joint connects the steering column to the steering gearbox. The steering column is designed to collapse in the event of an accident.

Frequently, when working on the suspension or steering system components, you may come across fasteners which seem impossible to loosen. These fasteners on the underside of the vehicle are continually subjected to water, road grime, mud, etc., and can become rusted or "frozen," making them extremely difficult to remove. In order to unscrew these stubborn fasteners without damaging them (or other components), be sure to use lots of penetrating oil and allow it to soak in for a while. Using a wire brush to clean exposed threads will also ease removal of the nut or bolt and prevent damage to the threads. Sometimes a sharp blow with a hammer and punch is effective in breaking the bond between a nut and bolt threads, but care must be taken to prevent the punch from slipping off the fastener and ruining the threads. Heating the stuck fastener and surrounding area with a

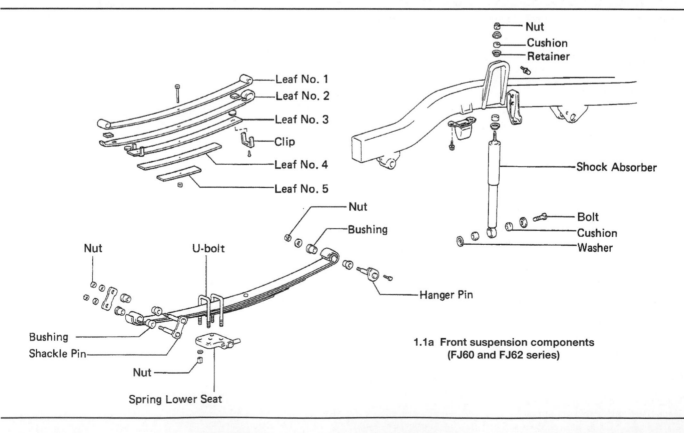

**1.1a Front suspension components
(FJ60 and FJ62 series)**

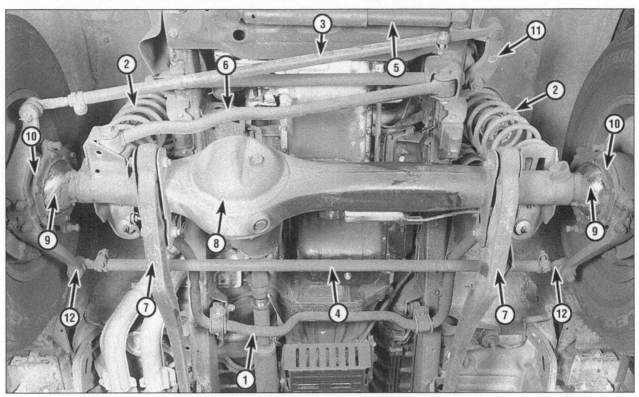

1.1b Front suspension and steering components (FJ80 series)

1	Stabilizer bar	5	Steering damper	9	Axle tube yokes
2	Coil spring	6	Track bar	10	Steering arm
3	Drag link	7	Suspension arm	11	Pitman arm
4	Tie-rod	8	Differential and axle housing	12	Tie-rod end

1.2a Rear suspension components (FJ80 series)

1	Shock absorber	3	Track bar	5	Differential	8	Lower suspension arm
2	Right coil spring (left coil spring not visible)	4	Differential and axle housing	6	Stabilizer bar	9	Upper suspension arm
				7	Stabilizer bar end link		

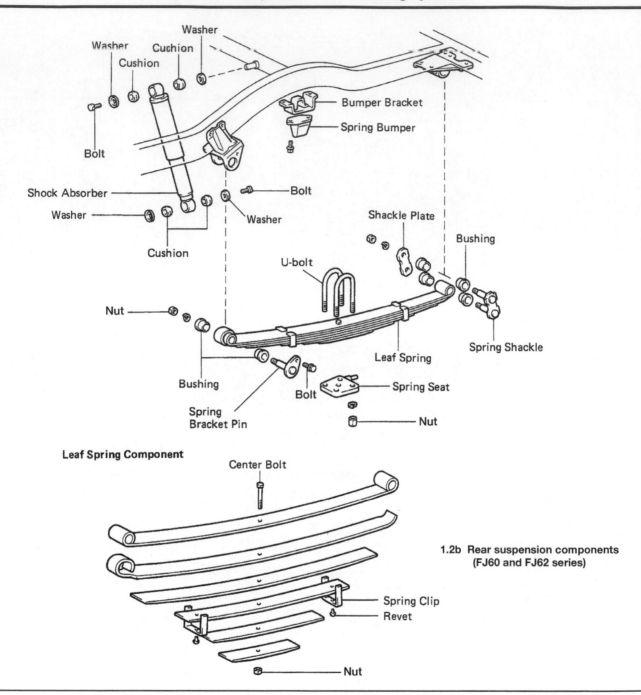

Washer
Washer
Cushion
Cushion
Bolt
Shock Absorber
Washer
Cushion
Nut
Bushing
Spring
Bracket Pin
Bolt
Washer
Bumper Bracket
Spring Bumper
Bolt
Shackle Plate
Bushing
U-bolt
Leaf Spring
Spring Shackle
Spring Seat
Nut

Leaf Spring Component

Center Bolt

1.2b Rear suspension components (FJ60 and FJ62 series)

Spring Clip
Revet
Nut

torch sometimes helps too, but isn't recommended because of the obvious dangers associated with fire. Long breaker bars and extension, or "cheater," pipes will increase leverage, but never use an extension pipe on a ratchet - the ratcheting mechanism could be damaged. Sometimes, turning the nut or bolt in the tightening (clockwise) direction first will help to break it loose. Fasteners that require drastic measures to unscrew should always be replaced with new ones.

Since most of the procedures that are dealt with in this Chapter involve jacking up the vehicle and working underneath it, a good pair of jackstands will be needed. A hydraulic floor jack is the preferred type of jack to lift the vehicle, and it can also be used to support certain components during various operation. **Warning:** *Never, under any circumstances, rely on a jack to support the vehicle while working on it. Whenever any of the suspension or steering fasteners are loosened or removed they must be inspected and, if necessary, be replaced with new ones of the same part number or of original equipment quality and design. Torque specifications must be followed for proper reassembly*

and component retention. Never attempt to heat or straighten any suspension or steering components. Instead, replace any bent or damaged part with a new one.

2   Stabilizer bar and bushings (front and rear) - removal and installation

Removal

Refer to illustrations 2.2a, 2.2b, 2.2c and 2.3

1   Apply the parking brake. Raise the vehicle and support it securely on jackstands.

2   Remove the stabilizer bar end links **(see illustrations)**. Note the order in which the upper and lower retainers and grommets are installed on the links on FJ60 and FJ62 models. If it is necessary to remove the links, simply unbolt them from the axle brackets.

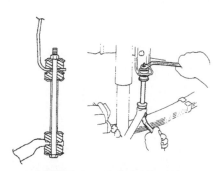

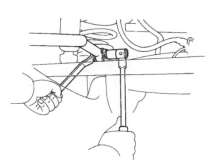

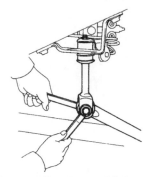

2.2a Front stabilizer bar end link mounting details (FJ60 and FJ62 series)

2.2b Front stabilizer bar end link mounting details (FJ80 series)

2.2c Rear stabilizer bar end link mounting details (FJ80 series)

2.3 To disconnect the stabilizer bar from the frame, remove the bushing clamp bolts (arrows); inspect the rubber bushings and replace them if they're hard, cracked or otherwise deformed

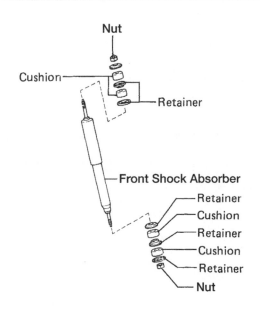

3   Remove the retaining bolts from the stabilizer bar bushing clamps **(see illustration)** and detach the bar from the vehicle.

4   Pull the bushings off the stabilizer bar and inspect them for cracks, hardness and other signs of deterioration. If the bushings are damaged, replace them. Inspect the bushings in the lower ends of the links, replacing them if necessary.

Installation

5   Position the stabilizer bar bushings on the bar.

6   Push the brackets over the bushings and raise the bar up to the frame. Install the bracket bolts but don't tighten them completely at this time.

7   Install the stabilizer bar end links in the reverse order of removal. Tighten the nuts to the torque listed in this Chapter's Specifications.

8   Tighten the bracket bolts to the torque listed in this Chapter's Specifications.

---

3   Shock absorbers (front and rear) - removal and installation

---

Removal

Refer to illustration 3.2

1   Loosen the wheel lug nuts, raise the vehicle and support it securely on jackstands. Apply the parking brake and remove the wheel. Support the axle with a floor jack positioned nearest the shock absorber to be replaced.

2   Remove the upper shock absorber nut(s) or bolts **(see illustration)**. On stud-mounted shock absorbers, use an open end wrench to keep the stem from turning. If the nut won't loosen because of rust, squirt some penetrating oil on the stud threads and allow it to soak in

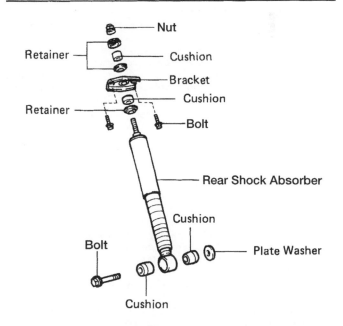

3.2 Front and rear shock mounting details (FJ80 series)

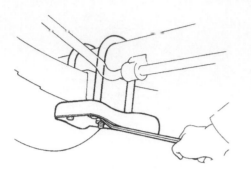

4.5 Remove the nuts, U-bolts and the leaf spring tie plate

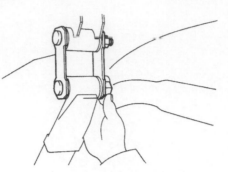

4.6 Remove the spring eye-to-shackle bolt

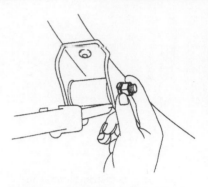

4.7 Remove the spring eye-to-frame bracket bolt

for awhile. It may be necessary to keep the stud from turning with a pair of locking pliers, since the flats provided for a wrench are quite small. **Note:** *On FJ60 and FJ62 models refer to illustrations 1.1a and 1.2b for the shock mounting details.*

3   Remove the lower shock mounting fastener(s) and pull the shock absorber out from the wheel well.

Installation

4   Follow any specific mounting instructions supplied with the new shock absorbers. If you're replacing the old shocks with gas-filled shock absorbers, do not cut the retaining strap on the new shock until the top of the shock has been firmly mounted.

5   Move the shock absorber back-and-forth to ensure the grommets are centered in the mount. Tighten the stem nut to the torque listed in this Chapter's Specifications, then guide the lower end into position while releasing the retaining strap.

6   Install the lower mounting fasteners and tighten them to the torque listed in this Chapter's Specifications.

7   Install the wheel and lug nuts. Lower the vehicle and tighten the lug nuts to the torque listed in the Chapter 1 Specifications.

---

4   Leaf springs and bushings (FJ60 and FJ62 series) - removal and installation

---

Bushing check

1   All models are equipped with a two-piece bushing that is pressed into each spring eye. The bushings should be inspected for cracks, damage and looseness indicating excessive wear. To check for wear, jack up the frame until the weight is removed from the spring bushing. Pry the spring eye up-and-down to check for movement. If there is considerable movement, the bushing is worn and should be replaced.

Spring removal

Refer to illustrations 4.5, 4.6 and 4.7

Note: *This procedure applies to the front and rear leaf spring assemblies. We recommend that you do one side at a time, to keep the axle under control and so you'll have one side to use as a "guide" to reassembly.*

2   Loosen the front or rear wheel lug nuts, raise the front or rear of the vehicle and support it securely on jackstands. Remove the wheel.

3   Position a floor jack under the axle and raise the jack just enough to support the axle.

4   Remove the lower shock mounting bolts (see Section 3) and move the shock absorber aside.

5   Unscrew the U-bolt nuts **(see illustration),** then remove the spring tie plate and the U-bolts from the axle.

6   Remove the spring-to-shackle bolt **(see illustration).**

7   Remove the spring eye-to-frame bracket bolt **(see illustration)** and remove the spring from the vehicle.

Bushing replacement

8   The leaf spring bushings have a two-piece design and can be removed by inserting a screwdriver into the inner diameter of the bushing half and pushing outward. If the bushings won't loosen because of dirt and rust, squirt some penetrating oil between the spring eye and bushing and allow it to soak in for awhile.

9   Coat the inside of the spring eye with a thin layer of chassis grease and press the new bushing into position.

Spring installation

10   Installation is the reverse of the removal procedure. Be sure to tighten the spring mounting bolts and the U-bolt nuts to the torque listed in this Chapter's Specifications. **Note:** *The vehicle must be sitting at normal ride height before tightening the front and rear mounting bolts.*

---

5   Coil spring (FJ80 series) - removal and installation

Front

Refer to illustrations 5.5 and 5.8

Warning: *This procedure requires a special coil spring compressor tool available at most auto part stores. Attempts to perform this procedure without the spring compressor tool may result in personal injury.*

1   Loosen the front wheel lug nuts, raise the front of the vehicle and support it securely on jackstands. Remove the wheels

2   Support the axle assembly with either a floor jack under the differential or, preferably, two jacks, one at each end of the axle (the latter option provides better balance).

3   Unbolt the stabilizer bar end links and the shock absorbers at the

5.5 Compress the coil spring evenly until it can be removed

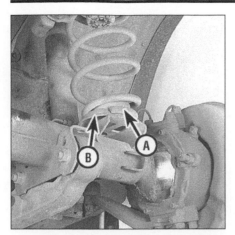

5.8 When installing the front coil springs, align the lower spring end (A) with the stop (B) in the lower spring seat

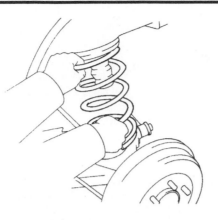

5.14 With the rear axle housing lowered, pull outward on the lower end of the coil spring to remove it from the vehicle

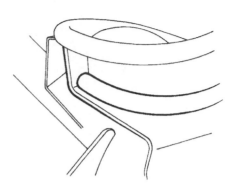

5.16 When installing the rear coil springs, align the lower spring end with the stop in the lower spring seat

front axle housing (see Sections 2 and 3).

4   Slowly lower the axle assembly until the coil springs are almost fully extended. If you are using only one jack, have an assistant support the right side of the axle as it's lowered.

5   Install the spring compressor tool and compress the coil spring enough to allow it to be removed **(see illustration)**.

6   Remove the coil spring from the vehicle. **Warning:** *Keep the ends of the spring pointing away from your body.*

7   Check the spring for deep nicks and corrosion, which will cause premature failure of the spring. Replace the spring if these or any other questionable conditions are evident.

8   Reinstall the spring compressor tool and compress the new coil spring. Guide the upper end of the coil spring into place and align the lower spring end with the stop in the lower spring seat **(see illustration)**.

9   The remainder of the installation is the reverse of removal. Tighten all fasteners to the torque listed in this Chapter's Specifications

Rear

Refer to illustrations 5.14 and 5.16

10   Loosen the front wheel lug nuts, raise the rear of the vehicle and support it securely on jackstands. Remove the wheels.

11   Support the axle assembly with either a floor jack under the differential or, preferably, two jacks, one at each end of the axle (the latter option provides better balance).

12   Unbolt the stabilizer bar brackets and the shock absorbers at the rear axle housing (see Sections 2 and 3).

13   Remove the lower bolt from the rear track bar, then position the track bar up and out of the way (see Section 6).

14   Slowly lower the axle assembly until the coil springs are fully extended. Grasp the lower end of the spring and remove the spring from the vehicle **(see illustration)**. **Caution:** *Be careful not to damage the rear brake hose.*

15   Check the spring for deep nicks and corrosion, which will cause premature failure of the spring. Replace the spring if these or any other questionable conditions are evident.

16   Installation is the reverse of removal. Make sure the coil spring is properly seated **(see illustration)**. Tighten all fasteners to the torque listed in this Chapter's Specifications.

6   Track bar (FJ80 series) - removal and installation

Refer to illustrations 6.2 and 6.3

Note: *This procedure applies to front and rear track bars.*

1   Raise the front or rear of the vehicle and support it securely on jackstands.

2   Remove the retaining nut and bolt from the frame rail bracket **(see illustration)**.

3   Remove the retaining nut and bolt from the axle tube bracket **(see illustration)**.

4   Remove the track bar.

5   Installation is the reverse of removal. Be sure to tighten the fas-

6.2 Remove the retaining nut and bolt securing the upper end of the track bar to the frame rail bracket

6.3 Remove the retaining nut and bolt securing the lower end of the track bar to the axle housing

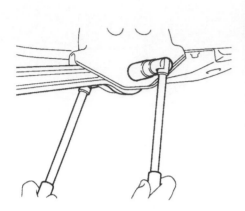

7.3 To remove the front suspension arm, unscrew the pivot nut and bolt from the frame rail bracket . . .

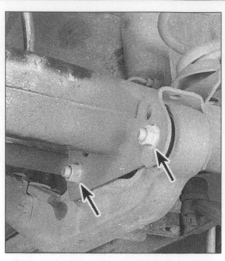

7.4 . . . and the nuts and bolts from the axle housing bracket

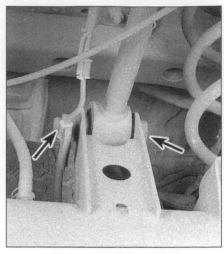

7.10a To remove the upper rear suspension arm, detach the nut and bolt (arrow) from the axle housing bracket . . .

teners to the torque listed in this Chapter's Specifications. **Note:** *The vehicle must be sitting at normal ride height before tightening the mounting fasteners.*

7 Suspension arms (FJ80 series) - removal and installation

Note: *Remove and install only one suspension arm at a time to avoid the possibility of the axle housing shifting out of position, which would make reassembly much more difficult. If it is absolutely necessary to remove more than one at a time, remove the coil springs (see Section 5) and support the axle with two floor jacks.*

Front

Refer to illustrations 7.3 and 7.4

1   Raise the front of the vehicle and support it securely on jack-stands.
2   Support the axle assembly with either a floor jack under the differential or, preferably, two jacks, one at each end of the axle (the latter option provides better balance).
3   Remove the retaining nut and bolt from the frame rail bracket **(see illustration)**.
4   Remove the retaining nuts and bolts from the axle housing

bracket **(see illustration)**.
5   Remove the front suspension arm. If it is difficult to remove the suspension arm, use a large pry bar to separate them from the frame and axle brackets.
6   Inspect the bushing in the suspension arm for cracking, hardness and general deterioration. If it is in need of replacement, reinstall the suspension arm and take the vehicle to a dealer service department or an automotive machine shop to have it replaced. Because of the special tools required to do this job, it can't be done at home.
7   Installation is the reverse of removal. Be sure to tighten the fasteners to the torque listed in this Chapter's Specifications. **Note:** *The vehicle must be sitting at normal ride height before tightening the mounting fasteners.*

Rear

Refer to illustrations 7.10a, 7.10b, 7.11a and 7.11b

8   Raise the rear of the vehicle and support it securely on jack-stands.
9   Support the axle assembly with either a floor jack under the differential or, preferably, two jacks, one at each end of the axle (the latter option provides better balance).
10   Remove the nuts and bolts securing the upper suspension arm to the axle housing bracket and to the frame bracket **(see illustrations)**. Remove the suspension arm from the vehicle.

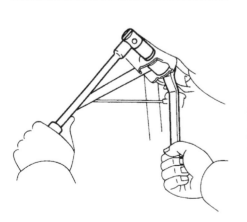

7.10b . . . and the frame bracket

7.11a To remove the lower rear suspension arm, detach the nut and bolt (arrows) from the axle housing bracket . . .

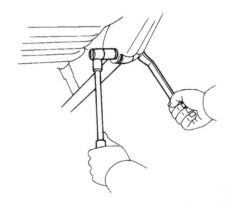

7.11b . . . and the frame bracket

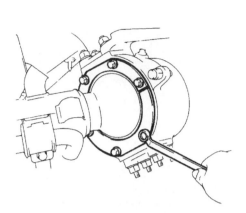

8.5 Remove the oil seal retainer bolts from the rear of the steering knuckle

8.6 Remove the nuts and cone washers (arrows) securing the steering knuckle arm

8.7a Remove the upper bearing cap retaining bolts . . .

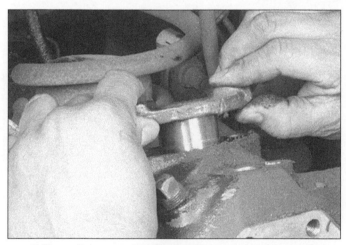

8.7b . . . then use a twisting motion to remove the upper bearing cap

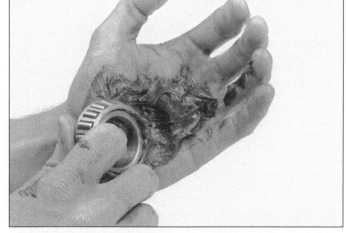

8.11 If a bearing packing tool is not available, work grease into the bearing rollers by pressing it against the palm of your hand

11   Remove the nuts and bolts securing the lower suspension arm to the axle housing bracket and to the frame bracket **(see illustrations)**. Remove the suspension arm from the vehicle.

12   Inspect the bushing in the suspension arm for cracking, hardness and general deterioration. If it is in need of replacement, reinstall the suspension arm and take the vehicle to a dealer service department or an automotive machine shop to have it replaced. Because of the special tools required to do this job, it can't be done at home.

13   Installation is the reverse of removal. Be sure to tighten the fasteners to the torque listed in this Chapter's Specifications. **Note:** *The vehicle must be sitting at normal ride height before tightening the mounting fasteners.*

8   Steering knuckle - removal and installation

Removal

Refer to illustrations 8.5, 8.6, 8.7a, 8.7b, 8.11, 8.13 and 8.17

1   Loosen the wheel lug nuts, raise the vehicle and support it securely on jackstands. Remove the wheel.

2   Disconnect the tie-rod from the steering knuckle (see Section 10).

3   Remove the brake components (see Chapter 9).

4   Remove the front axle axleshaft (see Chapter 8).

5   Unbolt the steering knuckle oil seal **(see illustration)**.

6   Remove the steering knuckle arm retaining nuts and cone washers **(see illustration)**. Grasp the lower steering arm assembly, then pull it downward to separate the arm from the steering knuckle.

7   Remove the upper bearing cap and shim (if equipped), then remove the steering knuckle **(see illustrations)**.

8   Remove the bearings and tag them (RH upper, RH lower or LH upper, LH lower) to avoid mixing them up.

9   Use solvent to remove all traces of the old grease from the bearings, axle tube yoke and the steering knuckle. A small brush may prove helpful; however make sure no bristles from the brush embed themselves inside the bearing rollers. Allow the parts to air dry.

10   Carefully inspect the bearings for cracks, heat discoloration, worn rollers, etc. Check the bearing races inside the axle tube yoke for wear and damage. If the bearing races are defective, drive the bearing race out of the axle tube yoke using a brass drift. Drive the new race into the tube yoke using the appropriate size bearing driver (inexpensive bearing driver sets are available at most automotive parts stores). Note that the bearings and races are replaced as matched sets; used bearings should never be installed on new races.

11   Use only high-temperature wheel bearing grease to pack the bearings. Inexpensive bearing packing tools are available at automotive parts stores, but not entirely necessary. If one is not available, pack the grease by hand completely into the bearings, forcing it between the rollers, cone and cage from the back side **(see illustration)**.

12   Apply a film of grease to each bearing race.

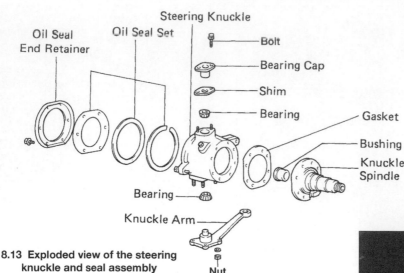

8.13 **Exploded view of the steering knuckle and seal assembly**

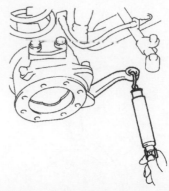

8.17 **Use a spring tension gauge to measure the steering knuckle bearing preload**

13   Install a new steering knuckle oil seal **(see illustration)**.

14   Install the upper bearing into the axle tube yoke, then hold the lower bearing in position on the tube yoke.

15   While holding the lower bearing in position, slide the steering knuckle over the axle tube yoke and install the upper bearing cap and shim (if equipped). Install the upper bearing cap bolts.

16   Install the steering knuckle arm, cone washers and retaining nuts. Tighten the upper and lower bearing cap retaining nuts and bolts to the torque listed in this Chapter's Specifications.

17   Measure the bearing preload with a spring tension gauge **(see illustration)**. If bearing preload is not as specified in this Chapter's Specifications purchase the correct size adjustment shim. Remove the upper bearing cap and reinstall the shim and check the preload again. **Note:** *Adding a shim will decrease (loosen) bearing preload while removing a shim will increase (tighten) bearing preload.*

18   The remainder of the installation is the reverse of removal

9   Steering wheel - removal and installation

Warning: *Some models are equipped with airbags. The airbag is armed and can deploy (inflate) anytime the battery is connected. To prevent accidental deployment and possible injury, disconnect the negative battery cable whenever working near airbag components. After the battery is disconnected, wait at least 90 seconds before*

beginning work. The system uses a capacitor as a back-up voltage source to the battery. This capacitor must fully discharge before the airbag module is actually disarmed. For more information, see Chapter 12.

Removal

Refer to illustrations 9.2a, 9.2b, 9.3a, 9.3b, 9.3c, 9.4, 9.5 and 9.6

1   Park the vehicle with the wheels pointing straight ahead. Disconnect the cable from the negative terminal of the battery. **Caution:** *If the*

9.2a **On airbag equipped models, pry open the small trim covers on each side of the steering wheel (left side shown , right cover identical) . . .**

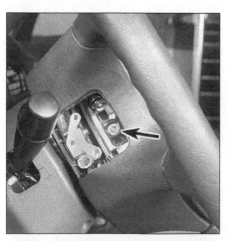

9.2b **. . . then remove the Torx screws (arrow) located behind each trim cover**

9.3a Remove the airbag module . . .

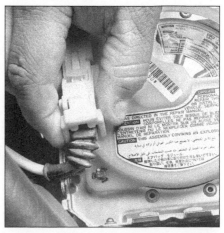

9.3b **. . . flip up the lock on the airbag module electrical connector . . .**

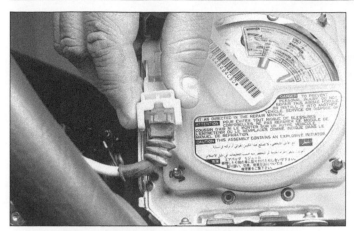

9.3c . . . and unplug the electrical connector

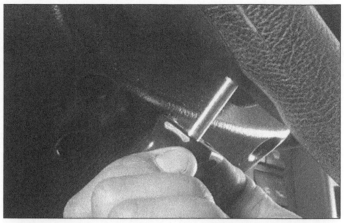

9.4 On models not equipped with airbags, remove the horn pad retaining screws located on the backside of the steering wheel

stereo in your vehicle is equipped with an anti-theft system, make sure you have the correct activation code before disconnecting the battery.

2   On airbag-equipped models, pry out the small covers on each side of the steering wheel and loosen the Torx head screws that attach the airbag module to the steering wheel **(see illustrations)**. Loosen each screw until the groove in the circumference of the screw catches on the screw case.

3   Detach the airbag module from the steering wheel and unplug the airbag module connector from the module **(see illustrations)**. Whenever handling the airbag module, always keep the airbag opening (the trim side) pointed away from your body. Never place the airbag module on a bench or other surface with the airbag opening (trim side) facing the surface. Always place the airbag module in a safe location with the airbag opening facing up.

4   On vehicles not equipped with airbags, remove the screws securing the horn pad **(see illustration)**.

5   Remove the steering wheel retaining nut, then mark the relationship of the steering shaft to the hub (if marks don't already exist or don't line up) to simplify installation and ensure steering wheel alignment **(see illustration)**. **Caution:** *Don't allow the steering shaft to turn while the steering wheel is removed.*

6   Use a puller to detach the steering wheel from the shaft **(see illustration)**. Don't hammer on the shaft to dislodge the steering wheel.

Installation

Refer to illustration 9.7

7   On airbag-equipped models, if the front wheels remained in the

9.5 Before removing the steering wheel, look for alignment marks between the steering wheel and the steering shaft; if there are none, use a sharp scribe or white paint to make your own marks

straight-ahead position throughout the procedure, the steering wheel can be installed now. If, however, the wheels were turned, the spiral cable for the airbag will have to be centered. To do this, place the front wheels in the straight-ahead position and turn the spiral cable hub counterclockwise by hand until it becomes harder to turn. Rotate the cable hub in the clockwise direction about three turns and align the two red pointers **(see illustration)**.

9.6 Remove the wheel from the shaft with a puller - DO NOT HAMMER ON THE SHAFT!

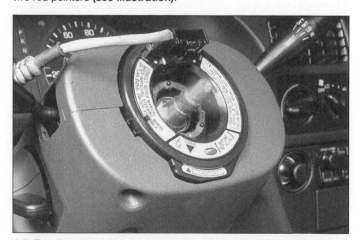

9.7 To align the spiral cable, turn the cable hub counterclockwise until it's harder turn, rotate it clockwise three turns and align the two red marks (the cable hub should be able to rotate about three turns in either direction when it's correctly centered)

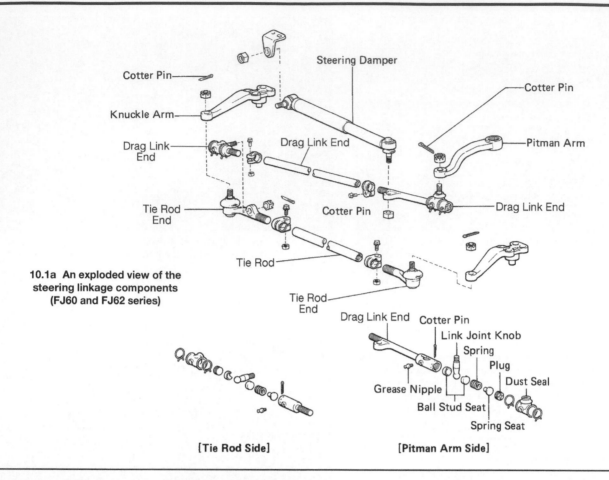

10.1a An exploded view of the steering linkage components (FJ60 and FJ62 series)

[Tie Rod Side]

[Pitman Arm Side]

8   To install the wheel, align the mark on the steering wheel hub with the mark on the shaft and slip the wheel onto the shaft. Install the nut and tighten it to the torque listed in this Chapter's Specifications.
9   Plug in the electrical connectors and install the airbag module (if equipped) or steering wheel pad.
10   Connect the negative battery cable.

10   Steering linkage - inspection, removal and installation

Inspection

Refer to illustrations 10.1a and 10.1b

1   The steering linkage (**see illustrations**) connects the steering gear to the front wheels and keeps the wheels in proper relation to each other. The linkage consists of the Pitman arm, fastened to the steering gear shaft, which moves the drag link back and forth. The back-and-forth motion of the drag link is transmitted to the steering knuckle. A tie-rod assembly connects the drag link to the steering knuckle. The tie-rod is made up of a tube, clamps and two tie-rod ends. A steering damper, connected between the drag link and the frame reduces shimmy and unwanted forces to the steering gear.
2   Set the wheels in the straight ahead position and lock the steering wheel.
3   Raise one side of the vehicle until the tire is approximately 1-inch off the ground.
4   Mount a dial indicator with the needle resting on the outside edge of the wheel. Grasp the front and rear of the tire and using light pressure, wiggle the wheel back-and-forth and note the dial indicator reading. The gauge reading should be less than 6.0 mm (0.236-inch). If the play in the steering system is more than specified, inspect each steering linkage pivot point and ballstud for looseness and replace parts if necessary.

5   Raise the vehicle and support it on jackstands. Check for torn ballstud boots, frozen joints and bent or damaged linkage components.

Removal and installation

Tie-rod ends

Refer to illustrations 10.7, 10.8 and 10.9
Note: *Tie-rod ends are located at each end of the tie-rod and drag link assemblies. This procedure applies to all four tie-rod ends.*
6   Loosen the wheel lug nuts, raise the vehicle and support it securely on jackstands. Apply the parking brake. Remove the wheel.
7   Remove the cotter pin and loosen, but do not remove, the castle nut from the ballstud (**see illustration**).
8   Using a small puller, separate the ballstud from the steering knuckle (**see illustration**). Remove the castle nut and pull the tie-rod end from the steering knuckle. **Caution:** *The use of a picklefork-type balljoint separator most likely will cause damage to the balljoint boot.*
9   If a tie-rod end must be replaced, count the number of threads showing and jot down this number to maintain correct toe-in during reassembly (**see illustration**). Loosen the adjuster tube clamp and unscrew the tie-rod end.
10   Lubricate the threaded portion of the tie-rod end with chassis grease. Screw the new tie-rod end into the adjuster tube and adjust the distance from the tube to the ballstud by threading the tie-rod into the adjuster tube until the same number of threads are showing as before (the number of threads showing on both sides of the adjuster tube should be within three threads of each other). Don't tighten the adjuster tube clamps yet.
11   To install the tie-rod end, insert the tie-rod end ballstud into the steering knuckle. Make sure the ballstud is fully seated. Install the nut and tighten it to the torque listed in this Chapter's Specifications. If a ballstud spins when attempting to tighten the nut, force it into the tapered hole with a large pair of pliers.

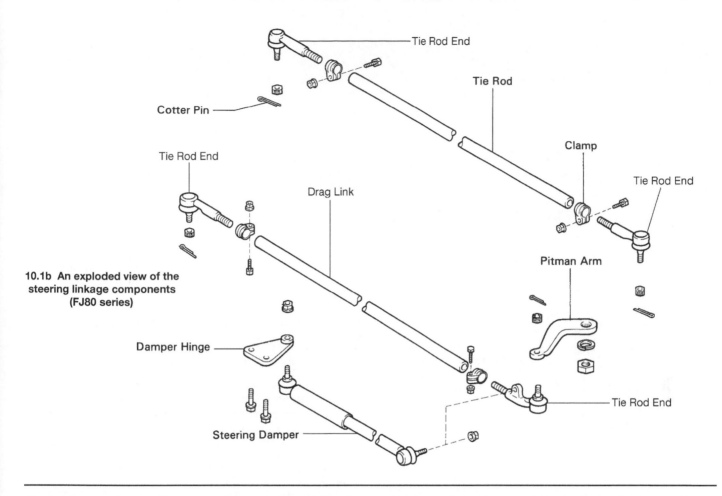

Tie Rod End

Tie Rod

Cotter Pin

Clamp

Tie Rod End

Tie Rod End

Drag Link

Pitman Arm

10.1b An exploded view of the steering linkage components (FJ80 series)

Damper Hinge

Tie Rod End

Steering Damper

12   Install new cotter pins. If necessary, tighten the nut slightly more to align a slot in the nut with the hole in the ballstud. **Note:** *DO NOT loosen the castle nut to align the cotter pin hole.*

13   Tighten the adjuster tube clamp nuts to the torque listed in this Chapter's Specifications.

14   Install the wheel and lug nuts, lower the vehicle and tighten the lug nuts to the torque listed in the Chapter 1 Specifications. Drive the

vehicle to an alignment shop to have the front end alignment checked and, if necessary, adjusted.

Drag link

15   Raise the front of the vehicle and support it securely on jackstands. Apply the parking brake.

16   Detach the steering damper from the drag link using the same

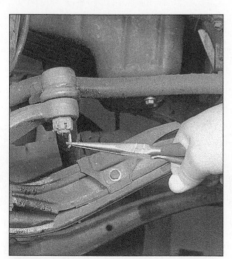

10.7 Remove the cotter pin from the tie-rod end castle nut, then loosen the nut a few turns

10.8 Use a two-jaw puller to detach the tie-rod end - notice the nut has been loosened but not removed; this will prevent the components from separating violently

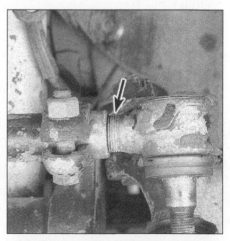

10.9 If the tie-rod end must be replaced, count the number of threads visible at the end of the adjuster tube and install the new tie-rod end with the same number of threads showing

10.22 Use a Pitman arm puller to detach the Pitman arm - notice the nut has been loosened but not removed; this will prevent the components from separating violently

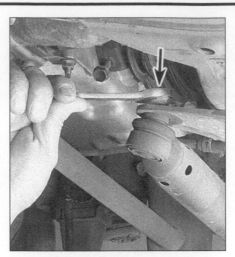

10.26 Remove the nut (arrow) to disconnect the steering damper from the frame bracket

11.2 Loosen these fittings (arrows) and disconnect the power steering fluid lines from the steering gearbox

technique described in Steps 7 and 8.

17   Separate the drag link ends (tie-rod ends) from the Pitman arm and the steering knuckle or tie-rod (FJ60 and FJ62 series) using the same technique described in Steps 7 and 8.

18   If the drag link ends (tie-rod ends) are in need of replacement, follow Steps 9 and 10.

19   Installation is the reverse of the removal procedure. If the ball-studs spin when attempting to tighten the nuts, force them into the tapered holes with a large pair of pliers. Be sure to tighten all fasteners to the torque listed in this Chapter's Specifications.

Pitman arm

Refer to illustration 10.22

20   Disconnect the drag link from the Pitman arm (see above).

21   Loosen - but don't remove - the Pitman arm retaining nut. Mark the relationship of the Pitman arm to the steering shaft.

22   Install a Pitman arm removal tool and pull off the Pitman arm **(see illustration)**.

23   Remove the Pitman arm retaining nut and remove the Pitman arm.

24   Installation is the reverse of removal. Make sure the marks you made before removal are aligned when installing the Pitman arm.

Steering damper

Refer to illustration 10.26

25   Separate the steering damper from the center link using the tech-

nique described in Step 16.

26   Unbolt the damper from the frame bracket **(see illustration)** and remove it from the vehicle.

27   Installation is the reverse of removal. Be sure to tighten the steering damper fasteners to the torque listed in this Chapter's Specifications.

11   Steering gear - removal and installation

Warning: *If the vehicle is equipped with an airbag, make sure the steering shaft is not turned while the steering gear is removed or you could damage the airbag system. To prevent the shaft from turning, place the ignition key in the LOCK position or thread the seat belt through the steering wheel and clip it into place.*

Removal

Refer to illustrations 11.2, 11.3 and 11.5

1   Raise the front of the vehicle and support it securely on jackstands. Apply the parking brake.

2   Place a drain pan under the steering gear (power steering only). Disconnect the line fittings **(see illustration)** and cap the ends to prevent excessive fluid loss and contamination. If available, use a flare-nut wrench to remove the hoses/lines.

3   Mark the relationship of the intermediate shaft lower universal joint to the steering gear input shaft. Remove the intermediate shaft

11.3 Mark the relationship of the intermediate shaft U-joint to the steering gear input shaft and remove the pinch bolt (arrow)

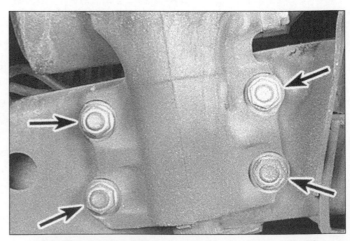

11.5 The steering gear is mounted to the frame rail with four bolts (arrows)

lower pinch bolt **(see illustration)**.

4   Detach the Pitman arm from the steering gear (see Section 10).

5   Support the steering gear and remove the mounting bolts **(see illustration)**. Lower the unit, separate the intermediate shaft from the steering gear input shaft and remove the steering gear from the vehicle.

Installation

6   Raise the steering gear into position and connect the intermediate shaft, aligning the marks.

7   Install the mounting bolts and washers and tighten them to the torque listed in this Chapter's Specifications.

8   Slide the Pitman arm onto the shaft. Make sure the marks are aligned. Install the washer and nut and tighten the nut to the torque listed in this Chapter's Specifications.

9   Install the intermediate shaft lower pinch bolt and tighten it to the torque listed in this Chapter's Specifications.

10   Connect the power steering hoses/lines to the steering gear and fill the power steering pump reservoir with the recommended fluid (see Chapter 1).

11   Lower the vehicle and bleed the steering system (see Section 13).

12 Power steering pump - removal and installation

Refer to illustration 12.2, 12.3a, 12.3b and 12.4

1   Loosen the pump drivebelt and slip the belt over the pulley (see Chapter 1).

2   Using a suction gun, withdraw as much power steering fluid from

12.2 Detach the power steering pump pressure line and return hose connections (1FZ-FE engine shown)

the reservoir as possible. Position a drain pan under the pump and disconnect the high pressure line and fluid return hose **(see illustration)**. Cap the ends of the lines to prevent excessive fluid leakage and the entry of contaminants.

3   On vehicles equipped with 2F and 3F-E engines, remove the mounting bolts from the front of the pump **(see illustration)**. On vehicles equipped with 1FZ-FE engines, remove the bolts attaching the

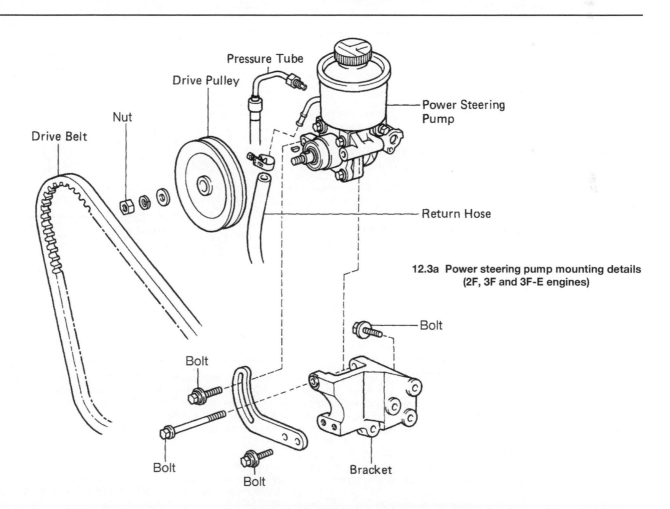

12.3a Power steering pump mounting details (2F, 3F and 3F-E engines)

12.3b On 1FZ-FE engines, remove the bolts securing power steering pump to the engine block

12.4 On 1FZ-FE engines, be sure to replace the O-ring between the engine block and the power steering pump

steering pump to the engine block, then lift the pump from the engine **(see illustration)**.

4   Installation of the power steering pump is the reverse of the removal procedure. On vehicles equipped with 1FZ-FE engines, be sure to replace the O-ring located between the engine block and power steering pump **(see illustration)**. Bleed the power steering system following the procedure in Section 13.

---

13 Power steering system - bleeding

1   Following any operation in which the power steering fluid lines have been disconnected, the power steering system must be bled to remove all air and obtain proper steering performance.

2   With the front wheels in the straight ahead position, check the power steering fluid level and, if low, add fluid until it reaches the Cold (C) mark on the dipstick.

3   Start the engine and allow it to run at fast idle. Recheck the fluid level and add more if necessary to reach the Cold (C) mark on the dipstick.

4   Bleed the system by turning the wheels from side-to-side, without hitting the stops. This will work the air out of the system. Keep the reservoir full of fluid as this is done.

5   When the air is worked out of the system, return the wheels to the straight ahead position and leave the vehicle running for several more minutes before shutting it off.

6   Road test the vehicle to be sure the steering system is functioning normally and noise-free.

7   Recheck the fluid level to be sure it is up to the Hot (H) mark on the dipstick while the engine is at normal operating temperature. Add fluid if necessary (see Chapter 1).

---

14 Wheels and tires - general information

Refer to illustration 14.1

Most vehicles covered by this manual are equipped with metric-sized fiberglass or steel-belted radial tires **(see illustration)**. Use of other size or type of tires may affect the ride and handling of the vehicle. Don't mix different types of tires, such as radials and bias belted, on the same vehicle as handling may be seriously affected. It's recommended that tires be replaced in pairs on the same axle, but if only one tire is being replaced, be sure it's the same size, structure and tread design as the other.

Because tire pressure has a substantial effect on handling and

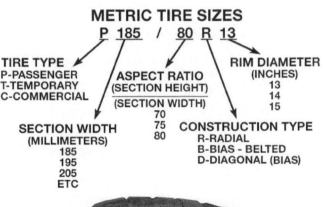

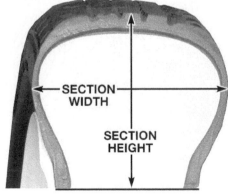

14.1 Metric tire size code

wear, the pressure on all tires should be checked at least once a month or before any extended trips (see Chapter 1).

Wheels must be replaced if they are bent, dented, leak air, have elongated bolt holes, are heavily rusted, out of vertical symmetry or if the lug nuts won't stay tight. Wheel repairs that use welding or peening are not recommended.

Tire and wheel balance is important to the overall handling, braking and performance of the vehicle. Unbalanced wheels can adversely affect handling and ride characteristics as well as tire life. Whenever a tire is installed on a wheel, the tire and wheel should be balanced by a shop with the proper equipment.

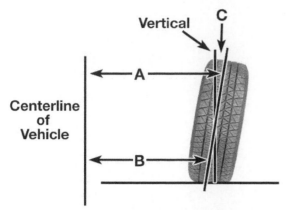

CAMBER ANGLE (FRONT VIEW)

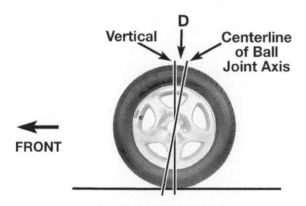

CASTER ANGLE (SIDE VIEW)

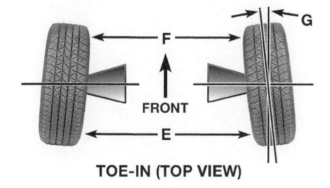

TOE-IN (TOP VIEW)

15.3 Front end alignment details

A minus B = C (degrees camber)
D = caster (expressed in degrees)
E minus F = toe-in (measured in inches)
G - toe-in (expressed in degrees)

15   Front end alignment - general information

Refer to illustration 15.3

A front end alignment refers to the adjustments made to the front wheels so they are in proper angular relationship to the suspension and the ground. Front wheels that are out of proper alignment not only affect steering control, but also increase tire wear. The only front end adjustment possible on these vehicles is toe-in. Caster and camber should be checked to determine if there are any damaged or worn parts.

Getting the proper front wheel alignment is a very exacting process, one in which complicated and expensive machines are necessary to perform the job properly. Because of this, you should have a technician with the proper equipment perform these tasks. We will, however, use this space to give you a basic idea of what is involved with front end alignment so you can better understand the process and deal intelligently with the shop that does the work.

Toe-in is the turning in of the front wheels **(see illustration)**. The purpose of a toe specification is to ensure parallel rolling of the front wheels. In a vehicle with zero toe-in, the distance between the front edges of the wheels will be the same as the distance between the rear edges of the wheels. The actual amount of toe-in is normally only a fraction of an inch. Toe-in adjustment is controlled by the position of the tie-rod ends on the tie-rod. Incorrect toe-in will cause the tires to wear improperly by making them scrub against the road surface.

Caster is the tilting of the top of the steering axis from the vertical. A tilt toward the rear is positive caster and a tilt toward the front is negative caster. Caster isn't adjustable on these vehicles.

Camber (the tilting of the front wheels from vertical when viewed from the front of the vehicle) is factory present and cannot be adjusted. If the camber angle isn't correct, either the wheel bearings need to be adjusted or the components causing the problem must be replaced. **Caution:** *Never attempt to adjust the camber angle by heating or bending the axle or any other suspension component!*

Notes

Chapter 11 Body

Contents

	Section		Section
Body - maintenance	2	Hinges and locks - maintenance	7
Body repair - major damage	6	Hood - removal, installation and adjustment	9
Body repair - minor damage	5	Hood release latch and cable - removal and installation	10
Bumpers - removal and installation	12	Instrument panel - removal and installation	25
Center console - removal and installation	24	Liftgate - removal, installation and adjustment	21
Dashboard trim panels - removal and installation	23	Outside mirrors - removal and installation	18
Door - removal, installation and adjustment	19	Radiator grille - removal and installation	11
Door latch, lock cylinder and handles - removal and installation	15	Seats - removal and installation	26
Door trim panel - removal and installation	14	Steering column covers - removal and installation	22
Door window glass - removal and installation	16	Tailgate - removal, installation and adjustment	20
Door window glass regulator - removal and installation	17	Upholstery and carpets - maintenance	4
Front fender - removal and installation	13	Vinyl trim - maintenance	3
General information	1	Windshield and fixed glass - replacement	8

1 General information

The vehicles covered by this manual are built with a body-on-frame construction. The frame is a ladder-type, consisting of two box steel side rails joined by crossmembers. These crossmembers are welded to the side rails, with the exception of the transmission crossmember which is bolted into place for easy removal. The vehicle bodies are secured to the chassis by rubber insulated mounts and can be completely removed from the chassis.

Certain components are particularly vulnerable to accident damage and can be unbolted and repaired or replaced. Among these parts are the body moldings, front fenders, doors, bumpers, the hood and tailgate and all glass. Only general body maintenance practices and body panel repair procedures within the scope of the do-it-yourselfer are included in this Chapter.

2 Body - maintenance

1   The condition of your vehicle's body is very important, because the resale value depends a great deal on it. It's much more difficult to repair a neglected or damaged body than it is to repair mechanical components. The hidden areas of the body, such as the wheel wells, the frame and the engine compartment, are equally important, although they don't require as frequent attention as the rest of the body.
2   Once a year, or every 12,000 miles, it's a good idea to have the underside of the body steam cleaned. All traces of dirt and oil will be removed and the area can then be inspected carefully for rust, damaged brake lines, frayed electrical wires, damaged cables and other problems. The front suspension components should be greased after completion of this job.
3   At the same time, clean the engine and the engine compartment with a steam cleaner or water-soluble degreaser.
4   The wheel wells should be given close attention, since undercoating can peel away and stones and dirt thrown up by the tires can cause

the paint to chip and flake, allowing rust to set in. If rust is found, clean down to the bare metal and apply an anti-rust paint.
5   The body should be washed about once a week. Wet the vehicle thoroughly to soften the dirt, then wash it down with a soft sponge and plenty of clean soapy water. If the surplus dirt is not washed off very carefully, it can wear down the paint.
6   Spots of tar or asphalt thrown up from the road should be removed with a cloth soaked in solvent.
7   Once every six months, wax the body and chrome trim. If a chrome cleaner is used to remove rust from any of the vehicle's plated parts, remember that the cleaner also removes part of the chrome, so use it sparingly.

3 Vinyl trim - maintenance

Don't clean vinyl trim with detergents, caustic soap or petroleum-based cleaners. Plain soap and water works just fine, with a soft brush to clean dirt that may be ingrained. Wash the vinyl as frequently as the rest of the vehicle. After cleaning, application of a high-quality rubber and vinyl protectant will help prevent oxidation and cracks. The protectant can also be applied to weatherstripping, vacuum lines and rubber hoses, which often fail as a result of chemical degradation, and to the tires.

4 Upholstery and carpets - maintenance

1   Every three months remove the floormats and clean the interior of the vehicle (more frequently if necessary). Use a stiff whisk broom to brush the carpeting and loosen dirt and dust, then vacuum the upholstery and carpets thoroughly, especially along seams and crevices.
2   Dirt and stains can be removed from carpeting with basic household or automotive carpet shampoos available in spray cans. Follow the directions and vacuum again, then use a stiff brush to bring back the "nap" of the carpet.

These photos illustrate a method of repairing simple dents. They are intended to supplement *Body repair - minor damage* in this Chapter and should not be used as the sole instructions for body repair on these vehicles.

1 If you can't access the backside of the body panel to hammer out the dent, pull it out with a slide-hammer-type dent puller. In the deepest portion of the dent or along the crease line, drill or punch hole(s) at least one inch apart . . .

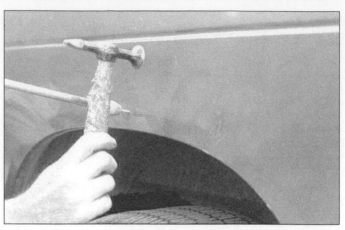

2 . . . then screw the slide-hammer into the hole and operate it. Tap with a hammer near the edge of the dent to help 'pop' the metal back to its original shape. When you're finished, the dent area should be close to its original contour and about 1/8-inch below the surface of the surrounding metal

3 Using coarse-grit sandpaper, remove the paint down to the bare metal. Hand sanding works fine, but the disc sander shown here makes the job faster. Use finer (about 320-grit) sandpaper to feather-edge the paint at least one inch around the dent area

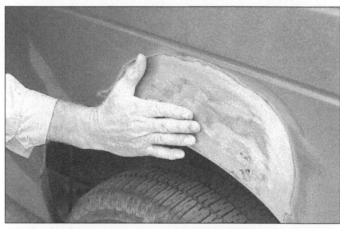

4 When the paint is removed, touch will probably be more helpful than sight for telling if the metal is straight. Hammer down the high spots or raise the low spots as necessary. Clean the repair area with wax/silicone remover

5 Following label instructions, mix up a batch of plastic filler and hardener. The ratio of filler to hardener is critical, and, if you mix it incorrectly, it will either not cure properly or cure too quickly (you won't have time to file and sand it into shape)

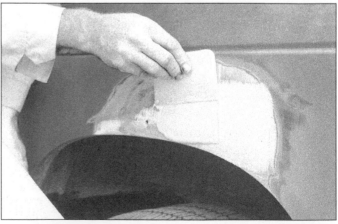

6 Working quickly so the filler doesn't harden, use a plastic applicator to press the body filler firmly into the metal, assuring it bonds completely. Work the filler until it matches the original contour and is slightly above the surrounding metal

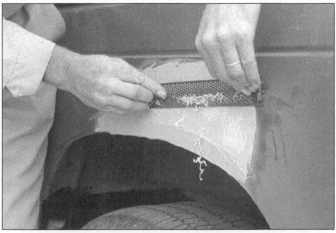

7 Let the filler harden until you can just dent it with your fingernail. Use a body file or Surform tool (shown here) to rough-shape the filler

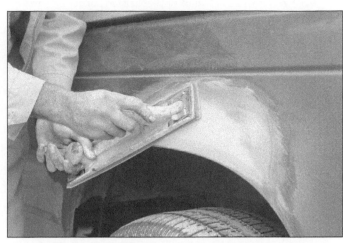

8 Use coarse-grit sandpaper and a sanding board or block to work the filler down until it's smooth and even. Work down to finer grits of sandpaper - always using a board or block - ending up with 360 or 400 grit

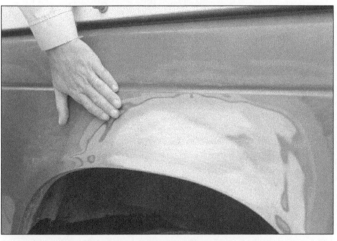

9 You shouldn't be able to feel any ridge at the transition from the filler to the bare metal or from the bare metal to the old paint. As soon as the repair is flat and uniform, remove the dust and mask off the adjacent panels or trim pieces

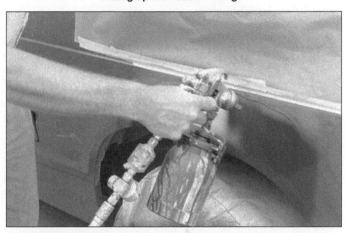

10 Apply several layers of primer to the area. Don't spray the primer on too heavy, so it sags or runs, and make sure each coat is dry before you spray on the next one. A professional-type spray gun is being used here, but aerosol spray primer is available inexpensively from auto parts stores

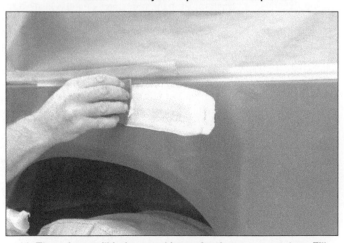

11 The primer will help reveal imperfections or scratches. Fill these with glazing compound. Follow the label instructions and sand it with 360 or 400-grit sandpaper until it's smooth. Repeat the glazing, sanding and respraying until the primer reveals a perfectly smooth surface

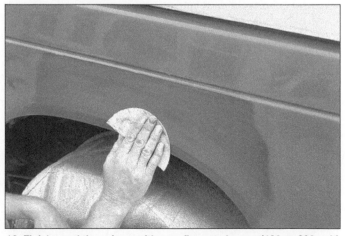

12 Finish sand the primer with very fine sandpaper (400 or 600-grit) to remove the primer overspray. Clean the area with water and allow it to dry. Use a tack rag to remove any dust, then apply the finish coat. Don't attempt to rub out or wax the repair area until the paint has dried completely (at least two weeks)

3   Most interiors have cloth or vinyl upholstery, either of which can be cleaned and maintained with a number of material-specific cleaners or shampoos available in auto supply stores. Follow the directions on the product for usage, and always spot-test any upholstery cleaner on an inconspicuous area (bottom edge of a back seat cushion) to ensure that it doesn't cause a color shift in the material.

4   After cleaning, vinyl upholstery should be treated with a protectant. **Note:** *Make sure the protectant container indicates the product can be used on seats - some products may make a seat too slippery.* **Caution:** *Do not use protectant on vinyl-covered steering wheels.*

5   Leather upholstery requires special care. It should be cleaned regularly with saddlesoap or leather cleaner. Never use alcohol, gasoline, nail polish remover or thinner to clean leather upholstery.

6   After cleaning, regularly treat leather upholstery with a leather conditioner, rubbed in with a soft cotton cloth. Never use car wax on leather upholstery.

7   In areas where the interior of the vehicle is subject to bright sunlight, cover leather seating areas of the seats with a sheet if the vehicle is to be left out for any length of time.

5   Body repair - minor damage

Repair of scratches

1   If the scratch is superficial and does not penetrate to the metal of the body, repair is very simple. Lightly rub the scratched area with a fine rubbing compound to remove loose paint and built-up wax. Rinse the area with clean water.

2   Apply touch-up paint to the scratch, using a small brush. Continue to apply thin layers of paint until the surface of the paint in the scratch is level with the surrounding paint. Allow the new paint at least two weeks to harden, then blend it into the surrounding paint by rubbing with a very fine rubbing compound. Finally, apply a coat of wax to the scratch area.

3   If the scratch has penetrated the paint and exposed the metal of the body, causing the metal to rust, a different repair technique is required. Remove all loose rust from the bottom of the scratch with a pocket knife, then apply rust inhibiting paint to prevent the formation of rust in the future. Using a rubber or nylon applicator, coat the scratched area with glaze-type filler. If required, the filler can be mixed with thinner to provide a very thin paste, which is ideal for filling narrow scratches. Before the glaze filler in the scratch hardens, wrap a piece of smooth cotton cloth around the tip of a finger. Dip the cloth in thinner and then quickly wipe it along the surface of the scratch. This will ensure that the surface of the filler is slightly hollow. The scratch can now be painted over as described earlier in this Section.

Repair of dents

See photo sequence

4   When repairing dents, the first job is to pull the dent out until the affected area is as close as possible to its original shape. There is no point in trying to restore the original shape completely as the metal in the damaged area will have stretched on impact and cannot be restored to its original contours. It is better to bring the level of the dent up to a point which is about 1/8-inch below the level of the surrounding metal. In cases where the dent is very shallow, it is not worth trying to pull it out at all.

5   If the back side of the dent is accessible, it can be hammered out gently from behind using a soft-face hammer. While doing this, hold a block of wood firmly against the opposite side of the metal to absorb the hammer blows and prevent the metal from being stretched.

6   If the dent is in a section of the body which has double layers, or some other factor makes it inaccessible from behind, a different technique is required. Drill several small holes through the metal inside the damaged area, particularly in the deeper sections. Screw long, self tapping screws into the holes just enough for them to get a good grip in the metal. Now the dent can be pulled out by pulling on the protruding heads of the screws with locking pliers.

7   The next stage of repair is the removal of paint from the damaged area and from an inch or so of the surrounding metal. This is easily done with a wire brush or sanding disk in a drill motor, although it can be done just as effectively by hand with sandpaper. To complete the preparation for filling, score the surface of the bare metal with a screwdriver or the tang of a file or drill small holes in the affected area. This will provide a good grip for the filler material. To complete the repair, see the Section on *filling and painting*.

Repair of rust holes or gashes

8   Remove all paint from the affected area and from an inch or so of the surrounding metal using a sanding disk or wire brush mounted in a drill motor. If these are not available, a few sheets of sandpaper will do the job just as effectively.

9   With the paint removed, you will be able to determine the severity of the corrosion and decide whether to replace the whole panel, if possible, or repair the affected area. New body panels are not as expensive as most people think and it is often quicker to install a new panel than to repair large areas of rust.

10   Remove all trim pieces from the affected area except those which will act as a guide to the original shape of the damaged body, such as headlight shells, etc. Using metal snips or a hacksaw blade, remove all loose metal and any other metal that is badly affected by rust. Hammer the edges of the hole on the inside to create a slight depression for the filler material.

11   Wire brush the affected area to remove the powdery rust from the surface of the metal. If the back of the rusted area is accessible, treat it with rust inhibiting paint.

12   Before filling is done, block the hole in some way. This can be done with sheet metal riveted or screwed into place, or by stuffing the hole with wire mesh.

13   Once the hole is blocked off, the affected area can be filled and painted. See the following subsection on *filling and painting*.

Filling and painting

14   Many types of body fillers are available, but generally speaking, body repair kits which contain filler paste and a tube of resin hardener are best for this type of repair work. A wide, flexible plastic or nylon applicator will be necessary for imparting a smooth and contoured finish to the surface of the filler material. Mix up only a small amount of filler on a clean piece of wood or cardboard (use the hardener sparingly). Follow the manufacturer's instructions on the package, otherwise the filler will set incorrectly.

15   Using the applicator, apply the filler paste to the prepared area. Draw the applicator across the surface of the filler to achieve the desired contour and to level the filler surface. As soon as a contour that approximates the original one is achieved, stop working the paste. If you continue, the paste will begin to stick to the applicator. Continue to add thin layers of paste at 20-minute intervals until the level of the filler is just above the surrounding metal.

16   Once the filler has hardened, the excess can be removed with a body file. From then on, progressively finer grades of sandpaper should be used, starting with a 180-grit paper and finishing with 600-grit wet-or-dry paper. Always wrap the sandpaper around a flat rubber or wooden block, otherwise the surface of the filler will not be completely flat. During the sanding of the filler surface, the wet-or-dry paper should be periodically rinsed in water. This will ensure that a very smooth finish is produced in the final stage.

17   At this point, the repair area should be surrounded by a ring of bare metal, which in turn should be encircled by the finely feathered edge of good paint. Rinse the repair area with clean water until all of the dust produced by the sanding operation is gone.

18   Spray the entire area with a light coat of primer. This will reveal any imperfections in the surface of the filler. Repair the imperfections with fresh filler paste or glaze filler and once more smooth the surface with sandpaper. Repeat this spray-and-repair procedure until you are satisfied that the surface of the filler and the feathered edge of the paint are perfect. Rinse the area with clean water and allow it to dry completely.

19   The repair area is now ready for painting. Spray painting must be carried out in a warm, dry, windless and dust free atmosphere. These conditions can be created if you have access to a large indoor work area, but if you are forced to work in the open, you will have to pick the

9.4 Mark around the hinge and remove the hood-to-hinge bolts - lift off the hood with the help of an assistant

9.10 To adjust the hood latch, loosen the retaining bolts (arrows), move the latch and retighten bolts, then close the hood to check the fit

day very carefully. If you are working indoors, dousing the floor in the work area with water will help settle the dust which would otherwise be in the air. If the repair area is confined to one body panel, mask off the surrounding panels. This will help minimize the effects of a slight mismatch in paint color. Trim pieces such as chrome strips, door handles, etc., will also need to be masked off or removed. Use masking tape and several thickness of newspaper for the masking operations.

20   Before spraying, shake the paint can thoroughly, then spray a test area until the spray painting technique is mastered. Cover the repair area with a thick coat of primer. The thickness should be built up using several thin layers of primer rather than one thick one. Using 600-grit wet-or-dry sandpaper, rub down the surface of the primer until it is very smooth. While doing this, the work area should be thoroughly rinsed with water and the wet-or-dry sandpaper periodically rinsed as well. Allow the primer to dry before spraying additional coats.

21   Spray on the top coat, again building up the thickness by using several thin layers of paint. Begin spraying in the center of the repair area and then, using a circular motion, work out until the whole repair area and about two inches of the surrounding original paint is covered. Remove all masking material 10 to 15 minutes after spraying on the final coat of paint. Allow the new paint at least two weeks to harden, then use a very fine rubbing compound to blend the edges of the new paint into the existing paint. Finally, apply a coat of wax.

6   Body repair - major damage

1   Major damage must be repaired by an auto body shop specifically equipped to perform body and frame repairs. These shops have the specialized equipment required to do the job properly.

2   If the damage is extensive, the body must be checked for proper alignment or the vehicle's handling characteristics may be adversely affected and other components may wear at an accelerated rate.

3   Due to the fact that all of the major body components (hood, fenders, etc.) are separate and replaceable units, any seriously damaged components should be replaced rather than repaired. Sometimes the components can be found in a wrecking yard that specializes in used vehicle components, often at considerable savings over the cost of new parts.

7   Hinges and locks - maintenance

Once every 3000 miles, or every three months, the hinges and latch assemblies on the doors, hood and trunk should be given a few drops of light oil or lock lubricant. The door latch strikers should also be lubricated with a thin coat of grease to reduce wear and ensure free movement. Lubricate the door and trunk locks with spray-on graphite lubricant.

8   Windshield and fixed glass - replacement

Replacement of the windshield and fixed glass requires the use of special fast-setting adhesive/caulk materials and some specialized tools and techniques. These operations should be left to a dealer service department or a shop specializing in glass work.

9   Hood - removal, installation and adjustment

Note: *The hood is heavy and somewhat awkward to remove and install - at least two people should perform this procedure.*

Removal and installation

Refer to illustration 9.4

1   Use blankets or pads to cover the cowl area of the body and fenders. This will protect the body and paint as the hood is lifted off.

2   Make marks or scribe a line around the hood hinge to ensure proper alignment during installation.

3   Disconnect any cables or wires that will interfere with removal.

4   Have an assistant support the hood. Detach the hood support struts (if equipped) and remove the hinge-to-hood nuts or bolts **(see illustration)**.

5   Lift off the hood.

6   Installation is the reverse of removal.

Adjustment

Refer to illustration 9.10

7   Fore-and-aft and side-to-side adjustment of the hood is done by moving the hinge plate slot after loosening the bolts or nuts.

8   Scribe a line around the entire hinge plate so you can determine the amount of movement.

9   Loosen the bolts or nuts and move the hood into correct alignment. Move it only a little at a time. Tighten the hinge bolts and carefully lower the hood to check the position.

10   If necessary after installation, the entire hood latch assembly can be adjusted up-and-down as well as from side-to-side on the radiator support so the hood closes securely and flush with the fenders. To make the adjustment, scribe a line or mark around the hood latch mounting bolts to provide a reference point, then loosen them and reposition the latch assembly, as necessary **(see illustration)**. Following adjustment, retighten the mounting bolts.

11   The hood latch assembly, as well as the hinges, should be periodically lubricated with white, lithium-base grease to prevent binding and wear.

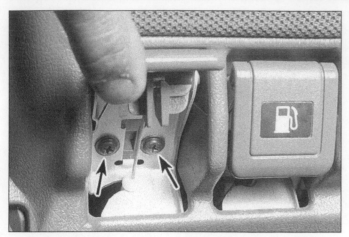

10.6 Lift the release lever and remove the lever retaining screws (arrows) (FJ80 series shown)

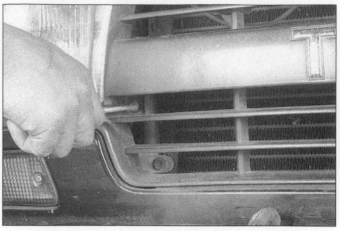

11.1 Remove the screws retaining the radiator grille (FJ80 series shown)

10 Hood release latch and cable - removal and installation

Latch

1   Scribe a line around the latch to aid alignment when installing, then detach the latch retaining bolts from the radiator support **(see illustration 9.10)**. Remove the latch.
2   Disconnect the hood release cable by disengaging the cable from the latch assembly.
3   Installation is the reverse of the removal procedure. **Note:** *Adjust the latch so the hood engages securely when closed and the hood bumpers are slightly compressed.*

Cable

Refer to illustration 10.6

4   Disconnect the hood release cable from the latch assembly as described above.
5   Attach a piece of stiff wire to the end of the cable, then follow the cable back to the firewall and detach all the cable retaining clips.
6   Working in the passenger compartment. Detach the screws securing the hood release lever **(see illustration).**
7   Pull the cable and grommet rearward into the passenger compartment until you can see the wire. Ensure that the new cable has a grommet attached, then remove the old cable from the wire and replace it with the new cable.
8   Working from engine compartment pull the wire back through

the firewall.
9   Installation is the reverse of the removal. **Note:** *Push on the grommet with your fingers from the passenger compartment to seat the grommet in the firewall correctly.*

11 Radiator grille - removal and installation

Refer to illustration 11.1

1   The radiator grille is held in place by clips and screws. Remove the screws and disengage the grille retaining clips with a small screwdriver **(see illustration)**.
2   Once all the retaining screws and clips are disengaged, pull the grille out and remove it.
3   Installation is the reverse of removal.

12 Bumpers - removal and installation

Refer to illustrations 12.3a and 12.3b

1   Disconnect any electrical connections that would interfere with bumper removal.
2   Support the bumper with a jack or jackstands. Alternatively, have an assistant support the bumper as the bolts are removed.
3   Working from the backside of the bumper, remove the retaining bolts securing the bumper bracket to the outside of each frame rail **(see illustrations)**. Remove the bumper from the vehicle.

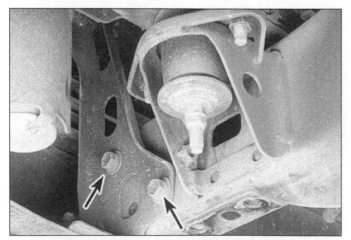

12.3a Typical front bumper bracket-to-frame rail bolt locations (arrows)

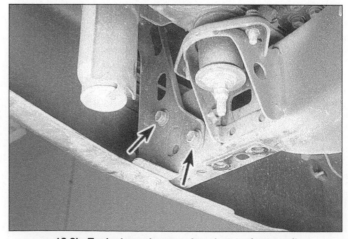

12.3b Typical rear bumper bracket-to-frame rail bolt locations (arrows)

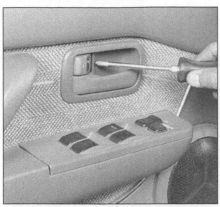

14.2 Detach the retaining screw and remove the inside door handle trim bezel

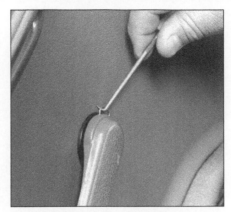

14.3 Use a hooked tool like this to remove the window crank retaining clip

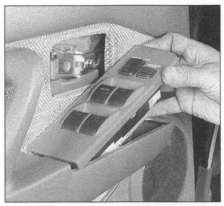

14.4 Use a small screwdriver to pry out the armrest switch control plate

4   If replacing the bumper fascia on FJ80 series vehicles, simply remove the nuts or bolts securing the bumper fascia to the bumper.
5   Installation is the reverse of removal.

13   Front fender - removal and installation

1   Loosen the front wheel lug nuts, raise the vehicle and support it securely on jackstands. Remove the wheel.
2   Remove the front bumper (see Section 12).
3   Remove the side marker and turn signal light assembly (see Chapter 12).
4   If removing the passenger side fender, remove the antenna (see Chapter 12).
5   Remove the fender mounting bolts and nuts.
6   Detach the fender. It's a good idea to have an assistant support the fender while it's being moved away from the vehicle to prevent damage to the surrounding body panels.
7   Installation is the reverse of removal.

14   Door trim panel - removal and installation

Refer to illustrations 14.2, 14.3, 14.4, 14.5 and 14.6

1   Disconnect the negative cable from the battery. **Caution:** *If the radio in your vehicle is equipped with an anti-theft system, make sure you have the correct activation code before disconnecting the battery.*
2   Remove the inside door handle trim bezel **(see illustration)**.
3   On manual window equipped models, remove the window crank using a hooked tool to remove the retainer clip **(see illustration)**. A

special tool is available for this purpose, but it's not essential. With the clip removed, pull off the handle.
4   On power window equipped models, pry out the armrest switch control plate **(see illustration)** and disconnect the electrical connections.
5   Detach the armrest pull handle retaining screws and remove the armrest **(see illustration)**.
6   Remove the remaining door panel retaining screws at the lower edge of the door panel. Insert a wide putty knife, a thin screwdriver or a special trim panel removal tool between the trim panel and the head of the retaining clip to disengage the door panel retaining clips **(see illustration)**.
7   Once all of the clips and screws are disengaged, detach the trim panel, disconnect any electrical connectors and remove the trim panel from the vehicle by gently pulling it up and out.
8   For access to the inner door remove the door panel support bracket (if equipped). Then peel back the watershield, taking care not to tear it. To install the trim panel, first press the watershield back into place. If necessary, add more sealant to hold it in place.
9   Installation is the reverse of removal.

15   Door latch, lock cylinder and handles - removal and installation

Door latch

Refer to illustration 15.4

1   Raise the window then remove the door trim panel and watershield (see Section 14).

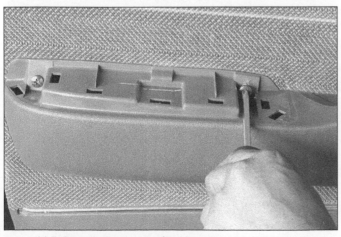

14.5 Detach the armrest/pull handle retaining screws

14.6 Insert a putty knife or trim removal tool between the door and the trim panel, then carefully pry the clips out

2   Working through the large access hole, disengage the outside door handle-to-latch rod, outside door lock-to-latch rod, the inside lock-to-latch rod, the inside handle-to-latch rod and the lock solenoid-to-latch rod (if equipped).

3   All door locking rods are attached by plastic clips. The plastic clips can be removed by unsnapping the portion engaging the rod and then by pulling the rod out of its locating hole.

4   Remove the screws securing the latch to the door **(see illustration)**, then remove the latch assembly from the door.

5   Installation is the reverse of removal.

Outside handle and door lock cylinder

Refer to illustration 15.7

6   To remove the outside handle and lock cylinder assembly, raise the window then remove the door trim panel and watershield as described in Section 14.

7   Working through the access hole, disengage the plastic clips that secure the outside handle to latch rod and the outside door lock to latch rod **(see illustration)**.

8   Remove the handle retaining bolts and the lock cylinder clip or bolt.

9   Remove the handle and lock cylinder assembly from the vehicle.

10   Installation is the reverse of removal.

Inside handle

Refer to illustration 15.12

11   Remove the door trim panel as described in Section 14 and peel away the watershield.

12   Unclip the door actuating rod guide, then remove the door handle retaining screws **(see illustration)**.

13   Pull the handle free from the door, then disconnect the actuating rod from the backside of the handle control and remove the handle from the door.

14   Installation is the reverse of removal.

---

16 Door window glass - removal and installation

Refer to illustration 16.4

1   Remove the door trim panel and the plastic watershield (see Section 14).

2   Lower the window glass all the way down into the door.

3   Carefully pry the inner weatherstrip out of the door window opening.

4   Raise the window just enough to access the window retaining bolts through the hole in the door frame **(see illustration)**.

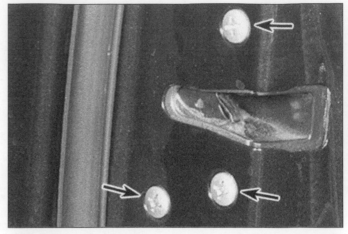

15.4 Remove the latch screws from the end of the door and pull the latch assembly through the access hole

15.7 The outside handle and door lock cylinder can be reached through the access hole in the door frame

5   Remove the vent window (if equipped).

6   Place a rag over the glass to help prevent scratching the glass and remove the two glass mounting bolts.

7   Remove the glass by pulling it up and out.

8   Installation is the reverse of removal.

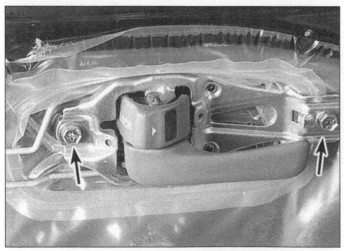

15.12 Remove the inside handle retaining screws (arrows), then rotate the handle outward and detach the actuating rods from the backside

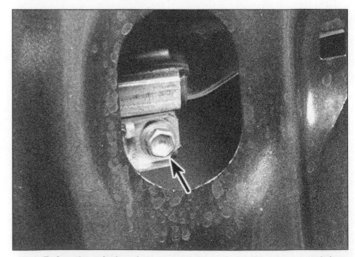

16.4 Raise the window just enough to access the glass retaining bolts (arrow) through the hole in the door frame

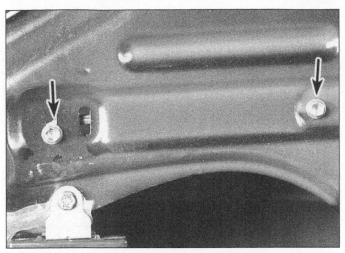

17.4a Detach the window equalizer mounting bolts (arrows) . . .

17.4b . . . then remove the window regulator bolts (arrows)

17   Door window glass regulator - removal and installation

Refer to illustrations 17.4a and 17.4b

1   Remove the door trim panel and the plastic watershield (see Section 14).
2   Remove the window glass assembly (see Section 16).
3   On power operated windows, disconnect the electrical connector from the window regulator motor.
4   Remove the equalizer arm bracket and the regulator mounting bolts **(see illustrations)**.
5   Pull the equalizer arm and regulator assemblies through the service hole in the door frame to remove it.
6   Installation is the reverse of removal.

18   Outside mirrors - removal and installation

Refer to illustration 18.2

1   Pry off the mirror trim cover.
2   Remove the mirror retaining bolts or screws and detach the mirror from the vehicle **(see illustration)**.
3   Disconnect the electrical connector from the mirror (if equipped).
4   Installation is the reverse of removal.

19   Door - removal, installation and adjustment

Refer to illustrations 19.6 and 19.8
Note: *The door is heavy and somewhat awkward to remove and install - at least two people should perform this procedure.*

Removal and installation

1   Raise the window completely in the door and then disconnect the negative cable from the battery. **Caution:** *If the radio in your vehicle is equipped with an anti-theft system, make sure you have the correct activation code before disconnecting the battery.*
2   Open the door all the way and support it on jacks or blocks covered with rags to prevent damaging the paint.
3   Remove the door trim panel and water deflector as described in Section 14.
4   Disconnect all electrical connections, ground wires and harness retaining clips from the door. **Note:** *It is a good idea to label all connections to aid the reassembly process.*
5   From the door side, detach the rubber conduit between the body and the door. Then pull the wiring harness through conduit hole and remove it from the door.
6   Remove the door stop strut center pin **(see illustration)**.
7   Mark around the door hinges with a pen or a scribe to facilitate realignment during reassembly.
8   With an assistant holding the door, remove the hinge-to-door bolts **(see illustration)** and lift the door off.
9   Installation is the reverse of removal.

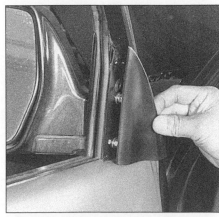

18.2 Pry off the trim cover and remove the mirror retaining bolts (FJ80 series shown)

19.6 Gently tap the pin for the door stop strut in the direction shown

19.8 A special door hinge wrench may be necessary to loosen the hinge bolts

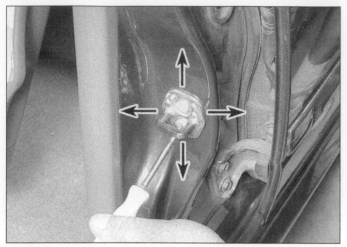

19.13 Adjust the door lock striker by loosening the mounting screws and gently tapping the striker in the desired direction (arrows)

20.4 Remove the tailgate support arm retaining bolts (arrow)

Adjustment

Refer to illustration 19.13

10   Having proper door to body alignment is a critical part of a well functioning door assembly. First check the door hinge pins for excessive play. Fully open the door and lift up and down on the door without lifting the body. If a door has 1/16-inch or more excessive play, the hinges should be replaced.

11   Door-to-body alignment adjustments are made by loosening the hinge-to-body bolts or hinge-to-door bolts and moving the door **(see illustration 19.8)**. Proper body alignment is achieved when the top of the doors are parallel with the roof section, the front door is flush with the fender, the rear door is flush with the rear quarter panel and the bottom of the doors are aligned with the lower rocker panel. If these goals can't be reached by adjusting the hinge-to-body or hinge-to-door bolts, body alignment shims may have to be purchased and inserted behind the hinges to achieve correct alignment.

12   To adjust the door closed position, scribe a line or mark around the striker plate to provide a reference point, then check that the door latch is contacting the center of the latch striker. If not adjust the up and down position first.

13   Finally adjust the latch striker sideways position, so that the door panel is flush with the center pillar or rear quarter panel and provides positive engagement with the latch mechanism **(see illustration)**.

20   Tailgate - removal, installation and adjustment

Note: *The tailgate is heavy and somewhat awkward to remove and install - at least two people should perform this procedure.*

Removal and installation

Refer to illustrations 20.4 and 20.5

1   Open the tailgate and remove the trim panels.

2   Disconnect all wiring harness connectors leading to the tailgate.

3   Cover the lower bumper area around the opening with pads or cloths to protect the painted surfaces when the tailgate is removed.

4   While an assistant supports the tailgate, detach the tailgate support arms **(see illustration)**.

5   Detach the hinge to tailgate bolts and remove the tailgate from the vehicle **(see illustration)**.

6   Installation is the reverse of removal.

Adjustment

Refer to illustration 20.9

7   Tailgate adjustments are made by loosening the hinge-to-body or

hinge-to-tailgate bolts and moving the tailgate. Proper alignment is achieved when the top of the tailgate is aligned with the top of the rear quarter panel. If these goals can't be reached by adjusting the hinge to body or hinge to tailgate bolts, body alignment shims may have to be purchased and inserted behind the hinges to achieve correct alignment.

8   To adjust the tailgate closed position, first check that the latch is contacting the center of the latch striker assembly. If not, remove striker assembly and add or subtract shims to achieve correct alignment.

9   Finally, adjust the latch striker assembly as necessary (up and down or sideways) to provide positive engagement with the latch mechanism **(see illustration)** and the outside of the tailgate is flush with rear quarter panel.

21   Liftgate - removal, installation and adjustment

Note: *The liftgate is heavy and somewhat awkward to hold - at least two people should perform this procedure.*

Removal and installation

Refer to illustration 21.4

1   Open the liftgate and support it securely.

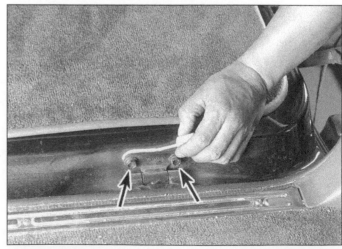

20.5 Before loosening the tailgate retaining bolts (arrows), draw a line around the hinge plate for a reinstallation reference

20.9 Adjust the tailgate lock striker by loosening the mounting screws and gently tapping the striker in the desired direction (arrows)

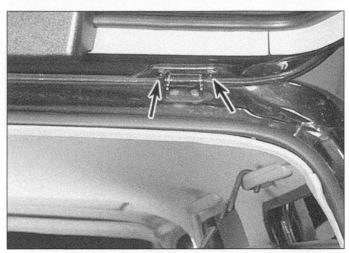

21.4 Loosen the hinge bolts (arrows) to remove the liftgate

2   Remove the liftgate trim panels and disconnect all wiring harness connectors leading to the liftgate.
3   While an assistant supports the liftgate, detach both ends of the support struts. Then pry or pull sharply to remove them from the vehicle.
4   Detach the hinge to liftgate bolts **(see illustration)**. and remove the liftgate from the vehicle.
5   Installation is the reverse of removal.

Adjustment

Refer to illustration 21.7
6   Adjustments are made by loosening the hinge-to liftgate bolts and moving the liftgate. Proper alignment is achieved when the edges of the liftgate are parallel with the rear quarter panel and the top of the tailgate.
7   Finally, adjust the latch striker assembly as necessary (up and down) to provide positive engagement with the latch mechanism **(see illustration)**.

22   Steering column covers - removal and installation

Refer to illustration 22.3
Warning: *Some models covered by this manual are equipped with airbags. Always disable the airbag system* (see Chapter 12) *before*

working in the vicinity of the impact sensors, steering column or instrument panel. Failure to follow these procedures may cause accidental deployment of the airbag, which could cause personal injury.
1   Remove the steering wheel (see Chapter 10).
2   Remove the steering column cover screws.
3   Separate the cover halves and detach them from the steering column **(see illustration)**.
4   Installation is the reverse of removal.

23   Dashboard trim panels - removal and installation

Refer to illustrations 23.1a and 23.1b
Warning: *Some models covered by this manual are equipped with airbags. Always disable the airbag system* (see Chapter 12) *before working in the vicinity of the impact sensors, steering column or instrument panel. Failure to follow these procedures may cause accidental deployment of the airbag, which could cause personal injury*
1   All of the dashboard trim panels are held in place by clips and screws **(see illustrations)**. Remove any screws or clips and disengage the trim panels as necessary.
2   Disconnect any wiring harness connectors which would interfere with removal.
3   Installation is the reverse of removal.

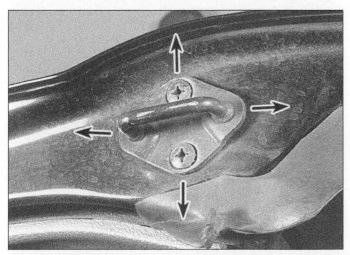

21.7 If the liftgate does not close properly, it will be necessary to loosen the screws (arrows) to adjust the striker plate

22.3 The steering column cover screws (arrows) are accessible from the bottom of the lower cover

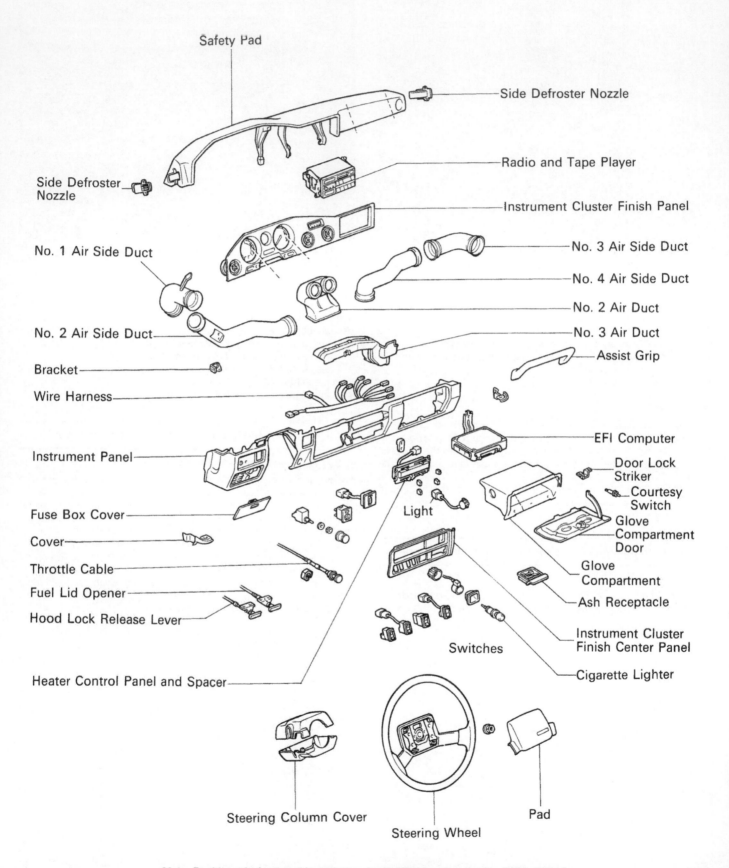

23.1a Dashboard trim panel installation details (FJ62 series shown, FJ60 similar)

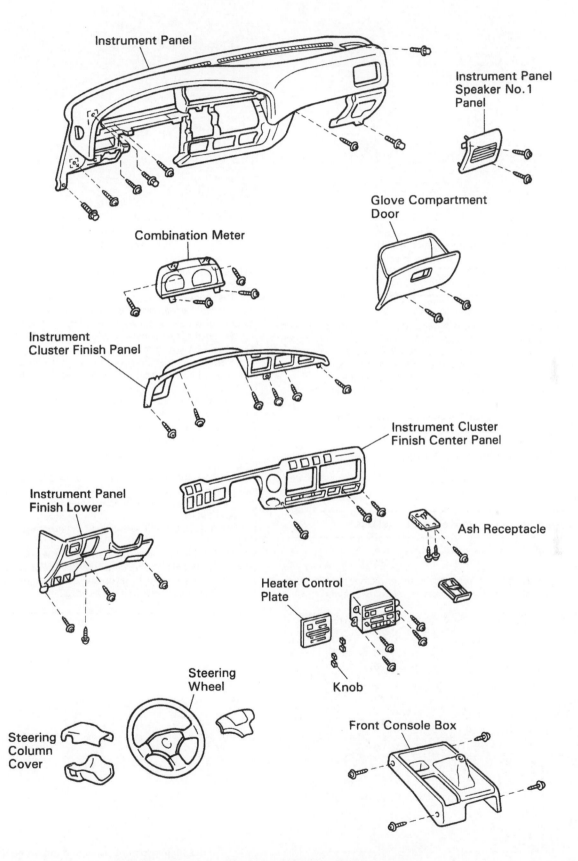

23.1b Dashboard trim panel installation details (FJ80 series)

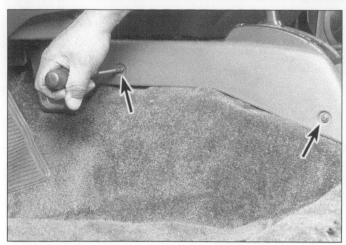

24.2a Detach the plastic trim caps (if equipped), then remove the retaining screws (arrows) from each side of the center console

24.2b Remove the screw (arrow) from the center of the console . . .

24 Center console - removal and installation

Refer to illustrations 24.2a, 24.2b and 24.2c

Warning: *Some models covered by this manual are equipped with airbags. Always disable the airbag system (see Chapter 12) before working in the vicinity of the impact sensors, steering column or instrument panel. Failure to follow these procedures may cause accidental deployment of the airbag, which could cause personal injury.*

1   Unscrew the shift knob(s) on manual transmission or transfer case shift levers.

2   Pry out the plastic trim caps (if equipped) and remove the console retaining screws **(see illustration)**.

3   Lift the console up and over the shift lever.

4   Disconnect any electrical connections and remove the console from the vehicle.

5   Installation is the reverse of removal.

25 Instrument panel - removal and installation

Refer to illustrations 25.10, 25.11a and 25.11b

Warning: *Some models covered by this manual are equipped with airbags. Always disable the airbag system (see Chapter 12) before working in the vicinity of the impact sensors, steering column or instru-*

ment panel. Failure to follow these procedures may cause accidental deployment of the airbag, which could cause personal injury.

1   Disconnect the negative battery cable. **Caution:** *If the radio in your vehicle is equipped with an anti-theft system, make sure you have the correct activation code before disconnecting the battery.*

2   Remove the steering wheel (see Chapter 10).

3   Remove the center floor console (see Section 24).

4   Remove all the dashboard trim panels such as the glove box, instrument cluster bezel, radio trim bezel and the driver side lower knee bolster.

5   Remove the radio and instrument cluster (see Chapter 12).

6   Remove the heater control panel.

7   Detach the nuts and bolts securing the fuse box and the hood release handle (see Section 10).

8   If equipped with fuel injection, remove the ECM from behind the glove box.

9   Remove the steering column mounting bolts and lower the steering column to the floor.

10   On FJ60 and FJ62 series vehicles remove the instrument panel pad **(see illustration)**.

11   Remove the bolts securing the instrument panel **(see illustrations)**.

12   Pull the instrument panel towards the rear of the vehicle and detach any electrical connectors interfering with removal.

13   Lift the instrument panel up and out to remove it from the vehicle.

14   Installation is the reverse of removal.

24.2c . . . then remove the screws (arrows) at the rear of the console

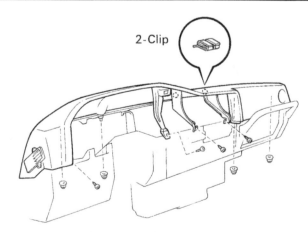

25.10 Instrument panel pad mounting details (FJ60 and FJ62 series)

26 Seats - removal and installation

Front seat

Refer to illustration 26.2

1   Position the seat all the way forward or all the way to the rear to access the seat retaining bolts.

2   Detach any bolt trim covers and remove the retaining bolts **(see illustration)**.

3   Tilt the seat upward to access the underneath, then disconnect any electrical connectors and lift the seat from the vehicle.

4   Installation is the reverse of removal.

Rear seat

Refer to illustrations 26.6

5   Detach the bolt trim covers and remove the seat cushion retaining bolts. Then lift up on the front edge and remove the cushion from the vehicle.

6   Detach the retaining bolts at the lower edge of the seat back **(see illustration)**.

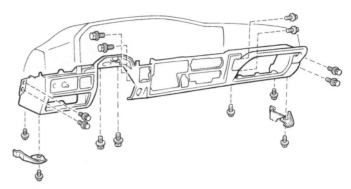

25.11a Instrument panel mounting details (FJ60 and FJ62 series)

7   Lift up on the lower edge of the seat back and remove it from the vehicle.

8   Installation is the reverse of removal.

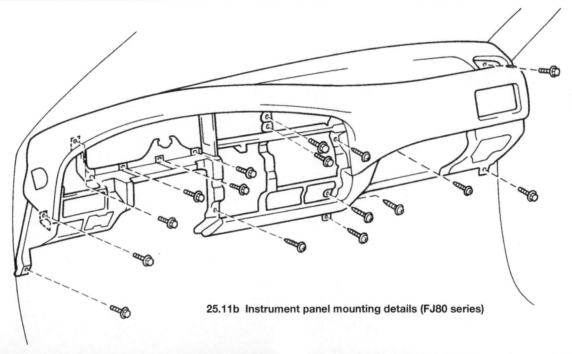

25.11b Instrument panel mounting details (FJ80 series)

26.2 Detach the trim covers to access the seat retaining bolts

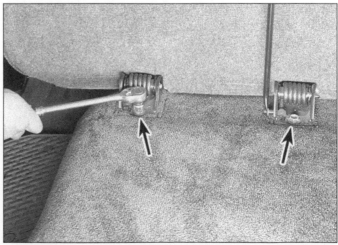

26.6 Remove the hinge bolts (arrows) from the rear seat back

Notes

Chapter 12
Chassis electrical system

Contents

	Section
Airbag - general information	26
Antenna - removal and installation	14
Bulb replacement	20
Circuit breakers - general information	5
Cruise control system - description and check	22
Daytime Running Lights (DRL) - general information	25
Electric side view mirrors - description and check	21
Electrical troubleshooting - general information	2
Fuses - general information	3
Fusible links - general information	4
General information	1
Headlight housing (Halogen type bulb) - removal and installation	18
Headlights - adjustment	17
Headlights - removal and installation	16
Horn - check and replacement	19
Ignition switch and lock cylinder - check and replacement	9
Instrument cluster - removal and installation	11
Instrument panel fuel, oil and temperature gauges - check	10
Power door lock system - description and check	24
Power window system - description and check	23
Radio and speakers - removal and installation	13
Rear window defogger - check and repair	15
Relays - general information and testing	6
Steering column switches - check and replacement	8
Turn signal and hazard flasher - check and replacement	7
Wiper motor - check and replacement	12
Wiring diagrams - general information	27

1 General information

The electrical system is a 12-volt, negative ground type. Power for the lights and all electrical accessories is supplied by a lead/acid-type battery which is charged by the alternator.

This Chapter covers repair and service procedures for the various electrical components not associated with the engine. Information on the battery, alternator, distributor and starter motor can be found in Chapter 5.

It should be noted that when portions of the electrical system are serviced, the cable should be disconnected from the negative battery terminal to prevent electrical shorts and/or fires. **Caution:** *If the stereo in your vehicle is equipped with an anti-theft system, make sure you have the correct activation code before disconnecting the battery.*

2 Electrical troubleshooting - general information

A typical electrical circuit consists of an electrical component, any switches, relays, motors, fuses, fusible links or circuit breakers related to that component and the wiring and electrical connectors that link the component to both the battery and the chassis. To help you pinpoint an electrical circuit problem, wiring diagrams are included at the end of this Chapter.

Before tackling any troublesome electrical circuit, first study the appropriate wiring diagrams to get a complete understanding of what makes up that individual circuit. Trouble spots, for instance, can often be narrowed down by noting if other components related to the circuit are operating properly. If several components or circuits fail at one time, chances are the problem is in a fuse or ground connection, because several circuits are often routed through the same fuse and ground connections.

Electrical problems usually stem from simple causes, such as loose or corroded connections, a blown fuse, a melted fusible link or a bad relay. Visually inspect the condition of all fuses, wires and connections in a problem circuit before troubleshooting it.

If testing instruments are going to be utilized, use the diagrams to plan ahead of time where you will make the necessary connections in order to accurately pinpoint the trouble spot.

The basic tools needed for electrical troubleshooting include a circuit tester or voltmeter (a 12-volt bulb with a set of test leads can also be used), a continuity tester, which includes a bulb, battery and set of test leads, and a jumper wire, preferably with a circuit breaker incorporated, which can be used to bypass electrical components. Before attempting to locate a problem with test instruments, use the wiring diagram(s) to decide where to make the connections.

Voltage checks

Voltage checks should be performed if a circuit is not functioning properly. Connect one lead of a circuit tester to either the negative battery terminal or a known good ground. Connect the other lead to a electrical connector in the circuit being tested, preferably nearest to the battery or fuse. If the bulb of the tester lights, voltage is present, which means that the part of the circuit between the electrical connector and the battery is problem free. Continue checking the rest of the circuit in the same fashion. When you reach a point at which no voltage is present, the problem lies between that point and the last test point with voltage. Most of the time the problem can be traced to a loose connection. **Note:** *Keep in mind that some circuits receive voltage only when the ignition key is in the Accessory or Run position.*

Finding a short

One method of finding shorts in a circuit is to remove the fuse and connect a test light or voltmeter in its place. There should be no voltage present in the circuit. Move the wiring harness from side to side while watching the test light. If the bulb goes on, there is a short to ground somewhere in that area, probably where the insulation has rubbed through. The same test can be performed on each component in the circuit, even a switch.

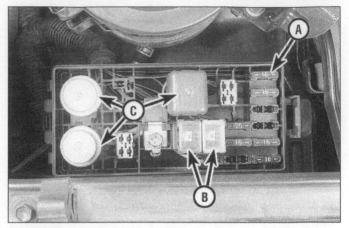

3.1a The interior fuse box is located on the lower half of the instrument panel, behind the fuse panel cover (FJ80 shown - FJ60 and FJ62 similar)

3.1b FJ80 series are equipped with an engine compartment fuse box which contains fuses (A) as well as fusible links (B) and relays (C)

Ground check

Perform a ground test to check whether a component is properly grounded. Disconnect the battery and connect one lead of a self-powered test light, known as a continuity tester, to a known good ground. Connect the other lead to the wire or ground connection being tested. If the bulb goes on, the ground is good. If the bulb does not go on, the ground is not good. **Caution:** *If the stereo in your vehicle is equipped with an anti-theft system, make sure you have the correct activation code before disconnecting the battery.*

Continuity check

A continuity check is done to determine if there are any breaks in a circuit - if it is passing electricity properly. With the circuit off (no power in the circuit), a self-powered continuity tester can be used to check the circuit. Connect the test leads to both ends of the circuit (or to the "power" end and a good ground), and if the test light comes on the circuit is passing current properly. If the light doesn't come on, there is a break somewhere in the circuit. The same procedure can be used to test a switch, by connecting the continuity tester to the power in and power out sides of the switch. With the switch turned On, the test light should come on.

Finding an open circuit

When diagnosing for possible open circuits, it is often difficult to locate them by sight because oxidation or terminal misalignment are hidden by the electrical connectors. Merely wiggling an electrical connector on a sensor or in the wiring harness may correct the open circuit condition. Remember this when an open circuit is indicated when troubleshooting a circuit. Intermittent problems may also be caused by oxidized or loose connections.

Electrical troubleshooting is simple if you keep in mind that all electrical circuits are basically electricity running from the battery, through the wires, switches, relays, fuses and fusible links to each electrical component (light bulb, motor, etc.) and to ground, from which it is passed back to the battery. Any electrical problem is an interruption in the flow of electricity to and from the battery.

3   Fuses - general information

Refer to illustrations 3.1a, 3.1b and 3.3

The electrical circuits of the vehicle are protected by a combination of fuses, circuit breakers and fusible links. Fuse blocks are located under the instrument panel and in the engine compartment depending on the model year of the vehicle **(see illustrations)**.

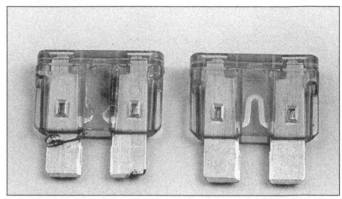

3.3 When a fuse blows, the element between the terminals melts - the fuse on the left is blown, the fuse on the right is good

Each of the fuses is designed to protect a specific circuit, and the various circuits are identified on the fuse panel cover.

Miniaturized fuses are employed in the fuse blocks. These compact fuses, with blade terminal design, allow fingertip removal and replacement. If an electrical component fails, always check the fuse first. The best way to check a fuse is with a test light. Check for power at the exposed terminal tips of each fuse. If power is present on one side of the fuse but not the other, the fuse is blown. A blown fuse can also be confirmed by visually inspecting it **(see illustration)**.

Be sure to replace blown fuses with the correct type. Fuses of different ratings are physically interchangeable, but only fuses of the proper rating should be used. Replacing a fuse with one of a higher or lower value than specified is not recommended. Each electrical circuit needs a specific amount of protection. The amperage value of each fuse is molded into the fuse body.

If the replacement fuse immediately fails, don't replace it again until the cause of the problem is isolated and corrected. In most cases, this will be a short circuit in the wiring caused by a broken or deteriorated wire.

4   Fusible links - general information

Some circuits are protected by fusible links. The links are used in circuits which are not ordinarily fused, such as the ignition circuit.

A conventional type of fusible link (described below) is used on the FJ60 and the FJ62 series. Cartridge type fusible links are used on the FJ80 series. Cartridge type fusible links are located in the engine and passenger compartment fuse blocks and are similar to a large

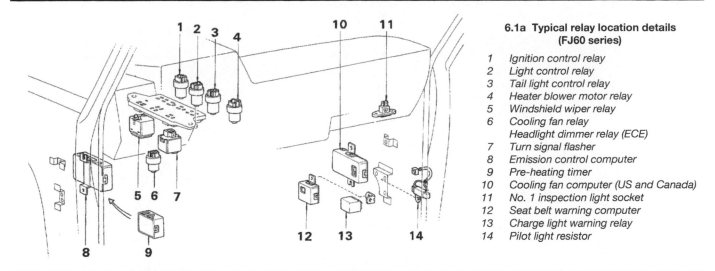

6.1a Typical relay location details (FJ60 series)

1 Ignition control relay
2 Light control relay
3 Tail light control relay
4 Heater blower motor relay
5 Windshield wiper relay
6 Cooling fan relay
   Headlight dimmer relay (ECE)
7 Turn signal flasher
8 Emission control computer
9 Pre-heating timer
10 Cooling fan computer (US and Canada)
11 No. 1 inspection light socket
12 Seat belt warning computer
13 Charge light warning relay
14 Pilot light resistor

fuse **(see illustration 3.1b)**, and, after disconnecting the negative battery cable, are simply unplugged and replaced by a unit of the same amperage. Some fusible links are held in place by a screw which must be loosened before removing the link. **Caution:** *If the stereo in your vehicle is equipped with an anti-theft system, make sure you have the correct activation code before disconnecting the battery.*

Conventional type fusible links cannot be repaired, a new link of the same size wire should be installed in its place. The procedure is as follows: **Caution:** *If the stereo in your vehicle is equipped with an anti-theft system, make sure you have the correct activation code before disconnecting the battery.*

a) *Disconnect the cable from the negative battery terminal.*
b) *Disconnect the fusible link from the wiring harness.*
c) *Cut the damaged fusible link out of the wiring just behind the connector.*
d) *Strip the insulation back approximately 1/2-inch.*
e) *Position the connector on the new fusible link and crimp it into place.*
f) *Use rosin core solder at each end of the new link to obtain a good solder joint.*
g) *Use plenty of electrical tape around the soldered joint. No wires should be exposed.*

h) *Connect the battery ground cable. Test the circuit for proper operation.*

5 Circuit breakers - general information

Circuit breakers protect components such as, power windows, power door locks and headlights.

On some models the circuit breaker resets itself automatically, so an electrical overload in a circuit breaker protected system will cause the circuit to fail momentarily, then come back on. If the circuit doesn't come back on, check it immediately. Once the condition is corrected, the circuit breaker will resume its normal function. Some circuit breakers must be reset manually.

6 Relays - general information and testing

General information
Refer to illustrations 6.1a, 6.1b and 6.1c

1 Several electrical accessories in the vehicle, such as the fuel

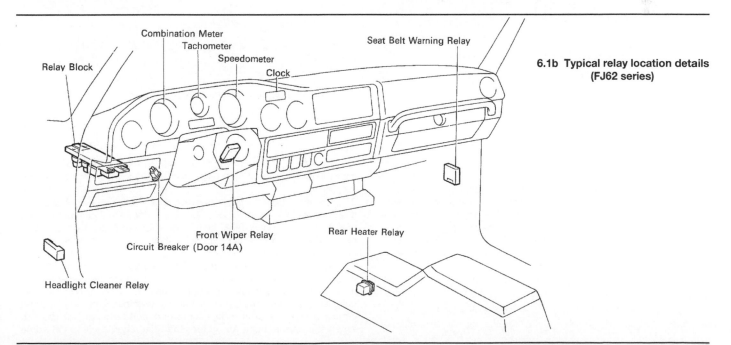

6.1b Typical relay location details (FJ62 series)

Combination Meter
Tachometer
Speedometer
Clock
Seat Belt Warning Relay
Relay Block
Front Wiper Relay
Circuit Breaker (Door 14A)
Rear Heater Relay
Headlight Cleaner Relay

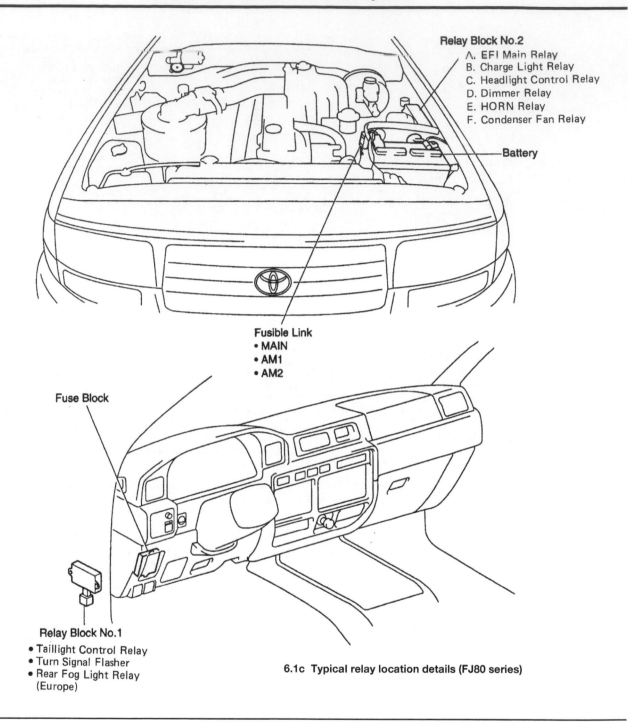

Relay Block No.2
- A. EFI Main Relay
- B. Charge Light Relay
- C. Headlight Control Relay
- D. Dimmer Relay
- E. HORN Relay
- F. Condenser Fan Relay

Battery

Fusible Link
- MAIN
- AM1
- AM2

Fuse Block

Relay Block No.1
- Taillight Control Relay
- Turn Signal Flasher
- Rear Fog Light Relay
  (Europe)

6.1c Typical relay location details (FJ80 series)

injection system, horns, starter, and fog lamps use relays to transmit the electrical signal to the component. Relays use a low-current circuit (the control circuit) to open and close a high-current circuit (the power circuit). If the relay is defective, that component will not operate properly. The various relays are mounted in engine compartment and several locations throughout the vehicle **(see illustrations)**. If a faulty relay is suspected, it can be removed and tested using the procedure below or by a dealer service department or a repair shop. Defective relays must be replaced as a unit.

Testing

2   It's best to refer to the wiring diagram for the circuit to determine the proper hook-ups for the relay you're testing. However, if you're not able to determine the correct hook-up from the wiring diagrams, you may be able to determine the test hook-ups from the information that follows.

3   On most relays, two of the terminals are the relay's control circuit (they connect to the relay coil which, when energized, closes the large contacts to complete the circuit). The other terminals are the power circuit (they are connected together within the relay when the control-circuit coil is energized).

4   The relays are usually marked as an aid to help you determine which terminals are the control circuit and which are the power circuit.

5   Connect a fused jumper wire between one of the two control circuit terminals and the positive battery terminal. Connect another jumper wire between the other control circuit terminal and ground. When the connections are made, the relay should click. On some

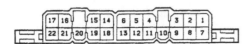

LIGHT CONTROL SWITCH

Switch position \ Terminal (Wire color)	10 EL (W)	11 T (W)	4 H (R)
OFF			
TAIL	O————O		
HEAD	O————O————O		

WIPER AND WASHER SWITCH

Switch position \ Terminal (Wire color)	20 +S (L-R)	21 +1 (L-R)	17 +B (L-W)	22 +2 (L-O)	19 C1 (Lg-R)	14 Ew (B)	15 W (L)
MIST		O————O					
OFF	O————O						
INT	O————O				O————O		
LO		O————O					
HI			O————O				
Washer OFF							
Washer ON						O————O	

HEADLIGHT DIMMER SWITCH

Switch position \ Terminal (Wire color)	13 ED (W-B)	6 HL (R-G)	5 Hu (R-Y)	12 HF (R-W)
Flash	O———————————O————O			
Low Beam	O————O			
High Beam	O———————————————O			

TURN SIGNAL AND HAZARD WARNING SWITCH

Switch position \ Terminal (Wire color)		9 TL (G-B)	3 TB (G-W)	8 TR (G-Y)	2 B1 (G-L)	7 F (G)	1 B2 (G-O)
Turn signal	L	O————O			O————O		
	N				O————O		
	R		O————O		O————O		
Hazard	ON	O————O————O				O————O

8.2a Combination switch connector identification and continuity chart (FJ60 and FJ62 series)

relays, polarity may be critical, so, if the relay doesn't click, try swapping the jumper wires on the control circuit terminals.

6   With the jumper wires connected, check for continuity between the power circuit terminals as indicated by the markings on the relay.

7   If the relay fails any of the above tests, replace it.

7   Turn signal and hazard flasher - check and replacement

1   The turn signal/hazard flasher are contained in a single small canister-shaped unit located in the relay center under the dash near the steering column **(see illustration 6.1a, 6.1b and 6.1c)**.

2   When the flasher unit is functioning properly, an audible click can be heard during its operation. If the turn signals fail on one side or the other and the flasher unit does not make its characteristic clicking sound, a faulty turn signal bulb is indicated.

3   If both turn signals fail to blink, the problem may be due to a blown fuse, a faulty flasher unit, a broken switch or a loose or open connection. If a quick check of the fuse box indicates that the turn signal fuse has blown, check the wiring for a short before installing a new fuse.

4   To replace the flasher, simply pull it out of the relay panel.

5   Make sure that the replacement unit is identical to the original. Compare the old one to the new one before installing it.

6   Installation is the reverse of removal.

8   Steering column switches - check and replacement

Warning: *Some models covered by this manual are equipped with airbags. Always disable the airbag system (see Section 26) before working in the vicinity of the impact sensors, steering column or instrument panel. Failure to follow these procedures may cause accidental deployment of the airbag, which could cause personal injury.*

Check
Refer to illustrations 8.2a and 8.2b

1   Trace the wire from the combination switch to the main wiring harness connector and disconnect the connectors.

2   Using an ohmmeter or self-powered test light and the accompanying diagrams, check for continuity between the indicated switch terminals with the switch in each of the indicated positions **(see illustrations)**. If the continuity isn't as specified, replace the switch.

Replacement
Refer to illustration 8.7

3   Disconnect the cable from the negative terminal of the battery. **Caution:** *If the stereo in your vehicle is equipped with an anti-theft system, make sure you have the correct activation code before disconnecting the battery.*

4   Remove the steering wheel (see Chapter 10).

5   Remove the steering column covers (see Chapter 11).

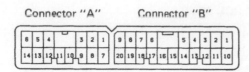

Connector "A"   Connector "B"

LIGHT CONTROL SWITCH

Terminal (Color) Switch position	A2 (W)	A11 (W)	A13 (R)
OFF			
TAIL	O————O		
HEAD	O————O————O		

HEADLIGHT DIMMER SWITCH

Terminal (Color) Switch position	A3 (R-G)	A9 (W-B)	A12 (R-Y)	A14 (R-W)
Flash		O————O————O		
Low beam	O————O			
High beam		O————O		

TURN SIGNAL SWITCH

Terminal (Color) Switch position	A1 (G-W)	A5 (G-B)	A8 (G-Y)
Left turn	O————O		
Neutral			
Right turn		O————O	

FRONT WIPER AND WASHER SWITCH

Terminal (Color) Switch position		B4 (L-R)	B7 (L-B)	B8 (L)	B13 (L-O)	B16 (B)	B18 (L-W)
Wiper	OFF		O				O
	MIST	O————O					
	LO		O				O
	HI				O————O		O
Washer	OFF						
	ON				O————O		

INTERMITTENT WIPER

Terminal (Color) Switch position		B4 (L-R)	B7 (L-B)	B8 (L)	B12 (Y-B)	13 (L-O)	B16 (B)	B18 (L-W)
Wiper	OFF	O————O						
	INT	O————O			O————O			
	LO			O				O
	HI					O————O		O
Washer	OFF							
	ON				O————O			

REAR WIPER AND WASHER SWITCH

Terminal (Color) Switch position		B1 (G)	B2 (V)	B10 (O)	B16 (B)
Washer	ON		O————O		
Wiper	OFF				
	INT			O————O	
	ON	O————O			
Washer	ON	O————O			

8.2b Combination switch connector identification and continuity chart (FJ80 series)

**8.7 Combination switch retaining screws (arrows)
(FJ80 series shown, FJ60 and FJ62 series similar)**

6   Unplug the electrical connectors from the combination switch.

7   Remove the retaining screws and pull the combination switch from the shaft **(see illustration)**.

8   Individual switches can now be removed from the switch body and replaced as necessary. Detach the retaining screws or pins and remove the defective switch.

9   Installation is the reverse the removal.

9   Ignition switch and lock cylinder - check and replacement

Warning: *Some models covered by this manual are equipped with airbags. Always disable the airbag system (see Section 26) before working in the vicinity of the impact sensors, steering column or instrument panel. Failure to follow these procedures may cause accidental deployment of the airbag, which could cause personal injury.*

Check

Refer to illustrations 9.2a, 9.2b and 9.2c

1   Trace the wire from the ignition switch to the main wiring harness connector and disconnect the connector.

Connector A

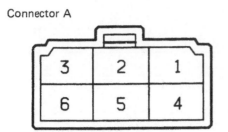

Ignition switch (Canada 60 Series)

Switch position	Terminal	A					
		1	3	6	4	2	5
LOCK						○	○
ACC		○	○			○	○
ON		○	○	○			
START		○		○	○		
Warning	Normal						
	Push					○	○

Connector B

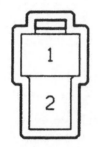

Ignition switch (Ex. Canada 60 Series)

Switch position	Terminal	A					B	
		1	3	6	4	2	1	2
LOCK								
ACC		○	○					
ON		○	○	○				
START		○		○	○	○		
Warning	Normal							
	Push						○	○

Connector B : Canada 70 Series only

9.2a Ignition switch connector identification and continuity chart (FJ60 series)

---

2   Using an ohmmeter or self-powered test light and the accompanying diagrams, check for continuity between the indicated switch terminals with the switch in each of the indicated positions **(see illustrations)**. If the continuity isn't as specified, replace the switch.

Replacement

Refer to illustrations 9.9, 9.11 and 9.12
3   Disconnect the cable from the negative terminal of the battery.

Caution: *If the stereo in your vehicle is equipped with an anti-theft system, make sure you have the correct activation code before disconnecting the battery.*
4   Place the ignition key in the ACC position.
5   Remove the steering wheel (see Chapter 10).
6   Remove the steering column cover (see Chapter 11).
7   Remove the combination switch from the steering column (see Section 8).

Ignition Switch

Switch position	Terminals	1	2	3	4	5	6	7	8
LOCK									
ACC				○	○				
ON			○	○	○			○	○
START		○	○		○	○		○	○

Key Unlock Warning Switch

Switch pin position	Terminals	9	10
Released (Ignition key removed)			
Pushed in (Ignition key set)		○	○

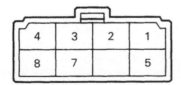

9.2b Ignition switch connector identification and continuity chart (FJ62 series)

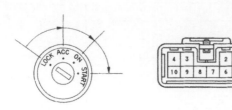

Terminal Switch position	2	3	4	6	7	9	10
LOCK							
ACC					○——○		
ON		○——○		○——○——○			
START	○——○——○			○		○——○	

9.2c Ignition switch connector identification and continuity chart (FJ80 series)

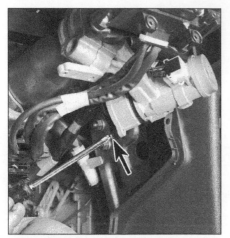

9.9 Detach the retaining screw (arrow) to remove the ignition switch from the steering column housing (FJ80 series shown, FJ60 and FJ62 series similar)

9.11 Detach the lock cylinder illumination ring (if equipped)

9.12 With the lock cylinder in the ACC position - use a small screwdriver to depress the retaining pin, then pull straight out to remove the lock cylinder (FJ80 series shown, FJ60 and FJ62 series similar)

Ignition switch

8   Unplug the ignition switch wiring harness connectors.
9   Detach the switch retaining screw **(see illustration)** remove the switch assembly from the steering column.
10   Installation is reverse of the removal.

Lock cylinder

11   Detach the lock cylinder illumination ring If equipped **(see illustration)**.
12   Using a small screwdriver or punch to depress the lock cylinder retaining pin, then pull straight out on the lock cylinder assembly to remove it from the vehicle **(see illustration)**.
13   To install the lock cylinder, depress the retaining pin and guide the lock cylinder into the steering column housing until the retaining pin to extends itself back into the locating hole in the steering column housing.
14   The remainder of the installation is the reverse of removal.

10   Instrument panel fuel, oil and temperature gauges - check

Warning: *Some models covered by this manual are equipped with airbags. Always disable the airbag system (see Section 26) before working in the vicinity of the impact sensors, steering column or instrument panel. Failure to follow these procedures may cause accidental deployment of the airbag, which could cause personal injury.*
1   All tests below require the ignition switch to be turned to ON position when testing.

2   If the gauge pointer does not move from the empty, low or cold positions, check the fuse. If the fuse is OK, locate the particular sending unit for the circuit you're working on (see Chapter 4 for fuel sending unit location, Chapter 2C for the oil sending unit location or Chapter 3 for the temperature sending unit location). Connect the sending unit connector to ground. If the pointer goes to the full, high or hot position replace the sending unit. If the pointer stays in same position, use a jumper wire to ground the sending unit terminal on the back of the gauge, if necessary, refer to the wiring diagrams at the end of this Chapter. If the pointer moves, the problem lies in the wire between the gauge and the sending unit. If the pointer does not move with the sending unit terminal on the back of the gauge grounded, check for voltage at the other terminal of the gauge. If voltage is present, replace the gauge.

11   Instrument cluster - removal and installation

Refer to illustrations 11.3a and 11.3b
Warning: *Some models covered by this manual are equipped with airbags. Always disable the airbag system (see Section 26) before working in the vicinity of the impact sensors, steering column or instrument panel. Failure to follow these procedures may cause accidental deployment of the airbag, which could cause personal injury.*
1   Disconnect the negative battery cable. **Caution:** *If the stereo in your vehicle is equipped with an anti-theft system, make sure you have the correct activation code before disconnecting the battery.*
2   Remove the instrument cluster bezel (see Chapter 11).
3   Remove the cluster mounting screws **(see illustrations)** and pull

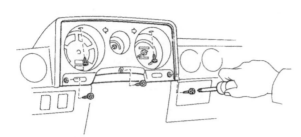

11.3a FJ60 and FJ62 instrument cluster mounting details

11.3b FJ80 instrument cluster mounting details

the instrument cluster towards the steering wheel.

4   Unscrew the speedometer cable from the backside of the instrument cluster if equipped.

5   Disconnect any electrical connectors that would interfere with removal.

6   Cover the steering column with a cloth to protect the trim covers then remove the instrument cluster from the vehicle.

7   Installation is the reverse of removal.

12   Wiper motor - check and replacement

Check

Note: *Refer to the wiring diagrams for wire colors and locations in the following checks. Keep in mind that power wires are generally larger in diameter and brighter colors, where ground wires are usually smaller in diameter and darker colors. When checking for voltage, probe a grounded 12-volt test light to each terminal at a connector until it lights; this verifies voltage (power) at the terminal.*

1   If the wipers work slowly, make sure the battery is in good condition and has a strong charge (see Chapter 1). If the battery is in good condition, remove the wiper motor (see below) and operate the wiper arms by hand. Check for binding linkage and pivots. Lubricate or repair the linkage or pivots as necessary. Reinstall the wiper motor. If the wipers still operate slowly, check for loose or corroded connections, especially the ground connection. If all connections look OK, replace the motor.

2   If the wipers fail to operate when activated, check the fuse. If the fuse is OK, connect a jumper wire between the wiper motor and ground, then retest. If the motor works now, repair the ground connection. If the motor still doesn't work, turn on the wipers and check for voltage at the motor. If there's no voltage at the motor, remove the motor and check it off the vehicle with fused jumper wires from the battery. If the motor now works, check for binding linkage (see Step 1 above). If the motor still doesn't work, replace it. If there's no voltage at the motor, check for voltage at the switch. If there's no voltage at the switch, check the wiring between the switch and fuse panel for continuity. If the wiring is OK, the switch is probably bad.

3   **If the wipers only work on one speed**, check the continuity of the wires between the switch and motor. If the wires are OK, replace the switch.

4   **If the interval (delay) function is inoperative**, check the continuity of all the wiring between the switch and motor. If the wiring is OK, replace the interval module.

5   **If the wipers stop at the position they're in when the switch is turned off (fail to park)**, check for voltage at the wiper motor when the wiper switch is OFF but the ignition is ON. If voltage is present, the limit switch in the motor is malfunctioning. Replace the wiper motor. If no voltage is present, trace and repair the limit switch wiring between the fuse panel and wiper motor.

6   **If the wipers won't shut off unless the ignition is OFF**, disconnect the wiring from the wiper control switch. If the wipers stop, replace the switch. If the wipers keep running, there's a defective limit

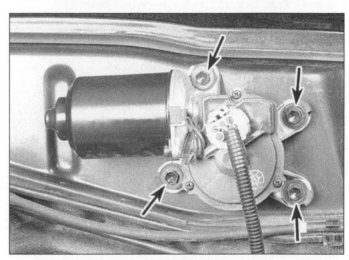

12.8 Front wiper motor retaining bolt locations (arrows)

switch in the motor; replace the motor.

7   **If the wipers won't retract below the hoodline,** check for mechanical obstructions in the wiper linkage or on the vehicle's body which would prevent the wipers from parking. If there are no obstructions, check the wiring between the switch and motor for continuity. If the wiring is OK, replace the wiper motor.

Replacement

Front wiper motor

Refer to illustration 12.8

8   Remove the wiper motor retaining bolts **(see illustration)**, then pull the motor out from the firewall.

9   Remove the motor spindle nut from the backside of the motor.

10   Disconnect the electrical connector and remove the motor from the vehicle.

11   Installation is the reverse of removal.

Rear wiper motor

Refer to illustration 12.16

12   Pull the rear wiper arm cover back to access the wiper arm retaining nut. Detach the nut and pull the wiper arm straight off the shaft to remove it.

13   Remove the drive spindle retaining nut.

14   Open the rear liftgate and remove the inside handle, tailgate trim panel and the tailgate access cover (see Chapter 11).

15   Disconnect the electrical connector from wiper motor.

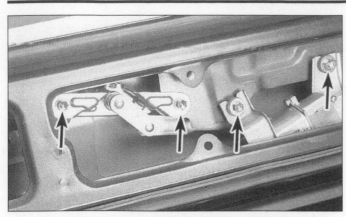

12.16 Rear wiper motor retaining bolt locations (arrows) (FJ80 series shown, FJ60 and FJ62 similar)

16 Detach the wiper motor mounting bolts, then remove the wiper motor and linkage assembly from the vehicle **(see illustration)**.
17 Installation is the reverse of removal.

13 Radio and speakers - removal and installation

Warning: *Some models covered by this manual are equipped with airbags. Always disable the airbag system (see Section 26) before working in the vicinity of the impact sensors, steering column or instrument panel. Failure to follow these procedures may cause accidental deployment of the airbag, which could cause personal injury.*
1 Disconnect the negative battery cable. **Caution:** *If the stereo in your vehicle is equipped with an anti-theft system, make sure you have the correct activation code before disconnecting the battery.*

Radio

Refer to illustration 13.4a, 13.4b and 13.4c
2 On earlier models, detach the radio control knobs. Use a deep socket to remove the control shaft nuts and washers, then remove the radio trim bezel.
3 On later models, remove the instrument cluster bezel (see Chapter 11).
4 Remove the retaining screws **(see illustrations)** and pull the radio outward to access the backside, then disconnect the electrical connectors and the antenna lead and lift the radio out of the vehicle **(see illustration)**.
5 Installation is the reverse of removal.

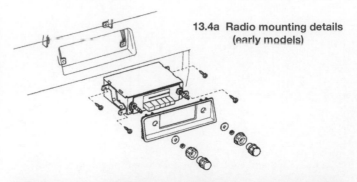

13.4a Radio mounting details (early models)

13.4b Radio mounting details (later models)

Speakers

Instrument panel mounted

6 On early models equipped with an instrument panel mounted front speaker, remove the radio as described above.
7 Remove the Instrument panel glove compartment (see Chapter 11).
8 Working through the radio and glove compartment opening detach the speaker retaining screws and the electrical connector, then withdraw the speaker from the instrument panel.

Door mounted

Refer to illustrations 13.10
9 On later models equipped with an door mounted front speakers,

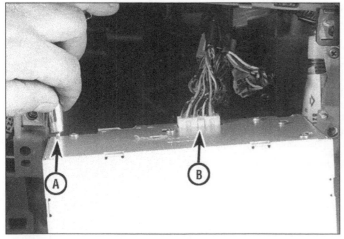

13.4c On all models, pull the radio forward, then disconnect the antenna lead (A) and the electrical connector (B)

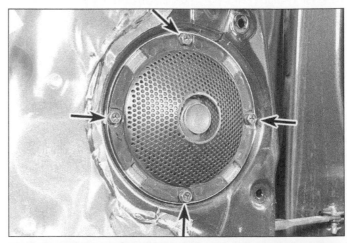

13.10 On vehicles equipped with door mounted speakers, remove the speaker retaining screws (arrows) and disconnect the electrical connector to remove the speaker from the vehicle

15.4 When measuring the voltage at the rear window defogger grid, wrap a piece of aluminum foil around the negative probe of the voltmeter and press the foil against the wire with your finger

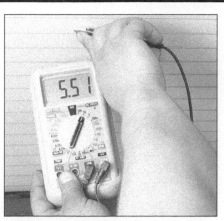

15.5 To determine if a heating element has broken, check the voltage at the center of each element - if the voltage is 6-volts, the element is unbroken - if the voltage is 12-volts, the element is broken between the center and the positive end - if there is no voltage, the element is broken between the center and ground

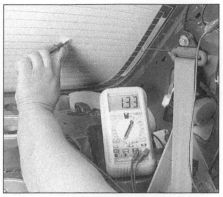

15.7 To find the break, place the voltmeter positive lead against the defogger positive terminal, place the voltmeter negative lead with the foil strip against the heating element at the positive terminal end and slide it toward the negative terminal end - the point at which the voltmeter reading changes abruptly is the point at which the element is broken

remove the door trim panels (see Chapter 11).

10   Remove the speaker retaining screws. Disconnect the electrical connector and remove the speaker from the vehicle **(see illustration)**.

11   Installation is the reverse of removal.

---

14   Antenna - removal and installation

1   Detach the radio and disconnect the antenna lead from the back-side of the radio (see Section 13). Attach a piece of stiff wire to the end of the antenna lead, then remove any retaining clips under the instrument panel securing the antenna lead.

2   On vehicles equipped with fixed type antennas, use a small wrench and remove the antenna mast retaining nut.

3   Using a pair of snap ring pliers or similar tool, remove the antenna base retaining nut.

4   On vehicles equipped with a power antenna, remove the antenna motor mounting screws from the inner fenderwell.

5   Pull the antenna up and out to remove it until you can see the wire, then remove the old antenna lead from the wire and replace it with the new antenna lead.

6   Working in the passenger compartment, pull the wire and the antenna lead rearward into the passenger compartment.

7   The remainder of installation is the reverse of removal.

---

15   Rear window defogger - check and repair

1   The rear window defogger consists of a number of horizontal elements baked onto the glass surface.

2   Small breaks in the element can be repaired without removing the rear window.

Check

Refer to illustrations 15.4, 15.5 and 15.7

3   Turn the ignition switch and defogger system switches to the ON position.

4   When measuring voltage during the next two tests, wrap a piece of aluminum foil around the tip of the voltmeter negative probe and press the foil against the heating element with your finger **(see illustration)**.

5   Check the voltage at the center of each heating element **(see illustration)**. If the voltage is 6-volts, the element is okay (there is no break). If the voltage is 12-volts, the element is broken between the center of the element and the positive end. If the voltage is 0-volts the element is broken between the center of the element and ground.

6   Connect the negative lead to a good body ground. The reading should stay the same.

7   To find the break, place the voltmeter positive lead against the defogger positive terminal. Place the voltmeter negative lead with the foil strip against the heating element at the positive terminal end and slide it toward the negative terminal end. The point at which the voltmeter deflects from zero to several volts is the point at which the heating element is broken **(see illustration)**.

Repair

Refer to illustration 15.13

8   Repair the break in the element using a repair kit specifically recommended for this purpose, such as Dupont paste No. 4817 (or equivalent). Included in this kit is plastic conductive epoxy.

9   Prior to repairing a break, turn off the system and allow it to cool off for a few minutes.

10   Lightly buff the element area with fine steel wool, then clean it thoroughly with rubbing alcohol.

11   Use masking tape to mask off the area being repaired.

12   Thoroughly mix the epoxy, following the instructions provided with the repair kit.

13   Apply the epoxy material to the slit in the masking tape, overlapping the undamaged area about 3/4-inch on either end **(see illustration)**.

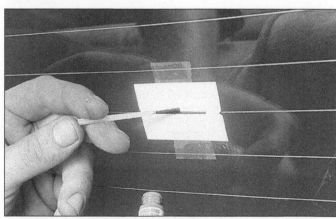

15.13 To use a defogger repair kit, apply masking tape to the inside of the window at the damaged area, then brush on the special conductive coating

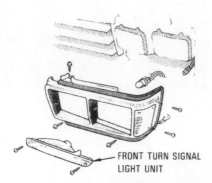

16.2 Headlight bezel mounting details (FJ62 series shown, FJ60 series similar)

16.4 Remove the retaining screws (don't confuse them with the adjustment screws) and lift the headlight ring off (rectangular headlight shown, round headlight similar)

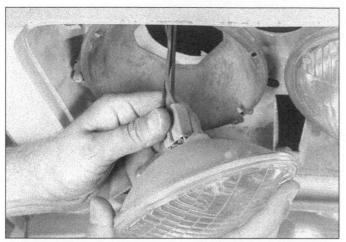

16.5 Pull the headlight forward and unplug the connector (round headlight shown, rectangular headlight similar)

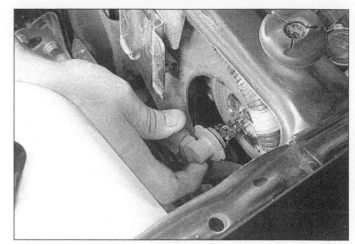

16.11 Rotate the bulb assembly out from the housing - remove the bulb from the bulb holder to replace it (Halogen type bulb)

14   Allow the repair to cure for 24 hours before removing the tape and using the system.

---

16 Headlights - removal and installation

1   Disconnect the negative cable from the battery. **Caution:** *If the stereo in your vehicle is equipped with an anti-theft system, make sure you have the correct activation code before disconnecting the battery.*

Sealed beam type

Refer to illustrations 16.2, 16.4 and 16.5

2   Remove the headlight trim bezel. Remove the turn signal assembly on FJ62 models **(see illustration)**.
3   Whenever replacing a headlight, be careful not to turn the spring-loaded adjusting screws of the headlight, as this will alter the aim.
4   Remove the screws which secure the retaining ring and withdraw the ring. Support the light as this is done **(see illustration)**.
5   Pull the headlight out slightly and disconnect the electrical connector from the rear of the light. Remove the light from the vehicle **(see illustration)**.
6   Position the new unit close enough to connect the electrical connector. Make sure that the numbers molded into the lens are at the top.
7   Install the retaining ring with its mounting screws and spring.
8   Check the headlights for proper operation and adjustment. If the adjusting screws were not altered, the new headlight will not need adjustment.
9   Install the radiator grille.

Halogen bulb-type

Refer to illustration 16.11
Warning: *Halogen gas filled bulbs are under pressure and may shatter if the surface is scratched or the bulb is dropped. Wear eye protection and handle the bulbs carefully, grasping only the base whenever possible. Do not touch the surface of the bulb with your fingers because the oil from your skin could cause it to overheat and fail prematurely. If you do touch the bulb surface, clean it with rubbing alcohol.*

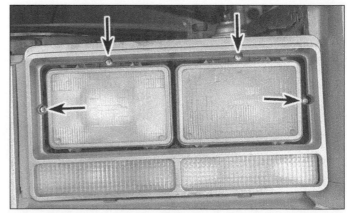

17.1a The headlight vertical adjustment screw is located at the top of the headlight and the horizontal screw is on the side of the headlight (arrows) (rectangular headlight shown, round headlight similar)

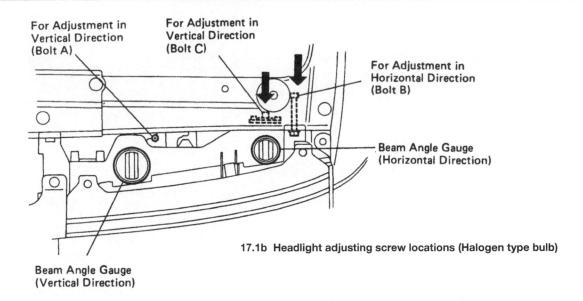

For Adjustment in Vertical Direction (Bolt A)

For Adjustment in Vertical Direction (Bolt C)

For Adjustment in Horizontal Direction (Bolt B)

Beam Angle Gauge (Horizontal Direction)

Beam Angle Gauge (Vertical Direction)

17.1b Headlight adjusting screw locations (Halogen type bulb)

10   Reach behind the headlight assembly and unplug the electrical connector.
11   Grasp the bulb holder securely and rotate it counterclockwise to remove it from the housing **(see illustration)**.
12   Pull straight out on the bulb to remove from it from the bulb holder. Without touching the glass with your bare fingers, insert the new bulb assembly into the headlight housing.
13   Plug in the electrical connector. Test headlight operation, then close the hood.

17   Headlights - adjustment

Refer to illustrations 17.1a, 17.1b and 17.3
Note: *The headlights must be aimed correctly. If adjusted incorrectly they could blind the driver of an oncoming vehicle and cause a serious accident or seriously reduce your ability to see the road. The headlights should be checked for proper aim every 12 months and any time a new headlight is installed or front end body work is performed. It should be emphasized that the following procedure is only an interim step which will provide temporary adjustment until the headlights can be adjusted by a properly equipped shop.*
1   Sealed beam headlights have two spring loaded adjusting screws, one on the top or bottom controlling up-and-down movement and one on the side controlling left-and-right movement **(see illustration)**. Halogen bulb type headlights have two adjustment screws located on the top of each headlight housing. Adjustments are made by turning the screws to center the level gauges located on top of the headlight housings **(see illustration)**.
2   There are several methods of adjusting the headlights. The simplest method requires masking tape, a blank wall and a level floor.
3   Position masking tape vertically on the wall in reference to the vehicle centerline and the centerlines of both headlights**(see illustration)**.
4   Position a horizontal tape line in reference to the centerline of all the headlights. **Note:** *It may be easier to position the tape on the wall with the vehicle parked only a few inches away.*
5   Adjustment should be made with the vehicle parked 25 feet from the wall, sitting level, the gas tank half-full and no unusually heavy load in the vehicle.
6   Starting with the low beam adjustment, position the high intensity zone so it is two inches below the horizontal line and two inches to the right of the headlight vertical line. Adjustment is made by turning the top or bottom adjusting screw to raise or lower the beam. The adjusting screw on the side should be used in the same manner to move the beam left or right.

7   With the high beams on, the high intensity zone should be vertically centered with the exact center just below the horizontal line.
Note: *It may not be possible to position the headlight aim exactly for both high and low beams. If a compromise must be made, keep in mind that the low beams are the most used and have the greatest effect on driver safety.*
8   Have the headlights adjusted by a dealer service department or service station at the earliest opportunity.

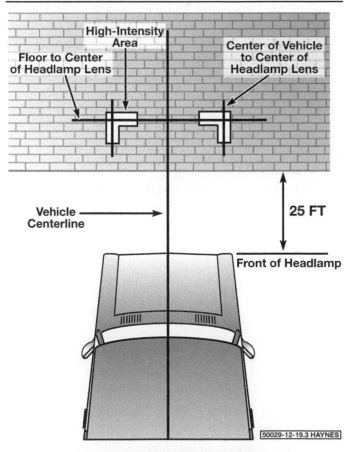

High-Intensity Area

Floor to Center of Headlamp Lens

Center of Vehicle to Center of Headlamp Lens

Vehicle Centerline

25 FT

Front of Headlamp

50029-12-19.3 HAYNES

17.3 Headlight adjustment details

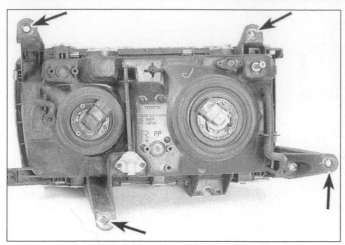

18.3 Headlight housing retaining screw (arrows) locations (Halogen type bulb)

18   Headlight housing (Halogen type bulb) - removal and installation

Refer to illustration 18.3
Warning: *These vehicles are equipped with halogen gas-filled headlight bulbs which are under pressure and may shatter if the surface is damaged or the bulb is dropped. Wear eye protection and handle the bulbs carefully, grasping only the base whenever possible. Do not touch the surface of the bulb with your fingers because the oil from your skin could cause it to overheat and fail prematurely. If you do touch the bulb surface, clean it with rubbing alcohol.*
1   Remove the headlight bulb (see Section 16).
2   Remove the radiator grille (see Chapter 11), side marker light (see Section 20) and the headlight trim bezel.
3   Remove the retaining bolts, detach the housing and withdraw it from the vehicle **(see illustration)**.
4   Installation is the reverse of removal.

19   Horn - check and replacement

Check

Refer to illustration 19.3
Note: *Check the fuses before beginning electrical diagnosis.*
1   Disconnect the electrical connector from the horn.

19.9 Disconnect the electrical connector, remove the bolt (arrows) and detach the horn

19.3 Check for power at the horn terminal with the horn button depressed

2   To test the horn, connect battery voltage to the two terminals with a pair of jumper wires. If the horn doesn't sound, replace it.
3   If the horn does sound, check for voltage at the terminal when the horn button is depressed **(see illustration)**. If there's voltage at the terminal, check for a bad ground at the horn.
4   If there's no voltage at the horn, check the relay (see Section 6). Note that most horn relays are either the four-terminal or externally grounded three-terminal type.
5   If the relay is OK, check for voltage to the relay power and control circuits. If either of the circuits is not receiving voltage, inspect the wiring between the relay and the fuse panel.
6   If both relay circuits are receiving voltage, depress the horn button and check the circuit from the relay to the horn button for continuity to ground. If there's no continuity, check the circuit for an open. If there's no open circuit, replace the horn button.
7   If there's continuity to ground through the horn button, check for an open or short in the circuit from the relay to the horn.

Replacement

Refer to illustration 19.9
8   Remove the radiator grille (see Chapter 11).
9   To replace the horn(s), disconnect the electrical connector and remove the bracket bolt **(see illustration)**.
10   Installation is the reverse of removal.

20   Bulb replacement

Front side marker lights

Refer to illustrations 20.1, 20.2a, and 20.2b
1   On early models, remove the screws that secure the side marker light lens **(see illustration)**. Lift the lens out to access the light bulbs.

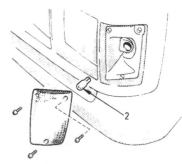

20.1 Front side marker light mounting details (FJ62 series shown, FJ60 series similar)

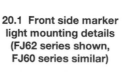

20.2a On FJ80 series, remove the retaining screw at the top of the housing assembly . . .

20.2b . . . then pull the housing outward to access the bulbs

20.4 Remove the lens retaining screws (arrows) to access the front turn signal bulbs (FJ62 and FJ80 series)

2   On later models, remove the screw that secures the top of the side marker light housing. Then pull the side marker light housing outward to access the bulbs located on the backside. rotate the bulb holder counterclockwise and pull the bulb out (see illustrations).
3   Installation is the reverse of removal.

Front turn signal light (FJ62 and FJ80 series)

Refer to illustration 20.4
4   Remove the screws securing the turn signal lens (see illustration).
5   Rotate the bulb counterclockwise to remove it.
6   Installation is the reverse of removal.

Rear tail light/brake light/turn signal

Refer to illustrations 20.7, 20.8a and 20.8b
7   On early models, remove screws and detach the tail light lens (see illustration). Push in and rotate the turn signal, brake light and tail light bulbs counterclockwise to remove them from the tail light housing.
8   On later models, lower the tailgate and remove tail light housing retaining screws located in the tailgate opening. Detach the remaining screw located in the center of the tail light lens and remove the tail light housing assembly (see illustrations).
9   Rotate the bulb holder counterclockwise and pull the bulb out to remove it.
10   Installation is the reverse of removal.

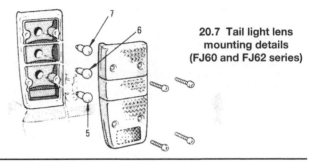

20.7 Tail light lens mounting details (FJ60 and FJ62 series)

License plate light

Refer to illustration 20.11
11   Remove the screws and detach the lens assembly (see illustration), then remove the bulb from the socket.
12   Installation is the reverse of removal.

Instrument cluster lights

Refer to illustration 20.14
13   To gain access to the instrument cluster illumination bulbs, the instrument cluster will have to be removed (see Section 11). The bulbs can then be removed and replaced from the rear of the cluster.

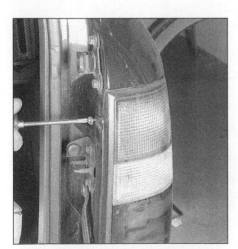

20.8a On FJ80 models, detach the lens retaining screws located in the tail gate opening . . .

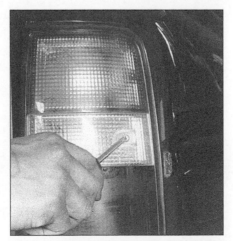

20.8b . . .then detach the screw located at the center of the lens and pull out the tail light housing to access the bulbs

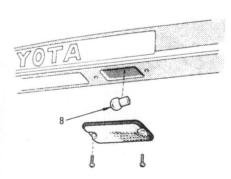

20.11 To remove the license plate light, detach the screws, then pull the lens down and out to access the bulb

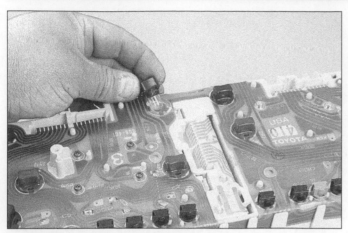

20.14 Remove the instrument cluster bulbs by rotating them 1/4-turn counterclockwise and pulling straight out

14   Rotate the bulb counterclockwise to remove it **(see illustration)**.
15   Installation is the reverse of removal.

Interior light

16   Pry the interior lens off the interior light housing.
17   Detach the bulb from the terminals. It may be necessary to pry the bulb out - if this is the case, pry only on the ends of the bulb (otherwise the glass may shatter).
18   Installation is the reverse of removal.

21   Electric side view mirrors - description and check

Refer to illustration 21.7
1   Most electric rear view mirrors use two motors to move the glass; one for up and down adjustments and one for left-right adjustments.
2   The control switch has a selector portion which sends voltage to the left or right side mirror. With the ignition ON but the engine OFF, roll down the windows and operate the mirror control switch through all functions (left-right and up-down) for both the left and right side mirrors.
3   Listen carefully for the sound of the electric motors running in the mirrors.
4   If the motors can be heard but the mirror glass doesn't move, there's probably a problem with the drive mechanism inside the mirror. Remove and disassemble the mirror to locate the problem.
5   If the mirrors do not operate and no sound comes from the mirrors, check the fuse (see Chapter 1).
6   If the fuse is OK, remove the mirror control switch from its mounting without disconnecting the wires attached to it. Turn the ignition ON and check for voltage at the switch. There should be voltage at one terminal. If there's no voltage at the switch, check for an open or short in the circuit between the fuse panel and the switch.
7   If there's voltage at the switch, disconnect it. Check the switch for continuity in all its operating positions **(see illustration)**. If the switch does not have continuity, replace it.
8   Re-connect the switch. Locate the wire going from the switch to ground. Leaving the switch connected, connect a jumper wire between this wire and ground. If the mirror works normally with this wire in place, repair the faulty ground connection.
9   If the mirror still doesn't work, remove the mirror and check the wires at the mirror for voltage. Check with ignition ON and the mirror selector switch on the appropriate side. Operate the mirror switch in all its positions. There should be voltage at one of the switch-to-mirror wires in each switch position (except the neutral "off" position).
10   If there's not voltage in each switch position, check the circuit between the mirror and control switch for opens and shorts.
11   If there's voltage, remove the mirror and test it off the vehicle with jumper wires. Replace the mirror if it fails this test.

22   Cruise control system - description and check

1   The cruise control system maintains vehicle speed with a electrically operated motor located in the engine compartment, which is connected to the accelerator pedal by a cable. The system consists of the cruise control unit, brake switch, control switches and vehicle speed sensor. Some features of the system require special testers and diagnostic procedures which are beyond the scope of this manual. Listed

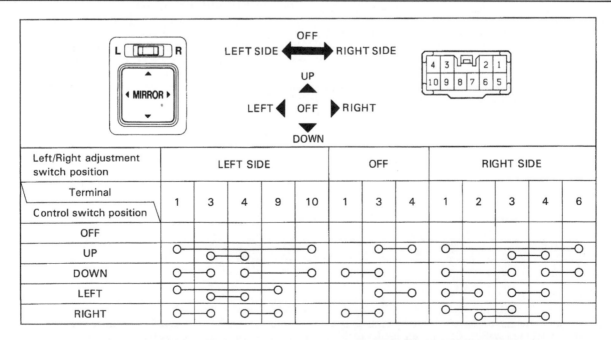

21.7 Power mirror switch terminal identification and continuity chart (FJ80 series)

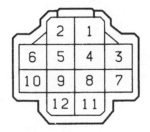

Window Lock Switch

Switch position \ Terminal	10	0
LOCK		
UNLOCK	O———	———O

Operation Window \ Switch position \ Terminal	Front · Left 2	10	1	3	Front · Right 5	10	7	4	Rear · Left 12	10	7	11	Rear · Right 9	10	7	8
UP	O—O				O—O	O—O			O—O	O—O			O—O	O—O		
OFF		O—			O—	—O—O			O—		—O—O		O—		—O—O	
DOWN		O—O			O—	O—O			O—		O—O		O—		—O—O	
DOWN-HOLD	O—O—O															

23.10a **Power window master control switch terminal identification and continuity chart (FJ60 and FJ62 series)**

below are some general procedures that may be used to locate common problems.

2   Locate and check the fuse (see Section 3).

3   Have an assistant operate the brake lights while you check their operation (voltage from the brake light switch deactivates the cruise control).

4   If the brake lights don't come on or stay on all the time, correct the problem and retest the cruise control.

5   Visually inspect the control cable between cruise control motor and the throttle linkage for free movement, replace if necessary.

6   The cruise control system uses a speed sensing device. The speed sensor is located in the speedometer. To test the speed sensor (see Chapter 6).

7   Test drive the vehicle to determine if the cruise control is now working. If it isn't, take it to a dealer service department or an automotive electrical specialist for further diagnosis

23 Power window system - description and check

Refer to illustrations 23.10a, 23.10b, 23.10c, 23.10d and 23.12

1   The power window system operates electric motors, mounted in the doors, which lower and raise the windows. The system consists of the control switches, relays, the motors, regulators, glass mechanisms and associated wiring.

2   The power windows can be lowered and raised from the master control switch by the driver or by remote switches located at the individual windows. Each window has a separate motor which is reversible. The position of the control switch determines the polarity and therefore the direction of operation.

3   The circuit is protected by a fuse and a circuit breaker. Each motor is also equipped with an internal circuit breaker, this prevents one stuck window from disabling the whole system.

4   The power window system will only operate when the ignition switch is ON. In addition, many models have a window lockout switch at the master control switch which, when activated, disables the switches at the rear windows and, sometimes, the switch at the passenger's window also. Always check these items before troubleshooting a window problem.

5   These procedures are general in nature, so if you can't find the problem using them, take the vehicle to a dealer service department or other properly equipped repair facility.

6   If the power windows won't operate, always check the fuse and circuit breaker first.

7   If only the rear windows are inoperative, or if the windows only operate from the master control switch, check the rear window lockout switch for continuity in the unlocked position. Replace it if it doesn't have continuity.

8   Check the wiring between the switches and fuse panel for continuity. Repair the wiring, if necessary.

9   If only one window is inoperative from the master control switch, try the other control switch at the window. **Note:** *This doesn't apply to the drivers door window.*

10   If the same window works from one switch, but not the other,

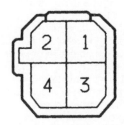

Switch position \ Terminal	1	2	4	3
UP		O—O	O—	—O
OFF				
DOWN	O—	O—O	—O	

23.10b **Power window switch terminal identification and continuity chart (FJ60 and FJ62 series)**

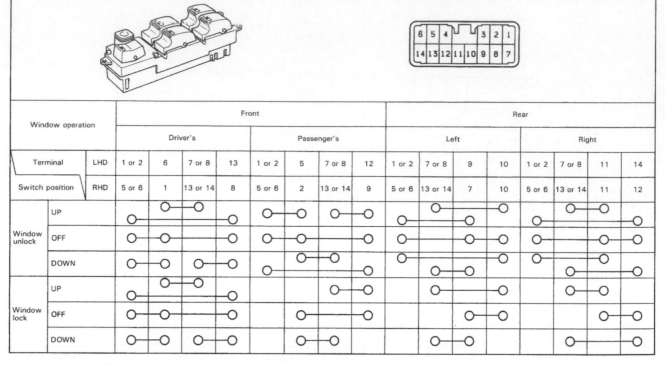

Window operation		Front								Rear							
		Driver's				Passenger's				Left				Right			
Terminal	LHD	1 or 2	6	7 or 8	13	1 or 2	5	7 or 8	12	1 or 2	7 or 8	9	10	1 or 2	7 or 8	11	14
Switch position	RHD	5 or 6	1	13 or 14	8	5 or 6	2	13 or 14	9	5 or 6	13 or 14	7	10	5 or 6	13 or 14	11	12
Window unlock	UP																
	OFF																
	DOWN																
Window lock	UP																
	OFF																
	DOWN																

23.10c Power window master control switch terminal identification and continuity chart (FJ80 series)

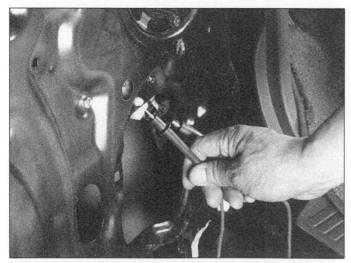

Terminal / Switch position	1	2	3	4	5
UP	○———————————————○		○———○		
OFF	○———○		○———○		
DOWN	○———○			○———○	

If continuity is not as specified, replace the switch.

23.10d Power window switch terminal identification and continuity chart (FJ80 series)

23.12 If no voltage is present at the motor with the switch depressed, check for voltage at the switch

check the switch for continuity (see illustrations). If the continuity is not as specified replace the switch.

11   If the switch tests OK, check for a short or open in the circuit between the affected switch and the window motor.

12   If one window is inoperative from both switches, remove the trim panel from the affected door and check for voltage at the switch and at the motor (see illustration) while the switch is operated.

13   If voltage is reaching the motor, disconnect the glass from the regulator (see Chapter 11). Move the window up and down by hand while checking for binding and damage. Also check for binding and damage to the regulator. If the regulator is not damaged and the window moves up and down smoothly, replace the motor. If there's binding or damage, lubricate, repair or replace parts, as necessary.

14   If voltage isn't reaching the motor, check the wiring in the circuit for continuity between the switches and motors. You'll need to consult the wiring diagram for the vehicle. If the circuit is equipped with a relay, check that the relay is grounded properly and receiving voltage.

15   Test the windows after you are done to confirm proper repairs.

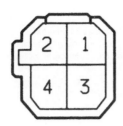

Switch position \ Terminal	2	1	4	3
LOCK	○———————○	○—————————○		
OFF				
UNLOCK	○	○———————○		○

24.3a Power door lock switch terminal identification and continuity chart (FJ60 and FJ62 series)

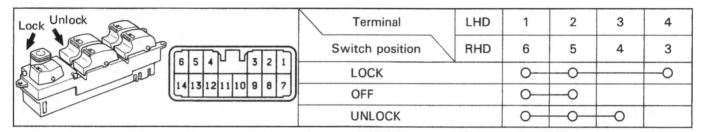

Terminal	LHD	1	2	3	4
Switch position	RHD	6	5	4	3
LOCK		○———○		○	
OFF		○———○			
UNLOCK		○———○———○			

If continuity is not as specified, replace the switch.

24.3b Power door lock switch terminal identification and continuity chart (FJ80 series)

24 Power door lock system - description and check

Refer to illustrations 24.3a, 24.3b and 24.6

The power door lock system operates the door lock actuators mounted in each door. The system consists of the switches, actuators, a control unit and associated wiring. Diagnosis can usually be limited to simple checks of the wiring connections and actuators for minor faults which can be easily repaired. Since this system uses an electronic control unit in-depth diagnosis should be left to a dealership service department. The door lock control unit is located behind the instrument panel, to the right of the fuse box.

Power door lock systems are operated by bi-directional solenoids located in the doors. The lock switches have two operating positions: Lock and Unlock. When activated, the switch sends a ground signal to the door lock control unit to lock or unlock the doors. Depending on which way the switch is activated, the control unit reverses polarity to the solenoids, allowing the two sides of the circuit to be used alternately as the feed (positive) and ground side.

Some vehicles may have an anti-theft systems incorporated into the power locks. If you are unable to locate the trouble using the following general Steps, consult your a dealer service department.

1   Always check the circuit protection first. Some vehicles use a combination of circuit breakers and fuses.
2   Operate the door lock switches in both directions (Lock and Unlock) with the engine off. Listen for the click of the solenoids operating.
3   Test the switches for continuity (**see illustrations**). Replace the switch if there's not continuity in both switch positions.
4   Check the wiring between the switches, control unit and solenoids for continuity. Repair the wiring if there's no continuity.
5   Check for a bad ground at the switches or the control unit.
6   If all but one lock solenoids operate, remove the trim panel from the affected door (see Chapter 11) and check for voltage at the solenoid while the lock switch is operated One of the wires should have voltage in the Lock position; the other should have voltage in the Unlock position (**see illustration**).
7   If the inoperative solenoid is receiving voltage, replace the solenoid.

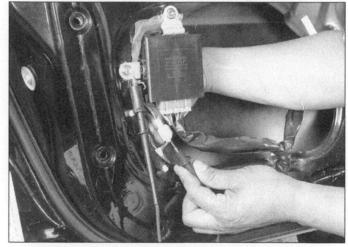

24.6 Check for voltage at the lock solenoid while the lock switch is operated

8   If the inoperative solenoid isn't receiving voltage, check the relay or for an open or short in the wire between the lock solenoid and the control unit. **Note:** *It's common for wires to break in the portion of the harness between the body and door (opening and closing the door fatigues and eventually breaks the wires).*

25 Daytime Running Lights (DRL) - general information

The Daytime Running Lights (DRL) system used on Canadian models illuminates the headlights whenever the engine is running. The only exception is with the engine running and the parking brake engaged. Once the parking brake is released, the lights will remain on as long as the ignition switch is on, even if the parking brake is later applied.

The DRL system supplies reduced power to the headlights so they won't be too bright for daytime use, while prolonging headlight life.

26 Airbag - general information

Some models are equipped with a Supplemental Restraint System (SRS), more commonly known as an airbag. This system is designed to protect the driver, and the front seat passenger, from serious injury in the event of a head-on or frontal collision. It consists of an airbag module in the center of the steering wheel and the right side of the instrument panel and a sensing/diagnostic module mounted in the center of the vehicle.

Airbag module

Steering wheel-mounted

The airbag inflator module contains a housing incorporating the cushion (airbag) and inflator unit, mounted in the center of the steering wheel. The inflator assembly is mounted on the back of the housing over a hole through which gas is expelled, inflating the bag almost instantaneously when an electrical signal is sent from the system. A coil assembly on the steering column under the module carries this signal to the module.

This coil assembly can transmit an electrical signal regardless of steering wheel position. The igniter in the air bag converts the electrical signal to heat and ignites the sodium azide/copper oxide powder, producing nitrogen gas, which inflates the bag.

Instrument panel-mounted

The airbag is mounted above the glove compartment and designated by the letters SRS (Supplemental Restraint System). It consists of an inflator containing an igniter, a bag assembly, a reaction housing and a trim cover.

The airbag is considerably larger that the steering wheel-mounted unit and is supported by the steel reaction housing. The trim cover is textured and painted to match the instrument panel and has a molded seam which splits when the bag inflates. As with the steering-wheel-mounted air bag, the igniter electrical signal converts to heat, converting sodium azide/iron oxide powder to nitrogen gas, inflating the bag.

Sensing and diagnostic module

The sensing and diagnostic module supplies the current to the airbag system in the event of the collision, even if battery power is cut off. It checks this system every time the vehicle is started, causing the "AIR BAG" light to go on then off, if the system is operating properly. If there is a fault in the system, the light will go on and stay on, flash, or the dash will make a beeping sound. If this happens, the vehicle should be taken to your dealer immediately for service.

Precautions

Warning: *Failure to follow these precautions could result in accidental deployment of the airbag and personal injury.*
Caution: *If the stereo in your vehicle is equipped with an anti-theft system, make sure you have the correct activation code before disconnecting the battery.*

Whenever working in the vicinity of the steering wheel, steering column or any of the other SRS system components, the system must be disarmed. To disarm the system:

a) *Point the wheels straight ahead and turn the key to the Lock position.*
b) *Disconnect the cable from the negative battery terminal.*
c) *Wait at least 90 seconds for the back-up power supply to be depleted.*

Whenever handling an airbag module, always keep the airbag opening pointed away from your body. Never place the airbag module on a bench of other surface with the airbag opening facing the surface. Always place the airbag module in a safe location with the airbag opening facing up.

Never measure the resistance of any SRS component. An ohmmeter has a built-in battery supply that could accidentally deploy the airbag.

Never use electrical welding equipment on a vehicle equipped with an airbag without first disconnecting the yellow airbag connector, located under the steering column near the combination switch connector (drivers airbag) and behind the glove box (passengers airbag).

Never dispose of a live airbag module. Return it to your dealer for safe deployment, using special equipment, and disposal.

27 Wiring diagrams - general information

Since it isn't possible to include all wiring diagrams for every year covered by this manual, the following diagrams are those that are typical and most commonly needed.

Prior to troubleshooting any circuits, check the fuse and circuit breakers (if equipped) to make sure they're in good condition. Make sure the battery is properly charged and check the cable connections (see Chapter 1).

When checking a circuit, make sure that all connectors are clean, with no broken or loose terminals. When unplugging a connector, do not pull on the wires. Pull only on the connector housings themselves.

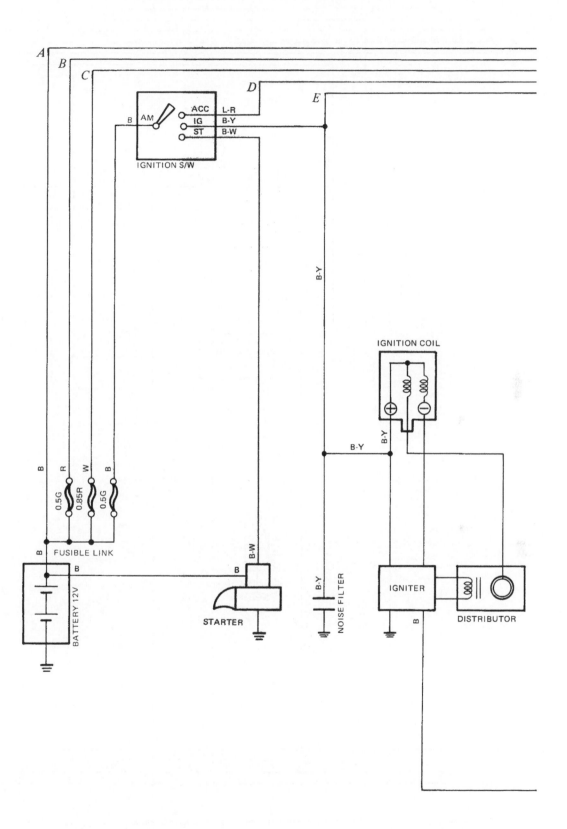

Typical starting, charging, ignition, emission control and engine compartment cooling fan systems (1980 through 1987 models) (1 of 3)

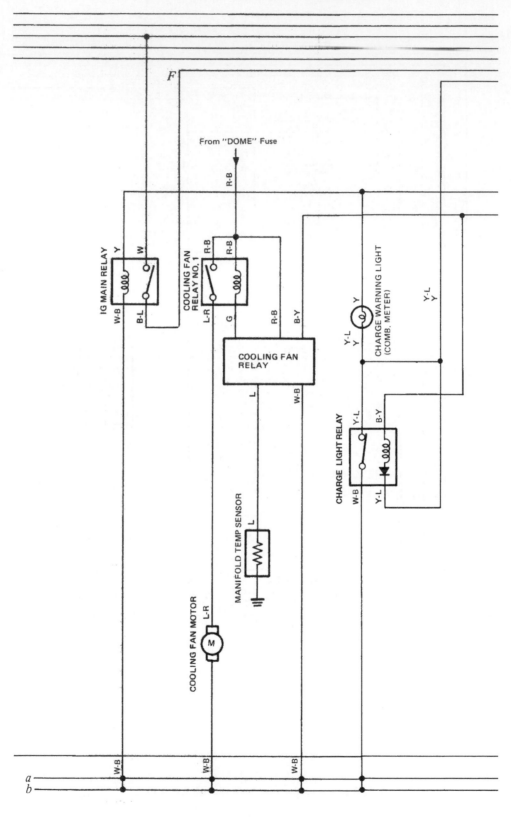

Ground points

a = Located on instrument panel under side of side fuse block

b = Located on instrument panel left side of side fuse block

Typical starting, charging, ignition, emission control and engine compartment cooling fan systems (1980 through 1987 models) (2 of 3)

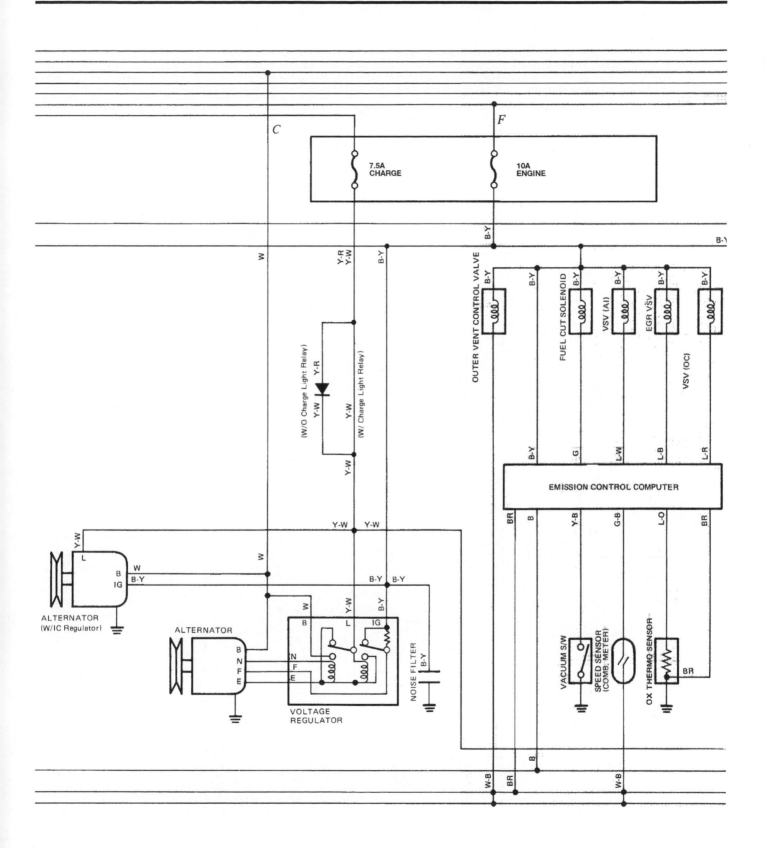

Typical starting, charging, ignition, emission control and engine compartment cooling fan systems (1980 through 1987 models) (3 of 3)

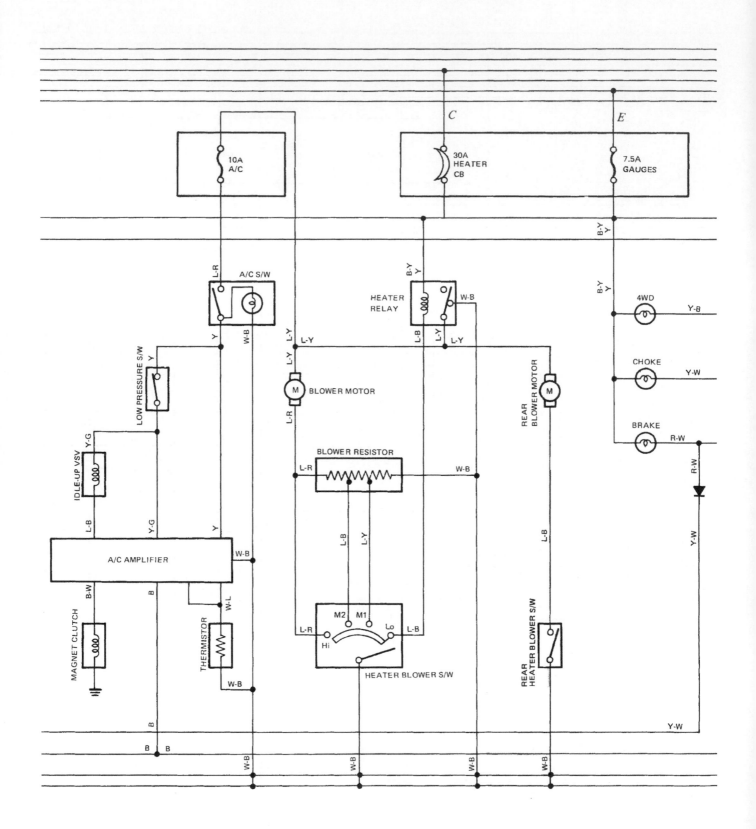

Typical heating, air conditioning, instrument panel and warning systems (1980 through 1987 models) (1 of 2)

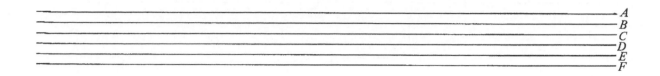

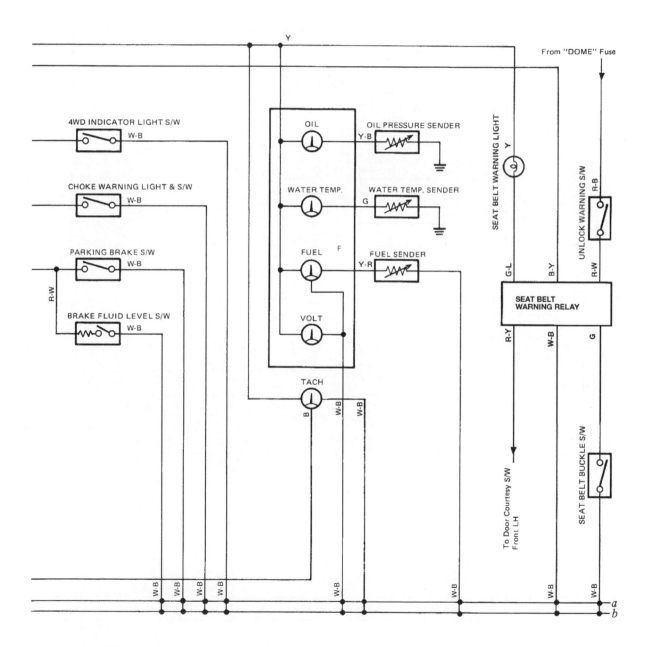

Typical heating, air conditioning, instrument panel and warning systems (1980 through 1987 models) (2 of 2)

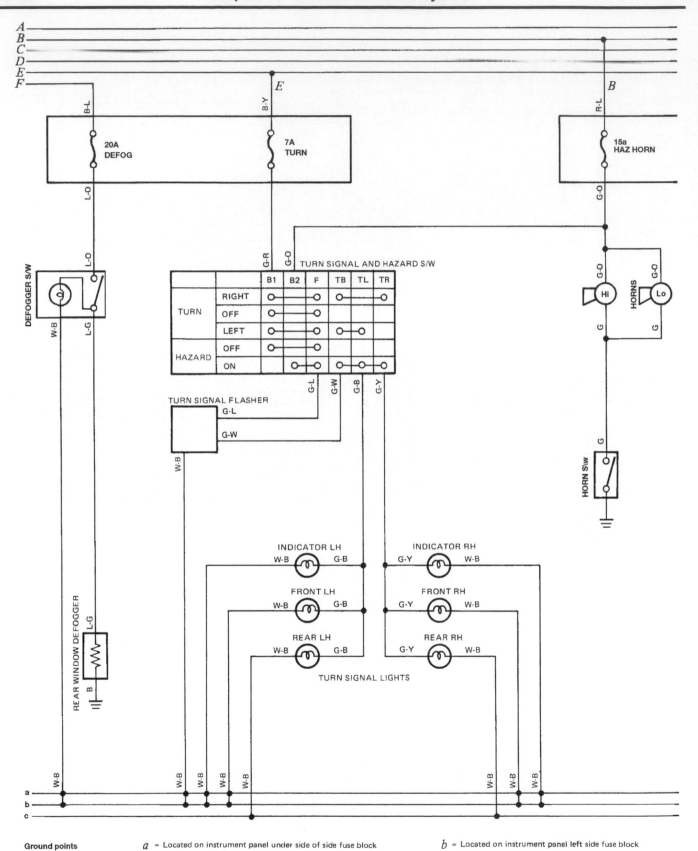

Ground points a = Located on instrument panel under side of side fuse block b = Located on instrument panel left side fuse block

**Typical rear window defogger, rear wiper and washer, turn signal, hazard, brake light and back-up lighting systems
(1980 through 1987 models) (1 of 2)**

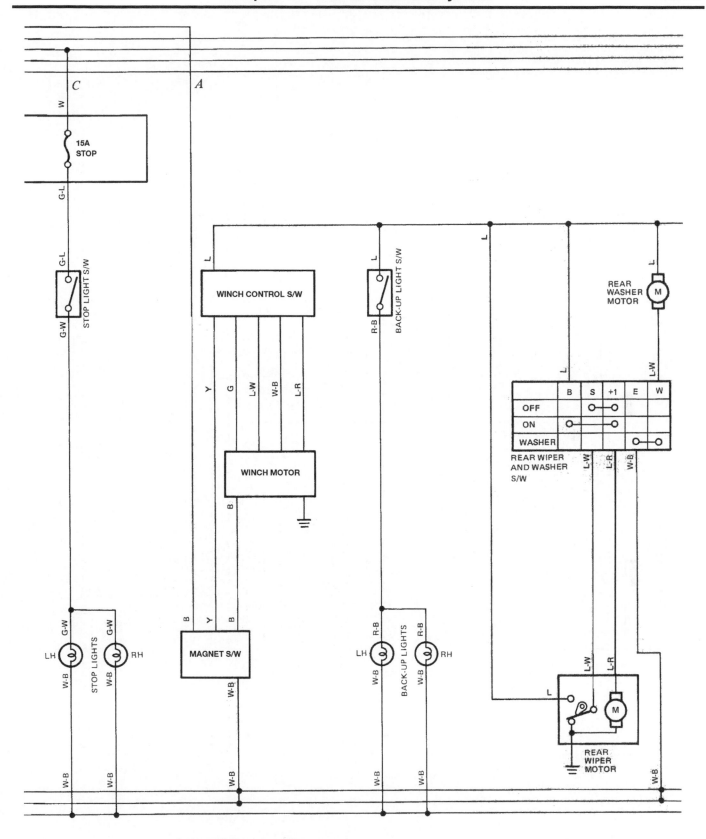

C = Located in frame center of last cross member

**Typical rear window defogger, rear wiper and washer, turn signal, hazard, brake light and back-up lighting systems
(1980 through 1987 models) (2 of 2)**

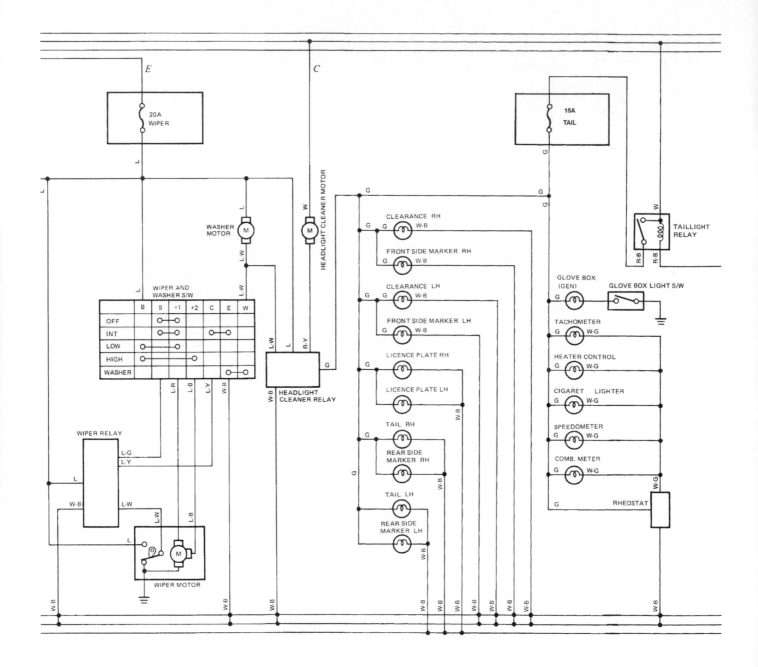

Typical windshield wiper and washer, tail lights, headlights and interior lighting systems (1980 through 1987 models) (1 of 2)

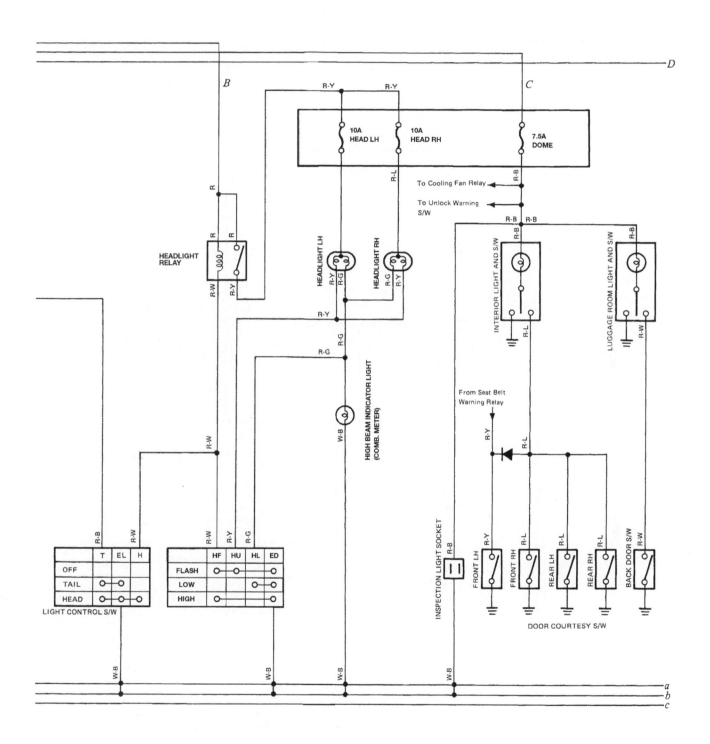

Typical windshield wiper and washer, tail lights, headlights and interior lighting systems (1980 through 1987 models) (2 of 2)

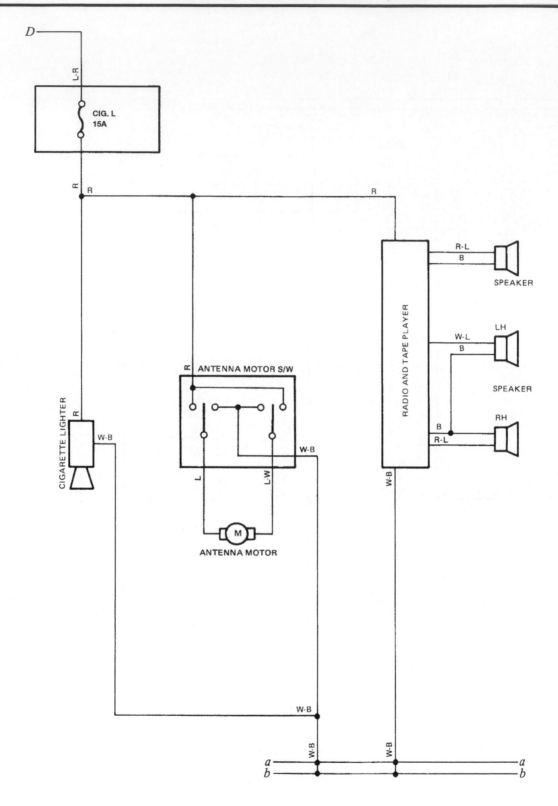

Ground points   *a* = Located on instrument panel under side of side fuse block
                 b = Located on instrument panel left side of side fuse block

Typical cigar lighter and audio systems (1980 through 1987 models)

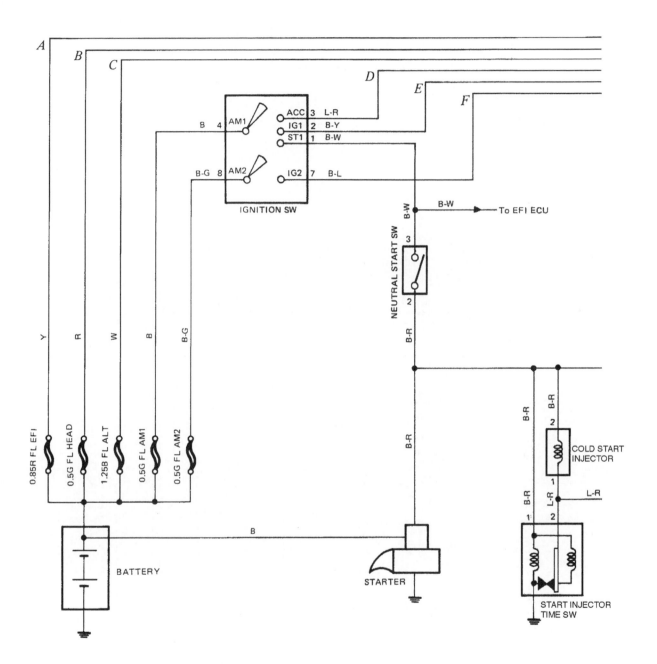

Typical starting system (1988 through 1990 models)

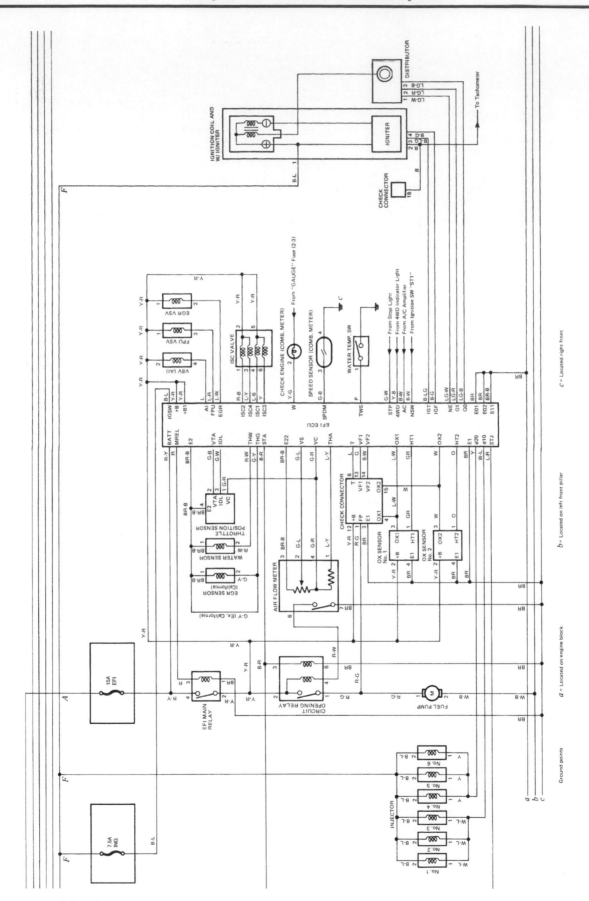

Typical engine control and ignition system (1988 through 1990 models)

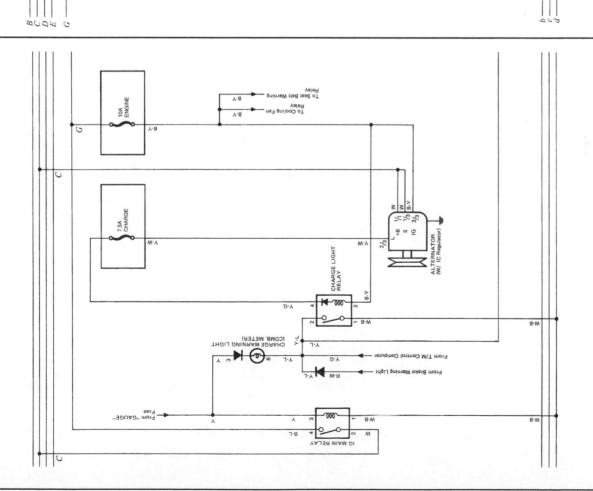

Typical heating and air conditioning system (1988 through 1990 models)

Typical charging system (1988 through 1990 models)

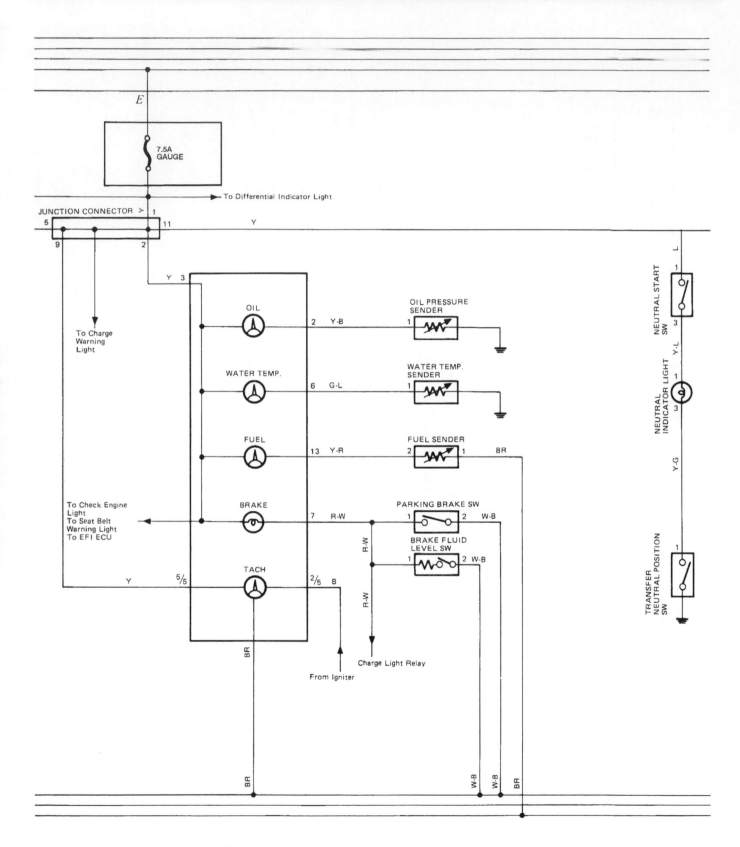

Typical engine warning system (1988 through 1990 models)

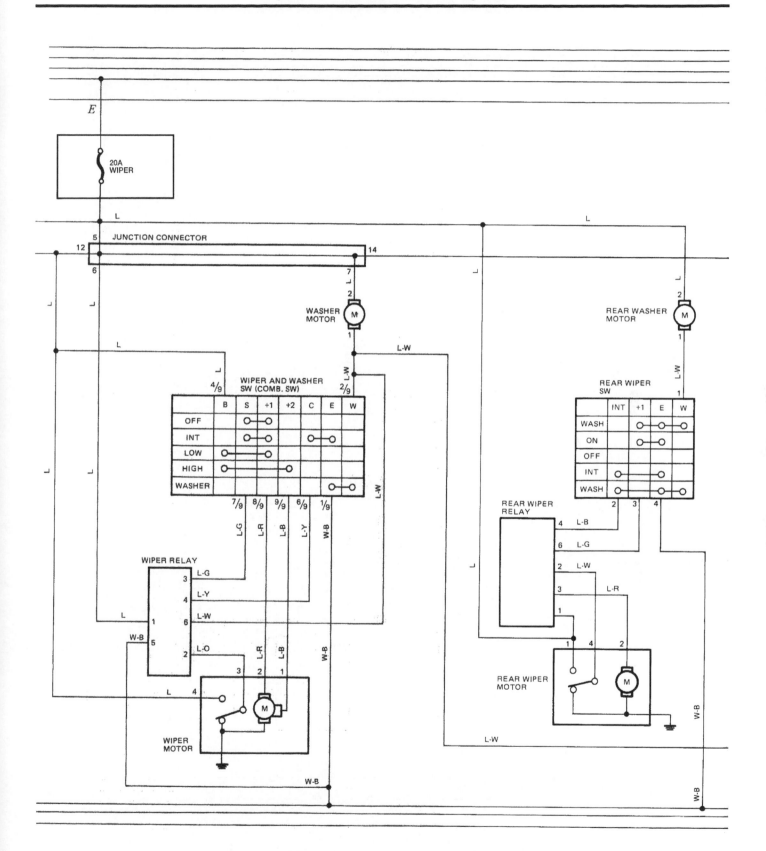

Typical windshield wiper and washer system (1988 through 1990 models)

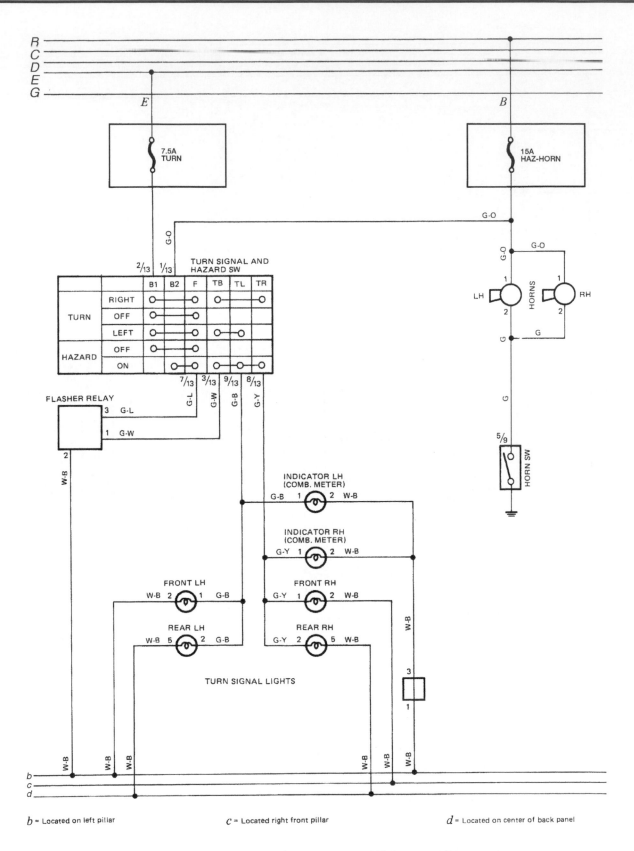

Typical turn signal and hazard lighting system (1988 through 1990 models)

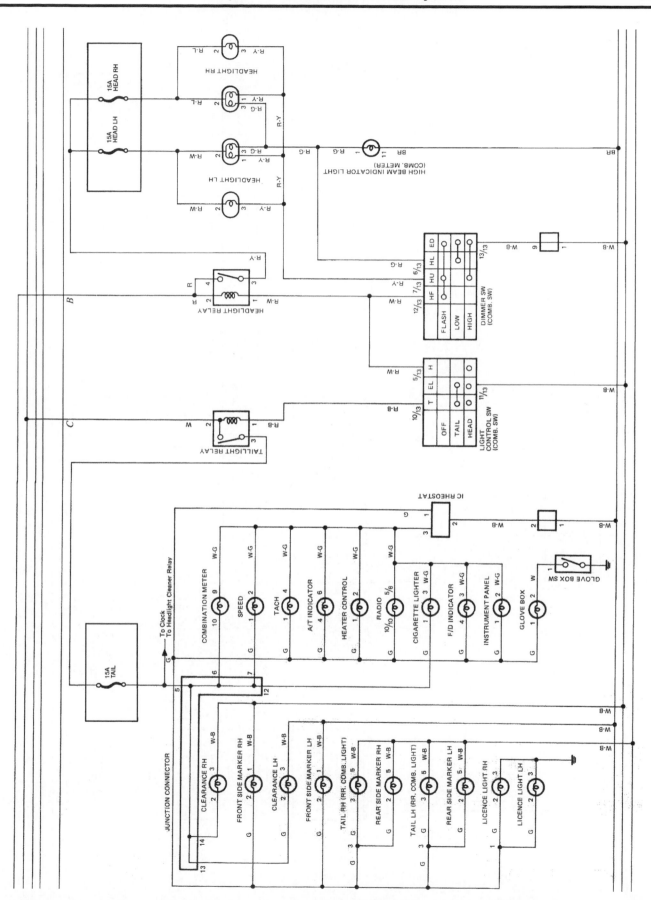

Typical tail light, headlight and instrument panel illumination system (1988 through 1990 models)

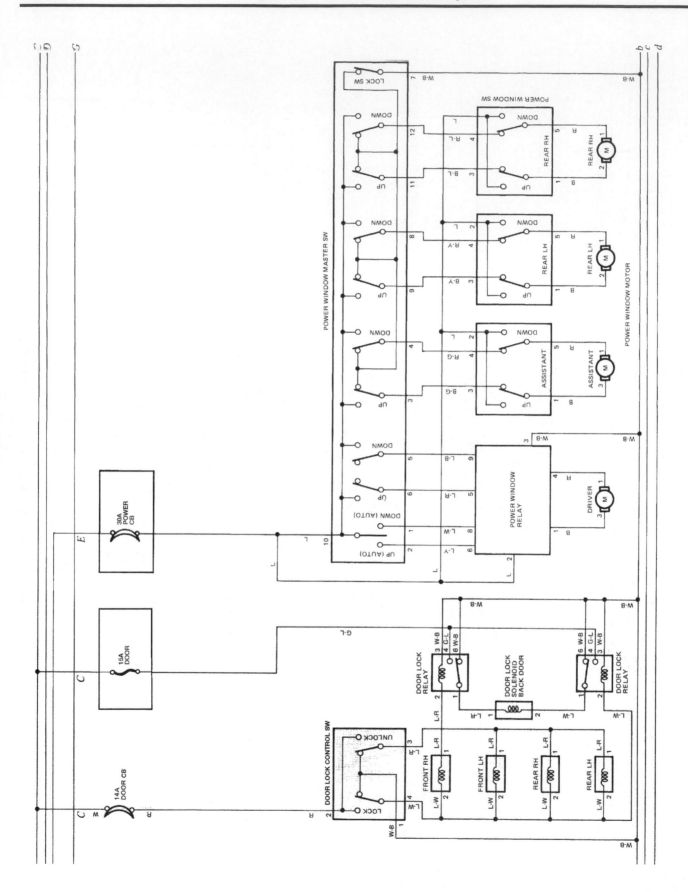

Typical power door lock and power window system (1988 through 1990 models)

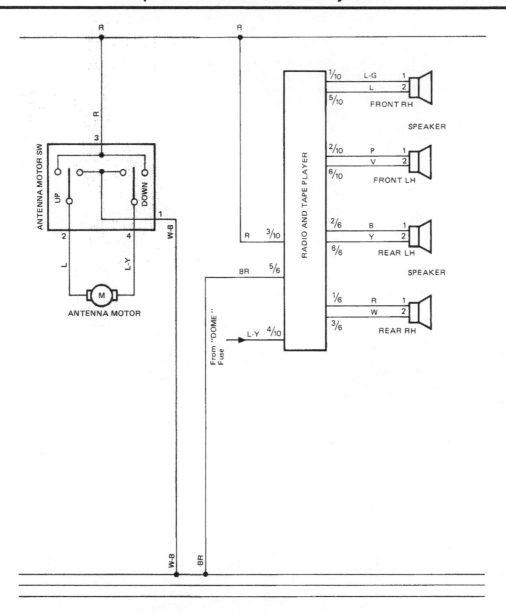

Typical audio system (1988 through 1990 models)

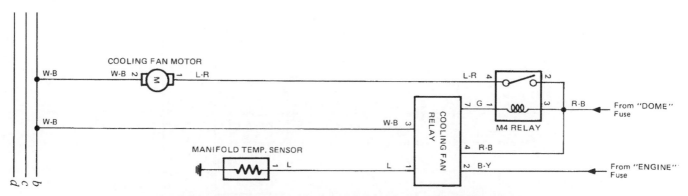

Typical engine compartment cooling system (1988 through 1990 models)

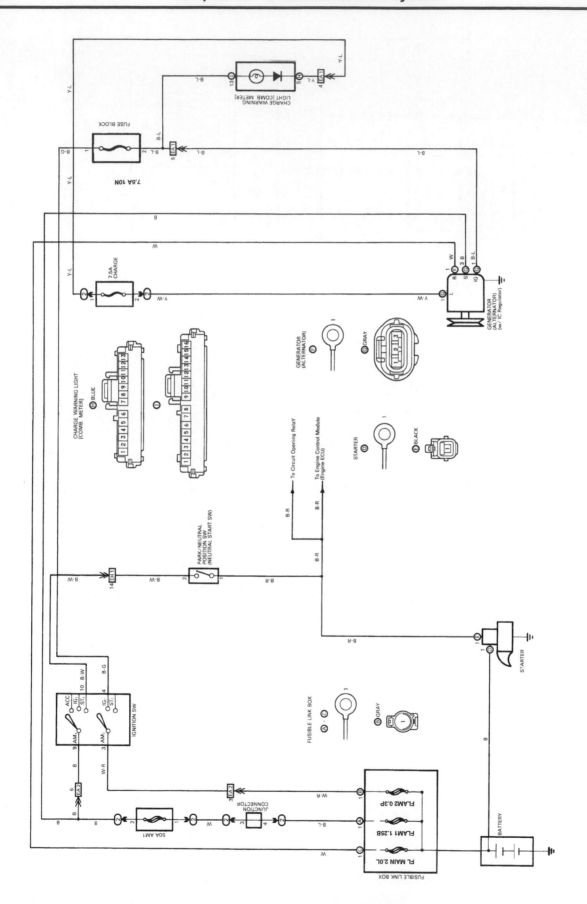

Typical starting and charging system (1991 and later models)

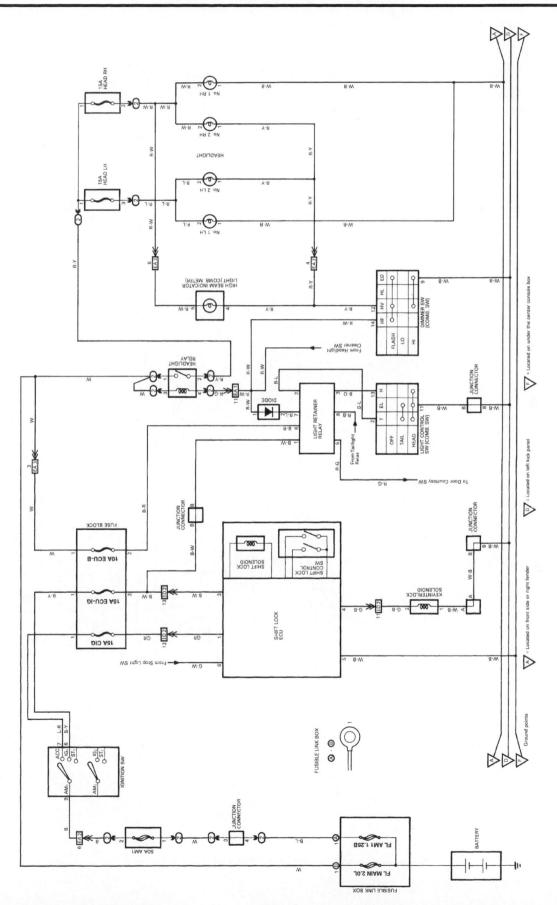

Typical headlight and shift lock system (1991 and later models)

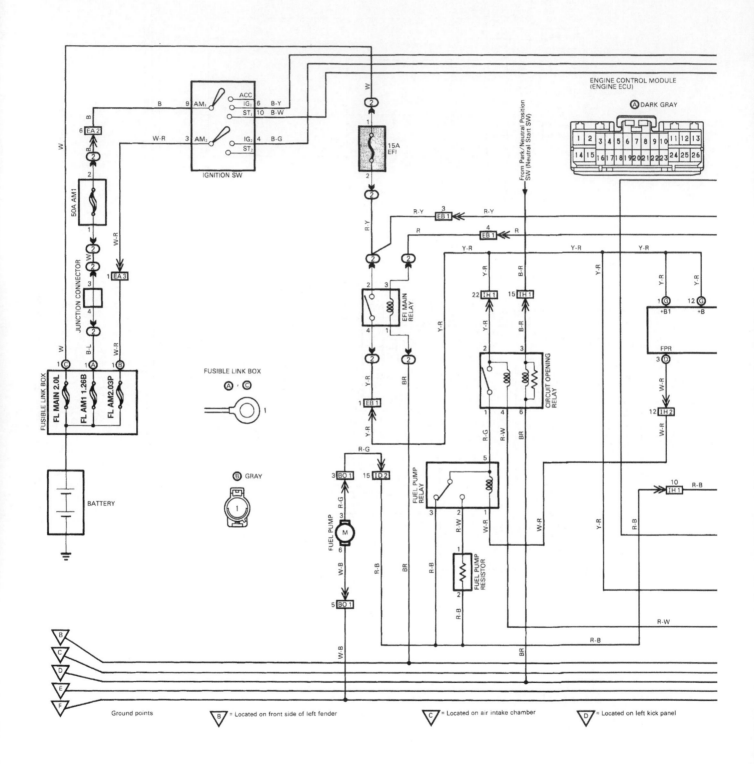

Typical engine control system (1991 and later models) (1 of 4)

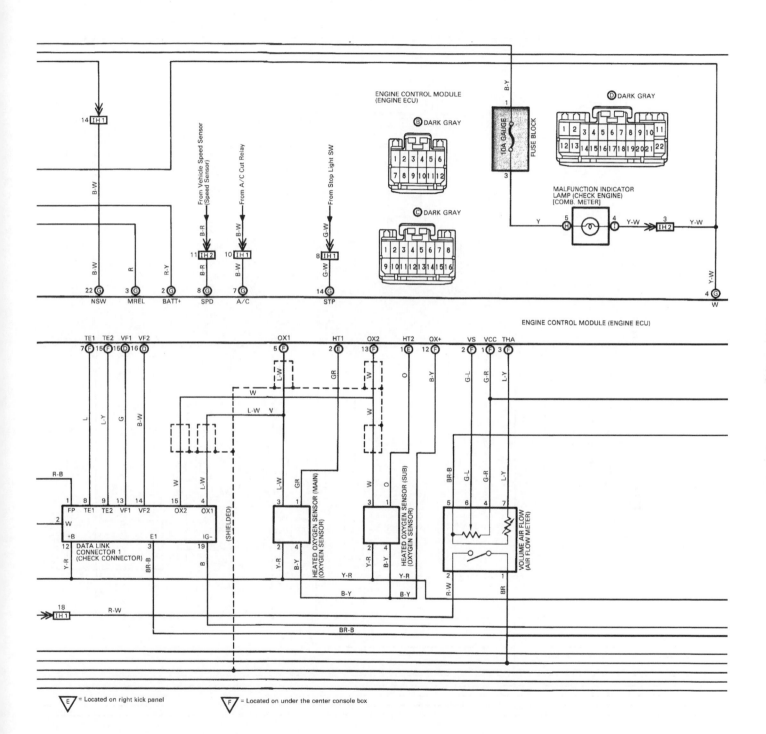

Typical engine control system (1991 and later models) (2 of 4)

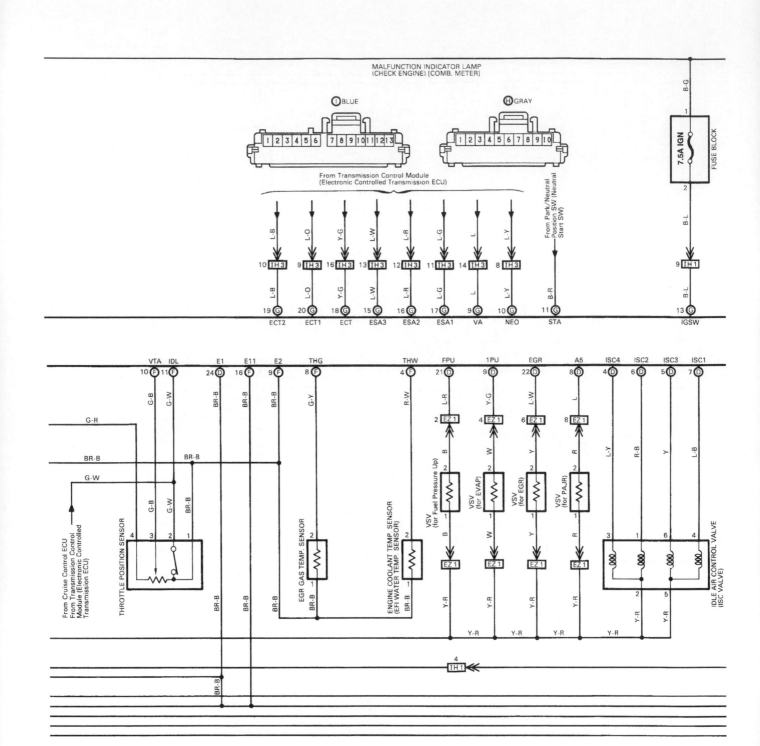

Typical engine control system (1991 and later models) (3 of 4)

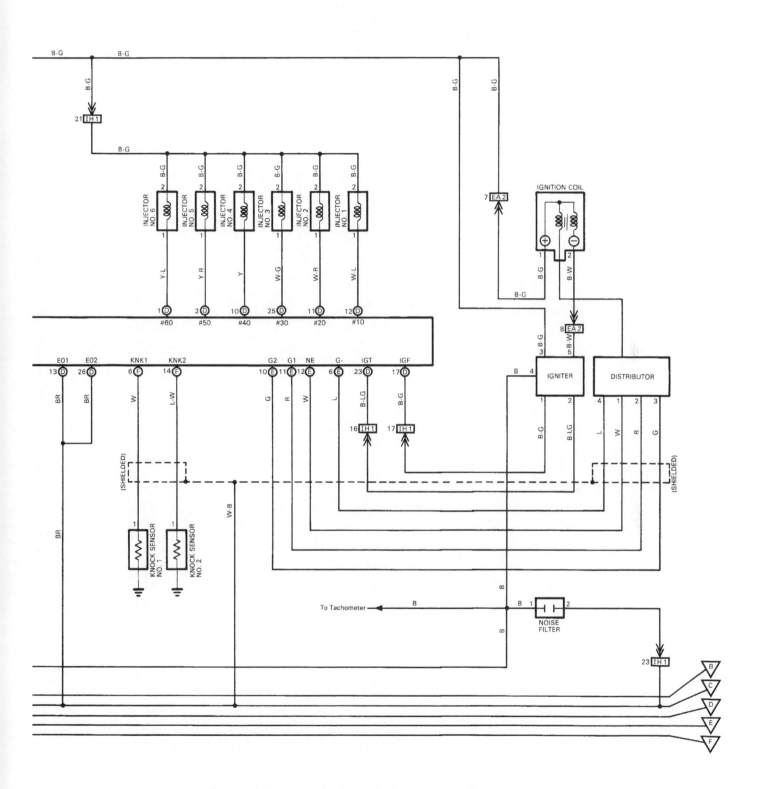

Typical engine control system (1991 and later models) (4 of 4)

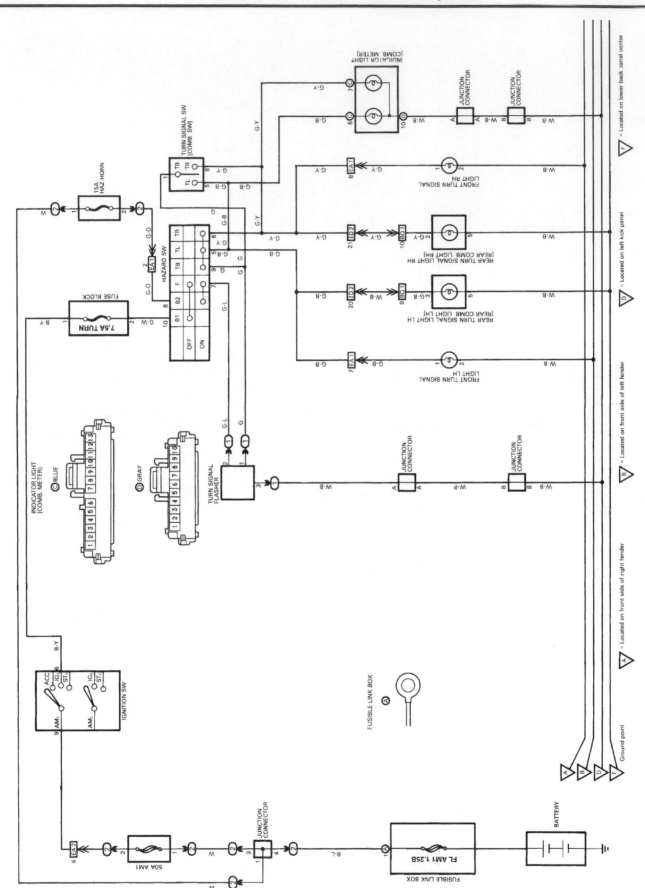

Typical turn signal and hazard flasher lighting system (1991 and later models)

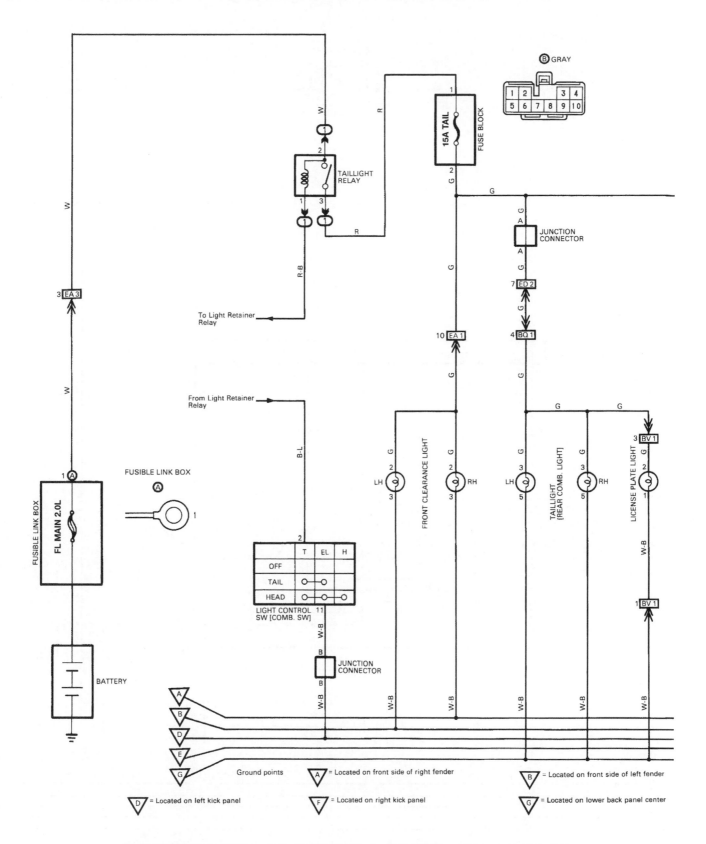

Typical tail light and instrument panel illumination system (1991 and later models) (1 of 3)

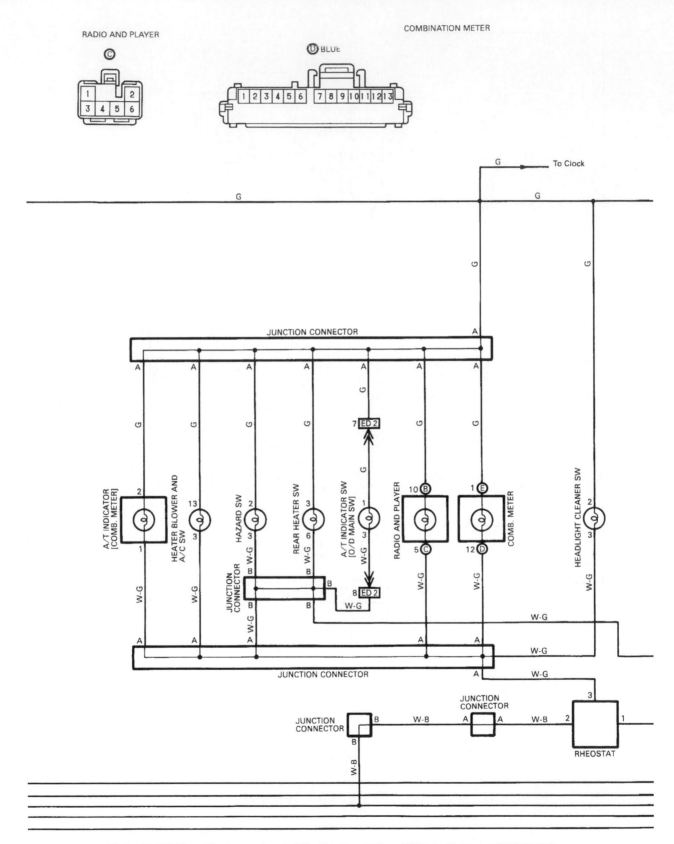

Typical tail light and instrument panel illumination system (1991 and later models) (2 of 3)

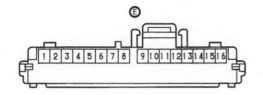

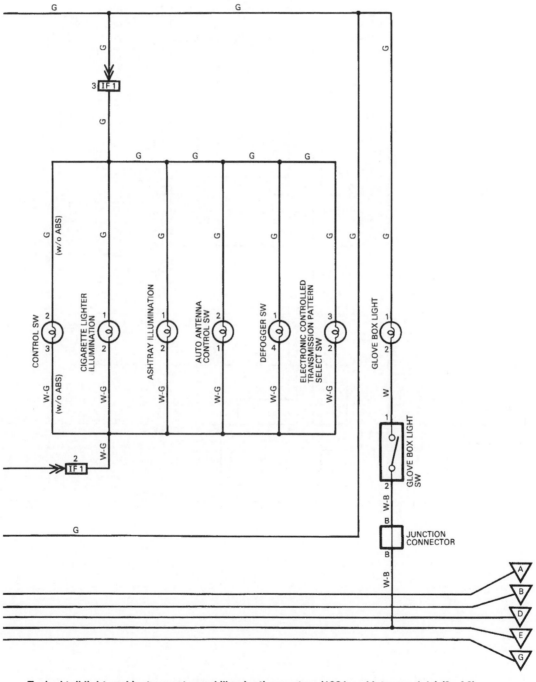

Typical tail light and instrument panel illumination system (1991 and later models) (3 of 3)

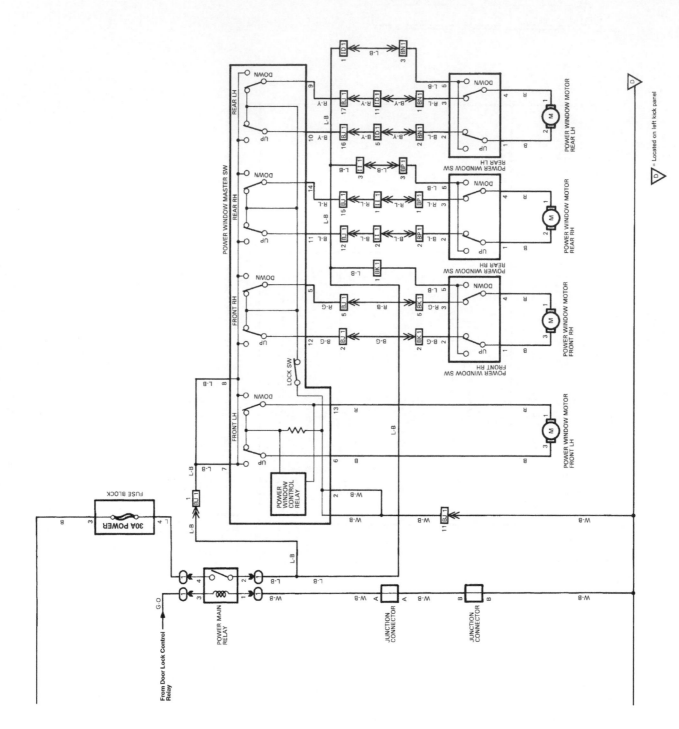

Typical power window system (1991 and later models)

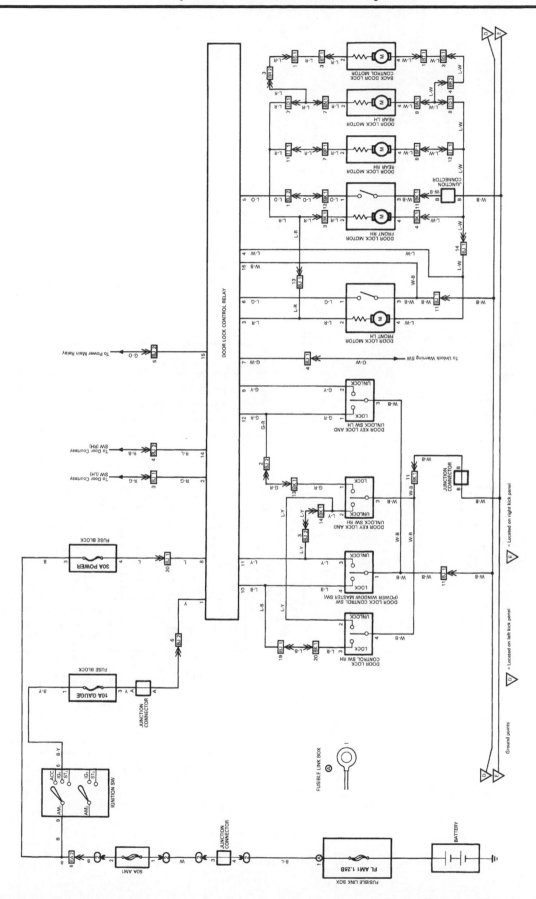

Typical power door lock system (1991 and later models)

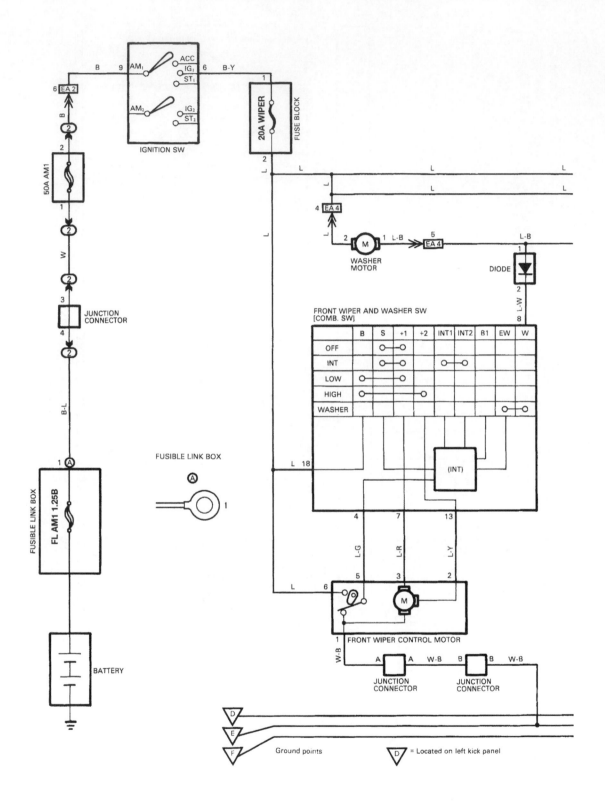

Typical windshield wiper and washer system (1991 and later models) (1 of 2)

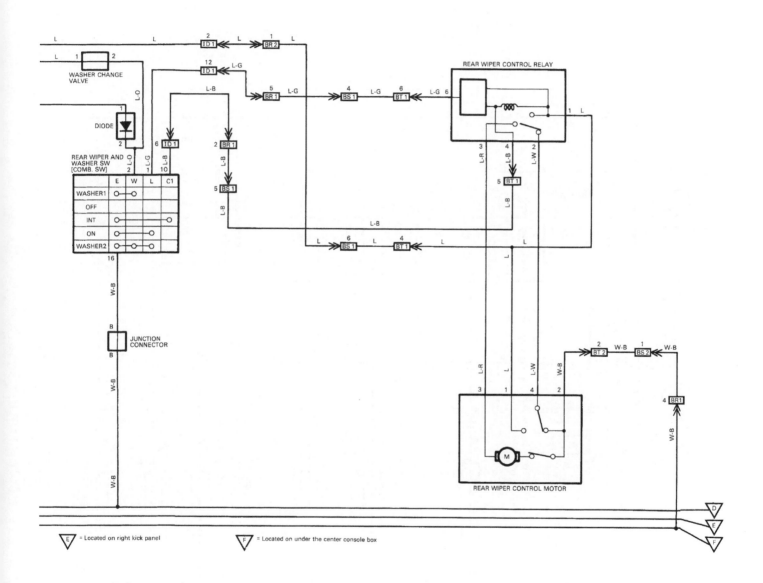

Typical windshield wiper and washer system (1991 and later models) (2 of 2)

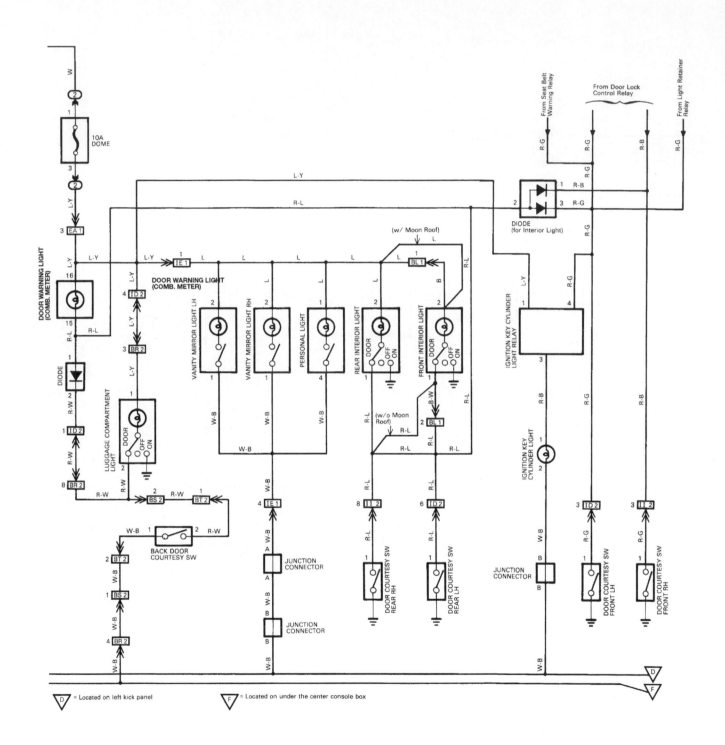

Typical interior lighting system (1991 and later models)

Index

A

About this manual, 0-5
Accelerator cable, removal, installation and
    adjustment, 4A-5, 4B-12
Air cleaner, removal and installation
    assembly, 4B-12
    housing, 4A-5
Air conditioning and heating system, check and
    maintenance, 3-14
Air conditioning system, parts removal and installation
    compressor, 3-17
    condenser, 3-18
    evaporator and expansion valve, 3-18
    receiver/drier, 3-17
Air filter check and replacement, 1-23
Air Injection Reactor (AIR) system, 6-3
Air intake plenum, 4B-21
Airbag, general information, 12-20
Airflow sensor, 6-19
Alignment, front end, general information, 10-17
Alternator
    components, check and replacement, 5-14
    removal and installation, 5-13
Antenna, removal and installation, 12-11
Antifreeze, general information, 3-2
Anti-lock Brake System (ABS), general information, 9-2
Automatic transmission, 7B-1 through 7B-8
    diagnosis, general, 7B-1
    fluid and filter change, 1-34
    fluid level check, 1-11
    neutral start switch, check and adjustment, 7B-6
    removal and installation, 7B-7
    shift linkage, check and adjustment, 7B-4
    shift lock system, description and check, 7B-5
    Throttle Valve (TV) cable, check and adjustment, 7B-5
    transmission mount, check and replacement, 7B-7
Automotive chemicals and lubricants, 0-17
Axle assembly, removal and installation, 8-14
Axles, driveshafts, differentials and, general information, 8-9
Axleshaft and oil seal, removal, overhaul and installation
    front, 8-12
    rear
        full-floating axle, 8-16
        semi-floating axle, 8-15

B

Battery
    cables, check and replacement, 5-2
    check, maintenance and charging, 1-14
    electrolyte, 1-9
    jump starting, 0-16
    removal and installation, 5-2
Blower unit, check and replacement, 3-11
Body, 11-1 through 11-16
    damage repair
        major, 11-5
        minor, 11-4
    maintenance, 11-1
Booster battery (jump) starting, 0-16
Brakes, 9-1 through 9-22
    Anti-lock Brake System (ABS), general information, 9-2
    booster, power, check, removal and installation, 9-18
    caliper, removal, overhaul and installation, 9-6
    check, 1-27
    disc, inspection, removal and installation, 9-8
    fluid, 1-9
    hoses and lines, inspection and replacement, 9-16
    hydraulic system, bleeding, 9-17
    light switch, removal, installation and adjustment, 9-21
    load sensing proportioning and by-pass valve, general
        information, 9-22
    master cylinder, removal, overhaul and installation, 9-14
    pads, disc, replacement, 9-4
    parking
        adjustment, 9-21
        cables, replacement, 9-21
        shoes (rear disc brakes only), inspection and replacement, 9-19
    shoes, drum, replacement, 9-9
    wheel cylinder, removal, overhaul and installation, 9-13
Bulb replacement, 12-14
Bumpers, removal and installation, 11-6
Buying parts, 0-8

C

Camshaft and bearings, inspection, 2A-15
Camshaft and timing gears (pushrod engines), removal and
    installation, 2A-13

Camshafts and valve lifters (OHC engine), removal, inspection and installation, 2B-5
Capacities, engine, 1-2
Carburetor
adjustments, 4A-7
air cleaner housing, removal and installation, 4A-5
choke check, 1-22
diagnosis and overhaul, 4A-5
engines, general information, 4A-1
removal and installation, 4A-8
Catalytic converter, 6-31
Center console, removal and installation, 11-14
Charcoal canister replacement, 6-25
Charging system
check, 5-12
general information and precautions, 5-9
Chassis electrical system, 12-1 through 12-54
Chassis lubrication, 1-19
Choke Breaker (CB) system, 6-12
Choke opener system, 6-8
Circuit breakers, general information, 12-3
Clutch and driveline, 8-1 through 8-18
Clutch
clutch/brake pedal height and freeplay adjustment, 1-29
components, removal and installation, 8-3
description and check, 8-2
hydraulic system, bleeding, 8-8
master cylinder, removal, overhaul and installation, 8-6
pedal assembly, removal and installation, 8-8
release bearing, removal, inspection and installation, 8-5
release cylinder, removal, overhaul and installation, 8-8
Coil spring (FJ80 series), removal and installation, 10-6
Cold start injector, 4B-20
Cold start injector time switch, 4B-20
Conversion factors, 0-18
Cooling system
blower unit, check and replacement, 3-11
check, 1-15
coolant temperature sending unit, check and replacement, 3-9
engine fan/clutch assembly, check, removal and installation, 3-4
oil cooler, removal and installation, 3-6
radiator and coolant reservoir, removal and installation, 3-5
refrigerant, adding, 3-16
servicing (draining, flushing and refilling), 1-32
thermostat, check and replacement, 3-3
Cooling, heating and air conditioning systems, 3-1 through 3-20
Crankshaft
front oil seal, replacement, 2A-12, 2B-15
inspection, 2C-20
installation and main bearing oil clearance check, 2C-24
pulley, removal and installation, 2A-11
removal, 2C-15
Crankshaft Position Sensor, 6-21
Cruise control system, description and check, 12-16
Cylinder
compression check, 2C-6
head
cleaning and inspection, 2C-11
disassembly, 2C-11
reassembly, 2C-13
removal and installation, 2A-9, 2B-10
honing, 2C-18

D

Dashboard trim panels, removal and installation, 11-11
Daytime Running Lights (DRL), general information, 12-19
Deceleration fuel cut system, 6-9

Diagnosis
general, 0-20
automatic transmission, general, 7B-1
Diagnostic Trouble Codes, 6-15
Differential lubricant
change, 1-36
level check, 1-20
Differentials, driveshafts, and axles, general information, 8-9
Disc brake
caliper, removal, overhaul and installation, 9-6
inspection, removal and installation, 9-8
pads, replacement, 9-4
Distributor, removal and installation, 5-6
Door
latch, lock cylinder and handles, removal and installation, 11-7
power lock system, description and check, 12-19
removal, installation and adjustment, 11-9
trim panel, removal and installation, 11-7
window glass
regulator, removal and installation, 11-9
removal and installation, 11-8
Drivebelt check, adjustment and replacement, 1-26
Driveline inspection, 8-9
Driveplate/flywheel, removal and installation, 2A-17, 2B-18
Driveshafts, differentials and axles, general information, 8-9
Driveshafts, removal and installation, 8-9
Drum brake shoes, replacement, 9-9

E

ECM removal and installation and Electronic Fuel Injection (EFI) system, general information, 6-13
EGR temperature sensor (California models), 6-20
Electrical troubleshooting, general information, 12-1
Electronic Fuel Injection (EFI) system
check, 4B-15
component check and replacement, 4B-15
general information, 4B-13
general information and ECM removal and installation, 6-13
Emissions and engine control systems, 6-1 through 6-32
Engine
block
cleaning, 2C-16
inspection, 2C-17
camshaft and bearings (pushrod engines), inspection, 2A-15
camshaft and timing gears (pushrod engines), removal and installation, 2A-13
camshafts and valve lifters (OHC engine), removal, inspection and installation, 2B-5
coolant, check, 1-8
cooling fan/clutch assembly, check, removal and installation, 3-4
crankshaft
front oil seal, replacement, 2A-12, 2B-15
inspection, 2C-20
installation and main bearing oil clearance check, 2C-24
pulley, removal and installation, 2A-11
removal, 2C-15
cylinder compression check, 2C-6
cylinder head
cleaning and inspection, 2C-11
disassembly, 2C-11
reassembly, 2C-13
removal and installation, 2A-9, 2B-10
cylinder honing, 2C-18
electrical systems, 5-1 through 5-20
exhaust manifold (pushrod engines), removal and installation, 2B-4
flywheel/driveplate, removal and installation, 2A-17, 2B-18
initial start-up and break-in after overhaul, 2C-27

intake and exhaust manifolds (pushrod engines), removal and
  installation, 2A-7
intake manifold (OHC engine), removal and installation, 2B-4
main and connecting rod bearings, inspection, 2C-20
mounts, check and replacement, 2A-18, 2B-20
oil and filter change, 1-12
oil pan, removal and installation, 2A-16, 2B-15
oil pump, removal and installation, 2A-16, 2B-17
oil type, 1-1
oil check, 1-7
overhaul
  disassembly sequence, 2C-8
  general information, 2C-5
  reassembly sequence, 2C-22
Overhead camshaft (OHC) engine, 2B-1
piston rings, installation, 2C-23
pistons/connecting rods
  inspection, 2C-18
  installation and rod bearing oil clearance check, 2C-25
  removal, 2C-14
pushrod
  cover, removal and installation, 2A-5
  engines, 2A-1
rear main oil seal installation, 2A-17, 2B-19, 2C-25
rebuilding alternatives, 2C-8
removal and installation, 2C-7
removal, methods and precautions, 2C-7
repair operations possible with the engine in the
  vehicle, 2A-3, 2B-3
rocker arms and pushrods, removal, inspection and
  installation, 2A-5
timing cover and chain, removal, inspection and installation, 2B-11
timing gear cover, removal and installation, 2A-12
Top Dead Center (TDC) for number one piston, locating, 2A-3
vacuum gauge diagnostic checks, 2C-5
valve
  cover, removal and installation, 2A-4, 2B-3
  lifters, removal, inspection and installation, 2A-10
  servicing, 2C-13
  spring, retainer and seals, replacement, 2A-6
Evaporative Emission Control (EVAP) system, 1-22, 6-23
Exhaust control valve (heat riser) lubrication and check, 1-21
Exhaust Gas Recirculation (EGR) system 1-22, 6-25
Exhaust manifold, removal and installation, 2B-4
Exhaust system
  check, 1-20
  servicing and general information, 4A-8

F

Fault finding, 0-20
Firing order and distributor rotation, 1-2
Fluid level checks, 1-7
Flywheel/driveplate, removal and installation, 2A-17, 2B-18
Freewheel hub, removal and installation, 8-10
Front end alignment, general information, 10-17
Front fender, removal and installation, 11-7
Front wheel bearing check and repack, 1-33
Fuel and exhaust systems
  carbureted engines, 4A-1 through 4A-8
  fuel-injected engines, 4B-1 through 4B-22
Fuel system
  check, 1-27
  filter replacement, 1-29
  level sending unit, check and replacement, 4A-4, 4B-11
  lines and fittings, inspection, repair and replacement, 4A-2, 4B-9
  pressure regulator, 4B-17
  pressure relief, 4B-2

pump, removal and installation, 4A-4, 4B-10
pump/fuel pressure check, 4A-1, 4B-3
rail and fuel injectors, 4B-19
tank
  cleaning and repair, general information, 4A-4
  removal and installation, 4A-3
Fuses, general information, 12-2
Fusible links, general information, 12-2

G

General engine overhaul procedures, 2C-1 through 2C-28

H

Headlight
  adjustment, 12-13
  housing (Halogen type bulb, removal and installation, 12-14
  removal and installation, 12-12
Heat Control Valve system, 6-11
Heater and air conditioning
  control assembly, check, removal and installation, 3-13
  system, check and maintenance, 3-14
Heater core, removal and installation, 3-12
High Altitude Compensation (HAC) system, 6-5
Hinges and locks, maintenance, 11-5
Hood release latch and cable, removal and installation, 11-6
Hood, removal, installation and adjustment, 11-5
Horn, check and replacement, 12-14
Hoses and lines (brake), inspection and replacement, 9-16
Hot Air Intake (HAI) system, 6-7
Hot Idle Compensation (HIC) system, 6-11
**Hub and wheel bearing assembly (rear) (full-floating axle),
  removal, installation and adjustment, 8-17**

I

Idle speed check and adjustment, 1-25
Idle Speed Control (ISC) or Idle Air Control (IAC) valve, 4B-18
Igniter, replacement, 5-7
Ignition system, 1-2
  check, 5-5
  coil, check and replacement, 5-6
  general information, 5-3
  switch and lock cylinder, check and replacement, 12-6
  timing check and adjustment, 1-31
Information sensors, 6-16
  airflow sensor, 6-19
  crankshaft position sensor, 6-21
  EGR temperature sensor, 6-20
  Intake Air Temperature (IAT) sensor, 6-20
  knock sensor, 6-21
  oxygen sensor, 6-17
  Throttle Position Sensor (TPS), 6-19
  vehicle speed sensor, 6-20
Initial start-up and break-in after overhaul, 2C-27
Instrument cluster, removal and installation, 12-8
Instrument panel
  fuel, oil and temperature gauges, check and replacement, 12-8
  removal and installation, 11-14
Intake Air Temperature (IAT) sensor, 6-20
Intake and exhaust manifolds, removal and installation, 2A-7
Intake manifold, removal and installation, 2B-4
Introduction to the Toyota Land Cruiser, 0-5

J

Jacking and towing, 0-16

K

Knock sensor, 6-21

L

Leaf springs and bushings (FJ60 and FJ62 series), removal and installation, 10-6
Liftgate, removal, installation and adjustment, 11-10
Load sensing proportioning and by-pass valve, general information, 9-22
Lock cylinder, handles and door latch, removal and installation, 11-7

M

Main and connecting rod bearings, inspection, 2C-20
Maintenance
    schedule, 1-6
    techniques, tools and working facilities, 0-8
Manual transmission, 7A-1 through 7A-2
    lubricant change, 1-36
    overhaul, general information, 7A-2
    removal and installation, 7A-1
Manual transmission/front differential lubricant level check, 1-20
Master cylinder
    brake, removal, overhaul and installation, 9-14
    clutch, removal, overhaul and installation, 8-6
Mirrors
    electric side view, description and check, 12-16
    outside, removal and installation, 11-9

N

Neutral start switch, check and adjustment, 7B-6

O

Oil cooler, removal and installation, 3-6
Oil pan, removal and installation, 2A-16, 2B-15
Oil pump, removal, inspection and installation, 2A-16, 2B-17
On Board Diagnosis (OBD) system, description and trouble code access, 6-14
Outside mirrors, removal and installation, 11-9
Overhead camshaft (OHC) engine, 2B-1 through 2B-20
Oxygen sensor, 6-17

P

Parking brake
    adjustment, 9-21
    cables, replacement, 9-21
    shoes (rear disc brakes only), inspection and replacement, 9-19
Pick-up coil, check and replacement, 5-7
Pilot bearing, inspection and replacement, 8-6
Pinion oil seal, replacement, 8-14
Piston rings, installation, 2C-23

Pistons/connecting rods
    inspection, 2C-18
    installation and rod bearing oil clearance check, 2C-25
    removal, 2C-14
Positive Crankcase Ventilation (PCV) system, 1-22, 6-28
Power brake booster, check, removal and installation, 9-18
Power door lock system, description and check, 12-19
Power steering system
    bleeding, 10-16
    fluid level check, 1-12
    pump, removal and installation, 10-15
Power window system, description and check, 12-17
Pulse Air Injection (PAIR) system, 6-29
Pushrod cover, removal and installation, 2A-5
Pushrod engines, 2A-1 through 2A-18

R

Radiator and coolant reservoir, removal and installation, 3-5
Radiator grille, removal and installation, 11-6
Radio and speakers, removal and installation, 12-10
Rear main oil seal
    installation, 2C-25
    replacement, 2A-17, 2B-19
Rear window defogger, check and repair, 12-11
Recommended lubricants and fluids, 1-1
Relays, general information and testing, 12-3
Repair operations possible with the engine in the vehicle, 2A-3, 2B-3
Rocker arms and pushrods, removal, inspection and installation, 2A-5

S

Safety first, 0-19
Seat
    belt check, 1-21
    removal and installation, 11-15
Shift diaphragm (early models), check and replacement, 7C-2
Shift linkage, check and adjustment, 7B-4
Shift lock system, description and check, 7B-5
Shift motor actuator (late models), check and replacement, 7C-3
Shock absorbers (front and rear), removal and installation, 10-5
Spark Control (SC) system, 6-2
Spark plug
    replacement, 1-23
    type and gap, 1-2
    wire, distributor cap and rotor check and replacement, 1-30
Stabilizer bar and bushings (front and rear), removal and installation, 10-4
Starter motor
    removal and installation, 5-19
    testing in-vehicle, 5-18
Starter solenoid (1980 through 1987 models), removal and installation, 5-19
Starting system, general information and precautions, 5-17
Steering system
    column
        covers, removal and installation, 11-11
        switches, check and replacement, 12-5
    gear, removal and installation, 10-14
    knuckle, removal and installation, 10-9
    linkage, inspection, removal and installation, 10-12
    power, bleeding, 10-16
    pump, power, removal and installation, 10-15
    wheel, removal and installation, 10-10
Suspension and steering systems, 1-18, 10-1 through 10-18
Suspension arms (FJ80 series), removal and installation, 10-8

T

Tailgate, removal, installation and adjustment, 11-10
Thermostat, check and replacement, 3-3
Thermostatically controlled air cleaner check, 1-21
Throttle body, 4B-16
Throttle position sensor (TPS), 4B-16, 6-19
Throttle Valve (TV) cable, check and adjustment, 7B-5
Timing cover and chain, removal, inspection and
    installation, 2B-11
Timing gear cover, removal and installation, 2A-12
Tire and tire pressure checks, 1-10
Tire rotation, 1-17
Tools, 0-11
Top Dead Center (TDC) for number one piston, locating, 2A-3
Towing the vehicle, 0-16
Track bar (FJ80 series), removal and installation, 10-7
Transfer case, 7C-1 through 7C-4
    general information, 7C-1
    lubricant change, 1-36
    lubricant level check, 1-20
    oil seals (early models), replacement, 7C-1
    removal and installation, 7C-3
    shift diaphragm (early models), check and replacement, 7C-2
    shift motor actuator (late models), check and replacement, 7C-3
Transmission
    automatic, 7B-1 through 7B-8
        diagnosis, general, 7B-1
        fluid and filter change, 1-34
        fluid level check, 1-11
        neutral start switch, check and adjustment, 7B-6
        removal and installation, 7B-7
        shift linkage, check and adjustment, 7B-4
        shift lock system, description and check, 7B-5
        Throttle Valve (TV) cable, check and adjustment, 7B-5
    manual, 7A-1 through 7A-2
        lubricant change, 1-36
        overhaul, general information, 7A-2
        removal and installation, 7A-1
        transmission mount, check and replacement, 7B-7
Trouble Codes, 6-15
Troubleshooting, 0-20
Tune-up and routine maintenance, 1-1 through 1-38
Turn signal and hazard flasher, check and replacement, 12-5

U

Underhood hose check and replacement, 1-16
Universal joint, replacement, 8-10
Upholstery and carpets, maintenance, 11-1

V

Vacuum gauge diagnostic checks, 2C-5
Valve
    clearance check and adjustment, 1-25
    cover, removal and installation, 2A-4, 2B-3
    lifters, removal, inspection and installation, 2A-10
    spring, retainer and seals, replacement, 2A-6
    servicing, 2C-13
Vehicle identification numbers, 0-6
Vehicle speed sensor, 6-20
Vinyl trim, maintenance, 11-1

W

Water pump
    check, 3-6
    replacement, 3-7
Wheel cylinder, removal, overhaul and installation, 9-13
Wheels and tires, general information, 10-16
Window
    defogger, check and repair, 12-11
    glass
        regulator, removal and installation, 11-9
        removal and installation, 11-8
    power system, description and check, 12-17
Windshield and fixed glass, replacement, 11-5
Windshield washer fluid, 1-9
Wiper blade inspection and replacement, 1-17
Wiper motor, check and replacement, 12-9
Wiring diagrams, general information, 12-20
Working facilities, 0-14

Haynes Automotive Manuals

NOTE: If you do not see a listing for your vehicle, please visit **haynes.com** for the latest product information and check out our **Online Manuals!**

ACURA
12020 Integra '86 thru '89 & Legend '86 thru '90
12021 Integra '90 thru '93 & Legend '91 thru '95
Integra '94 thru '00 - see HONDA Civic (42025)
MDX '01 thru '07 - see HONDA Pilot (42037)
12050 Acura TL all models '99 thru '08

AMC
14020 Concord/Hornet/Gremlin/Spirit '70 thru '83
14025 (Renault) Alliance & Encore '83 thru '87

AUDI
15020 4000 all models '80 thru '87
15025 5000 all models '77 thru '83
15026 5000 all models '84 thru '88
Audi A4 '96 thru '01 - see VW Passat (96023)
15030 Audi A4 '02 thru '08

AUSTIN-HEALEY
Sprite - see MG Midget (66015)

BMW
18020 3/5 Series '82 thru '92
18021 3-Series including Z3 models '92 thru '98
18022 3-Series including Z4 models '99 thru '05
18023 3-Series '06 thru '14
18025 320i all 4-cylinder models '75 thru '83
18050 1500 thru 2002 except Turbo '59 thru '77

BUICK
19010 Buick Century '97 thru '05
Century (front-wheel drive) - see GM (38005)
19020 Buick, Oldsmobile & Pontiac Full-size (Front wheel drive) '85 thru '05
19025 Buick, Oldsmobile & Pontiac Full-size (Rear wheel drive) '70 thru '90
19027 Buick LaCrosse '05 thru '13
Regal - see GENERAL MOTORS (38010)
Skyhawk - see GM (38015)
Skylark - see GM (38020, 38025)
Somerset - see GENERAL MOTORS (38025)

CADILLAC
21015 CTS & CTS-V '03 thru '14
21030 Cadillac Rear Wheel Drive '70 thru '93
Cimarron, Eldorado & Seville - see GM (38015, 38030, 38031)

CHEVROLET
10305 Chevrolet Engine Overhaul Manual
24010 Astro & GMC Safari Mini-vans '85 thru '05
24015 Camaro V8 all models '70 thru '81
24016 Camaro all models '82 thru '92
Cavalier - see GM (38016)
Celebrity - see GM (38005)
24017 Camaro & Firebird '93 thru '02
24018 Camaro '10 thru '15
24020 Chevelle, Malibu, El Camino '69 thru '87
24024 Chevette & Pontiac T1000 '76 thru '87
Citation - see GENERAL MOTORS (38020)
24027 Colorado & GMC Canyon '04 thru '12
24032 Corsica & Beretta all models '87 thru '96
24040 Corvette all V8 models '68 thru '82
24041 Corvette all models '84 thru '96
24042 Corvette all models '97 thru '13
24044 Cruze '11 thru '19
24045 Full-size Sedans Caprice, Impala, Biscayne, Bel Air & Wagons '69 thru '90
24046 Impala SS & Caprice and Buick Roadmaster '91 thru '96
Impala '00 thru '05 - see LUMINA (24048)
24047 Impala & Monte Carlo '06 thru '11
Lumina '90 thru '94 - see GM (38010)
24048 Lumina & Monte Carlo '95 thru '05
Lumina APV - see GM (38035)
24050 Luv Pick-up all 2WD & 4WD '72 thru '82
24051 Malibu '13 thru '19
24055 Monte Carlo all models '70 thru '88
Monte Carlo '95 thru '01 - see LUMINA (24048)
24059 Nova all V8 models '69 thru '79
24060 Nova/Geo Prizm '85 thru '92
24064 Pick-ups '67 thru '87 - Chevrolet & GMC
24065 Pick-ups '88 thru '98 - Chevrolet & GMC
24066 Pick-ups '99 thru '06 - Chevrolet & GMC
24067 Chevy Silverado & GMC Sierra '07 thru '14
24068 Chevy Silverado & GMC Sierra '14 thru '19
24070 S-10 & S-15 Pick-ups '82 thru '93
24071 S-10 & Sonoma Pick-ups '94 thru '04
24072 Chevrolet TrailBlazer, GMC Envoy & Oldsmobile Bravada '02 thru '09
24075 Sprint '85 thru '88, Geo Metro '89 thru '01
24080 Vans - Chevrolet & GMC '68 thru '96
24081 Chevrolet Express & GMC Savana Full-size Vans '96 thru '19

CHRYSLER
10310 Chrysler Engine Overhaul Manual
25015 Chrysler Cirrus, Dodge Stratus, Plymouth Breeze, '95 thru '00
25020 Full-size Front-Wheel Drive '88 thru '93
K-Cars - see DODGE Aries (30008)
25025 Chrysler LHS, Concorde & New Yorker, Dodge Intrepid, Eagle Vision, '93 thru '97
25026 Chrysler LHS, Concorde, 300M, Dodge Intrepid '98 thru '04
25027 Chrysler 300 '05 thru '18, Dodge Charger '06 thru '18, Magnum '05 thru '08 & Challenger '08 thru '18
25030 Chrysler & Plymouth Mid-size '82 thru '95
Rear-wheel Drive - see DODGE (30050)
25035 PT Cruiser all models '01 thru '10
25040 Chrysler Sebring '95 thru '06, Dodge Stratus '01 thru '06 & Dodge Avenger '95 thru '00
25041 Chrysler Sebring '07 thru '10, 200 '11 thru '17 & Dodge Avenger '08 thru '14

DATSUN
28005 200SX all models '80 thru '83
28012 240Z, 260Z & 280Z Coupe '70 thru '78
28014 280ZX Coupe & 2+2 '79 thru '83
300ZX - see NISSAN (72010)
28018 510 & PL521 Pick-up '68 thru '73
28020 510 all models '78 thru '81
28022 620 Series Pick-up all models '73 thru '79
720 Series Pick-up - see NISSAN (72030)

DODGE
30008 Aries & Plymouth Reliant '81 thru '89
30010 Caravan & Plymouth Voyager '84 thru '95
30011 Caravan & Plymouth Voyager '96 thru '02
30012 Challenger & Plymouth Sapporo '78 thru '83
30013 Caravan, Chrysler Voyager & Town & Country '03 thru '07
30014 Grand Caravan & Chrysler Town & Country '08 thru '18

EAGLE
Talon - see MITSUBISHI (68030, 68031)
Vision - see CHRYSLER (25025)

FIAT
34010 124 Sport Coupe & Spider '68 thru '78
34025 X1/9 all models '74 thru '80

FORD
10320 Ford Engine Overhaul Manual
10355 Ford Automatic Transmission Overhaul
11500 Mustang '64-1/2 thru '70 Restoration Guide
36004 Aerostar Mini-vans '86 thru '97
36006 Contour & Mercury Mystique '95 thru '00
36008 Courier Pick-up all models '72 thru '82
36012 Crown Victoria & Mercury Grand Marquis '88 thru '11
36016 Escort & Mercury Lynx '81 thru '90
36020 Escort & Mercury Tracer '91 thru '02
36022 Escape '01 thru '17, Mazda Tribute '01 thru '11 & Mercury Mariner '05 thru '11
36024 Explorer & Mazda Navajo '91 thru '01
36025 Explorer & Mercury Mountaineer '02 thru '10
36026 Explorer '11 thru '17
36028 Fairmont & Mercury Zephyr '78 thru '83
36030 Festiva & Aspire '88 thru '97
36032 Fiesta all models '77 thru '80
36034 Focus all models '00 thru '11
36035 Focus '12 thru '14
36045 Ford Fusion '06 thru '14 & Mercury Milan '06 thru '10
36048 Mustang V8 all models '64-1/2 thru '73
36049 Mustang II all 4-cylinder, V6 & V8 '74 thru '78
36050 Mustang & Mercury Capri '79 thru '93
36051 Mustang all models '94 thru '04
36052 Mustang '05 thru '14
36054 Pick-ups and Bronco '73 thru '79
36058 Pick-ups and Bronco '80 thru '96
36059 F-150 '97 thru '03, Expedition '97 thru '17, F-250 '97 thru '99, F-150 Heritage '04 & Lincoln Navigator '98 thru '17
36060 Super Duty Pick-up & Excursion '99 thru '10
36061 F-150 full-size '04 thru '10
36062 Pinto & Mercury Bobcat '75 thru '80
36063 F-150 full-size '15 thru '17
36064 Super Duty Pick-ups '11 thru '16
36066 Probe all models '89 thru '92
36070 Ranger & Bronco II gas models '83 thru '92
36071 Ranger '93 thru '11 & Mazda Pick-ups '94 thru '09
36074 Taurus & Mercury Sable '86 thru '95
36075 Taurus & Mercury Sable '96 thru '07
36076 Taurus '08 thru '14, Five Hundred '05 thru '07, Mercury Montego '05 thru '07 & Sable '08 thru '09
36078 Tempo & Mercury Topaz '84 thru '94
36082 Thunderbird & Mercury Cougar '83 thru '88
36086 Thunderbird & Mercury Cougar '89 thru '97
36090 Vans all V8 Econoline models '69 thru '91
36094 Vans full size '92 thru '14
36097 Windstar '95 thru '03, Freestar & Mercury Monterey Mini-van '04 thru '07

GENERAL MOTORS
10360 GM Automatic Transmission Overhaul
38005 Buick Century, Chevrolet Celebrity, Olds Cutlass Ciera & Pontiac 6000 '82 thru '96
38010 Buick Regal, Chevrolet Lumina, Oldsmobile Cutlass Supreme & Pontiac Grand Prix front wheel drive '88 thru '07
38015 Buick Skyhawk, Cadillac Cimarron, Chevrolet Cavalier, Oldsmobile Firenza Pontiac J-2000 & Sunbird '82 thru '94
38016 Chevrolet Cavalier/Pontiac Sunfire '95 thru '05
38017 Chevrolet Cobalt & Pontiac G5 '05 thru '11
38020 Buick Skylark, Chevrolet Citation, Olds Omega, Pontiac Phoenix '80 thru '85
38025 Buick Skylark & Somerset, Olds Achieva, Calais & Pontiac Grand Am '85 thru '98
38026 Chevrolet Malibu, Olds Alero & Cutlass, Pontiac Grand Am '97 thru '03
38027 Chevrolet Malibu '04 thru '12
38030 Cadillac Eldorado, Seville, Oldsmobile Toronado & Buick Riviera '71 thru '85
38031 Cadillac Eldorado, Seville, DeVille, Fleetwood, Oldsmobile Toronado & Buick Riviera '86 thru '93
38032 Cadillac DeVille '94 thru '05, Seville '92 thru '04 & Cadillac DTS '06 thru '10
38035 Chevrolet Lumina APV, Olds Silhouette & Pontiac Trans Sport all models '90 thru '96
38036 Chevrolet Venture, Olds Silhouette, Pontiac Trans Sport & Montana '97 thru '05
38040 Chevrolet Equinox '05 thru '17, GMC Terrain '10 thru '17 & Pontiac Torrent '06 thru '09

GEO
Metro - see CHEVROLET Sprint (24075)
Prizm - '85 thru '92 see CHEVY (24060), '93 thru '02 see TOYOTA Corolla (92036)
40030 Storm all models '90 thru '93
Tracker - see SUZUKI Samurai (90010)

GMC
Vans & Pick-ups - see CHEVROLET

HONDA
42010 Accord CVCC all models '76 thru '83
42011 Accord all models '84 thru '89
42012 Accord all models '90 thru '93
42013 Accord all models '94 thru '97
42014 Accord all models '98 thru '02
42015 Accord '03 thru '12 & Crosstour '10 thru '14
42016 Accord '13 thru '17

PONTIAC
79008 Fiero all models '84 thru '88
79018 Firebird V8 models except Turbo '70 thru '81
79019 Firebird all models '82 thru '92
79025 G6 all models '05 thru '09
79040 Mid-size Rear-wheel Drive '70 thru '07
Vibe '03 thru '10 - see TOYOTA Corolla (92037)
For other PONTIAC titles, see BUICK, CHEVROLET or GM listings.

PORSCHE
80020 911 Coupe & Targa models '65 thru '89
80025 914 all 4-cylinder models '69 thru '76
80030 924 all models including Turbo '76 thru '82
80035 944 all models including Turbo '83 thru '89

RENAULT
Alliance & Encore - see AMC (14025)

SAAB
84010 900 all models including Turbo '79 thru '88

SATURN
87010 Saturn all S-series models '91 thru '02
Saturn Ion '03 thru '07- see GM (38017)
87020 Saturn L-series all models '00 thru '04
87040 Saturn VUE '02 thru '09

SUBARU
89002 1100, 1300, 1400 & 1600 '71 thru '79
89003 1600 & 1800 2WD & 4WD '80 thru '94
89080 Impreza '02 thru '11, WRX '02 thru '14 & WRX STI '04 thru '14
89100 Legacy all models '90 thru '99
89101 Legacy & Forester '00 thru '09
89102 Legacy '10 thru '16 & Forester '12 thru '16

SUZUKI
90010 Samurai/Sidekick & Geo Tracker '86 thru '01

TOYOTA
92005 Camry all models '83 thru '91
92006 Camry '92 thru '96 & Avalon '95 thru '96
92007 Camry, Avalon, Solara, Lexus ES 300 '97 thru '01
92008 Camry, Avalon, Lexus ES 300/330 '02 thru '06 & Solara '02 thru '08
92009 Camry, Avalon & Lexus ES 350 '07 thru '17
92015 Celica Rear-wheel Drive '71 thru '85
92020 Celica Front-wheel Drive '86 thru '99
92025 Celica Supra all models '79 thru '92
92030 Corolla all models '75 thru '79
92032 Corolla rear-wheel drive models '80 thru '87
92035 Corolla front-wheel drive models '84 thru '92
92036 Corolla & Geo/Chevrolet Prizm '93 thru '02
92037 Corolla '03 thru '19, Matrix '03 thru '14, & Pontiac Vibe '03 thru '10
92040 Corolla Tercel all models '80 thru '82
92045 Corona all models '74 thru '82
92050 Cressida all models '78 thru '82
92055 Land Cruiser FJ40/43/45/55 '68 thru '82
92056 Land Cruiser FJ60/62/80/FZJ80 '80 thru '96
92060 Matrix '03 thru '11 & Pontiac Vibe '03 thru '10
92065 MR2 all models '85 thru '87
92070 Pick-up all models '69 thru '78
92075 Pick-up all models '79 thru '95
92076 Tacoma '95 thru '04, 4Runner '96 thru '02 & T100 '93 thru '08
92077 Tacoma all models '05 thru '18
92078 Tundra '00 thru '06, Sequoia '01 thru '07
92079 4Runner '03 thru '09
92080 Previa all models '91 thru '95
92081 Prius all models '01 thru '12
92082 RAV4 all models '96 thru '12
92085 Tercel all models '87 thru '94
92090 Sienna all models '98 thru '10
92095 Highlander '01 thru '19 & Lexus RX330/330/350 '99 thru '19
92179 Tundra '07 thru '19 & Sequoia '08 thru '19

TRIUMPH
94007 Spitfire all models '62 thru '81
94010 TR7 all models '75 thru '81

VW
96008 Beetle & Karmann Ghia '54 thru '79
96009 New Beetle '98 thru '10
96016 Rabbit, Jetta, Scirocco & Pick-up gas models '75 thru '92 & Convertible '80 thru '92
96017 Golf, GTI & Jetta '93 thru '98, Cabrio '95 thru 02
96018 Golf, GTI & Jetta '99 thru '05
96019 Jetta, Rabbit, GLI, GTI & Golf '05 thru '11
96020 Rabbit, Jetta, Pick-up diesel '77 thru '84
96021 Jetta '11 thru '18 & Golf '15 thru '19
96023 Passat '98 thru '05, Audi A4 '96 thru '01
96030 Transporter 1600 all models '68 thru '79
96035 Transporter 1700, 1800, 2000 '72 thru '79
96040 Type 3 1500 & 1600 '63 thru '73
96045 Vanagon Air-Cooled all models '80 thru '83

VOLVO
97010 120, 130 Series & 1800 Sports '61 thru '73
97015 140 Series all models '66 thru '74
97020 240 Series all models '76 thru '93
97040 740 & 760 Series all models '82 thru '88
97050 850 Series all models '93 thru '97

TECHBOOK MANUALS
10205 Automotive Computer Codes
10206 OBD-II & Electronic Engine Management
10210 Automotive Emissions Control Manual
10215 Fuel Injection Manual '78 thru '85
10225 Holley Carburetor Manual
10230 Rochester Carburetor Manual
10305 Chevrolet Engine Overhaul Manual
10320 Ford Engine Overhaul Manual
10330 GM and Ford Diesel Engine Repair Manual
10331 Duramax Diesel Engines '01 thru '19
10332 Cummins Diesel Engine Performance
10333 GM, Ford & Chrysler Engine Performance
10334 GM Engine Performance
10340 Small Engine Repair Manual, 5 HP & Less
10341 Small Engine Repair Manual, 5.5 thru 20 HP
10345 Suspension, Steering & Driveline Manual
10355 Ford Automatic Transmission Overhaul
10360 GM Automatic Transmission Overhaul
10405 Automotive Body Repair & Painting
10410 Automotive Brake Manual
10411 Automotive Anti-lock Brake (ABS) Systems
10425 Automotive Heating & Air Conditioning
10435 Automotive Tools Manual
10445 Welding Manual

HYUNDAI
43010 Elantra all models '96 thru '19
43015 Excel & Accent all models '86 thru '13
43050 Santa Fe all models '01 thru '12
43055 Sonata all models '99 thru '14

INFINITI
G35 '03 thru '08 - see NISSAN 350Z (72011)

ISUZU
Hombre - see CHEVROLET S-10 (24071)
47017 Rodeo '91 thru '02, Amigo '89 thru '94 & '98 thru '02 & Honda Passport '95 thru '02
47020 Trooper '84 thru '91 & Pick-up '81 thru '93

JAGUAR
49010 XJ6 all 6-cylinder models '68 thru '86
49011 XJ6 all models '88 thru '94
49015 XJ12 & XJS all 12-cylinder models '72 thru '85

JEEP
50010 Cherokee, Comanche & Wagoneer Limited all models '84 thru '01
50011 Cherokee '14 thru '19
50020 CJ all models '49 thru '86
50025 Grand Cherokee all models '93 thru '04
50026 Grand Cherokee '05 thru '19 & Dodge Durango '11 thru '19
50029 Grand Wagoneer & Pick-up '72 thru '91
50030 Wrangler all models '87 thru '17
50035 Liberty '02 thru '12 & Dodge Nitro '07 thru '11
50050 Patriot & Compass '07 thru '17

KIA
54050 Optima '01 thru '10
54060 Sedona '02 thru '14
54070 Sephia '94 thru '01, Spectra '00 thru '09, Sportage '05 thru '20
54077 Sorento '03 thru '13

LEXUS
ES 300/330 - see TOYOTA Camry (92007, 92008)
RX 300/330/350 - see TOYOTA Highlander (92095)

LINCOLN
Navigator - see FORD Pick-up (36059)
59010 Rear-Wheel Drive Continental '70 thru '87, Mark Series '70 thru '92 & Town Car '81 thru '10

MAZDA
61010 GLC (rear wheel drive) '77 thru '83
61011 GLC (front wheel drive) '81 thru '85
61012 Mazda3 '04 thru '11
61015 323 & Protegé '90 thru '03
61016 MX-5 Miata '90 thru '14
61020 MPV all models '89 thru '98
61030 Pick-ups '72 thru '93
Pick-ups '94 thru '09 - see Ford (36071)
61035 RX-7 all models '79 thru '85
61036 RX-7 all models '86 thru '91
61040 626 (rear-wheel drive) all models '79 thru '82
61041 626 & MX-6 (front-wheel drive) '83 thru '92
61042 626 '93 thru '01 & MX-6/Ford Probe '93 thru '02
61043 Mazda6 '03 thru '13

MERCEDES-BENZ
63012 123 Series Diesel '76 thru '85
63015 190 Series 4-cylinder gas models '84 thru '88
63020 230, 250 & 280 6-cylinder SOHC '68 thru '72
63025 280 123 Series gas models '77 thru '81
63030 350 & 450 all models '71 thru '80
63040 C-Class: C230/C240/C280/C320/C350 '01 thru '07

MERCURY
64200 Villager & Nissan Quest '93 thru '01
All other titles, see FORD listing.

MG
66010 MGB Roadster & GT Coupe '62 thru '80
66015 MG Midget & Austin Healey Sprite Roadster '58 thru '80

MINI
67020 Mini '02 thru '13

MITSUBISHI
68020 Cordia, Tredia, Galant, Precis & Mirage '83 thru '93
68030 Eclipse, Eagle Talon & Plymouth Laser '90 thru '94
68031 Eclipse '95 thru '05 & Eagle Talon '95 thru '98
68035 Galant '94 thru '12
68040 Pick-up '83 thru '96 & Montero '83 thru '93

NISSAN
72010 300ZX all models incl. Turbo '84 thru '89
72011 350Z & Infiniti G35 all models '03 thru '08
72015 Altima all models '93 thru '06
72016 Altima '07 thru '12
72020 Maxima all models '85 thru '92
72021 Maxima all models '93 thru '08
72025 Murano '03 thru '14
72030 Pick-ups '80 thru '97 & Pathfinder '87 thru '95
72031 Frontier, Xterra & Pathfinder '96 thru '04
72032 Frontier & Xterra '05 thru '14
72037 Pathfinder '05 thru '14
72040 Pulsar all models '83 thru '86
72042 Rogue all models '08 thru '20
72050 Sentra all models '82 thru '94
72051 Sentra & 200SX all models '95 thru '06
72060 Stanza all models '82 thru '90
72070 Titan pick-ups '04 thru '10 & Armada '05 thru '10
72080 Versa all models '07 thru '19

OLDSMOBILE
73015 Cutlass V6 & V8 gas models '74 thru '88
For other OLDSMOBILE titles, see BUICK, CHEVROLET or GM listings.

PLYMOUTH
For PLYMOUTH titles, see DODGE.

Over a 100 Haynes motorcycle manuals also available

10/22

Haynes North America, Inc. • (805) 498-6703 • www.haynes.com